T0145388

Smart Innovation, Systems and Technologies

Volume 83

About this Series

The Smart Innovation, Systems and Technologies book series encompasses the topics of knowledge, intelligence, innovation and sustainability. The aim of the series is to make available a platform for the publication of books on all aspects of single and multi-disciplinary research on these themes in order to make the latest results available in a readily-accessible form. Volumes on interdisciplinary research combining two or more of these areas is particularly sought.

The series covers systems and paradigms that employ knowledge and intelligence in a broad sense. Its scope is systems having embedded knowledge and intelligence, which may be applied to the solution of world problems in industry, the environment and the community. It also focusses on the knowledge-transfer methodologies and innovation strategies employed to make this happen effectively. The combination of intelligent systems tools and a broad range of applications introduces a need for a synergy of disciplines from science, technology, business and the humanities. The series will include conference proceedings, edited collections, monographs, handbooks, reference books, and other relevant types of book in areas of science and technology where smart systems and technologies can offer innovative solutions.

High quality content is an essential feature for all book proposals accepted for the series. It is expected that editors of all accepted volumes will ensure that contributions are subjected to an appropriate level of reviewing process and adhere to KES quality principles.

More information about this series at http://www.springer.com/series/8767

Suresh Chandra Satapathy
Amit Joshi
Editors

Information and Communication Technology for Intelligent Systems (ICTIS 2017) - Volume 1

Editors
Suresh Chandra Satapathy
Department of CSE
PVP Siddhartha Institute of Technology
Vijayawada, Andhra Pradesh
India

Amit Joshi
Sabar Institute of Technology for Girls
Ahmedabad, Gujarat
India

ISSN 2190-3018 ISSN 2190-3026 (electronic)
Smart Innovation, Systems and Technologies
ISBN 978-3-319-87610-8 ISBN 978-3-319-63673-3 (eBook)
DOI 10.1007/978-3-319-63673-3

Printed on acid-free paper

This Springer imprint is published by Springer Nature
The registered company is Springer International Publishing AG
The registered company address is: Gewerbestrasse 11, 6330 Cham, Switzerland

Preface

This SIST volume contains the papers presented at the ICTIS 2017: Second International Conference on Information and Communication Technology for Intelligent Systems. The conference was held during March 25 and 26, 2017, Ahmedabad, India, and organized combinedly by ASSOCHAM Gujarat Chapter, G R Foundation, Association of Computer Machinery, Ahmedabad Chapter and supported by Computer Society of India, Division IV, Communication, and Division V, Education and Research. It targeted state-of-the-art as well as emerging topics pertaining to ICT and effective strategies for its implementation for Engineering and Intelligent Applications. The conference had a large number of high-quality submissions from many academic pioneering researchers, scientists, industrial engineers, students from all around the world, and it provided a common provide a forum to researcher to interact and exchange ideas. After a rigorous peer-reviewed process with the help of program committee members and external reviewers, 147 (Vol-I: 74, Vol-II: 73) papers were accepted with an acceptance ratio of 0.39. The conference featured many distinguished personalities such as Dr. Pankaj L. Jani, Hon'ble Vice Chancellor, Dr. Babasaheb Ambedkar University; Shri Sunil Shah, President, Gujarat Innovation Society; Dr. S.C. Satapathy, Chairman, Division V, Computer Society of India; Shri Vivek Ogra, President, GESIA IT Association; and Dr. Nilesh Modi, Chairman, ACM Ahmedabad Chapter. Separate invited talks were organized in industrial and academic tracks in both days. The conference also hosted few tutorials and workshops for the benefit of participants. We are indebted to ACM Ahmedabad Professional Chapter, CSI Division IV, V for their immense support to make this conference possible in such a grand scale. A total of 15 sessions were organized as a part of *ICTIS 2017* including 12 technical, 1 plenary, 1 inaugural session, and 1 valedictory session. A total of 113 papers were presented in 12 technical sessions with high discussion insights. Our sincere thanks to all sponsors, press, print, and electronic media for their excellent coverage of this conference.

April 2017

Suresh Chandra Satapathy
Amit Joshi

Organization

Organizing Committee Chairs

Bharat Patel	COO, Yudiz Solutions
Durgesh Kumar Mishra	Division IV, CSI
Nilesh Modi	ACM Ahmedabad Chapter

Organizing Secretary

Mihir Chauhan	ACM Professional Member

Program Committee Chairs

Malaya Kumar Nayak	IT and Research Group, London, UK
Shyam Akashe	ITM University, Gwalior, India
H.R. Vishwakarma	VIT, Vellore, India

Advisory Committee

Z.A. Abbasi	Department of Electronics Engineering, AMU, Aligarh, India
Manjunath Aradhya	Department of MCA, SJCE, Mysore
Min Xie	Fellow of IEEE
Mustafizur Rahman	Endeavour Research Fellow, Institute of High Performance Computing, Agency for Science, Technology and Research
Chandana Unnithan	Deakin University, Melbourne, Australia
Pawan Lingras	Saint Mary's University, Canada

Mohd Atique	Amravati, Maharashtra, India
Hoang Pham	Department of Industrial and Systems Engineering, Rutgers University, Piscataway, NJ
Suresh Chandra Satapathy	Division V, CSI
Naeem Hannoon	Universiti Teknologi Mara, Malaysia
Hipollyte Muyingi	Namibia University of Science and Technology, Namibia
Nobert Jere	Namibia University of Science and Technology, Namibia
Shalini Batra	Computer Science & Engineering Dept, Thapar University, Patiala, Punjab, India
Ernest Chulantha Kulasekere	University of Moratuwa, Sri Lanka
James E. Fowler	Mississippi State University, Mississippi, USA
Majid Ebnali-Heidari	ShahreKord University, Shahrekord, Iran
Rajendra Kumar Bharti	Kumaon Engg College, Dwarahat, Uttarakhand, India
Murali Bhaskaran	Dhirajlal Gandhi College of Technology, Salem, Tamil Nadu
Pramod Parajuli	Nepal College of Information Technology, Nepal
Komal Bhatia	YMCA University, Faridabad, Haryana, India
Lili Liu	Automation College, Harbin Engineering University, Harbin, China
S.R. Biradar	Department of Information Science and Engineering, SDM College of Engineering and Technology Dharwad, Karnataka
A.K. Chaturvedi	Department of Electrical Engineering, IIT Kanpur, India
Margaret Lloyd	Faculty of Education School of Curriculum, Queensland University of Technology, Queensland
Jayanti Dansana	KIIT University, Bhubaneswar, Odisha
Desmond Lobo	Computer Engineering Department, Faculty of Engineering at KamphaengSaen, Kasetsart University, Thailand
Sergio Lopes	Industrial Electronics Department, University of Minho, Braga, Portugal
Soura Dasgupta	Department of TCE, SRM University, Chennai, India
Apurva A. Desai	Veer Narmad South Gujarat University, Surat, India
Abrar A. Qureshi	University of Virginia's College at Wise, One College Avenue

V. Susheela Devi	Department of Computer Science and Automation Indian Institute of Science, Bangalore
Subhadip Basu	The University of Iowa, Iowa City, USA
Bikash Kumar Dey	Department of Electrical Engineering, IIT Bombay, Powai, Maharashtra
Vijay Pal Dhaka	Jaipur National University, Jaipur, Rajasthan
Mignesh Parekh	Kamma Incorporation, Gujarat, India
Kok-Lim Low	National University of Singapore, Singapore
Chandana Unnithan	Victoria University, Australia

Technical Program Committee Chairs

Suresh Chandra Satapathy (Chair)	Division V, Computer Society of India
Vikrant Bhateja (Co-chair)	SRMGPC, Lucknow, India

Members

Dan Boneh	Computer Science Dept, Stanford University, California, USA
Alexander Christea	University of Warwick, London, UK
Aynur Unal	Stanford University, USA
Ahmad Al-Khasawneh	The Hashemite University, Jordan
Bharat Singh Deora	JRNRV University, India
Jean Michel Bruel	Departement Informatique IUT de Blagnac, Blagnac, France
Ngai-Man Cheung	University of Technology and Design, Singapore
Yun-Bae Kim	SungKyunKwan University, South Korea
Ting-Peng Liang	National Chengchi University, Taipei, Taiwan
Sami Mnasri	IRIT Laboratory Toulouse, France
Lorne Olfman	Claremont, California, USA
Anand Paul	The School of Computer Science and Engineering, South Korea
Krishnamachar Prasad	Department of Electrical and Electronic Engineering, Auckland, New Zealand
Brent Waters	University of Texas, Austin, Texas, USA
Kalpana Jain	CTAE, Udaipur, India
Avdesh Sharma	Jodhpur, India
Nilay Mathur	NIIT Udaipur, India
Philip Yang	Price water house Coopers, Beijing, China
Jeril Kuriakose	Manipal University, Jaipur, India

R.K. Bayal	Rajasthan Technical University, Kota, Rajasthan, India
Martin Everett	University of Manchester, England
Feng Jiang	Harbin Institute of Technology, China
Prasun Sinha	Ohio State University Columbus, Columbus, OH, USA
Savita Gandhi	Gujarat University, Ahmedabad, India
Xiaoyi Yu	National Laboratory of Pattern Recognition, Institute of Automation, Chinese Academy of Sciences, Beijing, China
Gengshen Zhong	Jinan, Shandong, China
Abdul Rajak A.R.	Department of Electronics and Communication Engineering, Birla Institute of Dr. Nitika Vats Doohan, Indore, India
Harshal Arolkar	CSI Ahmedabad Chapter, India
Bhavesh Joshi	Advent College, Udaipur, India
K.C. Roy	Kautilya, Jaipur, India
Mukesh Shrimali	Pacific University, Udaipur, India
Meenakshi Tripathi	MNIT, Jaipur, India
S.N. Tazi	Govt. Engineering College, Ajmer, Rajasthan, India
Shuhong Gao	Mathematical Sciences, Clemson University, Clemson, South Carolina
Sanjam Garg	University of California, Los Angeles, California
Garani Georgia	University of North London, UK
Hazhir Ghasemnezhad	Electronics and Communication Engineering Department, Shiraz University of Technology, Shiraz, Iran
Andrea Goldsmith	Stanford University, California
Cheng Guang	Southeast University, Nanjing, China
Venkat N. Gudivada	Weisberg Division of Engineering and Computer Science, Marshall University Huntington, Huntington, West Virginia
Rachid Guerraoui	I&C, EPFL, Lausanne, Switzerland
Wang Guojun	School of Information Science and Engineering of Zhong Nan University, China
Nguyen Ha	Department of Electrical and Computer Engineering, University of Saskatchewan, Saskatchewan, Canada
Z.J. Haas	School of Electrical Engineering, Cornell University, Ithaca, New York
Hyehyun Hong	Department of Advertising and Public Relations, Chung-Ang University, South Korea
Qinghua Hu	Harbin Institute of Technology, China

Marcos Roberto da Silva Borges	Federal University of Rio de Janeiro, Brazil
Soura Dasgupta	Iowa City, Iowa, USA
Gholamhossein Dastghaibyfard	College of Electrical & Computer Engineering, Shiraz University, Shiraz, Iran
Doreen De Leon	California State University, USA
Bartel Van de Walle	University Tilburg, Tilburg, Netherlands
David Delahaye	Saint-Martin, Cedex, France
Andrew G. Dempster	The University of New South Wales, Australia
Alan Dennis	Kelley School of Business Indiana University, Bloomington, Indiana, USA
Jitender Singh Deogun	Department of Computer Science and Engineering, University of Nebraska – Lincoln, Nebraska, USA
S.A.D. Dias	Department of Electronics and Telecommunication Engineering, University of Moratuwa, Sri Lanka
David Diez	Leganés, Spain, Europe
Zhang Dinghai	Gansu Agricultural University, Lanzhou, China
Ali Djebbari	Sidi Bel Abbes, Algeria
P.D.D. Dominic	Department of Computer and Information Science, Universiti Teknologi Petronas, Tronoh, Perak, Malaysia

Contents

Survey on Online Social Media Networks Facebook Forensics

V. Paul Selwin(✉)

TIFAC-CORE in Cyber Security, Amrita School of Engineering, Coimbatore, India
paul.selwin@gmail.com

Abstract. In this paper monitoring and surveying of data along with its protection and security parameters through the continuous evaluation of different OSNs are performed sequentially to identify any focused persons availability and activity in multiple OSN channels. A real time scenario in OSNs domain is initially monitored for data extraction of publicly available data. The available data is matched using semantic matching algorithm to ensure the link establishments, data consolidation and the data segregation which could possibly reveal some details and also used for the purpose of study, research and strategic report making. The main theme is the forensic evidence gathering and also enlisting the security parameters of different OSNs.

Keywords: Data collection · Monitoring · Extraction · E-mail ID · Forensic investigation · Posts

1 Introduction

In every area of online social media network connection and sharing there exists always a threat to the data followed by its preventive measures. The main deal of every social media is in protecting their user data integrity, availability, consistency and privacy. Most of the social media failed in protecting the user privacy policies as of the past reports. But nowadays, mostly every social media network tries to protect their user data from leaks and losses. They implement different innovative ideas to prove their security parameters.

Investigation of online social media network is termed to be the latest trend and a premium challenge in the cyber-crime and forensic wing. Most possible data gathering and analysis of social media networks turns to be negative due the cautious update and security patches of every social media networks every now and then. To make their social media network assured of user privacy, the data extracting methods used yesterday expires or fails to get the desired data today and so on. In this paper, we take consideration of the most preferred OSNs for the analysis of their security perimeters and documenting their data gathering possibilities.

The initial section begins with the monitoring of OSNs like LinkedIn, Twitter and Facebook.

S.C. Satapathy and A. Joshi (eds.), *Information and Communication Technology for Intelligent Systems (ICTIS 2017) - Volume 1*, Smart Innovation, Systems and Technologies 83, DOI 10.1007/978-3-319-63673-3_1

In the second section, analysis over the Facebook gives detailed exploration about how the security privacy is taken into consideration. The digital forensic investigation is done using several techniques with multiple angles of analysis on Facebook to find its security measure. Here we are able to get some idea about how communication and privacy terms are ensured. In general with that how Facebook uses the network traffic as a mode of transport.

In the third section of this paper, a scenario is taken, the process of real time monitoring and data collection of a particular group of unit or an organisation or an entity related pages over its Facebook cluster pages (example Amrita Universities related pages). The details about the different pages of the respective Facebook account are taken and analysed using the semantic matching algorithm and collected as evidence. Here only the Facebook OSN is taken into account because of its vast and well-structured nature. First the domain for investigation was as choose Amrita University Facebook pages. After finalising the domain the process moves with the peripheral pages or the different pages of Amrita University with respect to different year, department and location. Then the data is collected using two different methods, which includes extracting the data with and without logging in to the concerned user's account by using their login credentials. And after that s-match algorithm is used in analysis.

Then the conclusion of this paper, here all the details of a particular persons and his activities in a social media network like Facebook are retrieved and analysed for ratio of crime scene participation.

2 Monitoring of Networks

In this section the data traffic is monitored for the sensitive credentials. The monitoring and analysis is done for three different structured OSNs. The three OSNs are Facebook, Twitter and LinkedIn [1]. When the OSNs network traffic is considered before 2012 some credentials were always available over the network. This could be considered as data leakage when it is taken in terms of security privacy. And in this case data available open in the network may lead to a severe threat to the user. At any count the data over the channel should be secure. We can now see how the different OSNs are in securing their consistency in privacy ensuring.

2.1 Facebook

The monitoring of Facebook starts with, sniffing the online traffic using an effective sniffer multiple times at different instance. The sniffed data is the monitored with multiple iterations. Provided no illegitimate activity is done. The sniffed raw data reveals certain unrecognisable data and the ports details. With the available details we could be able frame some conclusion how the data handling happening and their security features.

Facebook forensics details the data availability over the traffic without the user's knowledge. A pattern of continuous sniffing is performed over the network of users using Facebook social media. The sniffed data is collected and checked for the details about the data as mentioned above. When we are looking for the data or the credentials, some chunk were found but not in a readable or meaningful format. So, we checked further for the port which is used by Facebook it was port 80 and 443. Port 80 is the http traffic which is an unsecure channel and 443 port opens the encrypted channel to avoid any eavesdropper or sniffer accessing the data (Fig. 1).

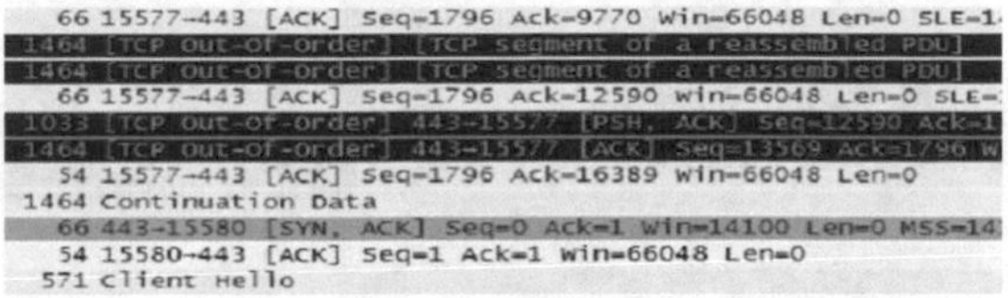

```
  66 15577→443 [ACK] Seq=1796 Ack=9770 Win=66048 Len=0 SLE=1
1464 [TCP Out-Of-Order] [TCP segment of a reassembled PDU]
1464 [TCP Out-Of-Order] [TCP segment of a reassembled PDU]
  66 15577→443 [ACK] Seq=1796 Ack=12590 Win=66048 Len=0 SLE=
1033 [TCP Out-Of-Order] 443→15577 [PSH, ACK] Seq=12590 Ack=1
1464 [TCP Out-Of-Order] 443→15577 [ACK] Seq=13569 Ack=1796 W
  54 15577→443 [ACK] Seq=1796 Ack=16389 Win=66048 Len=0
1464 Continuation Data
  66 443→15580 [SYN, ACK] Seq=0 Ack=1 Win=14100 Len=0 MSS=14
  54 15580→443 [ACK] Seq=1 Ack=1 Win=66048 Len=0
 571 Client Hello
```

Fig. 1. Data traffic detailed

Facebook tries to maintain the data integrity through encryption. It uses end–end encryption with public key cryptography. In the pair of keys one is designed to be shared widely and the other is the secret key. The Facebook chats are also encrypted using PGP encryption. Pretty good privacy, is an encryption process that let the people encrypt, decrypt emails and allows users to authenticate messages with digital signatures. In Facebook PGP is used as GnuPG to encrypt the chats and contacts sharing. As Facebook's user's data is snooped by the government, to ensure security and privacy Facebook have increased its level of security to a higher level. Facebook ensures the security encryption in both the communication ends between Facebook and the users using enigmail. And the chats are ensured with the encryption using GnuPG. Also not going beyond the legal bindings as allowed (public data is used). Forensic analysis is performed using multiple forensic frameworks. Most network forensic framework is performed to find how the data communication and function happens in the Facebook OSN.

2.2 Twitter

The Twitter data forensics reveals certain details like the previous monitoring report. The forensic investigation on Twitter uncovers the details like it uses 140 characters in tweets as limits. The Twitter API runs on standard web service ports: Port 80 for http and port 443 for SSL-based traffic (https). It uses RC4 for its session encryption. SSL verification fails often in Twitter and results in leaving the proxies unmonitored sometimes and network tampering. Sha256RSA signature algorithm is used. It uses public key RSA (2048 bits). In Twitter the extraction of electronic mail identity is found possible over the traffic. And also the transfer protocol error happening ratio is comparatively very less which reveals that the data communication function is with a well-built security boundaries (Figs. 2 and 3).

Currently, only SSL connections are allowed in Twitter application endpoints. Communicating over SSL preserves user privacy by protecting information between the user and the Twitter application as it spread across the Internet. O-Auth is mandated and protects the user from having their password captured in transit by substituting an O-Auth token for the user's credentials; it is not enough to ensure the complete privacy. According to Twitter, as much as 75 per cent of its internet traffic is already established using ECDHE; the remaining 25 per cent comes from older third-party clients that do not support the key agreement protocol. By November 2015, Twitter started their encryption procedures for the direct message (personal chats) to avoid snooping. Only tailored advertisements are allowed in Twitter to avoid misuse of Twitter account privileges.

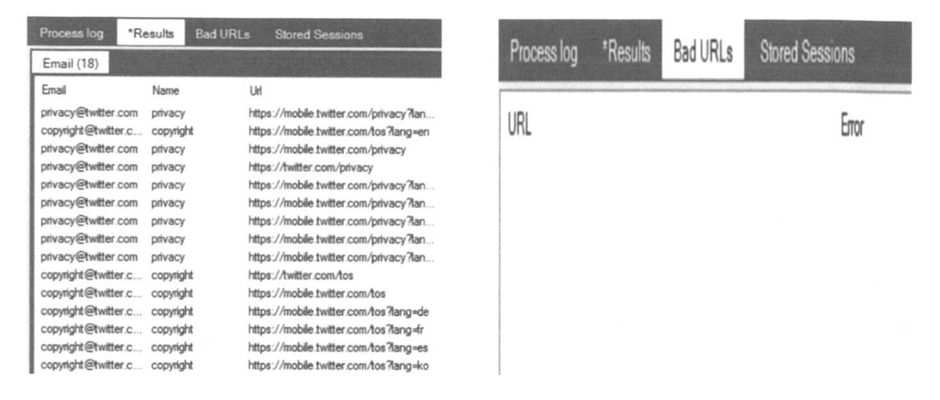

Email	Name	Url
privacy@twitter.com	privacy	https://mobile.twitter.com/privacy?lan...
copyright@twitter.c...	copyright	https://mobile.twitter.com/tos?lang=en
privacy@twitter.com	privacy	https://mobile.twitter.com/privacy
privacy@twitter.com	privacy	https://twitter.com/privacy
privacy@twitter.com	privacy	https://mobile.twitter.com/privacy?lan...
privacy@twitter.com	privacy	https://mobile.twitter.com/privacy?lan...
privacy@twitter.com	privacy	https://mobile.twitter.com/privacy?lan...
privacy@twitter.com	privacy	https://mobile.twitter.com/privacy?lan...
privacy@twitter.com	privacy	https://mobile.twitter.com/privacy?lan...
copyright@twitter.c...	copyright	https://twitter.com/tos
copyright@twitter.c...	copyright	https://mobile.twitter.com/tos
copyright@twitter.c...	copyright	https://mobile.twitter.com/tos?lang=de
copyright@twitter.c...	copyright	https://mobile.twitter.com/tos?lang=fr
copyright@twitter.c...	copyright	https://mobile.twitter.com/tos?lang=es
copyright@twitter.c...	copyright	https://mobile.twitter.com/tos?lang=ko

Fig. 2. E-mail ID extraction

Fig. 3. Bad URL report – EMPTY

2.3 LinkedIn

LinkedIn is a profession or business oriented OSN. Mobile version of LinkedIn application is not immunised to most of the vulnerabilities because it is used only for the purpose of job search, employee search, business partner search. Security criteria here are little out-dated since it doesn't share anything other than job profiles and sometimes advertisements and pictures. It provides a link with recruiter and the job seeker. Forensic analysis is performed using multiple forensic frameworks. Most network forensic framework is performed to find how the data communication and function happens in the LinkedIn OSN.

Process log | *Results | Bad URLs (211) | Stored Sessions

Email (8)

Email	Name	Url
privacy@slideshare...	privacy	https://www.linkedin.com/legal/priva...
press@linkedin.com	press	https://www.linkedin.com/about-us?t...
privacy@slideshare...	privacy	https://www.linkedin.com/legal/priva...
press@linkedin.com	press	https://www.linkedin.com/about-us
microsites-support...	microsites-support	https://forms.linkedin.com/formgen.p...
microsites-support...	microsites-support	https://forms.linkedin.com/formgen.p...
microsites-support...	microsites-support	https://forms.linkedin.com/formgen.p...
lcsfeedback@linke...	lcsfeedback	https://www.linkedin.com/uas/login-...

Fig. 4. Some data are not prevented from extraction and bad URL details

After this forensic analysis, some details about the LinkedIn is revealed which says that data from LinkedIn can be extracted directly which is not prevented (only page data). Figure 4 (bad URLs counts) reveals that the communication process error which gives an idea that LinkedIn is with a partly maintained data handling in the part of data transfer.

From the above three OSNs analysis we were able to get some strategic report. The report reveals that security ratings level drop down from Facebook, Twitter, LinkedIn to other OSNs.

3 Facebook Forensics: Conventional

Facebook forensic can be considered as conventional and non-conventional type. Some of the conventional type includes

1. Facebook API analysis.
2. Archival data forensics.
3. Facepager forensics.
4. Mobile hub forensics.

3.1 Facebook API Analysis is a Facebook availed facility to extract data directly from Facebook. Data includes the user account data like the photos, likes, views of the user, friend list and some personal descriptions. Provided the data extraction can be possibly done after the login verification by the account holders [4]. These APIs are roaming in the Facebook as an application which are used for getting some recreation analysis (example: like click here and share or like on Facebook to know about your character and so on). The method of creation is through the API interface of the Facebook and Facebook shares only a limited user data not like the sensitive data like the user chats (Fig. 5).

Fig. 5. Facebook API application creation.

When the account holder gives access to the application, the respective access token is shared with the application to justify its legitimate access process to the Facebook. Data extraction using this technique is discussed in the later leaves of this paper (Fig. 6).

```
54 6001→443 [ACK] Seq=3104 Ack=74380 Win=66048 Len=0
171 Application Data
1464 [TCP segment of a reassembled PDU]
54 6001→443 [ACK] Seq=3104 Ack=75907 Win=66048 Len=0
1284 Application Data
1464 [TCP segment of a reassembled PDU]
54 6001→443 [ACK] Seq=3104 Ack=78547 Win=66048 Len=0
1464 Application Data
231 Application Data
54 6001→443 [ACK] Seq=3104 Ack=80134 Win=66048 Len=0
1464 [TCP segment of a reassembled PDU]
171 Application Data
54 6001→443 [ACK] Seq=3104 Ack=81661 Win=66048 Len=0
```

Fig. 6. Port 443

Fig. 7. Data archival autopsy

Tunnelling traffic of Facebook data is analysed to find how the traffic is traversing. Through multiple iterations of data traffic analysis at different instance, it is found that the port used by the Facebook is 443 https port after the January 2016.

3.2 The Archival Data Forensics, as every OSN has a user access archival data, the data archived is retrieved and autopsied for details and found some details about the user activities which include the login schedules, post, status, etc., but the archival data could be extracted only through the login credentials and the with the knowledge of the Facebook. From the above analysis of data through archival data forensics techniques most of the details are revealed about the particular user. The details starting from the user details, user posts time, user's chat details, user's uploads, user's events to attend, user's posts, user valid saved data, user's contacts or the profile friends lists, user's managed and followed pages and groups along with the activities, user login detailed history, links shared, etc. These data are the evidences collected and analysed from the user after the legal warrant for accessing. This kind of analysis could be used in any account user profile only after getting the legal warrant from the legal department and proceeding with the chain of custody formats.

Figures 7 and 8 reveals the details and also the data of the user's photo uploads and share. The main use of this is ensuring the upload performed from this account or not and also the details about the upload like the upload date, time are also extracted. The other data interactions which are in the form of different formatted files are also possibly extracted from the user account as found in the Fig. 9. And also the particular file can be retrieved for use as evidence.

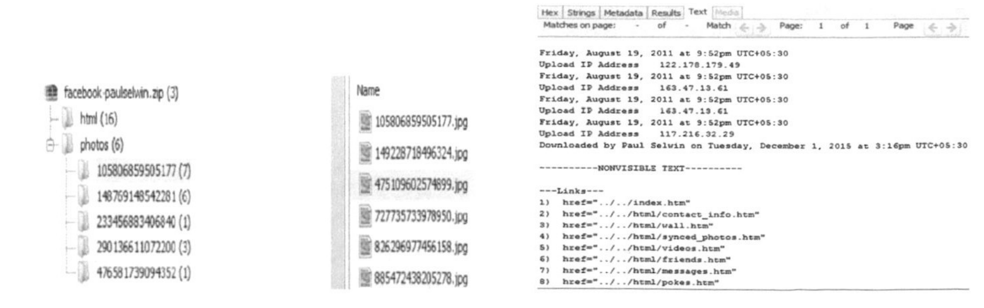

Fig. 8. Details of the photos upload.

Fig. 9. Uploaded photo details.

As per the Fig. 10, it is clear that the data and the details about the advertisement posted, contact details of friends, events organised and attended, friends list, messages details, linked mobile device details, voice chat details, video chat details, etc., are able to extract. And this also termed as an valuable evidence (Fig. 11).

Name	Location	Modified Time
ads.htm	/LogicalFileSet1/facebook-paulselwin.zip/html/ads.htm	2015-12-01 01:46:58 IST
contact_info.htm	/LogicalFileSet1/facebook-paulselwin.zip/html/contact_info.htm	2015-12-01 01:46:58 IST
events.htm	/LogicalFileSet1/facebook-paulselwin.zip/html/events.htm	2015-12-01 01:46:58 IST
friends.htm	/LogicalFileSet1/facebook-paulselwin.zip/html/friends.htm	2015-12-01 01:46:58 IST
messages.htm	/LogicalFileSet1/facebook-paulselwin.zip/html/messages.htm	2015-12-01 01:46:58 IST
mobile_devices.htm	/LogicalFileSet1/facebook-paulselwin.zip/html/mobile_devices.htm	2015-12-01 01:46:58 IST

Fig. 10. User interaction details

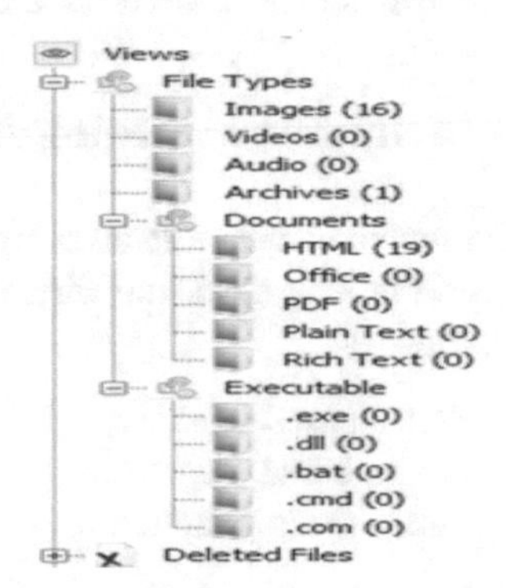

Fig. 11. All uploaded files

3.3 The Facepager forensics is a forensic platform availed legal access by the Facebook without compromising the user privacy by giving the access permission administration in the hands of the account holders. The Facepager analysis is performed over a page related to the Facebook using the login credentials or the appropriate access token. The data gathered are the publicly available data. From the above technique it is found that the Facebook is highly secure which gives tailored access to the forensic perspectives for the analysis. The forensic investigator needs to get the access token from the concerned user to get into access privilege. The details about the public profile are able to be retrieved. This data is possible only with the login credentials or the access token.

4 Real-Time Forensics over an Entity

The chosen domain Amrita Groups of University's related groups and pages are analysed for data. The dataset need to be extracted from the Facebook without getting logged in. Not breaking the legal bindings of the Facebook with the users, government and the domains. So, we are looking only for the public data [5]. Data set extracted are in .csv format with millions of URLs in it with the help of a forensic atomic mail hunter. Taking the data into consideration we are searching only for email-ID and their corresponding URLs. The theme of this extraction and analysis is that, to find the pages shared with email-IDs in their posts as a contact (for example in real-time scene like criminal weapon selling posts in a particular page with email-ID resembles the criminal link establishment with the group if not informed by the group members and also for inactive groups it is a different consideration, similar contact could also be searched in other pages, for the purpose of finding the crime links and also the availability).

In this work, data collection is performed sequentially. When the atomic mail extraction is made over the pages of Amrita, several thousands of URLs are extracted, along with the post related email-IDs. The extraction of the email-IDs from nearly different pages related to Amrita is performed. Facebook groups are also tried for data extraction and found not accessible at any point of approach. Then the data is classified using the semantic matching algorithm. Data is matched with different pages and searched for the similarities in the e-mail ID in more than one pages.

5 Semantic Matching Algorithm

The semantic matching algorithm is used in search engines like Google, Bing, etc. This algorithm is used for the wide range of search which comes in three different types of approaches.

1. Sense-based matcher,
2. String-based matcher,
3. Gloss-based matcher.

From the above we are apt to move with the string based matcher as a label matching algorithm for the initials and then with the other matchers. Therefore using this algorithm an efficient label and node matching is performed and the results are tabulated. The value of 'n' reveals the number of csv files as input is taken in account. The match operator takes two graph-like structures. In those structure two trees for each is graph structure is made available using the dataset. The trees are compared with each other in ensuring the similarity with the data. Then mapping is done between the nodes of the graphs that correspond semantically to each other.

The semantic matching algorithm approach [7] is based on two key notions, they are

1. Concept of a label specifies the set of input files that needs to be classified under a label.
2. Concept at a node specifies the set of files that is classified under a node, at certain position of its tree structure.

Semantic relations between the concepts at nodes of two schemas: equivalence (=) and non-equivalence (≠) as taken from [7]. The algorithm takes as input two schemas and computes as output a set of mapping elements in four macro steps

1. For all labels *L* in two trees, compute concepts of labels, *L*.
2. For all nodes N in two trees, compute concepts at nodes, N.
3. For all pairs of labels in two trees, compute relations among L's.
4. For all pairs of nodes in two trees, compute relations among N's.

Table 1 gives a detail about the availability of similar details in different pages (groups).

Table 1. Data after s-matching module 1

Value of n	Pages	FILE INPUT
5	amritaent hospital	CSV1
	Amrita International Program	CSV2
	Amrita School Business	CSV3
	Amrita Vishwa Vidhyapeetham-CIR	CSV4
	medical science and research amrita	CSV5

OUTPUT FILE		
E-Mail ID	Owner Name	URL
accommodations-ext@fb.com	accommodations-ext@fb.com	https://m.facebook.com/policies/?ref=pf
press@fb.com	press@fb.com	https://newsroom.fb.com/
privacy@groupm.com	privacy@groupm.com	http://www.groupm.com/privacy-policy
k_ganga@amrita.edu	k_ganga@amrita.edu	https://www.amrita.edu/center/tifac-core-cyber-security/people

Table 2 gives detail about a particular person's identity in any group or the file. And also how many time the particular person's identity is found in the form of post in several places.

Table 2. Data after s-matching module 2

SEARCH INPUT	CSV1	CSV2	CSV3	CSV4	CSV5
accommodations-ext@fb.com	1	1	1	1	1
press@fb.com	1	1	1	0	0
maileohye@gmail.com	0	0	0	0	1
k_ganga@amrita.edu	1	1	1	1	1
paul.selwin@gmail.com	0	0	0	0	0

6 API Data Extraction and Analysis

This is a process in which the details are extracted only with the help of Facebook and the user access permission. There are Facebook API explorers like the graph API explorer efficient in extracting the data about the users.

To start with the extraction here the first needed thing to be done is to get the access token by signing into the account using the user credentials [4]. This can be done to analyse own account, or the data analysis of a criminals account which is under charge of legal warrant. In the API explorer multiple data are extracted, viz., name, ID, events, pages liked, mutual friend's type links and posts. Likewise another user account data is also extracted. This additional account could be counted as the criminal's partner or co-worker in crime. Then the data extracted need to be analysed for the details. Here, the semantic matcher plays an important role in classifying the dataset into a confined data of need or demand. The semantic matching algorithm works as like the previous analysis in identifying the data with relations (Tables 3, 4, 5, and 6).

Table 3. Matching the profiles

Matching		T2			
		Photos	Likes	Posts	Events
T1	Photos	=	≠	≠	≠
	Likes	≠	=	≠	≠
	Posts	≠	≠	=	≠
	Events	≠	≠	≠	=

Table 4. Output FB/events (users availability)

Name	Place	City	Country	Latitude	Longitude	Street	Zip	ID
IT Careers in Madurai	Hotel Sangam, Alagar Koil Road	Madurai						204303979954809
Cyber Crime and Legal Enforcement	Bangalore	Bangalore	India	12.9833	77.5833			542899749213145
Riddles of the Sphinx	College of Engineering, Guindy	Chennai	India	13.010979	80.235426	Guindy	600025	1070500303002010
Ethical Hacking and Career Opportunities - CEH Demo	B9ITS - Hacking Trainer	Hyderabad	India	17.43143	78.44762	Show room	500073	436284166578634

Table 5. Output FB/pages_likes (similar identity)

Pages liked	Page ID	Created time
TIFAC - CORE in Cyber Security, Amrita Vishwa Vidyapeetham, Coimbatore	284013928423116	2016-07-16T15:05:01+0000
Cyber Research and Training Institute	778965892115473	2015-11-20T06:50:33+0000
Cyber Psychology Research Centre	8461826054.48504	2015-11-20T06:50:32+0000
Cyber Research Systems (CRS)	173774346002564	2015-11-20T06:50:30+0000
Cyber Research Academy	1496882713877330	2015-11-20T06:50:28+0000
Cyber Research Group - SDVOSB	1388011744772820	2015-11-20T06:50:25+0000
Cyber Research Systems	112008025491450	2015-11-20T06:50:24+0000
Cyber Umbrella Research and Development Corporation Pvt.	135769316631969	2015-11-20T06:50:23+0000

Table 6. Output FB/posts and shared post (Links)

User ID	Post ID	Created time
872261736193015	1004980412921140	2016-05-01T19:13:49+0000
872261736193015	1001719159913930	2016-04-26T12:44:25+0000
872261736193015	1001717946580720	2016-04-26T12:40:28+0000
872261736193015	999117890174065	2016-04-22T07:58:43+0000
872261736193015	999110076841513	2016-04-22T07:30:51+0000
872261736193015	989250904494097	2016-04-06T06:13:05+0000
872261736193015	985744834844704	2016-04-01T01:59:15+0000
872261736193015	968242943261560	2016-03- 13T17: 07:56+0000
872261736193015	966089840143537	2016-03- 10T03:57:47+0000

7 Conclusions

Thus from the entire forensic analysis the report reveals that Facebook is highly secure compared to their competitors and also well-structured in its approach of transfer of data. Data extraction is tougher in Facebook because of its security policies, immediate response to threats and vulnerabilities, most effective periodic upgrade in security measures. It reveals that most forensic techniques used effectively in past times are not able to do forensics now. This is a non-end case. In the organisations forensic analysis we were able to find some related e-mail IDs and their respective URLs are retrieved. This is found efficient in gathering details. This can be used in any suspicious page in-order to get the post's shared evidences. This is an efficient forensic technique. Any single person's post related e-mail IDs is also possibly extracted. When this analysis is performed over different pages of Facebook which is related to some illegal union or any other suspicious, we could possibly take out the post shared email ids and their related URLs. Using this we could possibly find out the availability of the same person's identity in multiple groups with the help of s-matching algorithm. When we take a person trying to sell weapon or any illegal details through post we could possibly trace him without the knowledge of Facebook or its respective user, accessed only using the public available data. The API data reveals the exact data how the user operates. Thus Facebook is a tailored OSN.

References

1. Gragna, G., Riva, G.: Social monitoring and understanding: an integrated mixed methods approach for the analysis of social media. Int. J. Web Based Communities **11**(1), 57–72 (2015)
2. Zhang, Y., Kolaczyk, E.D., Spencer, B.D.: Estimating network degree distributions under sampling: an inverse problem, with applications to monitoring social media networks. Ann. Appl. Stat. **9**, 166–199 (2015)
3. Huber, M., Mulazzani, M., Leithner, M., Schrittwieser, S., Wondracek, G., Weippl, E.: Social Snapshots: Digital Forensics for Online Social Networks. SBA Research, Vienna University of Technology, Vienna (2011)
4. Facebook. Graph API. https://developers.facebook.com/docs/reference/api/
5. The Washington Post. Facebook: a place to meet, gossip, share photos of stolen goods. http://www.washingtonpost.com/wp-dyn/content/article/2010/12/14/AR2010121407423pf:html. Accessed Dec 2010
6. https://www.facebook.com/notes/protect-the-graph/securing-email-communications-from-facebook/1611941762379302/
7. Giunchiglia, F., Yatskevich, M., Shvaiko, P.: Semantic Matching: Algorithms and Implementation. Department of Information and Communication Technology, University of Trento, Trento (2009)
8. Catanese, S., De Meo, P., Ferrara, E., Fiumara, G.: Analyzing the Facebook friendship graph: Department of Physics, Informatics Section. University of Messina, Department of Computer Sciences, Vrije Universiteit Amsterdam, Department of Mathematics. University of Messina

9. Gjoka, M.: Networked Systems UC Irvine, Maciej Kurant, School of Computer and Communication Sciences, EPFL, Lausanne, Carter T. Butts, Sociology Department, UC Irvine, Athina Markopoulou, EECS Department, UC Irvine, Walking in Facebook: A Case Study of Unbiased Sampling of OSNs
10. Karthika, S., Bose, S.: Department of computer Science and Engineering College of Engineering Guindy, Anna University, Chennai-600096, A Comparative Study of Social Networking Approaches In Identifying the Covert Nodes

Cooperative Biometric Multimodal Approach for Identification

Shaleen Bhatnagar(✉)

Computer Science and Engineering,
Alliance University, Bangalore 562106, India
shaleenbhatnagar@gmail.com

Abstract. Possibly the ultimate valuable utility of accurate personal detection is, providing security to authorized access systems from awful security attacks. Within all the currently present biometric schemes, fingerprint identification is most widely used from past few decades. Fingerprint identification scheme is very popular but there are some complexities in recognizing the patterns. Like: Greater dislocation or rotation in image which can causes overlap between fingerprint and create problem during matching, Distortion in fingerprint image, Skin condition and amount of pressure at the time of enrollment and matching, Feature extraction problems. This research paper proposes a cooperative biometric multimodal for fingerprint identification based on minutiae matching by addressing above mentioned problems. This research paper proposes a design a fingerprint identification technique which is expected to give better accuracy and acceptance rate. Main aim of this research work is to decrease False Acceptance Rate (FAR).

Keywords: Fingerprint identification scheme · Feature extraction · Minutiae matching · False acceptance rate

1 Introduction

Biometric scheme uses a person's physical features or behavioral feature to verify their identity. Biometric samples are very much unambiguous to every individual, effortlessly acquired non-intrusively, and remain mostly unchanged for entire life span and detectable without any specialized guidance. It has one of a kind impression, which is acceptable everywhere.

Before deciding the biometric technique we have to figure out the application domain as different domains have different requirements i.e. accuracy, cost, implementability and no single biometric feature can satisfy the entire requirement, which means no biometric is "optimal" [1]. There are two important phase in a biometric authentication technique, Data registration phase and data confirmation phase. During data registration phase, biometric data samples are collected and after applying appropriate processing on the samples they get stored in the database in form of template. During data confirmation phase, same processing will apply on the new data sample and testify the template characteristics to confirm the authenticity (Fig. 1).

S.C. Satapathy and A. Joshi (eds.), *Information and Communication Technology for Intelligent Systems (ICTIS 2017) - Volume 1*, Smart Innovation, Systems and Technologies 83, DOI 10.1007/978-3-319-63673-3_2

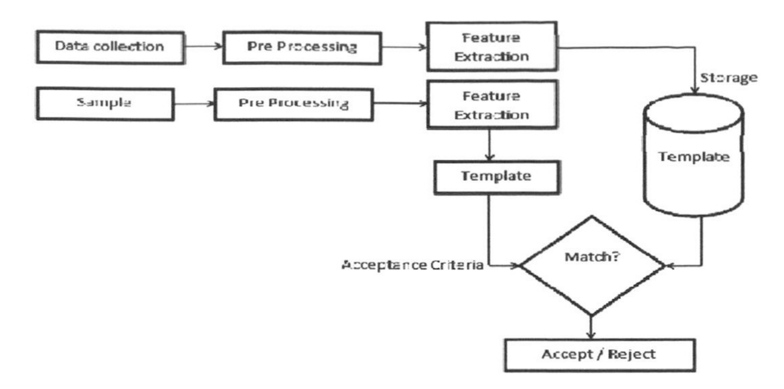

Fig. 1. General biometric model.

2 Fingerprints

Fingerprint recognition or fingerprint authentication deal with the automated method of verifying a comparison between two human fingerprints. Fingerprints are one of the biometrics forms, used to identify and verify the identity of an individual. Due to their uniqueness and consistency over time, fingerprints have been used from a long era, more recently becoming automated (i.e. a biometric) because of advancement in computing capabilities [2].

As shown in Fig. 2 Fingerprint has the highest average value. A short comparison of biometric techniques on the basis of seven factors provided in the below figure.

Biometrics	Universality	Uniqueness	Permanence	Collectability	Performance	Acceptability	Circumvention	Average
Face	100	50	75	100	50	100	50	75
Fingerprint	*75*	***100***	***100***	*75*	***100***	*75*	***100***	***89.3***
Hand geometry	75	75	75	100	75	75	75	78.6
Keystrokes	50	50	50	75	50	75	75	60.7
Hand veins	75	75	75	75	75	75	100	78.6
Iris	100	100	100	75	100	50	100	89.3
Retinal scan	100	100	75	50	100	50	100	82.1
Signature	50	50	50	100	50	100	50	64.3
Voice	75	50	50	75	50	100	50	64.3
Gait	75	50	50	100	50	100	75	71.4

Fig. 2. Average calculation on the basis of biometric system. (Source: A. Jain [3], U. Uludag [4] (the perception based on (High = 100, Medium = 75, Low = 50))

According to the data shown in the Fig. 2, use of each biometric technique is possible. As we know fingerprint are formed by various ridges and valleys on the surface of the finger. Upper skin layer segments of the finger patterns are known as Ridges and the lower segments are termed as valleys. These ridges and valleys, combined form a minutia point. There are many types of minutiae present in fingerprint, including dots, islands, ponds or lakes, spurs, bridges, and crossovers. By these minutiae points it is determined that how unique the fingerprint is.

3 Literature Review

Most of the available literature tends to categorize the fingerprint database on the basis of minutiae sets, singular points and other techniques [5, 6]. R.N. Verma and D.S. Chauhan investigates the major challenges in fingerprint recognition if the images are captured at varying pressure. Their work was directed towards removal of false minutiae points. Their Proposed algorithm starts with image capturing followed up by local histogram equalization, FFT, image Binarization, thinning, skeleton refinement, end-points, filtering and matching. The images achieved from these steps are passed through proposed m-margin algorithm. This algorithm shows improved minutiae extraction process [7]. Research by Dr. Neeraj Bhargava, Dr. Ritu Bhargava, Prafull Narooka, MinaxiCotia depicts that the fingerprints are the most widely used biometric feature for authentication. Their approach for implementation of a minutiae based fingerprint authentication and a minutiae based matching technique is the base for recent fingerprint recognition products [8] (Figs. 3 and 4).

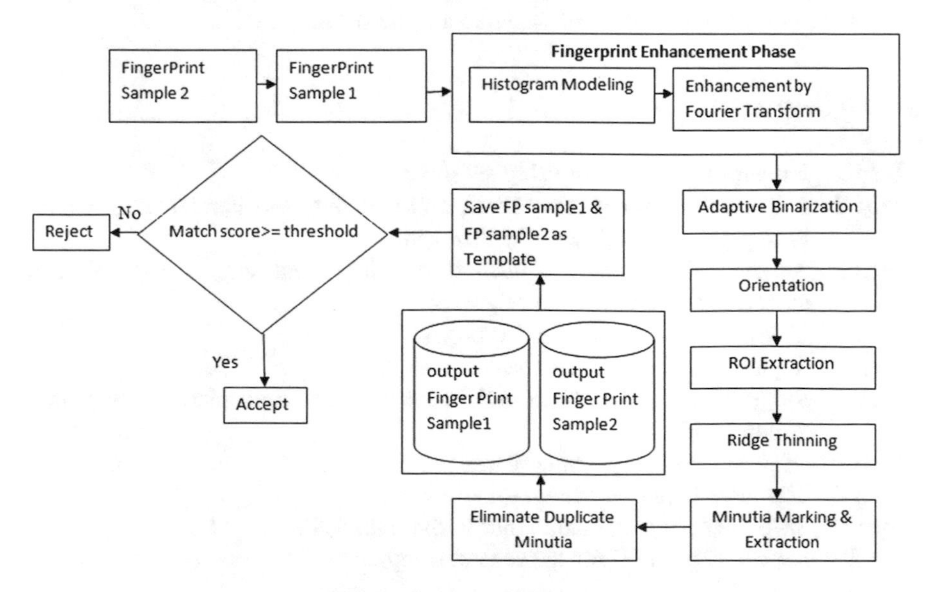

Fig. 3. System architecture.

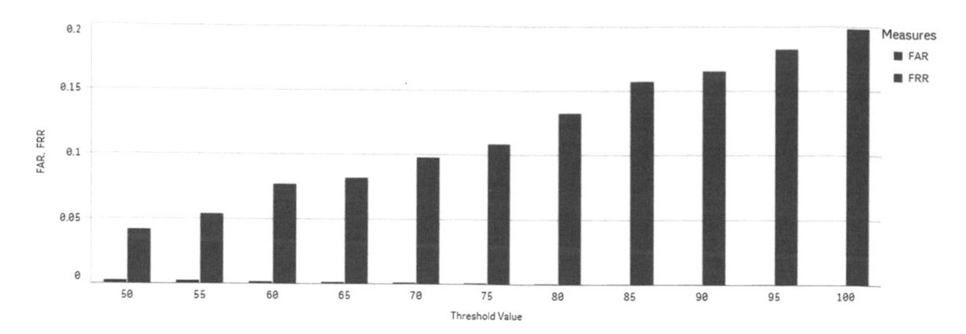

Fig. 4. Comparison graph between threshold value FAR and FRR.

4 Proposed Objective

This paper puts forward a cooperative biometric modal for fingerprint identification based on minutiae matching. Main aim of this research work is to decrease False Acceptance Rate (FAR). In this we are proposing a methodology for fingerprint matching, based on minutiae matching template. Though, like prevalent minutiae matching algorithms, proposed algorithm also works on region of interest (ROI) and line structures that present between minutiae pairs, but it also contains more structural information details of the fingerprint impression to be identified. This research paper is based on cooperative multimodal identification using two fingerprint images as input and using the algorithm it gives powerful assurance of matching minutiae against the given threshold. Results that we get from preprocessed images gives certainty that using such information provide us accurate and powerful matches.

5 Algorithm

Step 1: Fingerprint acquisition using sensor.
Step 2: Finger enhancement using Histogram modeling and Fourier transformation to improve the quality of the fingerprint.
Step 3: Adaptive Binarization to binarize the fingerprint sample to preserve the clarity and contrast of the ridges.
Step 4: Identify the orientation of the ridges.
Step 5: Region of interest extraction.
Step 6: Ridge Thinning to obtain and image having single pixel width with no discontinuities.
Step 7: Mark and extract Minutia points.
Step 8: Eliminate duplicate Minutia.
Step 9: Save the finalize minutia point in the output file.
Step 10: Repeat step 1 to 9 for the second sample.

Step 11: Merge both the output files and form a template and save in database.
Step 12: For authenticate repeat step 1 to 11.
Step 13: Generate Match score using template saved in Step 11 and Step 12.
Step 14: For identification compare match score against threshold value (minimum value acceptable for identification).

5.1 Database

For the experiments, we have used the databases FVC 2002 DB1, DB2, DB3 and FVC 2004 DB1, released on the web [9, 10].

5.2 Judgment Parameters for Fingerprint Authentication

Two parameters are widely approved to find out the accomplishment of a fingerprint authentication system:

FRR (false rejection rate) – Total number of times; a person rejected, instead identified positively. It is calculated by:

$$(\%)\mathrm{FRR} = (\mathrm{FR}/\mathrm{N}) * 100 \tag{1}$$

Where FR = number of incidents of false rejections, N = total number of finger print samples.

FAR (false acceptance rate) – Total number of times, a person positively matched, due to inaccuracy. It is calculated by:

$$(\%)\mathrm{FAR} = (\mathrm{FA}/\mathrm{N}) * 100 \tag{2}$$

FA = number of incidents of false acceptance, N = total number of finger print samples.

6 Result

Using the steps described in the Algorithm following results are obtained for Threshold value increase from 50 to 100 (Table 1).

The proposed algorithm provides 0 false acceptance rate guaranteeing authenticity of the user, when threshold value is 100.

Table 1. Experimental results of fingerprint authentication

Threshold value	False acceptance rate	False reject rate
50	0.22%	4.1%
55	0.19%	5.3%
60	0.14%	7.6%
65	0.10%	8.1%
70	0.8%	9.7%
75	0.5%	10.8%
80	0.3%	13.2%
85	0.2$	15.7%
90	0.1%	16.5%
95	0.01%	18.2%
100	0	19.8%

7 Conclusion

Using a cooperative multimodal biometric system along with fingerprint enhancement mechanism such as Histogram modeling and Fourier Transformation, Quality of the finger print sample is significantly enhanced. So as results obtained depict, very low false acceptance rate resulting in assured authenticity of the user.

8 Future Scope

Results can be further improved on improving image enhancement techniques along with better hardware to obtain the fingerprints. For enhancement in security fingerprint can also be stored with a secret key. To increase security Encryption methods can also be applied. Quality Estimation before enrollment phase can also decrease false reject rate of the fingerprint. Use of Fuzzy vaults can also make biometric system more reliable and secure.

References

1. Jain, L.C.: Intelligent Biometric Techniques in Fingerprint and Face Recognition. CRC Press, Boca Raton (1999)
2. Bhatnagar, S.: A reconsideration of schemes for fingerprint identification. Int. J. Adv. Res. Comput. Sci. **4**(3), 1 (2013)
3. Jain, A.K., Maltoni, D.: Handbook of Fingerprint Recognition. Springer, New York (2003)
4. Uludag, U., Pankanti, S., Prabhakar, S., Jain, A.K.: Biometric cryptosystems: issues and challenges. Proc. IEEE **92**, 948–960 (2004)
5. Jain, A.K., Ross, A.: Learning user specific parameters in a multibiometric system. In: Proceedings of International Conference on Image Processing (ICIP), New York, pp. 57–60 (2002)

6. Fingerprint Classification and Matching by Anil Jain (Department of Computer Science and Engineering, Michigan State University) & Sharath Pankanti (Exploratory Computer Vision Grp. IBM T.J. Watson Research Centre)
7. Wang, X., Li, J., Niu, Y.: Fingerprint matching using orientation codes and polylines. Pattern Recogn. **40**(11), 3164–3177 (2007)
8. Isenor, D.K., Zaky, S.G.: Fingerprint identification using graph matching. Pattern Recogn. **19**(2), 113–122 (1986)
9. http://bias.csr.unibo.it/fvc2002. Accessed 25 Mar 2010
10. http://bias.csr.unibo.it/fvc2004. Accessed 25 Mar 2010

A Hardware Based Technique with an Android Application to Avoid Road Accidents

Nikhat Ikram(✉) and Shilpa Mahajan

Department of CSE and IT, The NorthCap University, Gurugram, India
nk24786@gmail.com, shilpa@ncuindia.edu

Abstract. With the ever-increasing population of the world, increase the accidents at an alarming rate. Accidents are the most uninvited and unintentional miss-happening that causes a lot of injury and loss to human life. So there is an urgent need to stop these accidents to save mankind from destruction. Deaths by accidents have increased and have become the second largest reason for death in the world. Keeping in mind the status of accidents, a working hardware prototype has been proposed along with android based application to avoid road accident. A technique has been suggested where collision of vehicles can be avoided by alerting the driver with a buzzer and a message. Position of the approaching vehicles can be seen on an android application installed on a mobile device that aims to depict the exact location of the vehicles on Google map This whole proposed method is different from other methods as the methods that have been invented until date are mostly in-built techniques while the method proposed here is a prototype that can be implemented in real world.

Keywords: Accident · WSN · VANET · Vehicle

1 Introduction

Billions of people are losing their lives due to accidents. The accidents are the most uninvited, unintentional event that causes lot of damage and even cost person's life. Road accidents are the major integral part of accidents. Road accidents occur due to collision of vehicle with another vehicle or any object. Road accidents are responsible for almost 2.2 million of deaths all over the world each year. Death by accidents has increased to a greater level and has become the second largest reason for death after heart disease. Road injuries have resulted in death of many people in 1991; it was around 1.0 billion which has been increased up to 1.8 billion in 2013. Major reason for road accidents is driving at excess speeds, poorly under-developed highways, poor traffic rules, poor infrastructure and unattended zones. Road accidents occur daily in large number. Most prone areas are highways or sharp turns over road ends. By integrating VANETs with sensor networks, road accidents can be reduced to a greater level. The sensors will send their vehicle's information to the other vehicle using ZigBee. The

S.C. Satapathy and A. Joshi (eds.), *Information and Communication Technology for Intelligent Systems (ICTIS 2017) - Volume 1*, Smart Innovation, Systems and Technologies 83, DOI 10.1007/978-3-319-63673-3_3

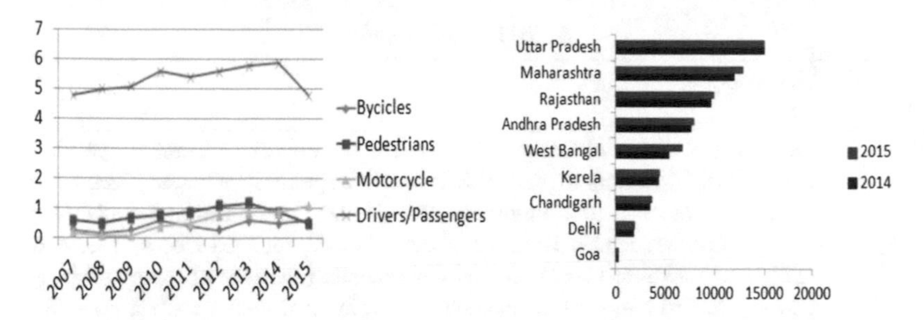

Fig. 1. (a) Traffic death per year (b) Fatality rate due to road accidents in various states in year 2014 and 2015

exchange of information will occur only when the two ZigBee embedded vehicles will fall within the range. Also the position of the nearby vehicle will be displayed on an android application. Figure 1(a) below shows the death rate over years. Figure 1(b) shows the fatality rate due to road accidents in various states of India in past two years.

2 Literature Review

See Table 1.

Table 1. Literature review

Published year	Description
2015	H. Qin et al. has raised an issue on lack of real time sensing of road conditions. A system has been proposed for avoiding road side accidents. In this paper, there are two types of nodes defined vehicle node and sensor nodes. Each vehicle node has two communication interfaces: Wi-Fi (for communication with other vehicle nodes) and ZigBee (for communication with roadside sensor nodes) Vehicle moves in clusters and one of the vehicle nodes will act as a cluster head, cluster head communicate with roadside sensors. Every access point periodically broadcast a beacon message, passerby cluster head hears this beacon message, it sends its registration request to that access point in return access point activates sensor nodes in forward direction, these sensors will monitor the road conditions. These sensors are active only when they find a cluster approaching towards them. If any dangerous condition is detected the sensors will send a warming message in backward direction. After simulation, it was concluded that integration of VANET-WSN will be better than using only VANET based systems [4]

(*continued*)

Table 1. (*continued*)

Published year	Description
2011	R. Alagu et al. stated that mobility models are important factor for performance of MANET. Authors explained Random Waypoint, Manhattan, Gauss-Markov and Random point group mobility models and two routing protocols namely Destination Sequenced Distance Vector (DSDV) and Ad-Hoc on Demand Distance Vector (AODV). Spatial dependency (measure of node mobility direction) and temporal dependency (velocity of different mobile node at different time slot) have been listed. The performances of these mobility models are compared on various parameters and have been found out that Random Point Group-AODV is better than DSDV [6]
2013	Anas Abu Taleb et al. stated that mobility model adopted by the mobile nodes in a network have a good impact on the performance of the whole WSNs. It reduces delay and enhances lifetime of the network and reduces the energy in sending and receiving data. Therefore, the routing protocol used for mobility in WSNs has a good impact on the network performance [7]
2013	Nisha Somani and Yask Patel in defined ZigBee Alliance which was established in 2001. It has two options for implementations: ZigBee (for smaller network) and ZigBee PRO (for larger networks) ZigBee devices are the combination of ZigBee logical (coordinator, router, end devices), application (such as lightening control) and ZigBee physical devices (full function devices and reduced function devices) Its low power consumption limits transmission distances to 10–100 meters line-of-sight. It can be used to transfer data over long distance by passing data through mesh network of intermediate devices. It supports star and tree network. It is built upon physical layer and media access control layer. In a ZigBee network there are 3 types of nodes- • **Coordinator:** its forms the root of network tree and helps to connect other networks. There exists only one coordinator in a network • **ZigBee Router(ZR):** It acts as an intermediate node, carrying data from one device to another. It helps to connect a device to an existing network • **ZigBee Sleepy End Devices:** These devices may be battery-oriented devices. They collect information from sensors [9]

3 Area of Concern

In this paper, a method is proposed to reduce number of collisions to reduce death rates. Collisions likely occur at blind turns, on highways during lane changing and on hilly areas. So an effort has been made to avoid accidents in these areas so as to make world a better and safe place to live Table 2.

Table 2. Various techniques along with their problems

Year	Technique	Problem
2012	Road accidents are increasing day by day. Considerable amount of work has been done in this area like eye blink and alcohol detection sensors. An intelligent car system was proposed by S.P. Bhumkar et al.	The Problem in this system is it cannot avoid collision of vehicle with other vehicles. If a vehicle is coming from back and is on the verge of colliding then this technique will fail [2]
2013	Another technique proposed by T.U. Anand Kumar and J. Mrudula, to avoid accidents that occur due to bad weather conditions	In case if accidents occur, its location is captured and its information is forwarded to a pre-defined numbers but this system also fails if there are no signals in the accident's location [1]
2014	An approach suggested by S. Sasikumar and Dr. J. Kalaivanan, in which NFC reader will be used to detect collision between vehicles. It will send an alert message to the driver if collision is likely to happen. If accident occurs, GSM sends message to server and server will send it to the hospital and nearby police station	The problem arises since signals will be affected by change in environment and also when the car is likely to collide there wouldn't be any time for driver to check his mobile device for the alert message and GPS data will be of little significance to the server if delay is more [3]

4 Proposed Work

This system works for accident prone areas like highways, dead ends and blind turns. The selection of area is the most crucial part of the implementation of the proposed work. The vehicle moving in the areas will act as sensor nodes. Users must have installed android application (developed in this work) on their mobile devices. This application will also help to track the other vehicles over Google Map. For the development of proposed model, the work has been divided into two phases. The first phase involves implementation of prototype and the next phase is to develop android based application for generation of alert signals and to trace location of vehicles on roads.

DETECTION stage: This stage first checks the vehicle is either standing or in motion. If standing, check if it is a blind turn (i.e. vehicle making angle $> 90°$ or $< 270°$ with the road), then collision is likely to occur. If the vehicle is moving, check if the approaching vehicle is within ZigBee's range. It is assumed that these vehicles have ZigBees installed in them and whenever a vehicle comes within this range, an alert message will be generated. The same scenario can be considered for vehicles coming from front as well as from back.

ALERT stage: An alert in the form of buzzer beep will be generated and a message will be flashed on the LCD screen installed inside the vehicle. After the driver gets alert, he then checks the android application to get the location of nearby heading vehicle on Google map. After getting the location, the driver acts accordingly to avoid collision. An algorithm for the scenario is defined below.

ALGORITHM

Notations

A: Area
R1: ZigBee Range R2:
IR Sensor Range V1:
Vehicle 1
V2: Vehicle 2

Step 1: /*Find Accident prone area*/
For all A such that A belongs to {Highways, Sharp turns, Dead ends, Hilly areas}
Step 2: Is vehicle equipped with IR sensor
Is there ZigBee installed in the vehicle
Eligible for application to run
successfully Else
Only detect the vehicles from
behind No detection possible
End

Step 3: Build an android tracking application
Step 4: Collision detection among vehicles
Vehicles belongs to areas, vehicles belongs to
{V1, V2, V3,.....} If |V1, V2| < R1, R1 belongs to
ZigBee Range
ALERT BUZZER (B1)
Driver gets alert and checks the android
application V2 location traced and COLLISION
AVOIDED
Else
No vehicle detected || No ZigBee installed GOTO Step
2 End

Step 5: No application found GOTO Step 3

Step 6: If (V1, V2) < R2, R2 belongs to IR Sensor
Range ALERT BUZZER (B2)
Else
No vehicle detection || No IR sensor installed GOTO Step
2 Else
No application found GOTO Step 3
End

5 Design Philosophy

5.1 Hardware Module

To verify the working of the proposed system, a hardware prototype is developed. A block diagram showing hardware module representing vehicle is shown in Fig. 2. In this, IR sensor is used as a proximity sensor as these sensors have minimum range. A ZigBee attach with the vehicle is used for communication purpose. This protocol has a pre-defined range which is 70–80 m. When a vehicle falls within the ZigBee's range of another vehicle, an alert warning in the form of buzzer will be generated and a message will be flashed on the LCD screen equipped inside the vehicle. Crystal in the model is used to avoid oscillations in the power supply. Microcontroller works at 5 V, the power supply that we are providing is via adapter or by 9 V battery.

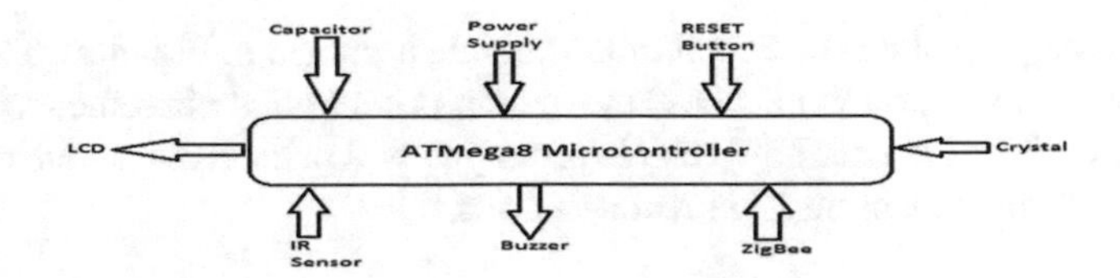

Fig. 2. Block diagram of a vehicle

Figure 3 given below shows the circuit diagram of a vehicle in detail:

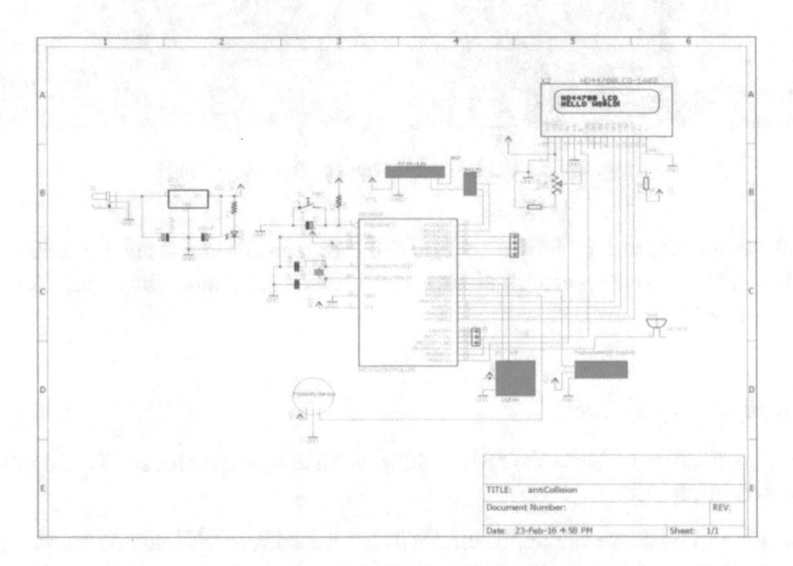

Fig. 3. Circuit diagram of a vehicle

- LED is connected with 5 V VCC along with a resistor, VCC is also connected to one end of 100 microFarad capacitor whose other end is grounded.

- There is an LCD whose pin VSS, RW and K is grounded, VDD is connected to 5 V VCC, VO is connected to a resistance of 10 k along with the load of 1 k, RS is connected to PB5 of microcontroller, E is connected to pin PB4 of microcontroller, DB4, DB5, DB6 and DB7 is connected to pin PD4, PD5, PD6, and PD7 of microcontroller respectively and A is connected to 5 V VCC along with the load of 1 k.
- A Tachometer has 3 pins out of which one is grounded, other one connected to pin PD2 of the microcontroller and the third one is connected to 5 V VCC. ZigBee has four pins one of which is ground, other is connected to 5 V VCC and rest two are input and output pins connected to PD0 and PD1 of microcontroller respectively.
- An RF Module has four pins connected to ground, 5 V VCC, ANT and TH12D respectively. TH12 takes one connection from RF Module and split it into two lines connected with pins PC0 and PC1 which are AC-to-Dc converter. A microcontroller having 28 pins.

A prototype presenting the circuit of the vehicle is shown in Fig. 4(a). The circuit is mounted over vehicle prototype whose movement is controlled via remote control. The prototype is designed to note how vehicle prototype will behave during the emergency, before actually implementing it on real scenarios.

Fig. 4. (a) Vehicle prototype (b) Remote control

The remote control is shown in Fig. 4(b). IR sensors are used for sending and receiving signals. These sensors control the vehicle movement in either forward or backward direction.

5.2 Android Application

Android application to track the approaching vehicle was developed. This application is executed in phases.

Phase1: In this phase the application will get the current vehicle's location via GPS on the display screen of the cellular device.
Phase2: The location captured is then transfer to the server where it is stored for later use. This transfer is done with the help of File Transfer Protocol.

Phase3: From server this application fetches location of all the other vehicles except for its own location. This process also takes place via File Transfer Protocol.
Phase4: All the location accessed by the application i.e. its own along with other vehicles are plotted over Google Map with different color markings(purple for its own and cyan for other vehicles) as shown in Fig. 6.

Figure 5 shows the four phases of android application.

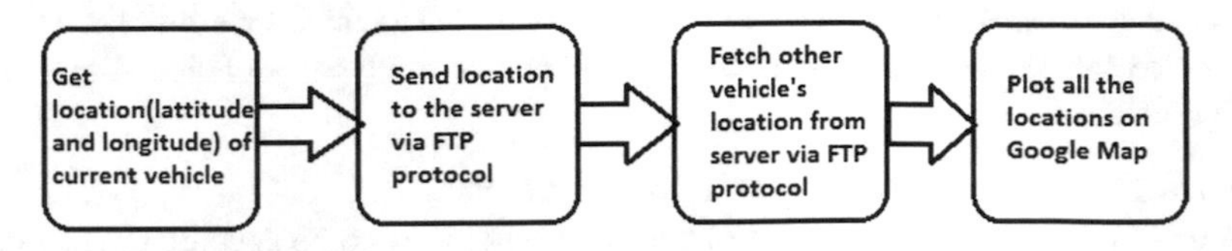

Fig. 5. Four phases of Android application

Android application code consists of five classes.

GPSTracker class make use of Location Manager which provide the location of vehicle.
Background Task class take the url of the server and POST the data onto the server.
Track class basically takes the location and separates them with a comma and then returns the values.
Maps Activity class takes latitude and longitude values from server and plot it into Google Map.
Main Activity class which is the main class, it makes use of Telephony Manager which is for getting the id of Mobile phone. This class calls all the above classes. Figure 6 below shows the screenshot of the android application.

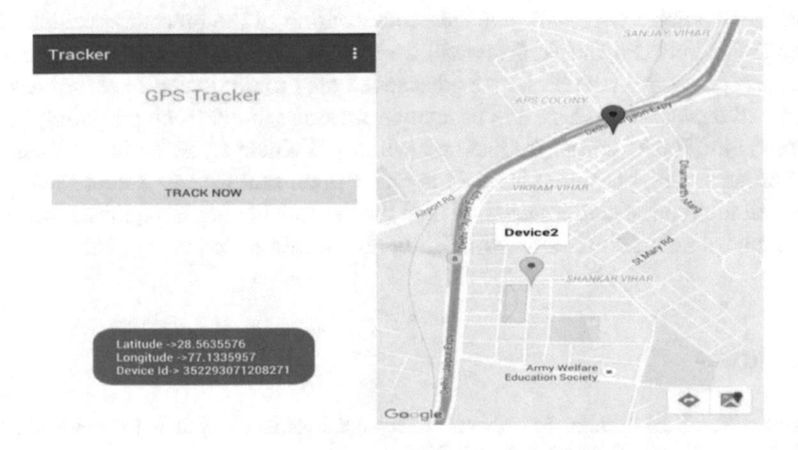

Fig. 6. Android application screenshots

6 Results

For evaluation of the above proposed model, different scenarios are considered. The results have been analyzed on monitoring the behavior of the vehicles.

Scenario 1: *Vehicles within range of each other*

When approaching vehicles comes within each other range, a buzzer will buzz and a message will be displayed on LCD screen "Vehicle Detected Nearby". Vehicle position can be seen on an android application installed on user's mobile. User can clearly see the position of approaching vehicle and will get sufficient time to take suitable action.

Scenario 2: *Vehicle moving infront of other vehicle*

When there is a vehicle in front of another vehicle. At some specific range a buzzer will ring to alert the driver and also a message will be displayed on LCD screen as "Vehicle in the Front Wake Up".

Scenario 3: *Vehicle moving at the back of other vehicle*

An IR sensor is also mounted at the back of the vehicle to alert the driver about approaching danger from back. A buzzer alert along the message on LCD "Vehicle at the Back Beware" will be signaled.

7 Conclusion and Future Work

The vehicle prototype mentioned in the paper is a defensive way of approaching the problem of road accidents by vehicle collision. Road accidents will be reduced by avoiding the collision of vehicles. There will be a buzzer alert inside the vehicle whenever the vehicle falls within the ZigBee range, a message will be displayed on the LCD of the vehicle stating "Vehicle Detected Nearby", the driver then will check the android application which will give the exact location of the preceding vehicle. In this way collision between the vehicles will be avoided. In addition to this, if there is a vehicle preceding from the back on a great speed the proximity sensor at the back of the vehicle will detect it and there will be an alert message in the front car stating "Vehicle Detected at the Back" and in the back car will be "Vehicle Detected in the Front Wake Up". Further if this prototype is made to work upon real time in future there will be changes in the type of sensors used, as then the sensors having large range will be used and the environment conditions will also be taken into account.

References

1. Kumar, T.A.S., Mrudula, J.: Advanced accident avoidance system for automobiles. Int. J. Comput. Trends Technol. **6**(2), 45 (2013)
2. Bhumkar, S.P., Deotare, V.V., Babar, R.V.: Accident avoidance and detection on highways. Int. J. Eng. Trends Technol. **3**(2), 247–249 (2012)

3. Sasikumar, S., Kalaivanan, J: Accident avoidance and device control using various intermediate devices in wireless technology. SSRG Int. J. VLSI Signal Process. (SSRG-IJVSP) **1**(1), (2014)
4. Qin, H., Li, Z., Wang, Y., Lu, X., Zhang, W. and Wang, G.: An integrated network of roadside sensors and vehicles for driving safety: concept, design and experiments. In: 2010 IEEE International Conference on Pervasive Computing and Communications (PerCom), pp. 79–87. IEEE
5. Festag, A., Hessler, A., Baldessari, R., Le, L., Zhang, W., Westhoff, D.: Vehicle-to-vehicle and road-side sensor communication for enhanced road safety, vol. 4 (2015)
6. Alagu Pushpa, R., Vallimayil, A., Sarma Dhulipala, R.: Impact of mobility models on mobile sensor networks. Int. J. Commun. Netw. Secur. **1** (2011)
7. Taleb, A.A., Alhmiedat, T., Hassan, O.A., Turab, N.M.: A survey of sink mobility models for wireless sensor network. J. Emerg. Trends Comput. Inf. Sci. **4**(9) (2013)
8. Rupinder, K., Gurpreet, S.: Survey of various mobility models in VANETs. Int. J. Eng. Comput. Sci. **3**(3) (2014)
9. Int. J. Eng. Comput. Sci. ISSN:2319-7242, Volume 3 Issue 3, March 2014
10. Somani, N.A., Patel, Y.: Zigbee: a low power wireless technology for industrial applications. Int. J. Control Theory Comput. Model. (IJCTCM) **2**, 27–33 (2012)
11. Prashanth, K.P, Padiyar, K., Naveen Kumar, P.H., Santhosh Kumar, K.: Road accident avoiding system using drunken sensing technique. Int. J. Eng. Res. Technol. (IJERT), **3**(10), (2014). ISSN: 2278-0181
12. Danquah, W.M., Altilar, T.D.: HYBRIST mobility model—a novel hybrid mobility model for VANET simulations. Int. J. Comput. Appl. 0975–8887, **86**(14), 15–21 (2014)
13. Vasanthi, V., Romenkumar, M., Ajithsingh, N., Hemalatha, M.: A detailed study of mobility models in wireless sensor networks. J. Theor. Appl. Inf. Technol. **33**(1), 7–14 (2011)
14. Bai, F., Sadagopan, N., Helmy, A.: The important framework for analyzing the impact of mobility on performance of routing protocols for Adhoc networks. AdHoc Netw. J. **1**, 383–403 (2003)

Hidden Decision Tree Based Pattern Evaluation Using Regression Models for Health Diagnosis

K. Chandra Shekar[1(✉)], K. Venugopala Rao[2], and Priti Chandra[3]

[1] IIIT – Rajiv Gandhi University of Knowledge Technologies, Basar, India
chandhra2k7@gmail.com
[2] G. Narayanamma Institute of Technology and Sciences, Hyderabad, India
kvgrao1234@gmail.com
[3] Advanced Systems Laboratory, DRDO, Hyderabad, India
priti_murali@yahoo.com

Abstract. Regression Models evolved since in Health Diagnosis contributes to predict the health condition through various approaches and analysis. These analytical approaches require statistical methodology for the better diagnosis in data mining applications, utilizing numerical predictions through regression coefficients' approximations. Here, we propose a model to diagnose the Health condition through error pattern discovery from the known and unknown databases; by designing a hidden layer to extract hidden patterns through Hidden Decision Tree approach. Our proposed model evaluates the database for (a) resulting the different health diagnosis, (b) patternising these to optimized queries based on returned results, (c) effective implementation and extraction of data being supplied by the end users, to handle data regularities and irregularities and (d) evaluate the pattern through HDT approach for the accuracy of the knowledge discovery. Proposed model can be implemented on a medical data to evaluate the error pattern for numeric additive quantities to obtain a better health diagnosis model.

Keywords: Pattern evaluation · Regression models · Hidden decision trees · Health diagnosis · Predictive accuracy

1 Introduction

Data Mining (DM) applications has been increasing for several years rapidly in variety of applications in public and private sectors; and it is still expected to grow in future as well. Application areas which adopted DM include banking, insurance, retail, telecom, pharmaceutics, health diagnosis, marketing, customer relationship management, fraud detection [4–6], product development, process planning and monitoring, information extraction, risk analysis, all sort of e-businesses and even in the area of Science and Technology. All these areas include DM analysis components: business area understanding, data understanding, preparation of data, data modelling, evaluation, application methods, testing and resulting to a DM analysis framework.

S.C. Satapathy and A. Joshi (eds.), *Information and Communication Technology for Intelligent Systems (ICTIS 2017) - Volume 1*, Smart Innovation, Systems and Technologies 83, DOI 10.1007/978-3-319-63673-3_4

Data mining is an attempt to make sense of the information explosion embedded in the huge volume of data. Data mining tasks can be descriptive ie., data interesting patterns or relationships, and predictive ie., data behaviour classification of the model based [1–3]. Data mining is a field of predicting outcomes and uncovering relationships of data. Data mining also embed with machine learning, statistics, artificial intelligence, databases, and visualizations [14, 15]. To improvise these, there is a need of more features and instances, algorithms and architectures, and automation tools to be developed through hybrid approaches. Data mining constitutes one or more of the following functions, namely classification, prediction - regression, clustering, characterization and discrimination, discovering association rules, functional dependencies and rule extraction.

Our proposed data mining approach, involves the following activities:

- extract efficiently the interested and meaningful patterns to the user from the given volume of data
- threshold the patterns to discover the needful data
- by subjective approach, examine the needful patterns for information about the data and
- deliver action on similarity and dissimilarity of information to the user

The purpose of regression model which can be linear or polynomial [10], is to find and map a data to a prediction variable. The variable is an attribute from the class containing the input attributes. The input to the regression model is a training dataset with several such attributes, which are holding dependable attributes, can be numerical in nature. The goal of the regression model is to distribute the several dependent variables in terms of the predictor attributes. These predicted attributes are assigned with the database for testing records to determine the dependent attribute [11, 12]. Or as an alternate, a numerical attribute which has continuous-valued with quantitative values, can be discretized to discrete values and symbolic values called as class labels and then proceed for classification.

Some systems running at fixed operating modes are linear in nature, resulting in to linear health diagnosis techniques. In case of nonlinear systems, which are not running at fixed operating modes, even the presence of well-established techniques utilized by linearized models cannot be employed. Further, even due the occurrence of nonlinear occurrences, the fixed operating models may deviate and as a result the linear models may increase in errors, run at false alarm instead of detecting the correct cause. Considering this, an idea of health diagnosis for nonlinear systems has been identified. Under this, several approaches has been introduced, including observer-based approaches, neural networks based approaches and fuzzy network based approaches. In this paper, we propose a Hidden Decision Tree (HDT) [7, 9, 13] based Pattern Regression Evaluation Model for nonlinear systems health diagnosis.

2 Pattern Regression Evaluation Model

The regression model-based methods require a relational model of the nonlinear system for the generation of response between modeled outputs and input relations [8]. Since various conditions in the system may cause different patterns in the relational model, an evaluation model is developed to process these various conditions. In order to properly select the input variables for nonlinear dynamic system, a linear multi-variant statistical regression model have to be designed to detect the health correlation between input and their relation variables in nonlinear systems.

In our proposed approach, a typical off-line procedure is considered, where data regarding abnormal and healthy conditions, were obtained from training. Health parameters were identified for health diagnosis, along with the threshold values. If the system response is within predetermined threshold values, the system is in a healthy condition. The regression evaluation model will withstand from variations in the model parameters estimated and digression of model parameters from normal are identified. The proposed data mining algorithm is an instantiation of the model-based-search components. The pattern regression evaluation model, generates sequential patterns, like in series analysis. The goal of the model is to state the generating patterns, extract and report deviations over the time.

Proposed Algorithm for Model Attribute:

Input: Let T = $a_1,a_2,\ldots,a_n$ denote finite text (character or symbols) data string
Let P = $b_1,b_2,,\ldots,b_m$ denote finite pattern data string to search in text T
where n and m are positive integers greater than 0
Let P = $\{P_1,P_2,,\ldots,P_k)$ is a finite set of sequence of patterns, where each sequence may result to equal or unequal alphabet.

Output: Model Attribute $y_{j,P}^{H}(v)$

Procedure:

1. Initiate the pattern weights using $\sum_{x=1}^{n} w_x a_x - t = 0$, defines the pattern decision boundary
2. Update the pattern weights according to $w_x(t+1) = w_x(t) + l(d-y)a_x$, where d is the desired pattern, t is the iteration threshold and l is learning rate
3. The regressive perception of pattern is given by, $y = \sum_{x=1}^{n} w_x(t+1)a_x - t$
4. The hidden pattern approach is, $y_j^{h+1} = \sum w_x^h(t+1)a_x^h - t_j^{h+1}$, where (h+1) are hidden changes, j is the hidden unit with total pattern input y_j^{h+1}
5. For h > 0 hidden layers, the output pattern is $y_{j,P}^{H}(v) = \frac{1}{1+e^{a_x^h}}$, where $y_{j,P}^{H}(v)$ is the output state obtained for node j in layer H for input pattern P^1

To explore the proposed decision tree, we take set of all references listed and we take one reference to create a node N for it and find the list reference. Once the list reference is built, the same N is repeated on references taken with offsets until whole decision tree is built. To select the pattern matching, the entries were checked with pattern reference entries with null label and not null label. Hence, checking of each label is listed and the best pattern is obtained as a resultant.

At that stage, we are left with no features, or data list is empty or tree has already reached maximum set length value. The following approximation model will provide the algorithm for developing a proposed decision tree. The proposed Hidden Layer Approximation Model Algorithm is as follows,

Proposed Algorithm for Hidden Layer Approximation Model:
Input: Approximation A, set of instances sI
Output: Pattern Instances Tree with approximated label node lN.
Procedure:

1. If Feature(A) return lN(A) else set the minimum set of instances (sImin) to zero
2. Approximate A into layers A1, A2,,…, AN according to the values of lN
3. If layers (A_1, A_2,,…, A_N) < sImin, sImin = layers (A_1, A_2,,…, A_N)
4. Approximate A into models Ai according to the model in F
5. If A_i not empty then A_i = layers (A_1, A_2,,…, A_N) else Ai is a lN (A)

3 Proposed Hidden Decision Tree Learning Algorithm

The basic function for hidden decision tree produce a response based on the input signal. The output response is formed through a weighted linear combination of the basic functions and system response. Learning a decision tree require knowledge about the each node with its node leaves and depth of the node tree.

Our proposed Hidden Decision Tree Learning Algorithm is tabulated below.

1. Check for the node root.
2. For each patternReference, find the Pattern by choosing the linkReference
3. Find the Feature of the pattern by the Condition predicted by C and valLeaf L.
4. For each pattern, by selecting the Approximation model, find the instances for minimum pattern recovery time.
5. Create a Decision Tree Node on the pattern chosen in above step.
6. Run above steps for remaining features decision tree.

4 Nonlinear Systems Health Diagnosis Based on Decision Tree Output

As we result, the proposed HDT output, we need to select some important features to diagnose with the large number of failures. Considering this, we use the following problem-solving approaches:

1. Node Tree leafs were not Health diagnosed because of zero Health results.
2. Filter: Filter the leaves which are less than or equal to percentage of c of the total number of failures, considering only large fraction of failures.
3. Merge: We merge the Node Trees by eliminating the preceding nodes.
4. Rank: Failure count will be prioritised based on the fail prediction.
   Considerable comparison between system response and its estimation is verified for system reliability.

In Fig. 1, the analytical model is illustrated. The method monitors the level of residual and when the signal reaches a given threshold, desired actions will be taken. Proposed FD algorithm is implemented using software on a processor controlled computer. Our proposed algorithm is a model based approach to improve the analytical redundancy in the system design. The proposed model concentrates on measuring the available inputs, output variables and evaluators; resulting towards health diagnosis. The proposed model consists of two stages.

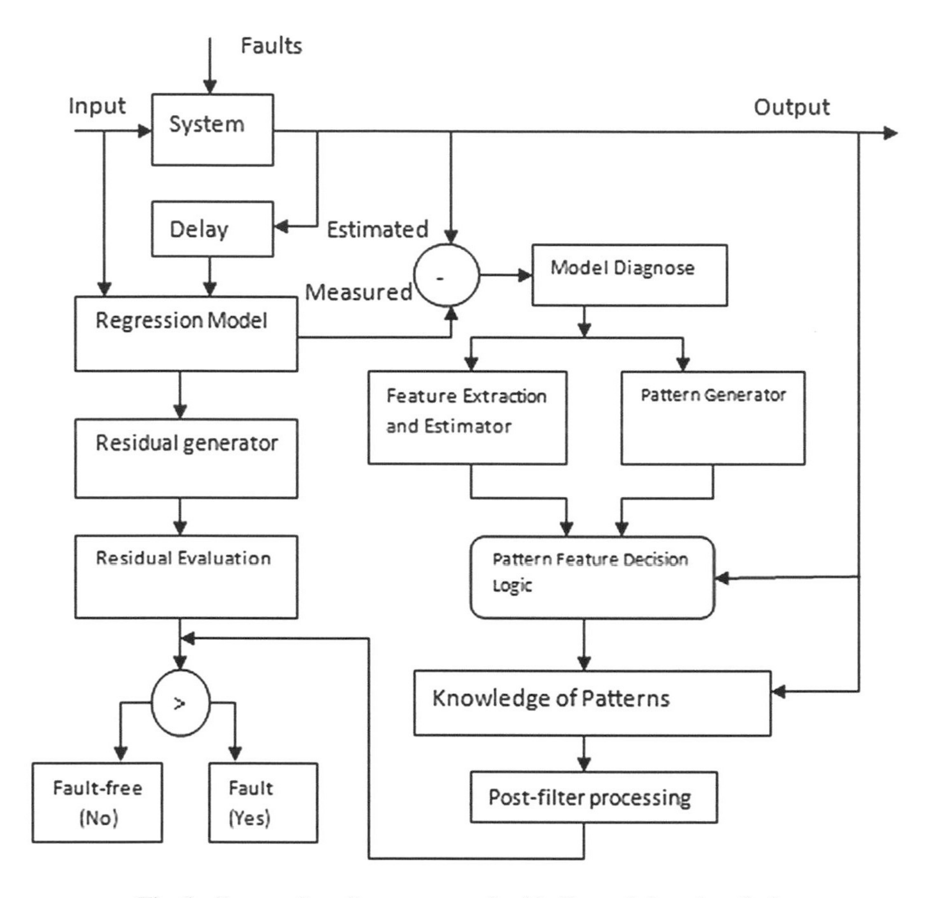

Fig. 1. Proposed nonlinear systems health diagnosis based on faults

In residual stage, the signal is generated by computing the outputs with its estimates. Since the output signal is affected by system noises and disturbances, we need to further process the results obtained at this stage. Our proposed model, represented in qualitative and quantitative behavior of the process, uses analytical model-based FD techniques for the purpose of residual generation. The Health result from residual evaluation will be in two states, either deviated from the normal function or where there is no functioning at all. In this stage, the residual model is processed through evaluated residual, which is compared with the threshold and if the value is out of range, an indication is provided for preventing from failure.

In Model stage, to detect the feature pattern, the deviation is obtained by comparing the estimated output with measured output, with their estimates. Based on the model selected, the features are extracted, and are estimated in parallel with the patterns generated. Model diagnosis is mainly used by the health control in order to estimate the measured states in the process, while the control is used for diagnosis purpose. The condition for process control is post-filter processing, which is used for health state estimation as well as FD control. This filtering process is robust against noises and disturbances, leading to an efficient health diagnostic system.

The proposed knowledge discovery learning procedure to determine the Health conditions and decision threshold settings in the given data set of linked error patterns, is provided below:

Proposed Algorithm: Proposed knowledge discovery algorithm (Health conditions, threshold)

Input: Health conditions and decision threshold settings
Output: filter processed patterns
Procedure:

1. identify the Health conditions and decision threshold settings to feature patterns.
2. identify the feature pattern vector [P_{v1}, $P_{v2,}$,…P_{vn}]
3. evaluate linear hidden layer equation $\sum_{y=1}^{n} R_y I_R$, defines the health decisions.
4. evalute the learning measure $\sum_{y} R(t,f_y) - s_y(t,f_y)$, where $s_y(t,f_y)$ is the Health condition state.
5. find the discovery weights $D_w(t,f_y) = \left[[R(t,f_y)]^T + \sum_{y} \omega(t)\alpha(t) \right]$, where $\omega(t)$ is weight of patterns discovered and $\alpha(t)$ is connected subset of weights.
6. patterns are filtered by $P_i = \sum D_w(t,f_y) N_{y+n}(t) * \omega(t)$, results the processed filter patterns.

5 Analytical Evaluation of Propose Model Behaviour

We have implemented the proposed algorithms for HDT in Java. We used the implementation of proposed algorithm and its associated rules using machine learning tools package Weka. The experiments are executed using JDK 1.4.2 on Intel(R) Core(TM) i5 @2.60 GHz Windows machine with 4 GB of RAM.

In our proposed method, we first classified the datasets into business and medical data. On the basis of meta data contained, the instances and attributes in datasets were selected. Weka Explorer enabled the proposed algorithm under trees classification root, which explores and implements the algorithm.

In our study, we considered 8 data-sets, collected from the UCI machine learning repository websites.

Table 1 summarizes the properties of the selected datasets.

Table 1. UCI datasets' characteristics

Dataset	Size	Attributes			No. of iterations	No. of instances	Missing values	No. of classes
		Total	Ordered	Unordered				
Balance-scale	103	5	5	0	53	625	0	3
Credit-rating	42	16	14	2	11	690	0	2
Mail database1	207	58	44	14	73	4601	0	2
Trains	3	33	30	3	4	10	0	2
Diabetes	39	9	8	1	9	768	0	2
Hepatitis	21	20	11	9	11	155	0	2
Liver-disorders	51	7	5	2	27	345	0	2
Lung-cancer	7	57	54	3	3	32	0	2

The algorithms are compared on equal Weka specific settings for each data set selected. We conducted experiments using 08 standard datasets from UCI repository benchmarks datasets used in patter learning, using our proposed HDT model approach. We evaluated the performance of proposed tree model by estimating the accuracy using the default 10-fold cross-validation, as shown in Table 2:

Table 2. Comparison of tree performance.

Dataset	No. of leaves	No. of inner classifiers	No. of hidden units	Execution time (in sec.)	RMSE
Balance-scale	52	489	136	0.01	0.2699
Credit-a	30	604	86	0.01	0.2313
Mail database1	104	4288	303	0.41	0.1562
Trains	2	10	0	0	0.2464
Diabetes	20	587	181	0.01	0.3463
Hepatitis	11	45	110	0.01	0.2630
Liver-disorders	26	247	88	0.01	0.4025
Lung-cancer	5	35	4	0.01	0.3394

UCI data set has been used for analysing performance of proposed HDT model approach. Table 3 lists the performance w.r.t. parameter analysis of the proposed HDT Model when run on UCI business and medical datasets.

Table 3. Parameter analysis of proposed HDT model

Dataset	TP rate	FP rate	Precision	Recall	F-Measure
Balance-scale	0.866	0.273	0.832	0.866	0.849
Credit-a	0.961	0.244	0.961	0.961	0.961
Mail database1	0.950	0.178	0.950	0.960	0.980
Trains	0.950	0.200	0.957	0.960	0.999
Diabetes	0.838	0.427	0.835	0.838	0.838
Hepatitis	0.939	0.558	0.925	0.939	0.925
Liver-disorders	0.787	0.456	0.783	0.787	0.880
Lung-cancer	0.881	0.324	0.868	0.881	0.866

6 Conclusions

In this paper, a HDT based Regression Model is presented for nonlinear systems. Hidden layer technique is applied for residual generation, a trained Pattern Evaluation Model is used for non-linear model of the system, proposed scheme presented in this paper is used to diagnose the problems in nonlinear system response and a HDT based decision tree is introduced to isolate the multiple faults based on decision thresholds. The developed algorithm approaches produced good pattern extraction, fault isolation and resulted in good diagnostic results. For nonlinear system diagnosis, the use of proposed decision tree reduced the computational efforts. The experiments conducted on 08 datasets from UCI, with proposed HDT decision tree achieved improved performance, by analyzing & removing irrelevant attributes, by pattern measures on the known and unknown datasets and by considering unique instances from the datasets. Finally, we conclude that the proposed hidden decision tree based regression model for pattern evaluation proves to be an effective method for predictive classification.

References

1. Chandra Shekar, K., Chandra, P., Venugopala Rao, K.: Fault diagnostics in industrial application domains using data mining and artificial intelligence technologies and frameworks. In: IEEE International Advance Computing Conference, pp. 538–543 (2014)
2. Pelillo, M., Siddiqi, K., Zucker, S.W.: Matching hierarchical structures using association graphs. IEEE Trans. Pattern Anal. Mach. Intell. **21**(11), 1105–1120 (1999)
3. Gehrke, J.E., Ramakrishnan, R., Ganti, V.: Rain-Forest—a framework for fast decision tree construction of large datasets. Data Min. Knowl. Discov. **4**(2/3), 127–162 (2000)
4. Rastogi, R., Shim, K.: PUBLIC: a decision tree classifier that integrates building and pruning. Data Min. Knowl. Discov. **4**(4), 315–344 (2000)
5. Chen, W., Saif, M.: Fault detection and isolation based on novel unknown input observer design. In: American Control Conference, p. 6, June 2006
6. Bennett, K.P., Cristianini, N., Shawe-Taylor, J., Wu, D.: Enlarging the margins in perceptron decision trees. Mach. Learn. **41**, 295–313 (2000)

7. Battula, B.P., Rama Krishna, K.V.S.S., Kim, T.-H.: An efficient approach for knowledge discovery in decision trees using inter quartile range transform. Int. J. Control Autom. **8**, 325–334 (2015)
8. Patton, R.J., Chen, J., Lopez-Toribio, C.J.: Fuzzy observers for nonlinear dynamic systems fault diagnosis. In: Proceedings of the 37th IEEE Conference on Decision and Control, vol. 1, pp. 84–89 (1998)
9. Ferdowsi, H., Jagannathan, S.: A unified model-based Health diagnosis scheme for non-linear discrete-time systems with additive and multiplicative faults. Trans. Inst. Meas. Control **35**, 742–752 (2013)
10. Demetriou, M.A.: A model-based fault detection and diagnosis scheme for distributed parameter systems: a learning systems approach. ESAIM Control Optim. Calc. Var. **7**, 43–67 (2002)
11. Isermann, R., Ball, P.: Trends in the application of model-based fault detection and diagnosis of technical processes. Control Eng. Pract. **5**(5), 709–719 (1997)
12. Patton, R.J., Chen, J.: Observer-based Health detection and isolation: robustness and applications. Control Eng. Pract. **5**(5), 671–682 (1997)
13. Chang, S.K., Hsu, P.L.: A novel design for the unknown input Health detection observer. Control Theory Adv. Technol. **10**, 1029–1051 (1995)
14. Ding, S.X.: Model-Based Health Diagnosis Techniques, Design Schemes, Algorithms, and Tools. Springer, Berlin (2008)
15. Chen, J., Patton, R.J.: Robust Model-Based Health Diagnosis for Dynamic Systems. Kluwer Academic Publisher, Boston (1999)

A Security Approach and Prevention Technique against ARP Poisoning

Sudhakar(✉) and R.K. Aggarwal

Department of Computer Engineering, National Institute of Technology, Kurukshetra, Haryana, India
sudhakarjnv@gmail.com, rkal5969@gmail.com

Abstract. Tenderfoot, presently clients who are utilizing the web however do not worry about the security issues. The information that is being transmitted on the system is not thought to be protected. There is such a variety of dangers like sniffing, ridiculing, phishing exits. With the assistance of a few devices like Wireshark, firewall and Microsoft disk operating framework, we can counter quantify the assaults. Here, in this paper we proposed an answer, which is greatly, improved the other proposed solutions based on the ARPWATCH and ARP central server (ACS).

Keywords: Spoofing · Sniffing · MITM · ARP poisoning · Ettercap · ARPWATCH

1 Introduction

The term MITM i.e. Man-in-the-middle assault which is gotten from the wicker container situation, where the players of a solitary group pass the ball to crate however other cooperative people's tries to seize them while doing the basket This is known as 'pail unit assaults' or 'Monkey-In-The Middle' attack. In this MITM assault, there might be a third individual that is mimicking the casualties between the client and server [1].

In this MITM assault, there is a typical situation, which includes two ends (victims) and outsider assailant. The aggressor can control the messages, which are getting traded on the communication channel. It is as shown in Fig. 1. Both the casualties attempt to instate the correspondence between them by trading their open/public keys (Message M1 and M2). Be that as it may, on the correspondence channel it will hinder by the interloper and the gate crasher or aggressor send its own open keys (Message M3 and M4) to the casualties. After that, casualty 1 scrambles their messages alongside the assailant's open/public keys and send the encoded message to the casualty/victim 2. The aggressor on the correspondence channel grabs the scrambled message and decodes it with his private key. After that, the aggressors encode the decoded message (plain content M6) with his own particular open/public key and send this scrambled message to casualty 2.

The outcomes that the aggressor persuades both the casualties that they are utilizing the protected channel however as a part of reality their messages are gotten to by the outsider know as interloper/intruder.

S.C. Satapathy and A. Joshi (eds.), *Information and Communication Technology for Intelligent Systems (ICTIS 2017) - Volume 1*, Smart Innovation, Systems and Technologies 83, DOI 10.1007/978-3-319-63673-3_5

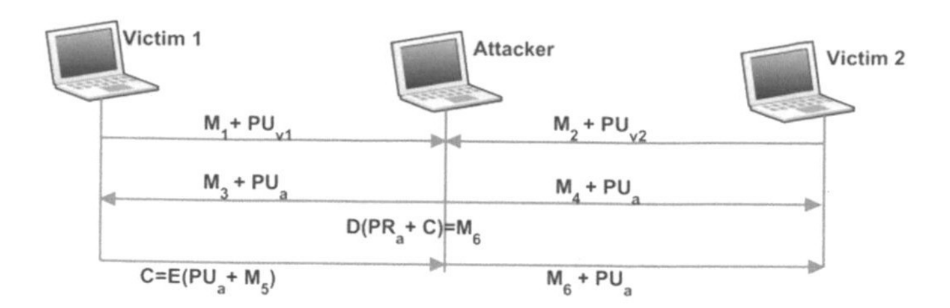

Fig. 1. Message exchanged in a MITM attack

1.1 ARP Poisoning Basic

Address resolution protocol (ARP) is stateless which means that it does not need any ARP request before ARP replying. So, ARP cache may be get infected with fraudulent MAC-IP associations. ARP protocol does not provide any reliable means for authentication, which leads it to some serious attacks like Session hijacking, DOSs or Man-In-The-Middle-Attacks. These all attacks may cause some serious loss or damage to the Local Area Network (LAN) [2].

Address Resolution Protocol (ARP) send requests and replies. In general, if a system wants to communicate with another system then it needs MAC-IP associations of both the communicating parties. So here, two systems like A and B wants to communicate on the network, it requires broadcasting an ARP request to fetch the MAC address of the other communicating entities. It is as shown in Fig. 2.

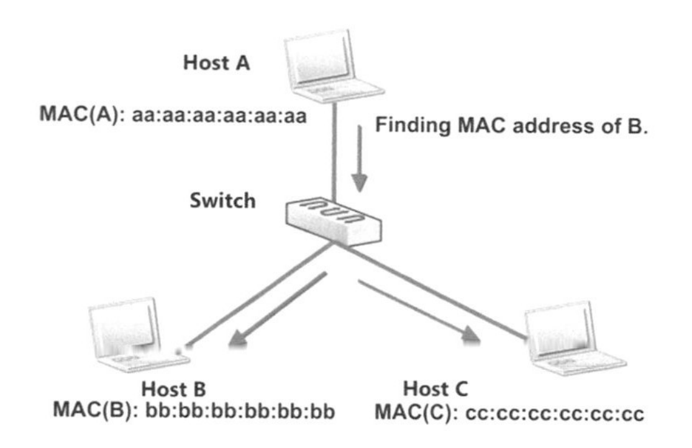

Fig. 2. Broadcasting the ARP request from Host A to Host C and Host C

After getting the ARP request, System B will send a unicast reply message with their MAC address. It is as shown in Fig. 3. When Host A will receive the B's reply, the communication process will proceed further and this MAC-IP association will be stored in ARP primary cache of Host A for a particular amount of time [3].

ARP is unable to authenticate the sender's identity so anyone can poison the ARP cache entry within the LAN. Here, it is possible for some other systems like system C to send fake ARP reply and impersonating as system B. Now, the traffic which is

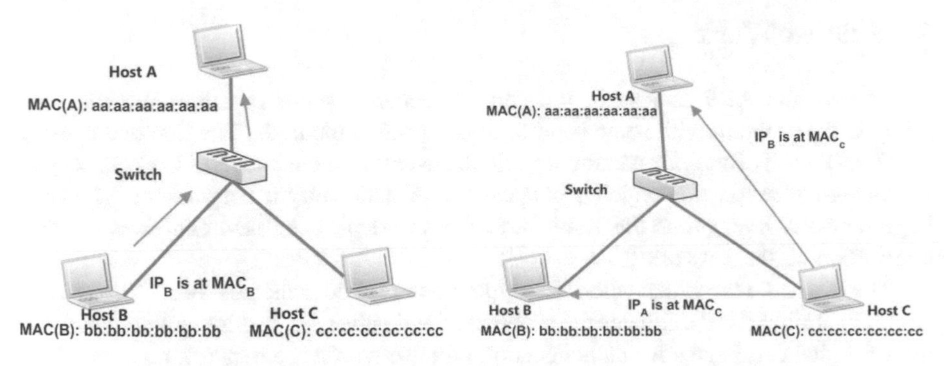

Fig. 3. Host B unicast message to Host A

Fig. 4. ARP Poisoning by host C on Host A and B

required by system B will receive by the system C as it have the intended MAC address of B [4]. It is as shown in Fig. 4.

1.2 ARP Centralized Server (ACS)

Actually, this ARP central server maintains two ARP cache table. The first one is Primary ARP cache and second one is Secondary ARP cache, which is maintained by the ACS. In case, if the primary ARP cache table get poisoned then the victim sends the request to the ACS for approval of MAC-IP binding. When the ACS receives the message from victim, if it mismatches with the secondary ARP cache table then it will send the correct MAC-IP binding.

This method allows the ACS to monitor the attempts of ARP poisoning and help to detect the ARP poisoning attacks on the network. This way ACS can counter measure the ARP poisoning [5].

1.3 ARPWATCH

ARPWATCH is an open source tool which is used to monitor the Ethernet traffic activity. It maintains a database of MAC-IP pairings with the timestamp. This helps us to WATCH carefully which Mac-IP association has been taken place for what period of time. It also has a facility for sending the report through email to the administrator. So, if there is unexpected pairing (like MAC-IP pairing changed or added) is found then it notices it and sends a report to the administrator.

Especially, the network administrator to keep Watching on the Ethernet traffic activity to detect poisoning of ARP cache table or unexpected MAC-IP bindings [6] utilizes this tool.

2 Related Work

To prevent the ARP attacks in the current scenario is not possible. We can only mitigate the problem with some handful recommended methods. The first one is using static ARP entry. But after making it static the user has to change the MAC gateway as per changing the location which is not possible. A static entry is not possible for a large huge network. Every time the administrator has to deploy the new entries as per new connection on the network [6].

The second recommendation is cryptography-based schemes like Secure ARP (S-ARP). This S-ARP mitigates the problem of authentication in ARP protocol with the use of digital certificates for authentication purpose within the network for every ARP replies. For public key distributions, Authoritative Key Distributor (AKD) server is used to different hosts. The implementation of this cryptography-based scheme requires changing the ARP standard specifications. This is such a high-level implementation cause's backward compatibility problem with the pre-existing network as well as AKD server is the central server for key distributions. So, it might cause a single point of failure problem [7].

S. Kumar et al. [5] proposed a brought together system for identification and counteractive action of ARP harming. In this procedure, an ARP Central Server (ACS) is utilized to approve the ARP table's entrances of all the host of the system. Clients likewise keep up an auxiliary long-term cache in this approach. Nonetheless, this procedure does not address the IP fatigue issue, which an attacker can make inside a system. In addition, this method is brought together in nature. Thus, the disappointment of ACS server leaves the system shaky. This also causes the problem of IP exhaustion like on the off chance that an MAC-IP in tosh mapping is not present in ACS reserve the aggressor can send ridiculed message with some irregular Macintosh address. Subsequently, the ACS will store this mapping into reserve and also a secondary table. Since, ACS itself contains this mapping, every single other host inside LAN will respect this mapping. The assailant can send numerous such ARP messages parodied with various IP address. This will prompt to IP exhaustion issue. Instead of this issue, it is very compatible to IP aliasing configuration. Since the approach permits mapping of a Macintosh address with more than one IP address, the approach is good with IP Associating setups. This strategy is in reverse good with the current system framework. Since the approach does not require any change in ARP specification.

The technique proposed by P. Pandey depends on ICMP bundles as test packets to approve the ARP messages [9]. This proposed show orders the aggressors under the accompanying three categories. The initial one is Weak Attacker second one Intermediate and the third one Strong attacker. The first weak attacker can create ARP satirize packets utilizing any software. However, they do not have traded off convention stack. The weak assailant has the force of creating fake packets yet it can't stop or control other host or system gadgets from doing their typical job. Assailant can't stop the other host or systems administration gadgets from producing the reaction or different packets. The second one is Intermediate attacker. This classification of an aggressor is an interfacing join between the Weak attackers and Strong attackers. An Intermediate attacker can produce mock packets and it can change its own convention stack with the end goal that it can create a reaction for any bundles which it gets as its interface. However, these attackers can't interfere with different other hosts in LAN. The third classification of an attacker is Strong attacker which is more capable than the Weak and Intermediate attacker. Such attacker can produce fake packets as well as can modify the protocol stack.

G.N. Nayak et al. [1] proposed two arrangements with a specific end goal to avoid ARP harming. The first sends ARPing demand messages to the default entryway at settled time interval. Nonetheless, this system is constrained for the checking of the host to entryway movement as it were. The activity between one host to another host is not examined for discovery and anticipation of ARP harming attack. The second strategy screens the ARP table at general interims. It, for the most part, checks what number of IP locations is connected with the Macintosh address of door. In the event that the number is more than 1, it cautions the client about conceivable harming. The hindrance is that the Macintosh address of the portal ought to be known ahead of time.

G. Jinhua et al. [10] proposed ICMP convention based identification algorithm for ARP spoofing. This algorithm gathers and dissects the ARP packets and afterward infuses ICMP echo request packets to test for the malicious host as indicated by its response packets. Nevertheless, the algorithm depends on a database accessible at Detection host. Along these lines, it causes the single point failure issue. Likewise, the algorithm does not address the issue of ARP harming utilizing fake ICMP resound demands. An assailant can send a fake ICMP to resound ask for with the mock source IP address of an honest to goodness have and the source Macintosh address of itself. Accordingly, when the victims have to get this message, it will overhaul its ARP store with ill-conceived restricting having aggressor's Macintosh deliver tie to a parodied authentic IP deliver assigned to another host inside the subnet. The assailant can send a similar sort of fake ICMP echo request to default entryway additionally to get the MITM position between the default door and the casualty (victim) have.

3 Problem Description

Here, in this Fig. 5, we are delineating the issue of ARPWATCH configured system. For our necessity, we have to do the IP aliasing in our network. But here at the time of listening to the network, ARPWATCH identify it as the ARP table get poisoned. After mismatching of the MAC-IP, the system will check the MAC-IP in current secondary ARP cache table. But, if the MAC-IP pairing is not present in the Secondary ARP cache table. It will generate the alarm otherwise it will treat as a legitimate user. So this IP associating prerequisite can make the issue which is illustrated with the help of flow diagram as below.

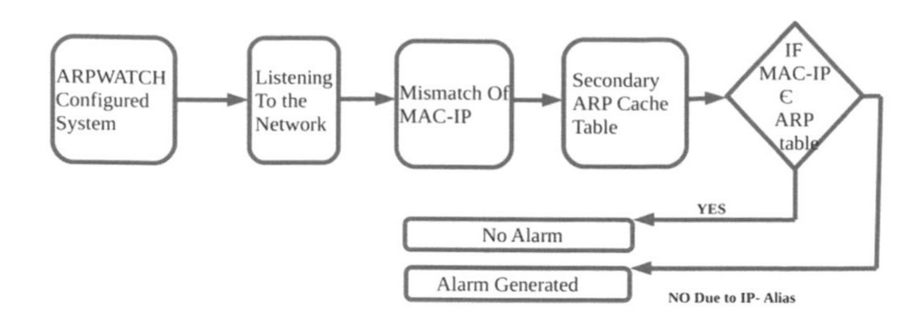

Fig. 5. Generating false alarm due to IP aliasing

4 Proposed Model

In this paper, we proposed a model which is made by the blend of ARPWATCH and Approval Server. Our work consolidate from the idea of DNS cache poisoning [12], ARPWATCH and Centralized detection tool. In this model, we utilize the ARP-WATCH as an identification apparatus on the system. Being conveyed in nature, the approach does not make a solitary purpose of disappointment issue. Be that as it may, the ARPWATCH is not good with the IP Associating setup. Since the location instrument raises the caution in the event that it sees an adjustment in Macintosh IP mappings, this approach will make the false alert if the IP Associating is designed for a portion of the Macintosh addresses. So the approach is not good with IP Associating arrangement.

Here, we concocted a thought that on the off chance that we join this approval server with the ARPWATCH apparatus along these lines, it can make good with the IP associating setup. This approval server is a sort of ACS server which approves the ARP tables' entries of all the host inside the network. This approach ought to keep up a long-term cache table at every client side. This way we can solve the problem of IP aliasing in the ARPWATCH (Fig. 6).

Here, the traffic, which is generated by the Attackers, is going to filter by the configured ARPWATCH and Centralized Server. So the problem of ARP poisoning

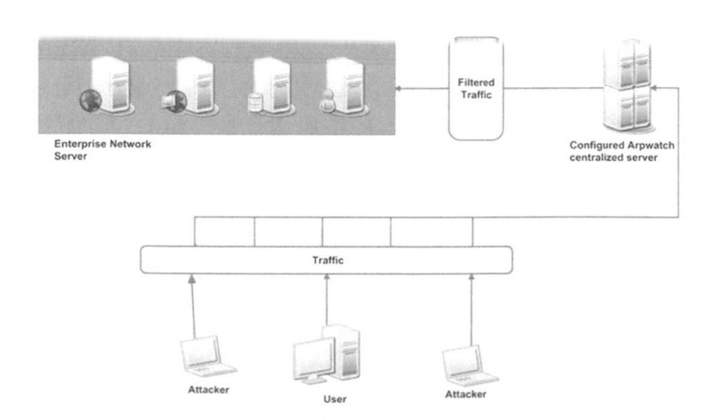

Fig. 6. ARPWATCH configured with ACS

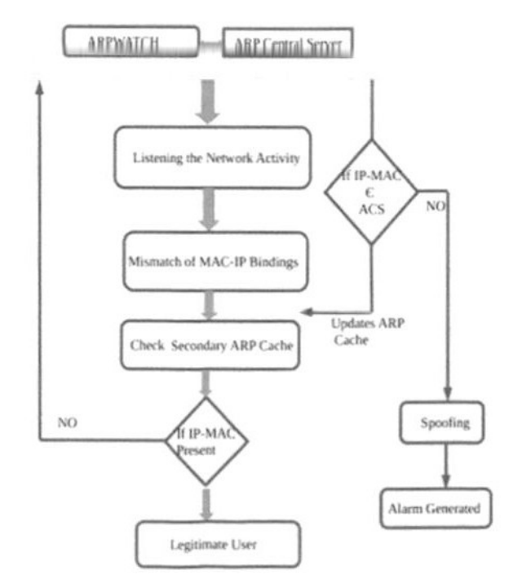

Fig. 7. Flow diagram to solvethe problem of IP Alias

can be tackled. Now, in this model, we are going to attach the ACS with configured ARPWATCH system so that the time generating the false alarm, it should check the MAC-IP associations with the secondary cache in ACS. If the secondary cache contains the associations of that MAC-IP, then it will not generate the false alarm. This can be explained in further flow diagram so that we can handle the mismatch of MAC-IP (IP Aliasing) which is as shown below.

To handle this issue, we are appending ACS with the ARPWATCH configured framework. In Fig. 7, we are outlining that at whatever point the ARPWATCH listen to any system interface. In the event that there is any bungle between the relationship of Macintosh IP. At that point, it will first go to the endorsement server, which has its own auxiliary ARP reserve table. If Macintosh IP partner has a place with that ARP cache table, then it will not create the caution else, it will. ACS will send a reply message to the Secondary ARP cache table. So that it can update its secondary ARP table and declare the user as a legitimate user for further future communications. Thusly our model can evacuate the issue of ARPWATCH at whatever point IP associating required in any system interface and handles the assault issue in the system.

4.1 ARPWATCH

Network administrator overseer screen ARP movement to distinguish ARP caricaturing, arrange flip-flops, changed and new stations and address reuse. ARPWATCH is cross-platform open source software and is released under the BSD permit.

ARPWATCH stores the just current condition of the system ETHERNET/IP pairings and permits to send email warning when a blending change happens. This is fine for little and rather static systems. In ARPWATCH case all the historical backdrop of matching is sent just organization post box. At the point when ARPWATCH is accustomed to checking at least dozen systems, it turns out to be difficult to monitor the notable address use data [8].

1. The commands to watch a particular interface on any system, with "I" alternative.
   # ARPWATCH -i eth0
2. Execute the commands to begin the administration service shown in Fig. 8
   # sudo /etc/init.d/arpwatch start

```
sud@sud-VirtualBox:~$ sudo /etc/init.d/arpwatch start
[sudo] password for sud:
Starting Ethernet/FDDI_station monitor daemon: (chown arpwatch /var/lib/arpwatch/eth0.dat) arpwatch-eth0.
```

Fig. 8. Showing ARPWATCH started

It creates a log of IP MAC pairing address alongside a timestamp when the IP MAC pairing showed on the system. Arpwatch uses packet capture library (pcap) to listen for arp packets on a local Ethernet interface. The pcap library gives a high-level interface to the systems. All packets on the network system, even those bounds for different hosts, are open through this mechanism.

3. To check the current ARP table

Address Resolution Protocol (ARP) is a convention for mapping a Web Convention address (IP address) to a physical machine address that is perceived in the local network. A table, more often than not called the ARP cache, is utilized to keep up a connection between's every Macintosh address and its comparing IP address. ARP gives the convention principles to making this relationship and giving location change in both headings (Fig. 9).

```
swastik@swastik:~$ arp -a
? (172.16.59.253) at 00:04:96:6c:f7:7e [ether] on eth0
? (172.16.58.217) at 08:9e:01:36:a6:c5 [ether] on eth0
swastik@swastik:~$
```

Fig. 9. ARP table

4. To check the version of ARPWATCH # nano version.c
5. To WATCH the syslog passages at "/var/log/syslog" or "/var/log/message" record says that there is another Macintosh or IP is changing the Macintosh address in the system (Fig. 10).

```
sud-VirtualBox arpwatch: flip flop 10.0.2.3 52:54:00:12:35:03 (08:00:27:6b:a9:45) eth0
sud-VirtualBox arpwatch: flip flop 10.0.2.3 08:00:27:6b:a9:45 (52:54:00:12:35:03) eth0
```

Fig. 10. Showing syslog entries

6. The commands to configure for listening the system. # sudo nano /etc/ARPWATCH.conf ARPWATCH.conf file sent to listen on eth0, and email root. After changing the file, restart ARPWATCH using: # sudo /etc/init. d/ARPWATCH restart (Fig. 11).

```
arpwatch.conf ×
# /etc/arpwatch.conf: Debian-specific way to watch multiple interfaces.
# Format of this configuration file is:
#
#<dev1> <arpwatch options for dev1>
#<dev2> <arpwatch options for dev2>
#...
#<devN> <arpwatch options for devN>
#
# You can set global options for all interfaces by editing
# /etc/default/arpwatch

# For example:

eth0    -m root
#eth1   -m root
#eth2   -m root

# or, if you have an MTA configured for plussed addressing:
#
#eth0   -m root+eth0
#eth1   -m root+eth1
#eth2   -m root+eth2
```

Fig. 11. ARPWATCH configured files

7. To send a caution to client mail id, we need to open the framework setup record "/document/sysconfig/ARPWATCH" and include the email address. The mail notice will be sent to the predetermined mail id with log points of interest. # OPTIONS=" -u ARPWATCH -e mailid@gmail.com -s 'root (ARPWATCH)'".
8. At the time of completion, we need to install the mailutils and configure it # sudo apt-get install mailutils- It is utilized to record the Hostname, IP address, Macintosh address, vendor name and timestamps.

4.2 The Basic Idea

ARPWATCH used to monitor the Ethernet activity, whenever our ARP cache table get poisoned then it generates the alarm to pay attention on the use of network. So that we can counter the assaults. But in case of IP aliasing it always generate the false alarm which is not required result.

4.2.1 IP aliasing [11]

IP associating will be partner more than one IP deliver to a system interface. Here, it is shown in figure (Fig. 12).

With this, one hub on a system can have various associations with a system, every filling an alternate need. The false alarm generated by ARPWATCH, when listening to the Ethernet network which is shown as below (Fig. 13).

```
root@sud-VirtualBox:~# ifconfig -a
eth0      Link encap:Ethernet  HWaddr 08:00:27:6b:a9:45
          inet addr:10.0.2.15  Bcast:10.0.2.255  Mask:255.255.255.0
          inet6 addr: fe80::a00:27ff:fe6b:a945/64 Scope:Link
          UP BROADCAST RUNNING MULTICAST  MTU:1500  Metric:1
          RX packets:53 errors:0 dropped:0 overruns:0 frame:0
          TX packets:164 errors:0 dropped:0 overruns:0 carrier:0
          collisions:0 txqueuelen:1000
          RX bytes:5177 (5.1 KB)  TX bytes:17926 (17.9 KB)

eth0:0    Link encap:Ethernet  HWaddr 08:00:27:6b:a9:45
          inet addr:10.0.2.117  Bcast:10.255.255.255  Mask:255.0.0.0
          UP BROADCAST RUNNING MULTICAST  MTU:1500  Metric:1

lo        Link encap:Local Loopback
          inet addr:127.0.0.1  Mask:255.0.0.0
          inet6 addr: ::1/128 Scope:Host
          UP LOOPBACK RUNNING  MTU:65536  Metric:1
          RX packets:531 errors:0 dropped:0 overruns:0 frame:0
          TX packets:531 errors:0 dropped:0 overruns:0 carrier:0
          collisions:0 txqueuelen:0
          RX bytes:47410 (47.4 KB)  TX bytes:47410 (47.4 KB)

root@sud-VirtualBox:~#
```

Fig. 12. IP Aliasing

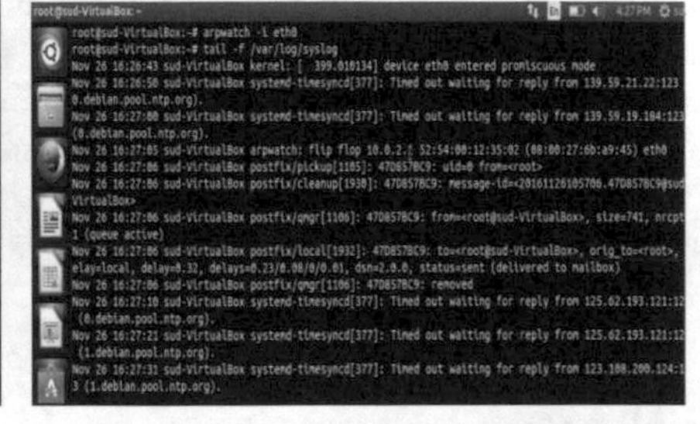

Fig. 13. False alarm by ARPWATCH

5 Result and Evaluation

Based on these three above parameter, we can easily observe from the table that how our approach is much better than the others proposed scheme are.

Here, we are using the integrated ARPWATCH along with the ACS, which is giving the confined result (Table 1).

Table 1. Concludes the comparison among different approaches for the mitigation of ARP poisoning.

Parameter	Existing technique ARPWATCH	Existing technique ACS	Proposed scheme ARPWATCH + ACS
Backward compatibility	YES	YES	YES
Comply with single point failure	YES	NO	YES
Affinity with IP-Aliasing	NO	YES	YES

Comply with single point of failure- The existing ACS technique is the centralized tool. It will fail if the ACS fails. Therefore, we are integrating with the ARPWATCH.

Affinity with IP-Aliasing- If our system is configured with the IP-aliasing. Individually, ARPWATCH will show the spoofing result. So to avoid this we are integrating with the Existing ACS technique.

The integrated system which is ARPWATCH along with the ACS will nullify the effects of the individual existing technique. So this way we can get better result.

6 Conclusion and Future Scope

The gave setup can introduce a conceivable answer for the ARP poisoning issue. It utilizes ARPWATCH and additionally ACS. This expels the irregularity from ARP entries from the system. Since the irregularities are expelled, the ARP harming can't be conceivable. This plan additionally permits the regressive similarity to existing systems and not helpless to central point of failure. Since the conveyance way of the ARPWATCH and there is no any adjustment in the specification of ARP convention. The IP fatigue assault is past the extent of this paper. We are just proposing the model not executing it in this paper. This is past the extent of this paper.

References

1. Nath Nayak, G., Samaddar, S.G.: Different flavours of man-in-the-middle attack, consequences and feasible solutions. In: Proceedings of 3rd IEEE International Conference on Computer Science Information Technology (ICCSIT), vol. 5, pp. 491–495 (2010)
2. ARP poisoning basics: Retrieved from http://www.ARPpoisoning.com/how-does-ARP-poisoning-work/. Accessed 22 Oct 2016
3. Tripathi, N., Mehtre, B.: Analysis of various ARP poisoning mitigation techniques: a comparison. In: International Conference on Control, Instrumentation, Communication and Computational Technologies (ICCICCT), pp. 125–132 (2014)
4. Khurana, S., kaur, R.: A security approach to prevent ARP poisoning and defensive tools. Int. J. Comput. Commun. Syst. Eng. (IJCCSE) **2**(3), 431–437 (2015)

5. Kumar, S., Tapaswi, S.: A centralized detection and prevention technique against ARP poisoning. In: 2012 International Conference on Cyber Security, Cyber Warfare and Digital Forensic (Cyberese), pp. 259–264 (2012)
6. Monitoring Ethernet. http://www.tecmint.com/monitor-ethernet-activity-in-linux/. Accessed Nov 2016
7. Bruschi, D., Ornaghi, A., Rosti, E.: S-ARP: a secure address resolution protocol. In: 2003 Proceedings of the 19th Annual on Computer Security Applications Conference, pp. 66–74. IEEE (2003)
8. ARP-s command. http://linux-ip.net/html/tools-arp.html. Accessed 22 Apr 2010
9. Pandey, P.: Prevention of ARP spoofing: a probe packet based technique. In: IEEE International Advance Computing Conference (IACC), pp. 147–153 (2013)
10. Jinhua, G., Kejian, X.: ARP spoofing detection algorithm using ICMP Protocol. In: 2013 International Conference on Computer Communication and Informatics (ICCCI), pp. 1–6. IEEE (2013)
11. IP Alias command. http://www.tldp.org/HOWTO/pdf/IP-Alias.pdf. Accessed Nov 2016
12. Antonakakis, M., Dagon, D., Luo, X., Perdisci, R., Lee, W., Bellmor, J.: A centralized monitoring infrastructure for improving DNS security. In: International Workshop on Recent Advances in Intrusion Detection. pp. 18–37. Springer, Berlin, September 2010

A High-Speed Image Fusion Method Using Hardware and Software Co-Simulation

Rudra Pratap Singh Chauhan[1](✉), Rajiva Dwivedi[2], and Rishi Asthana[3]

[1] Uttarakhand Technical University, Dehradun, UK, India
chauhanrudra72@gmail.com
[2] Galgotia University, Greater Noida, UP, India
dr.rajivadwivedi@gmail.com
[3] IMS Engineering College, Ghaziabad, UP, India
asthanarishi1973@gmail.com

Abstract. The process of adding significant information of two source images obtained from various sources into one image is called image fusion. Large volumes of data informations are obtained from various remote sensors. These informations are useful for image diagnosis through image fusion. Thus image fusion is the promising area of research. Many methods of image fusion have been suggested by the previous authors to produce a fused image having higher spatial resolution, but due to large amounts of data calculations, it is a time-consuming process. Therefore, a reconfigurable hardware system having high speed such as Field-programmable Gate Array (FPGA) is used for solving complex algorithm with reduced computation time to achieve parallel operation with high-speed characteristics. This paper describe the design and implementation of improved speed discrete wavelet transform based multisensor image fusion process with its implementation on hardware. MATLAB 2016a Simulink tools are used to integrate the Xilinx System generator with averaging method for image fusion. Algorithm design has been synthesized in Xilinx ISE 14.1and the same is implemented on ML 605 Virtex-6 FPGA kit. From the result, it is observed that the design consumes a total power of 4.36 W and operates at a maximum frequency of 851. 06 MHz.

Keywords: DWT · Image fusion · Xilinx System Generator (XSG) · Field Programmable Gate Array (FPGA)

1 Introduction

The process of adding important pictorial information of two source images obtained from various sources into one image is called image fusion. The source images may be taken from various satellite sensors, obtained at different times, or having different spatial and spectral characteristics. The prime objective of this method is to preserve the most relevant features of each source image. Image fusion process required large amount of data computation, and sometimes it is needed to store a large volume of data and process it with less computation time. This task required to perform complex algorithms with significant amount of computation [2, 4, 5]. A hardware/software

S.C. Satapathy and A. Joshi (eds.), *Information and Communication Technology for Intelligent Systems (ICTIS 2017) - Volume 1*, Smart Innovation, Systems and Technologies 83, DOI 10.1007/978-3-319-63673-3_6

co-design are the right choice to solve complex algorithms with less computing time using FPGAs. Designing a system and use of appropriate algorithm will reduce the computation time and provide an efficient solution to this problem. Given this, a reconfigurable system and software co-simulation method are proposed [6].

Recently, FPGAs has emerged as a useful reconfigurable programming device which is highly popular as compare to available other programmable devices like Programmable Array Logic (PAL) and Complex Programmable Logic Device (CPLDs). The reconfigurable property of the FPGA is its ability to change the functionality of this device. The internal architecture of gate array makes re-programming possible. Because of this flexibility of FPGA, it has a vital application in the field of image fusion. In this paper, we use FPGA for the hardware implementation of a picture fusion algorithm using pixel averaging methods and CDF 9/7 filter transforms. The rapid prototyping of image fusion algorithm is possible due to model-based design.

2 Hardware and Software Co-Simulation Method of Image Fusion

Figure 1 shows the wavelet transform based image fusion programming model developed by using Xilinx System Generator (XSG) of Xilinx 14.1 design tool which is configured with MATLAB R2016a. The MATLAB R2016a has a unique feature to integrate the MATLAB, Simulink environment with Xilinx block set and support Virtex-6 FPGA [1]. The MATLAB is a powerful high-level language used for technical computing and scientific calculation. The important feature of MATLAB is its integration ability with another language.

In Fig. 1 we use two input images PAN and MS images for the fusion process. The DWT is used to decompose the PAN and MS images into different frequency sub-bands by using the pixel averaging method the various subband images are fused to retain the relevant information of the picture. The inverse wavelet transforms IDWT

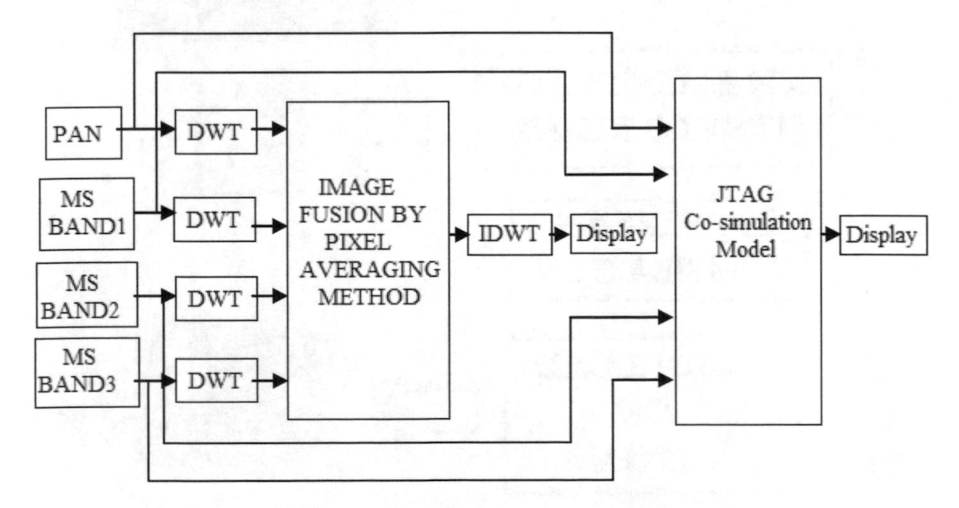

Fig. 1. Wavelet transform based image fusion programming model

used to reconstruct the fused image and finally applied to display unit to display the resultant fused image. Once the model is designed properly, it is useful for hardware and software co-simulation by using JTAG Co-simulation model.

3 Implementation of Design Flow

Figure 2 shows the implementation of image fusion algorithm and hardware software co-simulation design flow. The first step of design flow uses MATLAB, Simulink for the development of an image fusion algorithm shown in Fig. 3. Xilinx block sets

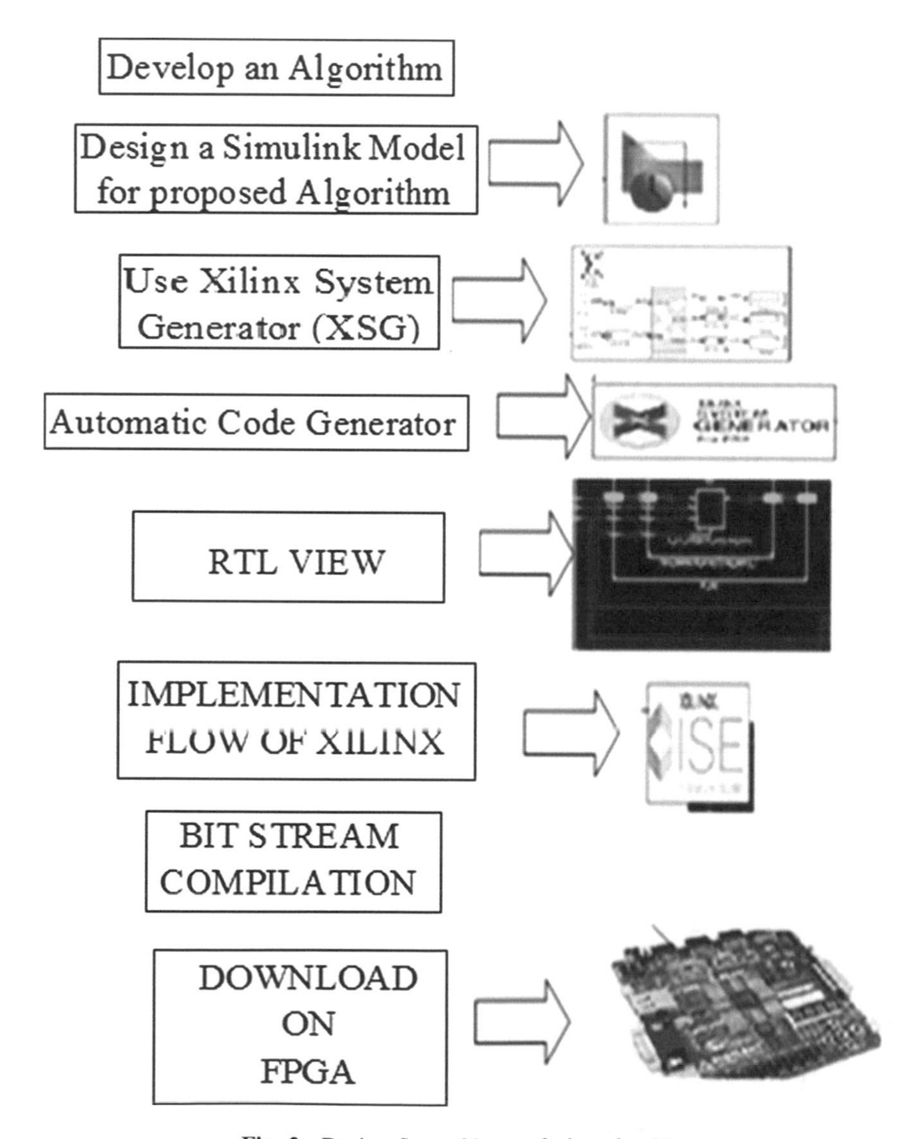

Fig. 2. Design flow of image fusion algorithm

library is used to model the picture fusion algorithm. The input images are now transformed into vector form with the help of Xilinx fixed point format, which is applied to Xilinx model. With suitable simulation period of time, model is simulated into MATLAB Simulink environment. System generator token is configured on FPGA Virtex-6 board. To integrate the design running on an FPGA directly into a Simulink environment, system generator is used for hardware implementation.

The model is then implemented for JTAG hardware co-simulation when input, output clock planning is done. When we compile the design, it will generate a netlist and Xilinx ISE accessible programming in Verilog HDL [3]. The proposed image fusion model is tested for behavioral syntax, and it is synthesized for the implementation on FPGA. It is the feature of the Xilinx system generator to configure user constraint file (UCF), test bench and test vector for the architecture testing. An FPGA bit file is created by bit stream compilation that is suitable as an input to FPGA and for the implementation of the Virtex-6 ML605 target device.

4 Implementation of Hardware and Software Co-Simulation for Image Fusion

Figure 3 shows the proposed design algorithm for hardware implementation of image fusion method of two multi-sensors images. We have considered here two registered multi-sensors images as an input. By using RGB to intensity converter block, the PAN

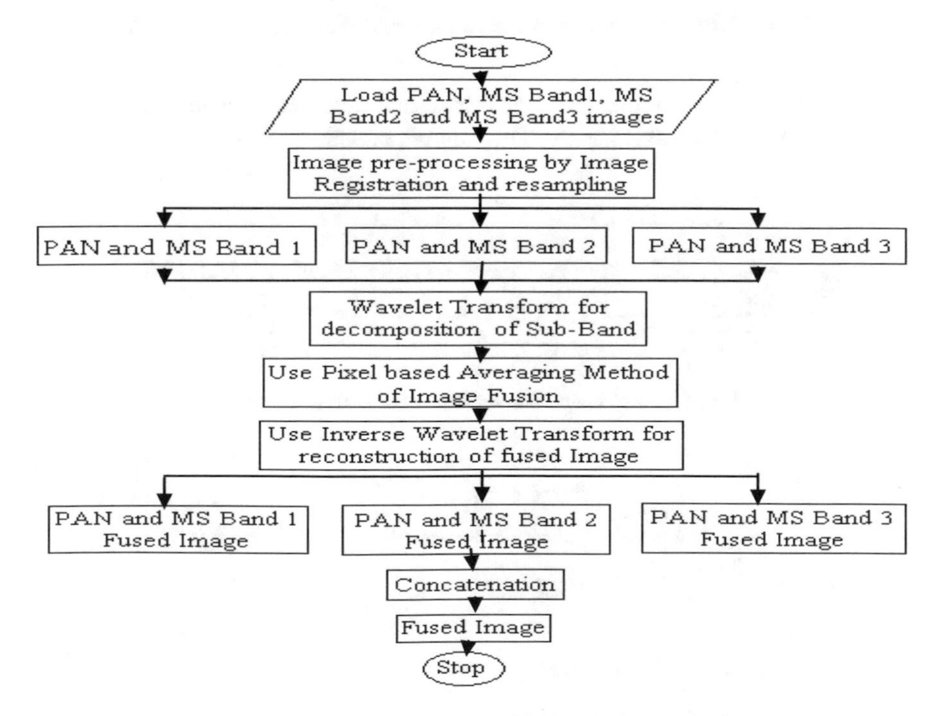

Fig. 3. Algorithm of proposed image fusion method

images of wavelengths 0.5 μm to 0.8 μm are converted into an intensity image from RGB. Due to memory limitation in FPGA, it is necessary to resize the multi-sensor images (MS and PAN) of size "256X256" into "128X128". Now we use 2D-1D Simulink block sets to convert, resize image "128X128" from two-dimensional image data into the one-dimensional bit stream. To implement in FPGA, this bit stream is applied as inputs to system generator model.

5 The Design of SUB Blocks

Figure 4 shows a schematic diagram for the conversion of the input image into its corresponding serial bit stream [8].

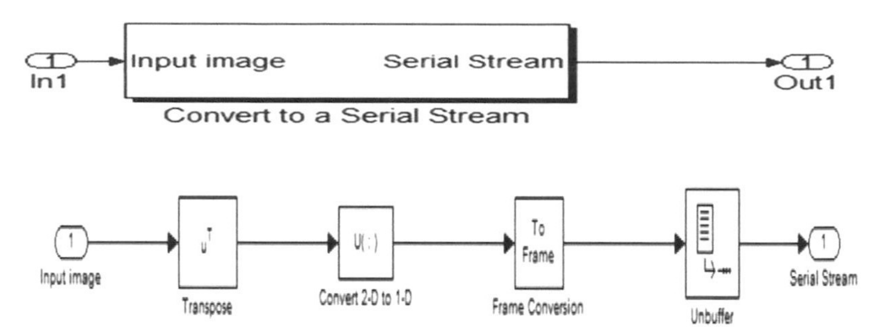

Fig. 4. Schematic diagram of image conversion from 2D-1D bit stream

The block diagram representation of Figs. 5 and 6 shows a transformation process of PAN and MS band images from 2D to 1D bit stream.

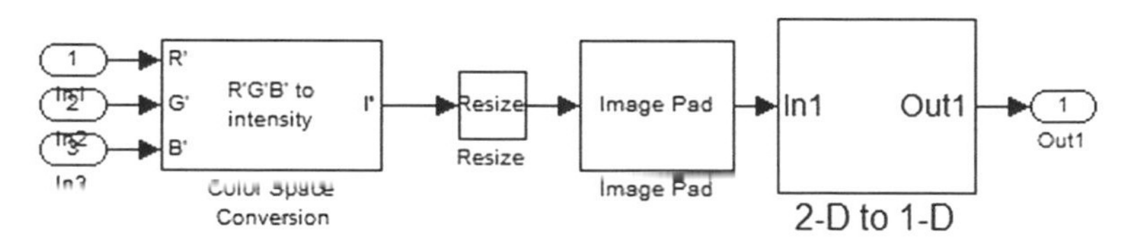

Fig. 5. PAN image conversion process from 2D to 1D

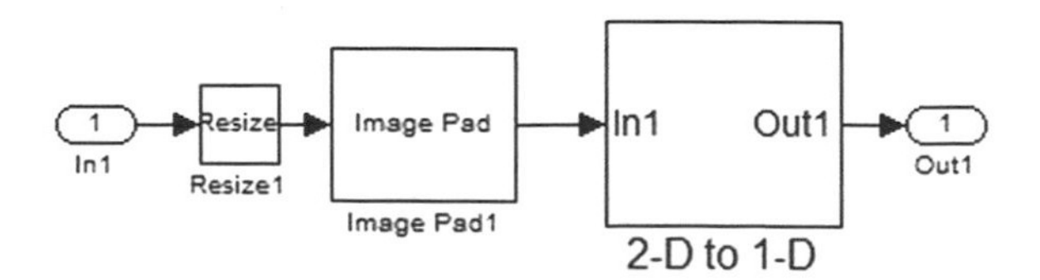

Fig. 6. MS Band image conversion process from 2D to 1D

This image bit stream data is used as inputs for system generator block set of DWT analysis filters through input gateway. These filters are used to separate each bit stream into the approximation and detail coefficients [7]. The resulting coefficients are produced by convolving the input values with the low-pass filter to give approximation and with high-pass filter value to provide detail coefficients. By doing this, we get a collection of sub-bands having smaller bandwidth and slower sample rates [1]. The four subbands HH, HL, LH and LL are produced by the 2-level decomposition of DWT. The produce resultant fused image we use the sub-bands HH, HL, LH and LL of both the images. The pixel averaging method of image fusion is carried out in two steps. In the first step, we add the image coefficients, and in second steps we multiply the result of the first step by factor 0.5 to calculate the average. To reconstruct the fused image bit stream the fused coefficient after averaging is applied to the synthesis filter using inverse wavelet transform IDWT. The fused image is 1-D data stream it is converted into a 2-D image by using 1-D to 2-D block sets [8] which are shown in Fig. 7.

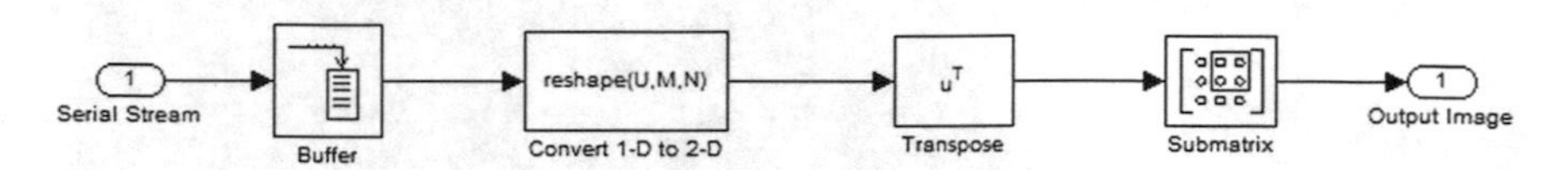

Fig. 7. Conversion of 1D to 2D of fused bit stream of image fusion

6 Result and Discussion

The Simulation result of input images of Fig. 8 is obtained from system generator, and FPGA are shown in Figs. 9(a) and (b). Figure 10 shows the simulated output of input and fused images for proposed averaging method. In synthesis process, we convert the Register Transfer Level (RTL) into a design implementation regarding logic gates [9]. With the help of Xilinx Synthesis Technology (XST), we perform the synthesis of Verilog code. The XST tool is a part of the Xilinx ISE software. The results of device utilization of averaging method are tabulated in Table 1 given below (Table 2).

(a)

(b)

Fig. 8. (a) Multispectral low-resolution input image (b) Panchromatic high-resolution input image.

Fig. 9. (a) Fused image using averaging method and (b) FPGA output for averaging method

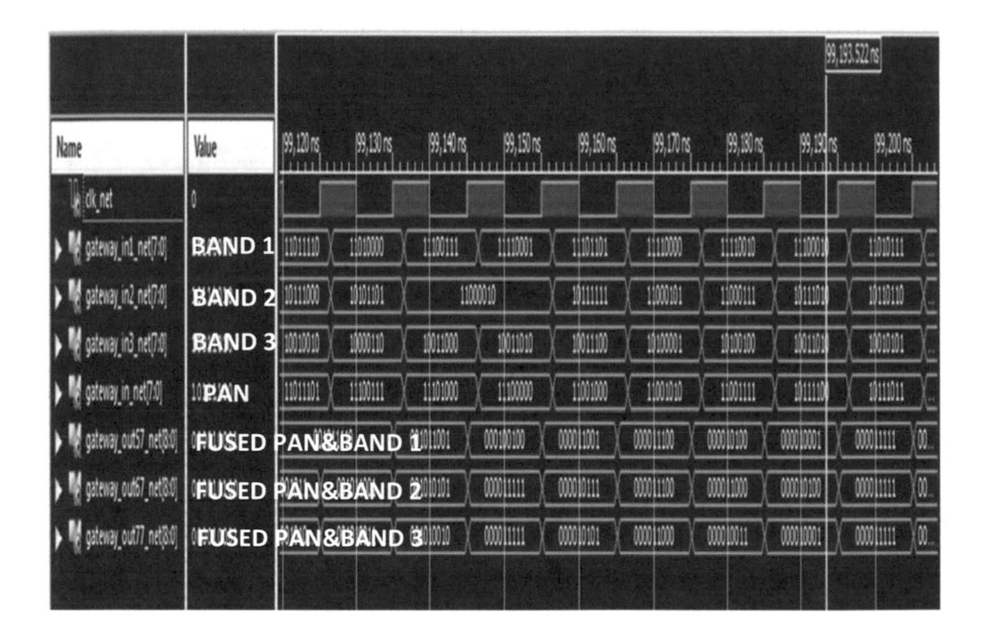

Fig. 10. Simulation results of input and fused images for averaging method

Table 1. Device utilization of averaging method of image fusion

Logic utilization	Used	Available	Utilized
Slice register	218	385,600	0%
Slice LUTs	215	197500	0%
Completely used LUT-FF pairs	215	218	98.62%
Bonded IOBs	145	600	24.16%
BUFG/BUFGCTRLs	1	32	3.12%

Table 2. Time required for averaging method

Speed grade	3
Lowest period	1.175 nsec (operating frequency 851.06 MHz)
Smallest input clock	0.437 nsec
Highest output time after clock	0.562 nsec
Highest path delay	0.448 nsec

Table 3. Speed performance of FPGA

FPGA family	Device properties	Minimum period	Maximum frequency
Virtex 6	XC6VSX315T-3FF1156	1.175 nsec	851.06 MHz

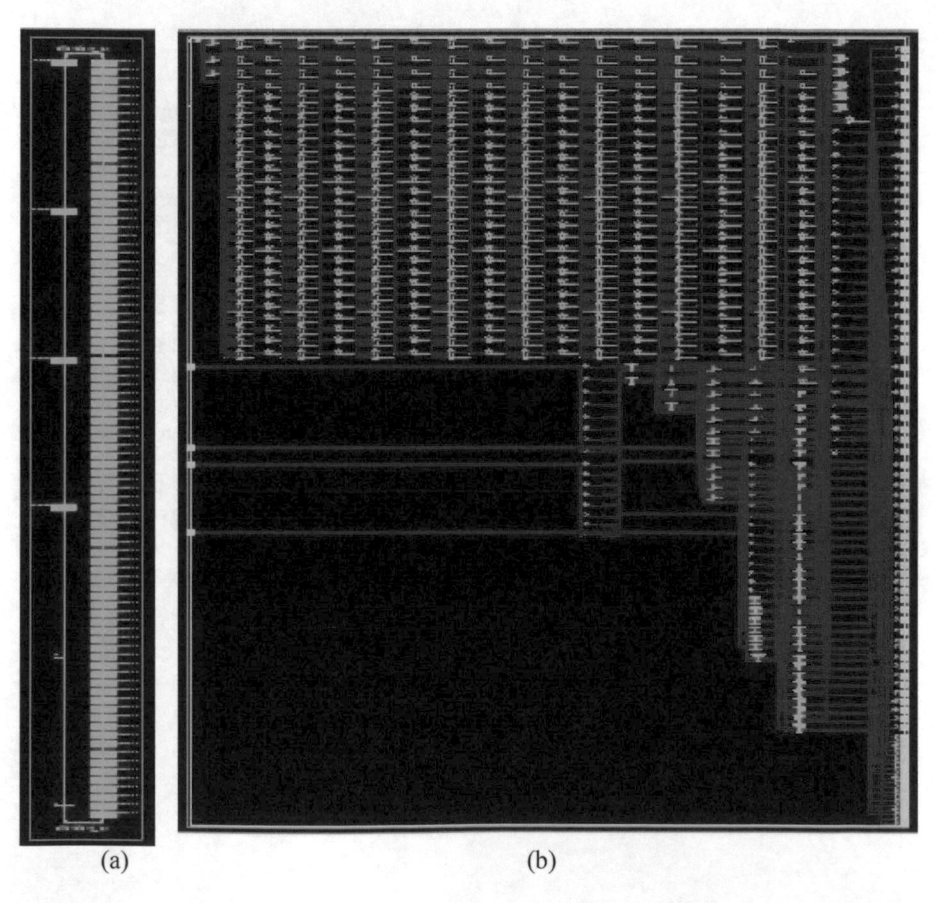

(a) (b)

Fig. 11. (a) Top level RTL diagram (b) RTL internal schematic of averaging method

7 Conclusion

It has been observed from Table 3 that by using Virtex 6, we have designed efficient hardware architecture of the satellite image fusion method using pixel averaging. The Virtex 6 operates on operating frequency 851.06 MHz at less time 1.175 nsec. It has been concluded that the proposed method is successfully developed wavelet-based hardware and software co-simulation using CDF 9/7 filter structure implemented on FPGA (Fig. 11).

Acknowledgment. We thanks to all the respective authors who had contributed their important research in the field of image fusion with its hardware implementation and have suggested novel approach of image fusion techniques.

References

1. Sulochana, T., Dilip Chandra, E., Manvi, S.S., Rasheed, I.: Design and testing of DWT based image fusion system using MATLAB-simulink. Int. J. Adv. Res. Electr. Electron. Instrum. Eng. **3**(3), 8177–8184 (2014)
2. Gupta, A.: Hardware software co-simulation for traffic load computation using Matlab Simulink model blockset. Int. J. Comput. Sci. Inf. Technol. **1**(2), 1–12 (2013)
3. Suthar, A.C., Vayada, M.D., Patel, C.B., Kulkarni, G.R.: Int. J. Comput. Sci. Issues **9**(2), 560–562 (2012)
4. Burkule, A., Borole, P.B.: Int. J. Adv. Res. Comput. Sci. Softw. Eng. **3**(1), 532–536 (2013)
5. Chandrashekar, M., Naresh Kumar, U., Sudershan Reddy, K., Nagabhushan Raju, K.: FPGA implementation of high speed infrared image enhancement. Int. J. Electron. Eng. Res. **1**(3), 279–285 (2009)
6. Johnston, C.T., Gribbon, K.T., Bailey, D.G.: Institute of Information Sciences and Technology, Massey University, pp. 118–124 (2004)
7. Anbumozhi, S., Manoharan, P.S.: Performance analysis of high efficient and low power architecture for fuzzy based image fusion. Am. J. Appl. Sci. **11**(5), 769–781 (2014)
8. Xilinx system generator reference guide
9. Chaithra, N.M., Ramana Reddy, K.V.: Int. J. Eng. Adv. Technol. **2**(6), 243–247 (2013)

Personalized Indian Bschool Counsellor System: A Rational Approach

Prajwal Eachempati(✉) and Praveen Ranjan Srivastava

Indian Institute of Management Rohtak, Rohtak, India
{fpm03.007, praveen.ranjan}@iimrohtak.ac.in

Abstract. With the proliferation of so many Bschools in India, it is becoming more and more difficult to choose a good Bschool. Over the decades since management education rankings first appeared, numerous debates have surfaced about their methodologies and objectivity. Although there has been significant research, especially by the various coaching institution magazines, about the ways in which rankings might be improved, there has been less research on providing the rankings according to the preferences of a person. In addition, there has been scant research on how rankings may impact students' access to industry interaction, and their selection of particular colleges. The objective of this paper is to provide a general ranking and a preferential ranking of top 28 Bschools in India. The approach is to provide accurate rankings by researching on a lot of existing data and by taking students preferences into consideration.

Keywords: AHP · TOPSIS · CR ratio · Weightage · Ranking · IIM · PGP · FPM

1 Introduction

The importance of education for the development of a country must not be underestimated because education is the tool which alone can inculcate national and cultural values and also liberate people of false prejudice, ignorance and representations. Education provides them required knowledge, technique, skill, information and enables them to know their rights and duties towards their family, society and towards their motherland at large. Just as a face is the mirror to the heart of a person, level of education reflects the status of a nation and is the key to development especially for developing countries like India.

Today, in India, there are many avenues for career growth of which management education is assuming a lot of importance. This is because though, people with technical skills and certifications are still valued, a management degree takes them forward by helping in providing a finishing touch to a person and by making the person a complete individual.

Earlier in India if one had to apply for a management degree (post-graduate or doctoral) there were only a handful of IIMs to choose from and entry into those institutes was very difficult due to the high quality of education [1] there and high cut-offs. But today due to government policy of opening new IIMs in almost 19 states and due to multiple private organizations opening new private business schools, there is a plethora of options for management aspirants to choose from, leading to more

S.C. Satapathy and A. Joshi (eds.), *Information and Communication Technology for Intelligent Systems (ICTIS 2017) - Volume 1*, Smart Innovation, Systems and Technologies 83, DOI 10.1007/978-3-319-63673-3_7

confusion among candidates as to which option to finalize as they have to examine many factors and make tradeoffs among those factors to come to a wise decision.

In order to resolve this confusion in the mind of the students, this paper comes up with a new innovative personalized Bschool Counsellor System that has two main functionalities: one to recommend an appropriate Bschool to the candidate based on his profile and his preferences based on a weightage- ranking system and the other to suggest to the lower ranked Bschools as to which parameters to strengthen for improving their rankings and hence to enhance their visibility in the market.

2 Existing Bschool Ranking Systems

Most of the institutions who provide the ranking of Bschools are the coaching institutions which help students to prepare for the management tests. The ranking of MBA-colleges has become a very subjective issue; as different coaching institutions give different weightage to different parameters without taking the candidate preferences into account. Also, according to the NIRF [2] website the ranking of Bschools is provided only on the basis of teaching and research making the ranking biased and not holistic. A student becomes confused when he sees a lot of ranking provided by various institutions and doesn't know which one is the authentic one due to lack of transparency in existing systems. Moreover, how these institutions have arrived to this ranking, there methodology, importance given to different parameters are not known. Apart from this these rankings are much generalized, which does not suit to the need and specific preferences of each aspirant. Each student might have different priorities and expectation from a MBA college but these ranking may not serve the purpose. For ex- A student who is more oriented towards research and is looking for A Bschool that would provide him with more research work and orient him towards the doctoral FPM programme. There are some websites like Outlook India [3] and Business Today [4] that may provide rankings based on research work but they are less in number and also less authentic. That is why this paper would try to address these issues also by suggesting personalized Bschools to both MBA and FPM aspirants who are confused. Currently, the Bschool aspirants are very confused as to whether they should apply for old IIMs or should they go for new IIMs or should they apply for the private Bschools but their confusion is not getting resolved through existing websites or recommender systems. The next section proposes a new, holistic and personalized Bschool Counsellor System as shown in Fig. 2 to help resolve their confusion.

3 Proposed System

The limitations and biasness in the ranking procedure of existing Bschool ranking systems, a new model has been proposed to resolve the confusion of the confused Bschool aspirant as shown in Fig. 1 by providing the ideal solution using AHP-TOPSIS tool.

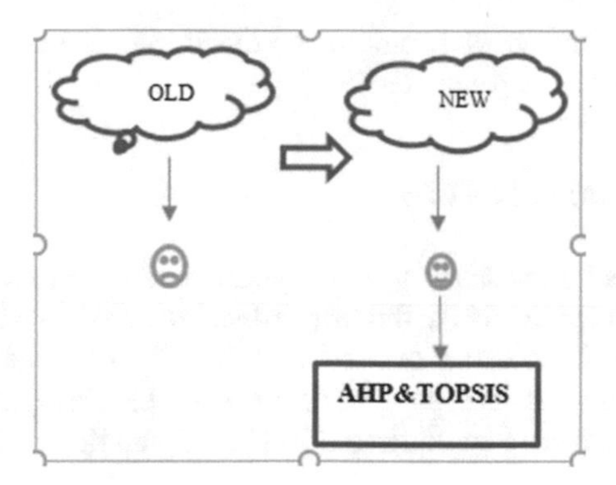

Fig. 1. Approach of the new system

The below proposed model of the new Bschool recommender system considers these techniques by using AHP to generate weightages and TOPSIS to give rankings of Bschools.

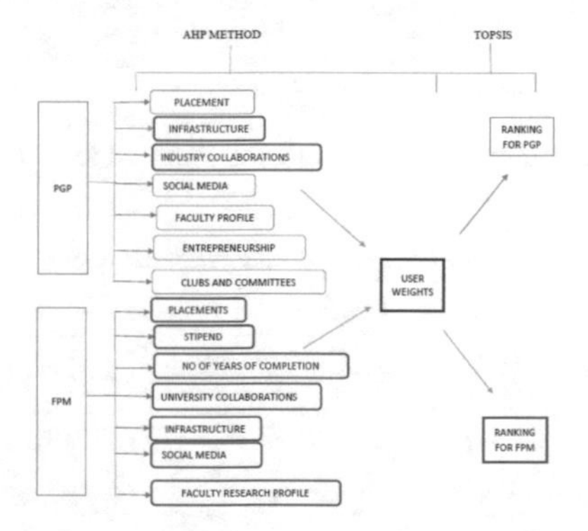

Fig. 2. Proposed model for AHP-TOPSIS method

The above model in Fig. 2 indicates broadly two main modules: one for the PGP and the other for FPM aspirants.

In the first module PGP parameters are taken into account where a user enters his/her weightages for all the PGP parameters which has been further analyzed by AHP method for identifying weightages and priorities. In this regard, the key parameters were found to be Placement, Infrastructure and Industrial Collaboration highlighted in the figure. Then TOPSIS method is adopted to find ranking of the Bschools based on the weightages provided by the user to the parameters. Similarly, for the FPM module the FPM parameters are taken into account where a user enters his weightages for all the FPM parameters which has been further analyzed by AHP method for identifying weightages and priorities. In this regard, the key parameters here found to be Placement, Stipend and No of years completed highlighted in the figure. Then TOPSIS method is again adopted to find ranking of the Bschools based on the weightages provided by the user to the parameters.

4 Methodology and Results

The input data was collected in the form of questionnaires (300 samples). The surveys were floated in social media, where the respondents were asked which parameters were important to them while selecting an MBA College for both the post-graduate MBA and doctoral FPM programmes on a scale of 1–10 i.e., how much weightage they assigned to these parameters as illustrated in Figs. 3 and 4.

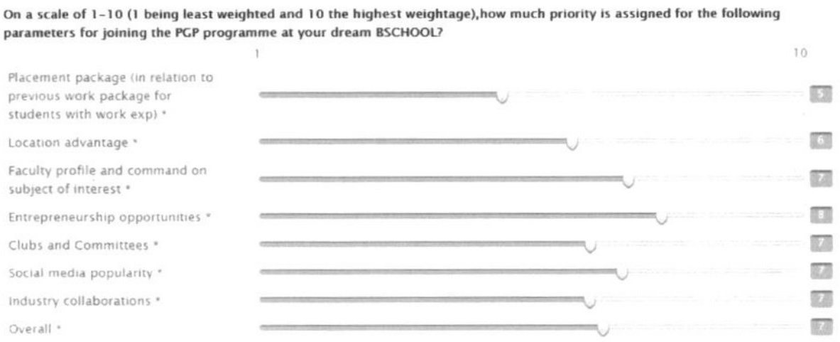

On a scale of 1-10 (1 being least weighted and 10 the highest weightage),how much priority is assigned for the following parameters for joining the PGP programme at your dream BSCHOOL?

Placement package (in relation to previous work package for students with work exp) *
Location advantage *
Faculty profile and command on subject of interest *
Entrepreneurship opportunities *
Clubs and Committees *
Social media popularity *
Industry collaborations *
Overall *

Fig. 3. Parameters evaluated for PGP students (sample)

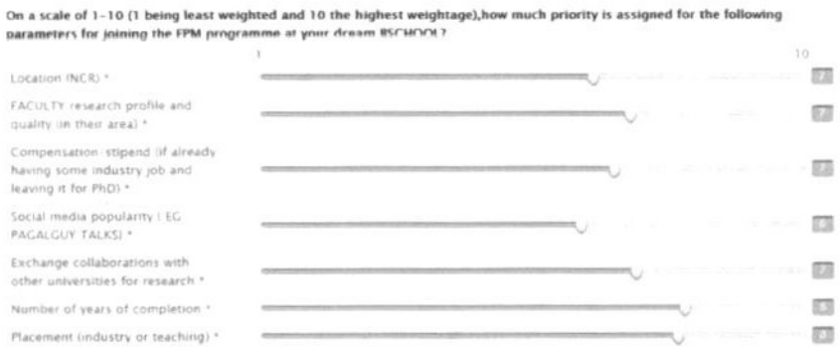

Fig. 4. Parameters evaluated for FPM students (sample)

The various parameters that both PGP and FPM students considered while making decision regarding choice of Bschool are:

For the **PGP** students: Placement package and opportunities, Faculty profile, Entrepreneurship opportunities, Industry Collaborations, Infrastructure, Social media popularity and Clubs and Committees.

For the **FPM** students: Placement, Faculty research profile, Infrastructure, Stipend offered, Exchange Collaborations with Universities, Number of years for completion and Social media popularity. The reason for choosing different parameters is that PGP is a post-graduate program that equips students with skills for a manager's job while FPM prepares students for a career of research hence the difference in criteria chosen.

After floating this surveys, an aggregate of 300 responses (150 responses for the PGP and 150 for the FPM) was obtained as shown in Figs. 5 and 6. Final weights for

Response ID	Age group	Qualification	Work experie	On a scale of 1-10 , weightage given to parameters for PGP programme						
				Placement	Faculty profil	Entrepreneu	Clubs and Co	Social media	Industry (	Infrastructure
1	1	PGDM	1	10	9	3	7	7	10	9
2	1	Graduate	1	10	7	5	9	8	10	10
3	2	Bachelor of Technology	3	10	7	6	8	7	8	10
4	1	MBA	2	10	8	5	10	10	10	8
5	1	MBA	1	10	7	6	7	7	9	10
6	1	Mba	1	10	7	7	7	7	9	10
7	2	B Tech	3	10	10	4	4	4	8	8
8	1	Graduation	1	10	5	2	6	5	8	7

Fig. 5. Sample responses for PGP Students

Response ID	Age group	Educational	Work Experience	On a scale of 1-10 weightage given to parameters for FPM programme						
				FACULTY	Compensatio	Social media	Exchange coll	No of years of	Placement (in	Infrastructure
1	1	FPM	2	7	10	1	6	4	10	9
2	1	MBA	1	7	7	3	6	8	3	10
3	1	MBA	1	8	7	1	8	6	9	10
4	1	B.E (ECE)	1	9	8	6	6	9	9	8
5	1	B. Tech.	1	10	10	8	10	8	10	10
6	1	B.Tech	1	8	7	2	9	9	8	10
7	1	B.Com. M.Com	1	10	4	5	8	9	9	8
8	1	Doctoral student	1	9	8	3	5	8	7	7

Fig. 6. Sample responses for FPM Students

each of these 7 parameters was calculated by taking the average of all the responses. The weights were normalized to reduce the error while calculation [5].

The aim of this research paper is to provide a ranking of top 28 Bschools in India. Primary research was done on various websites and then these colleges were assigned a score from 28 to 1 based on their performance on each of the 7 parameters. For example, taking a particular parameter of placements, the college with the best placement was given the score of 28 while the college performing worst on placement was given a score of 1. This process was repeated for all the 7 parameters. The data for these calculations was researched exhaustively through various rankings available across different platforms. These scores were further normalized to reduce the error in calculation. Primarily AHP-TOPSIS [6, 7] method was adopted because for large dataset with multiple parameters AHP efficiently provides weightages and TOPSIS ranks the alternatives as depicted in below Tables 1, 2, 3 and 4.

4.1 AHP and TOPSIS Method

The ranking system shown in Table 1 is a result of the following weightages and normalized weightages as shown in Tables 2 and 3 calculated by AHP process.

Table 1. Ranking system for PGP and FPM using AHP-TOPSIS

College	Ranking in terms of PGP and FPM	
	PGP	FPM
IIM Ahmedabad	*1*	*2*
IIM Calcutta	*2*	*3*
ISB Hyderabad	*3*	*4*
XLRI Jamshedpur	*4*	*17*
IIM Bangalore	*5*	*1*
IIM Kozhikode	*6*	*5*
IIT DELHI	*7*	*6*
IIM Lucknow	*8*	*9*
MDI Gurgaon	*9*	*12*
NITIE	*10*	*10*
IRMA	*11*	*11*
IIM-Ind(Mum)	*12*	*25*
JBIMS	*13*	*13*
XIMB	*14*	*15*
IIFT DELHI	*15*	*16*
IIM Shillong	*16*	*18*
IIFT K	*17*	*27*
TISS	*18*	*19*
NMIMS	*19*	*14*
Symbiosis	*20*	*8*
FMS Delhi	*21*	*20*
IIT Bombay	*22*	*21*
IIM Udaipur	*23*	*22*
IIM Kashipur	*24*	*23*
IIM Indore	*25*	*7*
IIM Rohtak	*26*	*24*
IIM Trichy	*27*	*26*
IIM Raipur	*28*	*28*

Table 2. Weightages assigned to parameters according to AHP-TOPSIS [PGP]

Criteria [AHP]	Job package	Faculty profile	Ecell	Clubs-committees	Social media	Industry collaboration	Infrastructure
Avg weight	*8.32*	*8.68*	*8.14*	*8.2*	*8.74*	*8.3*	*8.58*
Final weight	*0.148*	*0.215*	*0.163*	*0.166*	*0.241*	*0.19*	*0.205*

From Tables 2 and 3, the key highlighted parameters are Job package and Infrastructure for PGP while stipend and duration are key parameters for FPM [8, 9]. In order to validate both the results, a Consistency Ratio was computed for the data and compared with expected result as in Table 4.

Table 3. Weightages assigned to parameters according to AHP-TOPSIS [FPM]

Criteria [AHP]	Faculty research profile	Stipend	Social media	Exchange program	Duration	Placement	Infrastructure
Avg weight	*7.34*	*9.23*	*8.23*	*8.25*	*8.85*	*9.45*	*8.21*
Final weight	*0.132*	*0.225*	*0.167*	*0.169*	*0.199*	*0.255*	*0.165*

Table 4. Validation for ranking systems-PGP and FPM

Parameters	PGP	FPM
LAMBDA-MAX	*7.29002969*	*6.08577*
Ci	*0.048338282*	*0.017155*
Ri	*1.32*	*1.24*
Cr	*0.03661991*	*0.013835*

5 Analysis and Recommendations

Based on the results it can be inferred that when weightages given to parameters are varied in both the methods the rankings are also updated accordingly, this clearly indicates that the ranking procedure developed is a personalized ranking system that is taking into account the preferences of the user and is suggesting the appropriate Bschool enabling him/her to take a wise decision.

But now considering the second functionality of the recommender system i.e. to suggest improvements to the lower ranked Bschools there is a need to analyze the Bschools considered in the study parameter-wise and examine in which parameters the lower ranked Bschools [10] are lacking in order to be able to suggest corrective measures to improve the position of those Bschools. For this study the IIM Rohtak has been taken as the Bschool to suggest improvements as the author is a stakeholder of this institution. Given below are some graphical representations of Bschools against the parameters which they are compared with.

The above Figs. 7 and 8 show that for strengthening the PGP program at IIM Rohtak the above two parameters i.e. Placement, Infrastructure have to be worked upon.

The above Figs. 9 and 10 show that for strengthening the FPM program at IIM Rohtak the above two parameters i.e. Duration and Compensation/stipend and have to be worked upon. If these parameters are strengthened within the next 5–10 years, IIM Rohtak can be seen as one of the most sought after Bschools by prospective candidates.

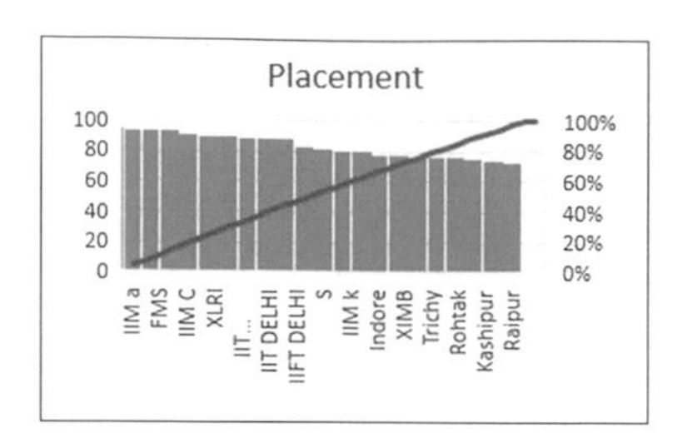

Fig. 7. Bschools for placement (PGP)

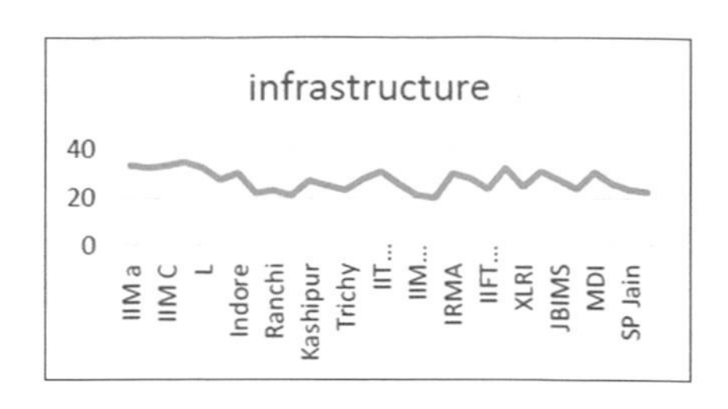

Fig. 8. Bschools for infrastructure (PGP)

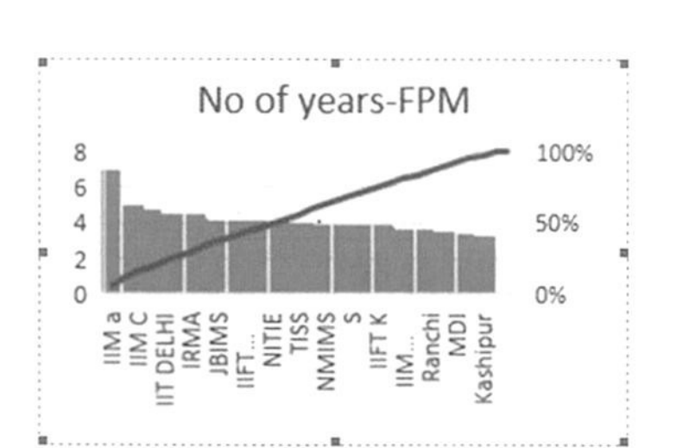

Fig. 9. Bschools-duration (FPM)

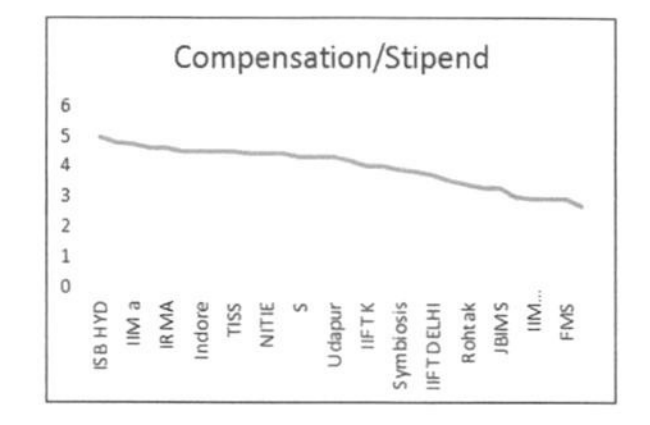

Fig. 10. Bschools-Stipend

6 Conclusion

This paper is an attempt to devise a personalized counsellor system that has a dual role of giving rankings and suggestions to improve Bschool rankings. AHP-TOPSIS method was devised for the purpose of comparing the impact on weightages and hence on the different ranking systems observed only to get multiple perspectives for ranking the Bschools. It is hoped that this attempt will help in resolving the confusion and dilemma of PGP and FPM aspirants [10] across the country while choosing Bschools.

There are other Multi-criteria decision making methods like PURE TOPSIS method and FUZZY AHP method for further optimizing the results and other categorical parameters can also be considered for presenting a more holistic ranking system.

Disclaimer: The Authors and Publishers disclaim any legal or other responsibility for opinions expressed and these may be considered as an example only and not to suggest any realistic ratings to the esteemed Institutions mentioned here. Readers who are keen to test the model may enter their data and observe the outcome purely for academic interest only.

References

1. McMahon, M., Patton, W., Watson, M.: Creating career stories through reflection: an application of the systems theory frame-work of career development. Aust. J. Career Dev. **13** (3), 13–17 (2004)
2. NIRF India. https://www.nirfindia.org/mgmt
3. Outlook-India. http://www.outlookindia.com/magazine/story/indias-best-schools-in-2015/295392
4. Business Today. http://bschools.businesstoday.in/
5. Olson, D.L.: Comparison of weights in TOPSIS models. Math. Comput. Model. **40**(7), 721–727 (2004)
6. Jahanshahloo, G.R., Lotfi, F.H., Izadikhah, M.: Extension of the TOPSIS method for decision-making problems with fuzzy data. Appl. Math. Comput. **181**(2), 1544–1551 (2006)
7. Behzadian, M., Otaghsara, S.K., Yazdani, M., Ignatius, J.: A state-of the-art survey of TOPSIS applications. Expert Syst. Appl. **39**(17), 13051–13069 (2012)
8. Whitaker, R.: Validation examples of the analytic hierarchy process and analytic network process. Math. Comput. Model. **46**(7), 840–859 (2007)
9. Ho, W.: Integrated analytic hierarchy process and its applications–a literature review. Eur. J. Oper. Res. **186**(1), 211–228 (2008)
10. Zolfani, S.H., Ghadikolaei, A.S.: Performance evaluation of private universities based on balanced scorecard: empirical study based on Iran. J. Bus. Econ. Manag. **14**(4), 696–714 (2013)

Proposed Model for an Expert System for Diagnosing Degenerative Diseases – Using Digital Image Processing with Neural Network

Mittal N. Desai[1(✉)], Vishal Dahiya[2], and A.K. Singh[3]

[1] Faculty of IT and Computer Science, Parul Institute of Computer Application, Parul University, Vadodara, India
bhattmittal2008@gmail.com
[2] Indus University, Ahmedabad, India
[3] Space Applications Centre, ISRO, Ahmedabad, India

Abstract. In the era of any information on fingertip or on one click, medical diagnosis is context in which wrong diagnosis should be avoided using extensive information related to patients and symptoms. There should be an efficient system in diagnosis in terms of expert diagnostic opinion within short span of time, so that disease should be prevented to become chronic. To streamline this expert diagnostic opinion process to the patients, in daily routine, Expert System (ES) using artificial neural network can be employed. It is the method which can simulate two very important characteristics of humans, learning and generalization. Using ANN algorithms various types of medical data are handled and output is achieved with defining various relations between that data. Radiology is one of the branches of medical science in which various medical imaging techniques are used to diagnose difference internal medical problems. Digital Image Processing is the science of processing various digital images: such that important information will be generated. An Expert System is also an efficient tool from which diagnosis can be made. Integrating outcomes of neural network from diseased X-ray, to the knowledge based expert system; an expert opinion of diagnosing disease can be generated. In this paper a model is proposed for diagnosing, seven lower lumbar problems as degenerative diseases.

Keywords: Radiology · Medical imaging techniques · Artificial neural network · Lower lumbar diagnosis · Expert system

1 Introduction

Peter O'Sullivan (2005) noticed that, eighty-five percent of Chronic Low Back Pain (CLBP) disorders have no known diagnosing under specific classification that creates gap between diseases and its management [10]. So there is requirement of an intelligent system that play diagnostic role in order to generate an opinion, which fills the gap of precise diagnosing in time. An Expert System is the solution for that provides advantages like forward chaining, backward chaining, coping uncertainty and explanation about its own reasoning solve the problem of diagnosing such diseases.

S.C. Satapathy and A. Joshi (eds.), *Information and Communication Technology for Intelligent Systems (ICTIS 2017) - Volume 1*, Smart Innovation, Systems and Technologies 83, DOI 10.1007/978-3-319-63673-3_8

1.1 Artificial Neural Network (ANN)

In our human body, neurons are computing elements of the brain. The main function of this neuron is to process input signals from various body part and take decision in various situation, on the basis of past experiences from memory.

ANN is the branch of Intelligent Information Processing Systems using the same analogy of neurons like processing elements [1]. It possesses "learning" and "generalization" ability of human behavior so that it can give output in terms of various relationships among nonlinear variables. It is made up of series of neurons like processing elements and they all are arranged in various layers. Each neuron in the layer is connected to each neuron in the next layer with weighted vector.

In learning Process: In this, ANN is formed by, input vector $X_{im} = (x_{i1}, x_{i2}, \ldots, x_{im})$ and output vector $Y_{in} = (y_{i1}, y_{i2}, \ldots, y_{in})$. The objective of the learning process is to approximate the function f, between the vectors X_{im} and Y_{in}. This is achieved through changing weights on connections, and deriving error from expected outcome Vs current outcome.

In between input layer and output layer, there are some layers which does intermediate processing in order to achieve expected outcome, are known as hidden layers. Neurons on the input layer receive input data and transfer them into next hidden layer through weighted connections. Here data are processed mathematically and send to next hidden or output layer. The mathematical calculations are done by calculating weighted sum of all input data and by adding bias term $(\theta)_j$,

$$\text{sum} = \sum_{i=1}^{m} (\text{input}) \times (\text{weights}) + \theta j \ (j = 1, 2, \ldots.., n)$$

After calculating weighted sum, it should be transferred to next neuron using various transferring functions like, Binary Threshold, Bipolar Threshold, Linear, Linear Threshold, Sigmoid, Hyperbolic tangent, Gaussian, Stochastic etc., out of which sigmoid is most commonly used,

$$f(x) = 1/1 + e^{-x}$$

After that learning is done by calculating difference between expected out come to current outcome, change in error with respect to weights will be calculated and multiplied with learning constant.

1.2 Importance of ANN in Medicine

Artificial Neural Network is used in situation, where typical algorithmic solution will not work. The problems are too complex and needs to predict output on the basis of historical data, for current inputs. These characteristics appear in the medicine and applied in different areas of medicine like: diagnostic systems, biomedical analysis, image analysis etc.

1.3 Radiology Based Medical Diagnosis

Radiology refers to medical imaging techniques, generates visual representation of internal body parts may be, bones, tissues etc. Using Digital Image Processing these all visual representations can process and analyzed, can produce valuable information from which diagnosis is possible. There is an extensive need of computer aided systems that can generate expert opinion on the basis of some imaging technology as well as patient's symptoms. Out of so many diseases Spinal problems are very common and lower lumbar problems are even more common as it falls under the category of degenerative diseases.

2 Model of Radiology Based Diagnosis of Lower Lumbar Problems using ANN

For development of systems in which system will give you diagnostic opinion based on X-ray as well as patient symptoms, following model is proposed:

As mentioned in the figure above, model is made up of two modules, (i) from X-ray images, using supervised learning neural network outcomes generated and (ii) that outcomes are integrated in the system so that from the knowledge base, system lead towards an expert opinion for diagnosing degenerative diseases (Figs. 1 and 2).

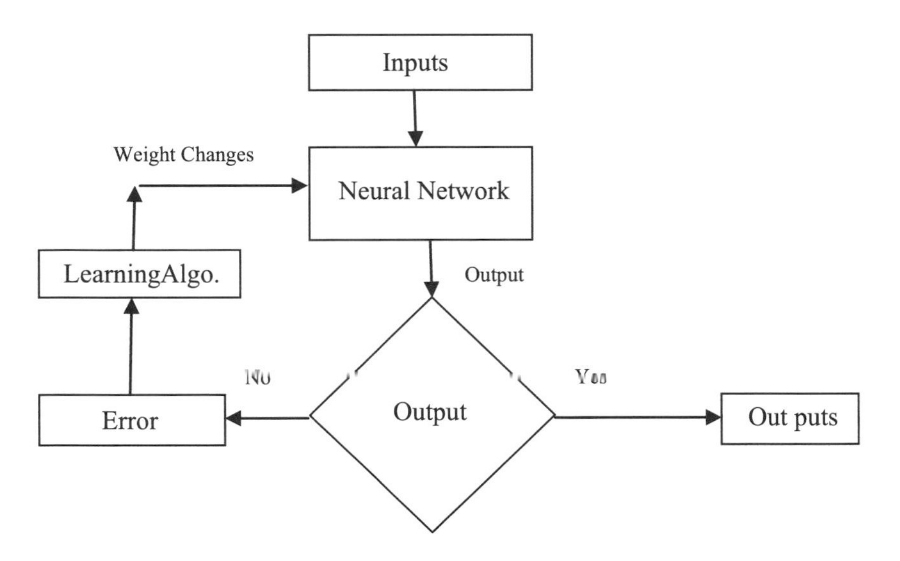

Fig. 1. Adaptive nature of neural network

2.1 First Module: Trained Neural Network

X-ray image is taken needs to be diagnosed by ANN and once the NN is trained to diagnose such X-rays, should be integrated with expert system. After image processing,

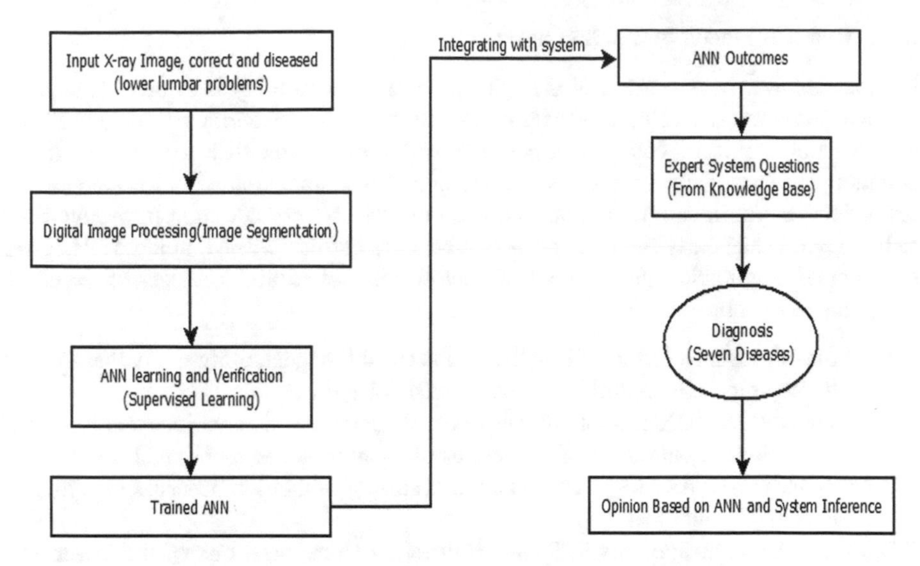

Fig. 2. Proposed model of medical diagnosis system for lower lumbar problems

system should ask questions to patients to know state of diseases, like early stage, middle stage or at chronic stage.

Artificial Neural Network (ANN): As mentioned above, ANN is possessing two important characteristics learning and generalization. In our proposed model, learning of NN leads to train NN to diagnose various medical abnormalities in terms of degenerative diseases, disk size is reduced or distance between two disks changed etc. From the given set of parameters like age, pain intensity, symptoms etc for six degenerative diseases will be given to NN and accordingly, it will be trained to classify them.

After getting out comes of NN it will be integrated into Expert System, so the system will ask questions to patients in order to leads to an expert opinion. Neural Network is not a statistical model based on assumptions but it is complex mathematical model, which gives accurate results in many areas of medicine like diagnosis and imaging [12].

2.2 An Expert System (ES)

It is a composition of several elements like, knowledge base, inference engine, user interface, such that an intelligent system evolved in direction of achieving same goal. As its name suggest, knowledge base is not just classical data base but it is more of a storing expert knowledge generally in terms of several rules.

Various applications of ES were observed like diagnosing, interpreting, predicting and instructing especially in medical field. It is not just a decision support system but, using NN, complex relationships between various attributes can establish by incorporating expert knowledge as rule based system, leads to an accurate and timely opinion.

2.3 Diseases Needs to be Diagnosed

Lumbar is lower back portion of the spine, and its responsibility is to provide support and stabilize majority of body's weight. Overtime as body undergoes repetitive lifting and twisting motions, causes compression and pressure on disk and joints. If the situation continuous, then it creates various spine conditions, fall under the category of lower lumbar problems. As it is degeneration in disk, is very common now days need to be diagnosed in early stages to prevent it become chronic. Above model is designed to generate diagnostic opinion for such below mentioned six diseases, observed as lower lumbar problems.

2.3.1 Spondylosis Deformns: Growth of bone around degenerating intervertebral disc in the spine also referred as osteoarthritis of spine.
2.3.2 Intervertebral Degeneration: Disc gets degenerated due various reasons like, overweight, continuous lifting and twisting motions as mentioned above.
2.3.3 Spondylosis: As a degeneration process stiffing or fixation observes in vertebrae is called spondylosis.
2.3.4 Lumbar Non-Sponlolytic Spondylosthesis: In this case, one of the lower vertebrae displaced either forward or backward on the below vertebrae.
2.3.5 Ankylosing Spondylosis: spine gets fused due to calcification of ligaments in areas where they attach to your spine.
2.3.6 Lumbar Spinal Canal Stenosis: Spinal space gets reduced, so it limits amount of space for spinal cord and nerves, can cause various symptoms.

References

1. Liao, S.-H.: Expert system methodologies and applications—a decade review from 1995 to 2004. Expert Syst. Appl. **28**(1), 93–103 (2005)
2. Hanan, S.A., Manza, R.R., RamTeke, R.J.: Generallized regression neural network and radial basis function for heart disease diagnosis. Int. J. Comput. Appl. (0975-8887) **7**(13), 7–13 (2010)
3. Amato, F., Lopez, A., María Peña-Méndez, E.M., Vanhara, P., Hampl, A., Havel, J.: Artificial neural network in medical diagnosis. J. Appl. Biomed. **11**, 47–58 (2013)
4. Jose, J., Emilio, G.: A combined neural network and decision tree model for prognosis breast cancer relapse. Artif. Intel. Med. **27**(1), 45–63 (2003)
5. Karabatak, M., Ince, M.C.: An expert system for detection of breast cancer based on association rules and neural network. Expert Syst. Appl. **36**(2), 3465–3469 (2009)
6. Cheng, H.D., Shan, J., Ju, W., Guo, Y., Zang, L.: Automated breast cancer detection and classification using ultrasound images: a survey. Pattern Recogn. **43**(1), 299–317 (2010)
7. Zhou, Z.-H., Jiang, Y., Yang, Y.-B., Chen, S.-F.: Lung cancer cell identification based on artificial neural network ensembles. Artif. Intel. Med. **24**(1), 25–36 (2002)
8. Naser, S.S.A., Ola, A.Z.A.: An expert system for diagnosing eye diseases using CLPS, JATIT (2005–2008)
9. Mandal, I.: Developing new machine learning ensembles for quality spine diagnosis. Knowl. Based Syst. **73**, 298–310. Science Direct (2015)
10. O' Sullivan, P.: Diagnosing and classification of chronic low back pain disorders: maladaptive movement and motor control impairments as underlying mechanism, July 2005. ScienceDirect.com

11. Lin, L., Hu, P.J.H., Sheng, O.R.L.: A decision support system for lower back pain diagnosis: uncertainty management and clinical evaluations. Decis. Support Syst. 1152–1169. Science Direct, Elsevier (2006)
12. Wei, J.T., Zhang, Z., Branhill, S.D., Madyastha, K.R., Zhang, H., Osterling, J.E.: Understanding artificial neural networks and exploring their potencial applications for practicing urologist. Clin. Rev. Elsevier (1998)
13. Lisboa, P.J.G.: A review evidence of health benefit from artificial neural networks in medical intervention. Neural Netw. **15**(1), 11–39 (2002). Pergamon
14. Tayade, M.C., Wankhede, S.V., Bhamare, S.B., Sabale, B.B.: Review article: role of image processing technology in health care sector. Int. J. Healthcare Biom. Res. **3**(3), 8–11 (2014)
15. Kurniawan, R.: Expert system for self diagnosing of eye diseases using Naïve Bayes. In: International Conference on Advance Informatics: Concepts, Theory and Applications, UIN Sultan Syarif Kasim Riau, Falulti Teknologi & Sains Maklumat (2014)

An Obscure Method for Clustering in Android Using k-Medoid and Apriori Algorithm

Amar Lalwani, Sriparna Banerjee, Manisha Mouly Kindo, and Syed Zishan Ali(✉)

Bhilai Institute of Technology, Raipur, Chhattisgarh, India
amarlalwani0202@gmail.com,
sriparnabanerjee15@gmail.com, mouly.kindo@gmail.com,
zishan786s@gmail.com

Abstract. In today's scenario, there is quick evolution in each field which contains majority and distinctive sorts of information. In order to differentiate sample data from the other, the amalgamation of data mining techniques with other useful algorithms is done. Android development is one of the major arena where there is tremendous need to execute these calculations. Combining frequent pattern calculation with clustering is extremely efficacious for android. In this paper the work is done in two levels, initial stage concentrates on generation of clusters and final stage deals with finding the frequent patterns.

Keywords: Android · Clustering · Itemsets

1 Introduction

Clustering is a unique multi-objective optimization problem solving technique of generating set of objects that are similar to one another but dissimilar from the objects belonging to the other set. The term "clustering" is used in several research communities to describe methods for grouping of unlabeled data. Clustering is significant in various exploratory pattern- analysis, decision-making, grouping, machine-learning situations, including data mining, pattern classification document retrieval and image segmentation. Clustering techniques [1, 2, 4–6, 12] are enforced as it is tough for humans to intuitively perceive knowledge in multi dimensional space. Here a brief overview of each of the techniques is presented and later on the methodology along with the main concept will be delineated.

2 Algorithms

Finding frequent patterns among various datasets requires the implementation of Algorithms. The Algorithms is applied according to the work and objects utilized in the subject. Below are the algorithms which is utilized in this work:

S.C. Satapathy and A. Joshi (eds.), *Information and Communication Technology for Intelligent Systems (ICTIS 2017) - Volume 1*, Smart Innovation, Systems and Technologies 83, DOI 10.1007/978-3-319-63673-3_9

2.1 k-Medoid Clustering

Medoids [9, 10, 15] are representative objects of a data set or a cluster with a data set which are similar in concept to means or centroids, but medoids are always members of a data set.when a mean or centroid cannot be defined medoids are used such as 3-D trajectories. *k*- medoids algorithm breaks the dataset into groups and attempts to minimize the list and between points labeled to be in clusters. A point is designated as the centre of particular cluster. In k-medoids, data points are chosen as centres (medoids or exemplars) that works with an arbitrary metrics of distances between them.

Objects are taken in account to represent an individual cluster. For each of the remaining objects the value is assigned to the cluster and based on the distance between the objects the minimum cost is compared and then evaluated. This algorithm [11, 14] is very efficacious in generating clusters which is required in most of the pragmatic cases.

The objective function is defined as

$$j = \sum_{\substack{I=1 \\ p \in C_i}}^{K} \sum_{p \in C_i} |p - o_i| \tag{1}$$

where,

E: the sum of absolute error for all objects in the data set
p: the data point in the space representing an object
$o_{i:}$ is the representative object of cluster C_i

2.2 Apriori Algorithm

The apriori algorithm [3] is one of the most broadly used tools for association rule mining. It uses priori knowledge of frequent itemset property for association rule mining. The algorithm makes use of downward closer property and is a bottom up search which moves upward level wise in the lattice. The main idea is to produce candidate itemsets of a given size and then verify if the count is actually large. Here k-itemsets are used to explore (k+1)-itemset. Therefore this is an iterative process in which we can generate the candidate of any pass by joining frequent itemsets of previous pass.

3 Methodology

The foremost concept implemented in this paper is based on the detection of frequent patterns in a particular cluster. Each object in the clusters have a distinct position that may vary depending upon the points obtained and the distinct clusters are formed by applying k-medoid clustering method along with apriori algorithm.

Step 1 Initially inputs are given as coordinates for in order to create set of clusters. The points are then imported to the RStudio [7, 8] for further plotting of the graph Table 1.

Table 1. Number of objects

Shapes		
○	+	▲
2	-	-
-	7	-
-	-	6

Step 2 Based on this information, the values of input provided for the location of the object are assigned with two centres of the cluster. Then the cost is evaluated using the Minkoswki or Manhattan distance.

Step 3 Implementing the procedure of *k*-medoid clustering [13] method, such that the minimum cost of traversal is obtained.

Step 4 The medoids using which the minimum cost is obtained is utilised in further steps.

Step 5 Final clusters will be formed after applying Apriori algorithm to find the frequent object or itemset.

Step 6 The number of objects in the corresponding clusters is highlighted by various shapes Fig. 1.

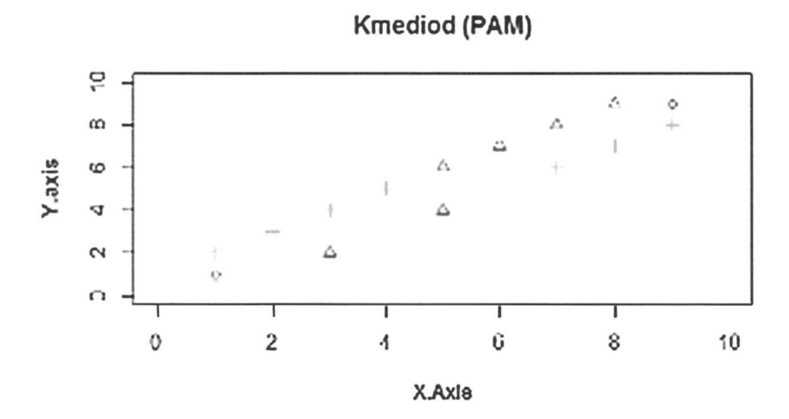

Fig. 1. Classification of objects

In the above figure there are different objects assumed, having non-identical items in the individual datasets. It is clearly displayed that the objects (triangle, star and circle) is capable of representing any living being like human, animals which is dynamic in nature Fig. 2.

Step 1. Input the number of objects to be clustered..

Step 2. Feed the coordinate values for the objects.

Step 3. Calculate the value of absolute error for all objects, then compare the value of minimum cost 'S'. .

Step 4. If S<0 then continue else the program will terminate .

Step 5. Assign the number of clusters to be formed.

Step 6. Final clusters are plotted along with objects.

Step 7.The frequency of the objects in the respective clusters is calculated.

Fig. 2. Flow of algorithm

This concept in successfully implemented in java environment as shown below: Figs. 3 and 4

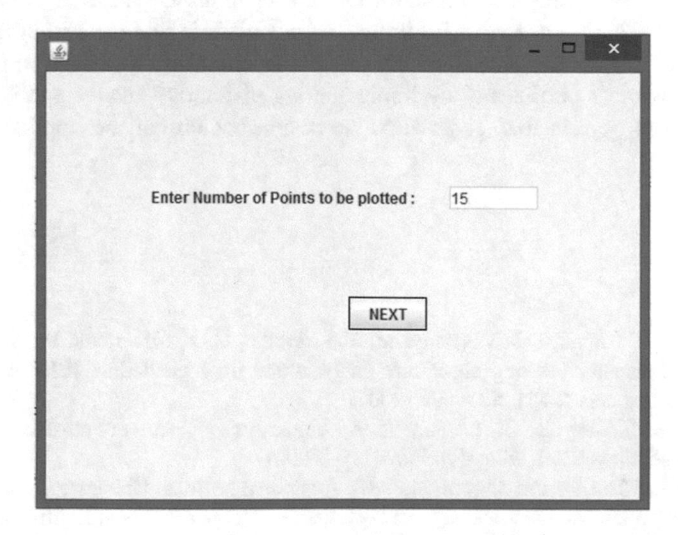

Fig. 3. Coordinate points

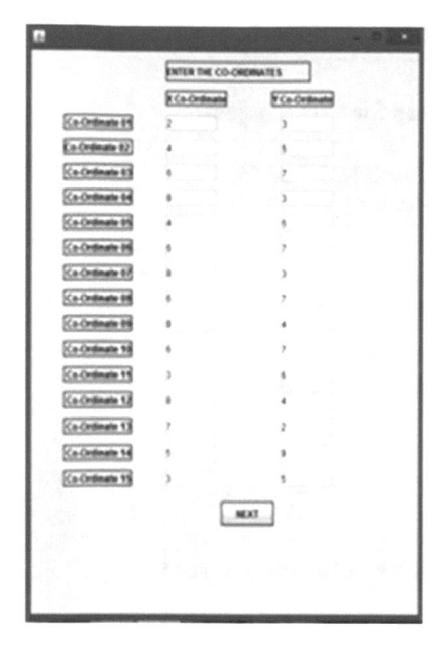

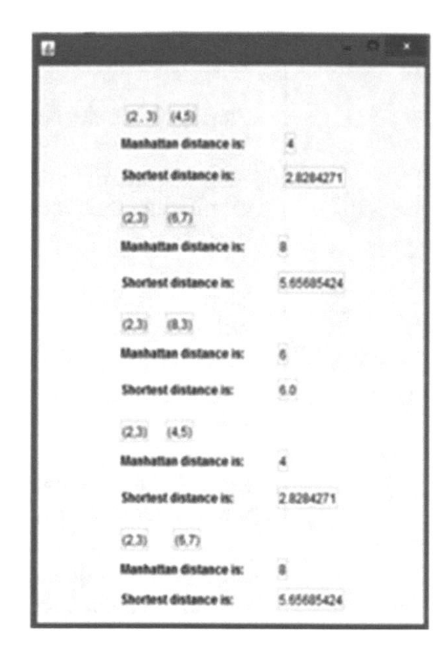

Fig. 4. Feeding coordinate values and calculating distance

4 Conclusion

As observed through the experimental analysis, a procedure for clustering in android has helped in identifying the maximum number of itemsets and their existence in an individual cluster. One of the major impact of this work is that it helps in representing the filtered objects in the cluster which can be beneficial for Android and other platforms. This work is carried out by implementing distinctive shapes and colors representing various objects that gives effective results which can be applied in Google Maps.

References

1. Kanungo, T., Mount, D.M., Netanyahu, N.S., Piatko, C.D., Silverman, R., Wu, A.Y.: An efficient k-means clustering algorithm: analysis and implementation. IEEE Trans. Pattern Anal. Mach. Intell. **24**(7), 881–892 (2002)
2. Steinbach, M., Karypis, G., Kumar, V.: A comparison of document clustering techniques. KDD. Workshop. Text. Min. **400**, 109–110 (2000)
3. Hegland, M.: The Apriori Algorithm CMA. Australian National University, Canberra (2005)
4. Han, J., Kamber, M.: Data Mining Concepts and Techniques. Morgan Kaufmann Publishers, Burlington (2006)
5. Singh, R.V., Bhatia, M.S.: (2011) Data clustering with modified k-means Algorithm. In: Proceedings of the IEEE, pp. 717–721, June 2011. ISBN: 978-1-4577-0588-5
6. Xu, R.: Survey of clustering algorithms. IEEE Trans. Neural Netw. **16**(3), 45–678 (2005)

7. Matloff N The Art of R Programming. (http://heather.cs.ucdavis.edu/~matloff/132/NSPpart.pdf)
8. Grolemund G Hands-on programming with R, Garrett (http://droppdf.com/v/KoQlw)
9. Kaufman, L., Rousseeuw, P.J.: Finding Groups in Data: An Introduction to Cluster Analysis. Wiley, New Jersey (2005)
10. Velmurugan, T., Santhanam, T.: Computational complexity between k-means and k-medoids clustering algorithms for normal and uniform distributions of data points. J. Comput. Sci. **6**(3), 363–368 (2010)
11. Patel, V.R., Mehta, R.G.: Clustering algorithms: A Comprehensive Survey. In: Proceedings of the International Conference on Electronics Information and Communication Systems Engineering, Jodhpur (2011)
12. Oyelade, O.J., Oladipupo, O.O., Obagbuwa, I.C.: Application of k-Means Clustering algorithm for prediction of Students' Academic Performance. Int. J. Comput. Sci. Inf. Secur. **7**, 1002 (2010)
13. Ng, R.T., Han, J.: CLARANS: a method for clustering objects for spatial data mining. IEEE Transact. Knowl. Data Eng. **14**(5), 1003–1016 (2002)
14. Sharmila, M.R.C.: Performance evaluation of clustering algorithms. Int. J. Eng. Trends Technol. **4**(7), 2231 (2013)
15. Revathi, NalinI, s: T performance comparison of various clustering algorithm. Int. J. Adv. Res. Comput. Sci. Softw. Eng. **3**(2), 67–72 (2013)

Extended BB84 Protocol Using Lucas Series and Identity Based Encryption

AmrinBanu M. Shaikh[1(✉)] and Parth D. Shah[2]

[1] U. & P.U. Patel Department of Computer Engineering, Chandubhai S. Patel Institute of Technology, CHARUSAT, Changa, Anand 388421, India
amrinbanushaikh.ce@charusat.ac.in

[2] Department of Information Technology, Chandubhai S. Patel Institute of Technology, CHARUSAT, Changa, Anand 388421, India
parthshah.ce@charusat.ac.in

Abstract. In 1984 Bennett and Brassard proposed a Quantum Key Distribution (QKD) protocol known as BB84 protocol to distribute a random and frequently changed key using quantum mechanism. A major problem in this Protocol is to prove authentication. One of a solution of this problem has already been proposed in [4]. But in presence of Hardware Fault or Interception, above protocol is not applicable. In This paper Proposed System, Key Distillation has been added to overcome Hardware Fault or Interception with minor changes were not available in [4] and was available in original algorithm. It may increase performance of the Proposed System.

Keywords: Quantum key distribution (QKD) · BB84 · Identity key · Hybrid key

1 Introduction

BB84 Protocol is having mainly three steps: Raw Key Exchange, Key Shifting and Key Distillation. In Raw key exchange, Alice chooses randomly selected bit value with randomly selected bases from rectilinear or diagonal bases which results in four quantum states which are exchanged with Bob through quantum channel. The only way for Bob to derive any information from the incoming quantum states is to measure them against a randomly selected sequence of bases of his own. In Key Shifting if Alice/Bob selects the same base, which was used for decoding/encoding, then the result is determined to be correct. When the bases are different, then the result of this measurement is in deterministic [6]. After Key Shifting, key should be free of errors but it is possible only if no interception or no Hardware Fault is there. Key Distillation is consists of two steps. The first step corrects all the errors in the key, by using a Classical Error Correction protocol to precisely estimate the actual error rate. With this error rate, it is possible to accurately calculate the amount of information the eavesdropper may have on the key. The second step is called privacy amplification which compresses the key by an appropriate factor to reduce the information of the eavesdropper. The compression factor depends on the error rate. The higher the error rate, the more information an

S.C. Satapathy and A. Joshi (eds.), *Information and Communication Technology for Intelligent Systems (ICTIS 2017) - Volume 1*, Smart Innovation, Systems and Technologies 83, DOI 10.1007/978-3-319-63673-3_10

eavesdropper might have on the key and the more it must be compressed to be secure [1]. (For more information on how BB84 protocol works please refer [2, 3].) To calculate error rate Alice will repeatedly picks up a random position and check the bit value stored in that position with Bob then both will calculate error rate. Then after checked positions and their bit values are going to be deleted. But still BB84 is not able to prove authentication so it is possible that Alice is communicating with Eve. For that, Key Distillation Step should be removed. To know how BB84 protocol works without Key distillation step please refer [4]. But it is possible that because of Hardware Fault or Interception, keys at both ends are not same. In that case [4] is not applicable. For that Key Distillation step is in need. Outcome of above written introduction is, now we have to keep original BB84 protocol by considering another aspect. Proposed system is a combination of BB84 protocol [2] and "BB84 and Identity Based Encryption (IBE) based a novel Symmetric Key Distribution Algorithm" [4] with some changes which may improve performance. And here, rather than selecting random positions for Key Distillation step Lucas series [5] is used which is as same as Fibonacci series but the starting two numbers are chosen by Alice and Bob. The Advantage of Lucas series is that instead of sharing all random position numbers, Alice or Bob have to share only starting two numbers.

2 Proposed Algorithm

Step: 1 Raw Key Exchange

Alice Encodes randomly selected bit value by randomly selecting bases (Rectilinear or Diagonal) and generate polarized photon states and send to Bob through Quantum Channel and keep base-value combination with her in digital memory.
Bob measures the polarized photons in one of two set of bases (Rectilinear or Diagonal).

- If Bob selects same base as Alice result will be correct for that bit
- Else result will be random for that bit.

[Keep record for result and bases used in digital memory.]

Step: 2 Key Shifting

Bob tells Alice which Bases he used where bases are encrypted with public key of Alice via authenticated classical channel.
Alice tells Bob Which Bases she used where bases are encrypted with public key of Bob via authenticated classical channel.
Both will compare their Bases and keep only those values where they both used the same base and discard others let N be the number of remaining bits.

[Known as raw key named k1]

Step: 3 Transfer Challenge

Alice will send (nonce encrypted with k1A (hash value of k1 (at Alice side)), known as challenge) and encrypt this message with KB (public key of Bob) and send it to Bob.
Bob will decrypt the message with his private key and decrypt the challenge with k1B (hash value of k1 (at Bob side)) and send message back to Alice by encrypting it with KA (public key of Alice).
Alice will verify whether they both are having same k1 by verifying the received challenge with sent challenge. If yes then they will execute step:7 otherwise they will continue by executing step: 4.

Step: 4 Error Estimation

Alice and Bob will confidentially choose two numbers, and that will be starting numbers of Lucas Series.
Lucas series values are behaving like position numbers.
Alice will calculate the MAC (Message Authentication Code) by inserting position number, bit value and a secret key [Assume Secret key is confidentially chosen] and transfer it to Bob until Lucas series, position number exceed then the length of k1 and Bob will verify whether error is there or not and communicate where it didn't match and discard all Lucas Series Positioned bits.
Alice and Bob both are going to calculate error rate e. If error rate is higher than threshold they have to abort otherwise continue.

[Assume error rate e = 0.2 because it could be possible that it didn't get the eroded position because proposed system is using Lucas series.]

Step: 5 Reconciliation

It is a parity based protocol. To estimate error it generates a block and checks parity for that block.
Initial Block size K0 = $\left(\frac{1}{e}\right) + \left(\frac{1}{(4+e)}\right)$
Alice and Bob both are going to Check parity for that block and if it didn't match discard the whole block otherwise keep it as it is.
Increase Block size by K i + 1 = 2*Ki
Repeat until Block Size K i + 1 exceeds $\frac{1}{4}$ of all bits (length of key at step 1).

[After this step Alice and Bob both will have one key k2. If length (k2) < length (message) then abort.] [3]

Step: 6 Transfer Challenge

Alice will send (nonce encrypted with k2A (hash value of k2 (at Alice side)), known as challenge) and encrypt this message with KB (public key of Bob) and send it to Bob.
Bob will decrypt the message with his private key and decrypt the challenge with k2B (hash value of k2 (at Bob side)) and send message back to Alice by encrypting it with KA (public key of Alice).
Alice will verify whether they both are having same k2 by verifying the received challenge with sent challenge. If yes then they will continue to communicate by executing step: 7 otherwise they will abort [4].

Step: 7 Generate Identity Key

Assume Bob has already exchange his image (photo) to Alice by entity Authentication. Here, Bob can share his image after or before Quantum key established. Alice and Bob both will generate Image key (Ik) from image [4].

[This step could be performed first]

Step: 8 Generate Hybrid key

Both will repeat Ik until its length becomes same as k1(or k2) length and find final key k by XOR operation.

Step: 9 Encryption

Alice will Encrypt message using k and send it to Bob. Here, Alice encrypt message using public key of Bob.

Step: 10 Decryption

Bob knows his private key, can only decrypt the message then after Alice and Bob both will Exit.

Assumptions
Private keys of Alice and Bob are not known to Eve.
Identity (Image) is confidentially shared.
Lucas Series starting two numbers are confidentially shared.
Single polarized photons are transferred.
Length of Raw Key Exchange = 8 Message length in bits.*

3 Proposed Algorithm Flow

Below Figure shows flow of Proposed Algorithm (Fig. 1)

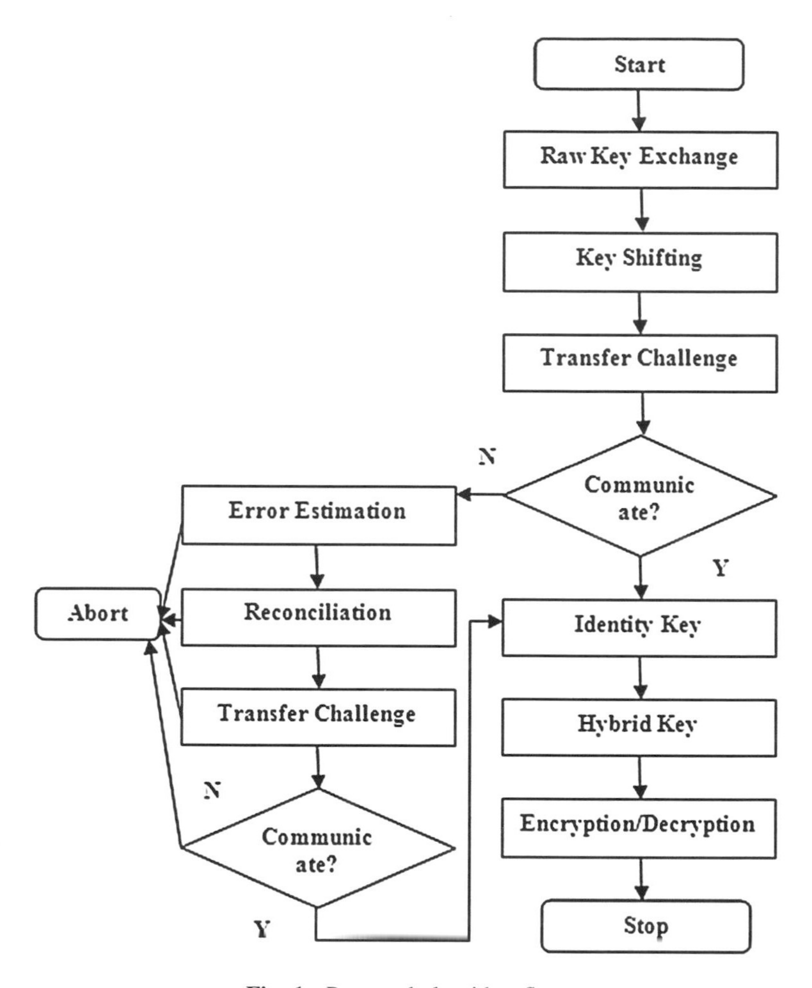

Fig. 1. Proposed algorithm flow

4 Conclusion

In this Paper the Proposed System has used Lucas Series to remove the number of assumptions and disadvantage of this series is that if starting two numbers are known then whole series is known. If Eve Knows Lucas series then interception becomes easy. So, rather than communicating with bit value Alice and Bob will communicate through MAC (Message Authentication Code). At the end, Alice or Bob will check whether keys are same or not by proving the given challenge and communicate if challenge has been proved. After adding Key Distillation Step, it improves the performance in presence of Hardware Fault or Interception. It Transfers challenge twice to overcomes cycle problem.

Acknowledgement. We Express our Auspicious thank to ARTCom2013 conference who has motivated us to think extra on our previous proposed system by selecting our paper as an extended paper. And we have never worked with QKD hardware so whether this combination is going to work or not that we can't answer.

References

1. http://swissquantum.idquantique.com/?Key-Distillation
2. Cobourne, S.: Quantum key distribution protocols and applications (2011). http://www.ma.rhul.ac.uk/static/techrep/2011/RHUL-MA-2011-05.pdf
3. Implementation of the BB84 QKD Protocol. http://www.cki.au.dk/experiment/qrypto/doc/QuCrypt/bb84prot.html
4. Shaikh, A.M., Shah, P.D.: BB84 and identity based encryption based a novel symmetric key distribution algorithm. ARTCom2013 (2013)
5. http://www.maths.surrey.ac.uk/hosted-sites/R.Knott/Fibonacci/lucasNbs.html
6. Scharitzer, G.: Basic quantum cryptography. Version 0.9, Vienna University of Technology Institute of Automation (2003)

Performance Analysis of WSN Routing Protocols with Effective Buffer Management Technique

T. Padmashree(✉), N.K. Cauvery, and Soniya Chavan

R.V. College of Engineering, Bengaluru, India
{padmashreet, cauverynk}@rvce.edu.in,
soniyachavan080@gmail.com

Abstract. Routing in WSN has dependably been a serious issue of concern mainly because of a few case studies which extend from unfriendly deployment conditions, network topology that change over and over, network failures, resource constraints at each sensor hub to issues in designing of routing protocols. Accordingly, the implementation of routing protocol is influenced by a few fundamental elements which must be thought of before any attempt at designed routing are implemented. Two major protocols used in WSN are Dynamic-Source Routing (DSR) and Destination Sequenced Distance Vector Routing (DSDV). DSR protocol is mainly source based routing protocol and implemented to limit the bandwidth utilized by packets in WSN by avoiding the regular messages transmitted to update table in table-driven approach. The proposed system is Multilayer Buffer Management DSR (MBMDSR) where multilayer buffer management mechanism is implemented in existing DSR protocol and a considerable increase in performance was noted.

Keywords: DSR · WSN · Multilayer buffer management · DSDV

1 Introduction

Wireless network is a gathering of wireless sensor hubs with no fixed structure and communicates with each other through remote connections. Applications for wireless sensor networks include different domains, such as medical monitoring [1–3], environmental monitoring, home surveillance, military operations, and machine monitoring in the industries. For recent many years analysts have concentrated on major routing protocols such as, DSDV and DSR and their working mechanisms. DSR protocol is based on source, utilizing intermediate nodes for information exchange from source hub to destination hub. In DSR protocol problem arises due to fast movement of nodes which causes collision between the nodes which in turn causes flooding in the network. Efficient buffer management policy aims at maximizing the overall throughput by reducing the number of packets that are retransmitted due to packet loss. Network Simulator-2 (NS2) is used for the simulation of the scenario in the network. The comparative analysis is carried out between the existing DSR and DSDV. A Multilayer Buffer Management policy is introduced to DSR and compared with DSR.

S.C. Satapathy and A. Joshi (eds.), *Information and Communication Technology for Intelligent Systems (ICTIS 2017) - Volume 1*, Smart Innovation, Systems and Technologies 83, DOI 10.1007/978-3-319-63673-3_11

The results show that the loss of relevant packets is less in MBMDSR when compared to the existing DSR protocol.

2 Literature Survey

Akyildiz et al. [4] stated that the characteristics of sensor networks such as fault tolerance, flexibility, high sensing fidelity, rapid deployment and low cost creates new application areas for remote sensing. However, sensor networks needs to fulfill the constraints presented by components, for example, adaptation to internal failure, adaptability, cost, hardware, topology change, environment and power utilization.

V. Jacobson [5] presented that computer systems have encountered an unstable development in the course of recent years and with that development have come extreme congestion issues. For instance, it is presently consistent to see web gateways drop 10% of the incoming packets on account of local buffer overflows. Examination of some of these issues has demonstrated that a significant part of the cause lays in transport protocol executions. The "self-evident" approaches to actualize a window-based transport protocol can bring about precisely the wrong behavior because of network congestion.

J. Postel [6] depicted that currently available buffer management mechanisms are divided into two categories mainly congestion avoidance and congestion control. Congestion avoidance mechanism first identifies the congestion in the network or avoids it from happening whereas congestion control mechanism focuses on recovery of a packet loss in the network. The main disadvantage of congestion avoidance mechanism is that it does not fit to the type of arrangement where several hubs sends their reading to a particular node.

Zafar Mahmood, Muhammad Awais Nawaz [8] and other authors studied and compared the behavior of AODV, DSR, and DSDV routing protocols. Analysis was carried out using NS2 as a computer simulator tool and results of analysis were depicted in graphical format. The graph is represented with respect to the pause time which depicts DSR protocol is improved than other two protocols, it is mainly because the system was fewer intense and fewer stressful DSR was better in packet delivery ratio with respect to packet sent when compared to the AODV and DSDV with minimum routing load, but DSDV have highest value for the performance metrics like average end to end delay.

Review of above literature survey indicates that most of the authors worked on routing protocols in WSN, studied their behavior, compared many routing protocols with each other and determined which routing protocol is better in terms of performance metrics by using the NS2 simulation tool. Also work has been carried out on buffer management policies as the WSN has limited memory and power. To overcome this limitation of WSN, authors have carried out survey on buffer management policies. This survey clearly indicates that if multilayer buffer management policy is implemented in routing protocol the loss of relevant packets was decreased and the throughput was gradually increased. The main objective of this work was to study and compare the DSDV and DSR routing protocol and to check which protocol is better in terms and performance metrics. Further, implement the routing protocol MBMDSR in order to increase the throughput and decrease the loss of relevant packets.

3 Existing System (Existing DSR)

The Route Maintenance protocols do not fix a broken connection. The broken connection is just conveyed to source. The DSR protocol is just productive in Mobile System with less than 200 hubs. Problem occurs due to rapid movement of nodes in a network. Flooding packets in the network may lead to collision among the packets. Additionally there is a little time delay at the start of another association with a new node on basis that the initiator should first discover the route to the objective.

4 Proposed System–MBMDSR

Multilayer buffer management technique is implemented to the existing DSR protocol (MBMDSR). The process includes four modules: DSR route discovery, packet classification, buffer partitioning and a discard policy.

It finds the multiple reliable paths from source to destination so that if one link fails the information is sent through other alternate paths. The MBMDSR protocol is efficient for large number of nodes up to 500. There is less chance for flooding the network which reduces the collisions between the packets. As the main buffer is divided into multiple buffers the relevant data can be saved and successfully transferred to the destination and drop of packets is less at destination side when compared to the existing DSR.

4.1 DSR Route Discovery

Source node Send request (RREQ) to intermediate nodes if respond (RREP) from the true destination is received by the source, it begins to transfer data packet. Otherwise exceeded dynamic hop discovery again source node send RREQ and the process repeats. Data packets are transmitted to destination through shortest path.

When path is not up and if target finds that packet delivery ratio drastically falls to the threshold, the detection scheme is activated to recognize the constant maintenance and real time reaction efficiency. The threshold is a differing value in the range [85%, 95%] that can be balanced by current system efficiency. The initial threshold value is set to 90%. This can be done by utilizing a dynamic threshold calculation that control the time when packet delivery ratio falls under the same threshold.

4.2 Packet Classification

Every node divides the arriving packets into three distinctive types and subsequently every packet is said to be of type *i*, $1 \leq i \leq 3$. First type of packets is relevant packets that include relevant and important data. Second type of packet is irrelevant packets that include diverse type of information that is not related to the receiver information.

Last type of packets is normal packets that incorporate hello packets and regular packets those created at consistent time interval. Regardless, it ensures that there is no loss of important packets and disregard the loss of other types of packets.

4.3 Buffer Partitioning

Buffer partitioning indicates measure of storage space accessible to specified queue and characterizes how memory is shared between distinctive queues. In network situation, every hub comprises of a total buffer size B, shared by T diverse kind of queue. Entire buffer is divided into T queues as per expected incoming packet type. Main buffer is divided in three queues (relevant, irrelevant and normal) and every queue accepts packets with corresponding type only. The memory space of relevant, irrelevant and normal type of queues is L, M and N respectively. Consequently, total capacity of three queues should not exceed total limit of main buffer, $L + M + N \leq B$. Every hub can recognize the kind of received packets by the data represented in packet header.

4.4 Discard Policy

Discard Policy primarily manages the policy that incorporates tolerating or dismissing of arriving packets and moreover pushing out a previously stored packets to make space for an incoming packets. The judgment is made in perspective of the type of incoming packets. Arrived packets are explicitly divided in three types as described previously. When the main buffer is full, discard policy is implemented and it executes as below.

If incoming packet is of normal type and if there are a couple of packets in normal queue, then it replaces the oldest packet in normal queue with the newly arrived packet. If the length of normal queue is zero, i.e. there are no current normal packets to be dropped or evacuated, it drops the incoming packet. If the incoming packet is important, it drops oldest packet either from normal or relevant queue to make a space for new incoming relevant packet. If the incoming packet is irrelevant, it drops the irrelevant packets.

Comprehensive analysis and comparison is carried out between existing DSR and MBMDSR.

5 Experimental Analysis and Results

Simulation (By default the X-axis is always taken as simulation time period) has been conducted to compare DSDV with DSR and it was noted that DSR performed well as compared to DSDV in terms of Packets dropped, Packet delivery ratio and Throughput as depicted in the Table 1.

Table 1. Comparative results of DSDV with DSR

Simulation time (milliseconds)	Parameter	DSDV	DSR
0–17	Packets dropped	65	15
6–18	Packet delivery ratio	0.95	1.000
15	Throughput	85	90

With respect to the above results, since DSR performed well compared to DSDV, the proposed buffer management technique was introduced to obtain the results as shown in Table 2.

Table 2. Comparative results of DSR with MBMDSR (considering no. of nodes)

Number of nodes	Parameter	DSR	MBMDSR
500	Packets dropped	8	3
300	Packets dropped	6	2
100	Packets dropped	3	0

It is observed that the packets dropped gradually decreased with increase in number of nodes.

Simulation results were also tabulated for Packet Rate Vs. Packet Dropped as shown in Table 3.

Table 3. Comparative results of DSR with MBMDSR (considering packet rate)

Packet rate	Parameter	DSR	MBMDSR
5500	Packets dropped	3	0
3500	Packets dropped	1.8	0
1500	Packets dropped	0	0

It is observed that as the Packet rate increases the Drop of packets were NIL in MBMDSR.

The result of simulation is depicted in the graphical format. Simulation of DSR protocol and Multilayer Buffer Management DSR (MBMDSR) is performed for 100–500 nodes and the result is depicted in graphical format shown.

Figure 1 shows the Loss of relevant packets vs the Buffer size for three layer WSN. Packet loss is gradually decreasing for three layer WSN, which means as the buffer size increased its capacity to accommodate the packets is also increased. Hence the maximum packets can be accommodated in buffer. As the buffer size increases the number of packets accommodating is more, as a result the loss of relevant packets is less.

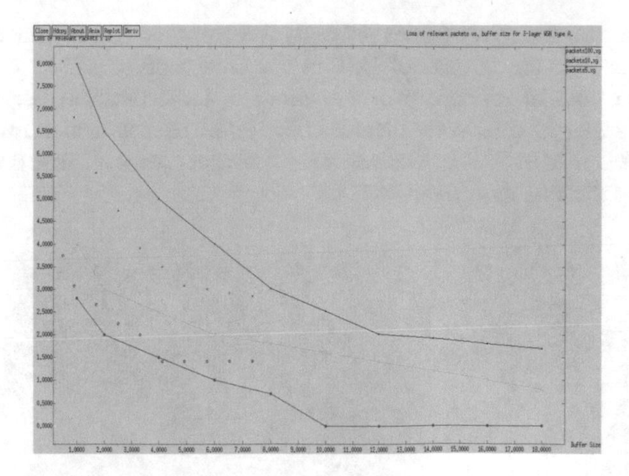

Fig. 1. Loss of relevant packets vs Buffer size

Figure 2 shows the loss of relevant packets vs number of nodes for three layer WSN. Loss of relevant packets is less in MBMDSR when compared to DSR. The graph depicts that loss of relevant packets is gradually decreasing with respect to increase in number of nodes in case of MBMDSR.

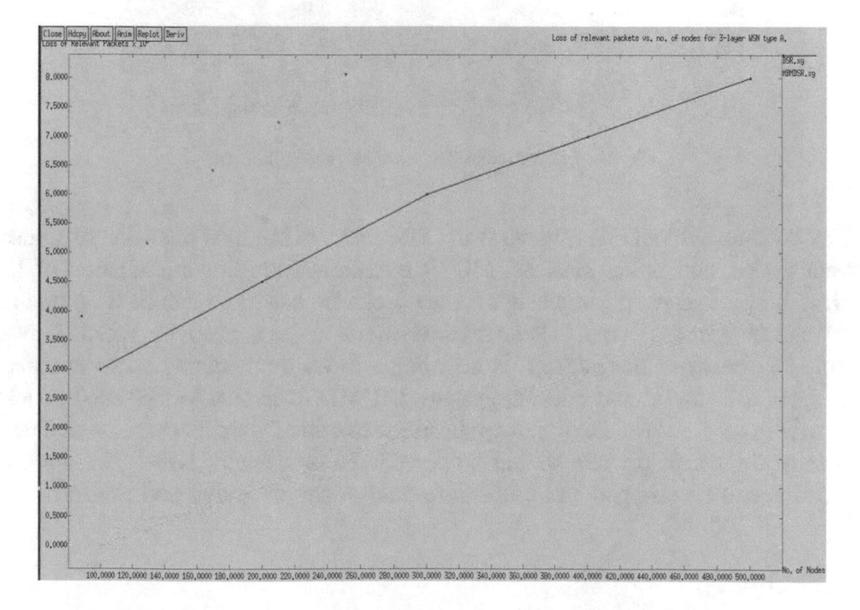

Fig. 2. Loss of relevant packets vs number of nodes

Figure 3 shows the loss of relevant packets vs packet rate for three layer WSN. Loss of packets is more in case of DSR when compared to MBMDSR. The graph depicts that the loss of relevant packet is more in DSR protocol because when the packet moves faster in a network, there is collision of packets which results in loss of packets. But when MBMDSR is considered, it overcomes with that problem as it is efficient during the fast movement of nodes.

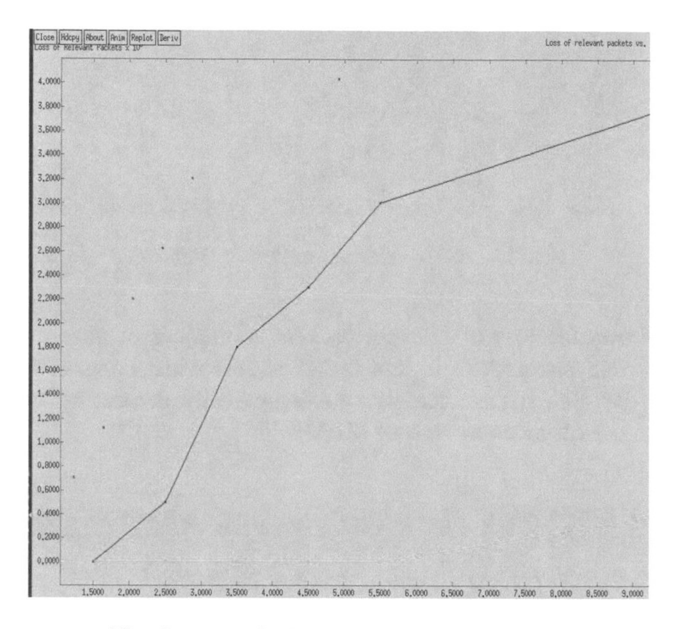

Fig. 3. Loss of relevant packets vs packet rate

To conclude, MBMDSR is better than DSR protocol in terms of packet rate, loss of relevant packet and buffer size. MBMDSR overcomes the major limitation of DSR which is Loss of relevant packets and hence it can be used in network were there are more number of nodes. It can also be used when the nodes are moving fast as it avoids the collision between the packets. It also helps in accommodating more number of packets and also useful and relevant packets. MBMDSR avoids loss of useful packets to a great extent. As WSN has major applications in critical domains such as healthcare, the loss of important packets should be considered as a major issue. The unwanted information can be dropped and useful information can be stored and utilized.

6 Conclusion

A. The main focus of the work is on routing protocols with respect to their performance in the wireless sensor network. And implementing the multilayer buffer management technique to one of the routing protocol is a step towards achieving a network with better Quality of Service.

B. MBMDSR protocol is efficient for large number of nodes.
C. There is less chance for flooding the network which reduces the collusions between the packets.
D. As the main buffer is divided into multiple buffers relevant data can be saved and successfully transferred to the destination. For MBMDSR drop of packets is less at destination side when compared to the existing DSR.
E. By implementing the buffer management technique in DSR protocol it is possible to reduce the loss of relevant packets in WSN.

References

1. Kidd, C.: The aware home: a living laboratory for ubiquitous computing research. In: Second International Workshop on Cooperative Buildings (CoBuild) (1999)
2. Intille, S.: Designing a home of the future. IEEE Pervasive Comput. **1**(2), 76–82 (2002)
3. Schwiebert, L., Gupta, S., Weinmann, J.: Research challenges in wireless networks of biomedical sensors. In: Proceedings of the Seventh Annual International Conference on Mobile Computing and Networking (MobiCom) (2001)
4. Akyildiz, F.: Wireless sensor networks: a survey. Comput. Netw. **38**, 393–422 (2002)
5. Jacobsn, V.: Congestion avoidance and control. In: IEEE/ACMSIGCOMM, pp. 314–329 (1988)
6. Postel, J.: Transmission control protocol specification. In: SRI International CA (1981)
7. Abdala, T.M., Daud, N., Sanam, E., Ahmed, M.S., Abdalla, A.A., Aboghsesa, S.M.: Performance tradeoffs of routing protocols in wireless sensor networks. In: International Conference on Network Security and Computer Science (ICNSCS-15), 8–9 Febuary 2015
8. Mahmood, Z., Nawaz, M.A., Iqbal, M., Khan, S., Haq, Z.U.: Varying pause time effect on AODV, DSR and DSDV performance. Int. J. Wirel. Microwave Technol. **1**, 21–33 (2015). doi:10.5815/ijwmt.2015.01.02. MECS. http://www.mecs-press.net

Comparison of Accelerator Coherency Port (ACP) and High Performance Port (HP) for Data Transfer in DDR Memory Using Xilinx ZYNQ SoC

Rikin J. Nayak[1(✉)] and Jaiminkumar B. Chavda[2]

[1] E&C Department, Chandubhai S. Patel Institute of Technology, Charotar University of Science and Technology, Changa 388421, Gujarat, India
rikinnayak.ec@charusat.ac.in

[2] IT Department, Chandubhai S. Patel Institute of Technology, Charotar University of Science and Technology, Changa 388421, Gujarat, India
jaiminchavda.it@charusat.ac.in

Abstract. ZYNQ 7000 Embedded Processing Platform SOC is chips includes ARM dual core A9-MPCore Processor Processing System-(PS-Microprocessor) along with Xilinx Programmable Logic (PL)-Artix 7 FPGA on a single die. ZYNQ SoC provides the high performance and computing throughput at low power using PS along with the flexibility of PL. ZYNQ SoC incorporates independent interfaces for communication of data control signals between PL and PS in various configurations to access the system resources. This paper describes the performance evaluation of such interfaces in terms of resource utilization and power consumption. Here, in this paper, data transfer from PL to PS using low speed AXI GP port and high speed bus like AXI HP port and ACP (Accelerator Coherency) port is discussed. The paper includes design, implementation and testing results on Zynq-7000 SoC based Avnet Zed board.

Keywords: Field programmable gate array · ZYNQ-7000 SoC · System on chip · High speed data acquisition

1 Introduction

As embedded systems become popular in multimedia gaming to space applications, it is highly required to meet critical constraints like time, power and others. Such system provides high performance with operation completion within deadlines [1]. For such high performance tasks, System on Chip (SoC) is highly used having advantages of utilization of end product as an IP (intellectual properties). Some of the SoC supports configurable hardware such as Field programmable gate array (FPGA), which offers high flexibility with soft hardware and can be reconfiguring whenever required.

S.C. Satapathy and A. Joshi (eds.), *Information and Communication Technology for Intelligent Systems (ICTIS 2017) - Volume 1*, Smart Innovation, Systems and Technologies 83, DOI 10.1007/978-3-319-63673-3_12

1.1 ZYNQ SoC Architecture

Xilinx ZYNQ-7000 is a programmable SoC which incorporates Artrix-7 FPGA (PL) and Dual core ARM Cortex MP-Core Processor (PS) on a single die with footprint as small as 13 mm × 13 mm and it is benefitted from the 28 nm fabrication technology [2]. The architecture of ZYNQ SoC contains Processing System (PS), Programmable Logic (PL) and PS-PL Communication interfaces using AXI bus switches for internal communication.

ZYNQ-PS consist of Application processing unit (APU) and Interface Controllers for DDR Memory, RS232-UART, SPI, USB,CAN, SDIO, Gigabyte Ethernet, etc. PS system contains ARM Cortex A9 dual-core processor which is based on the ARM V7-A architecture and supports virtual memory, which could run 32-bit/64-bit ARM instructions, 16&32-bit Thumb instructions, and Jazelle state 8 Java byte code. NEON Coprocessor is used to accelerate media and signal processing algorithms for audio, video and visualization graphics instruction [5]. APU and system components are connected by AMBA switches. ZYNQ-PL is based on either Xilinx Kintex-7/Artix-7 FPGA with programmable resources like configurable logic blocks, block RAMs, digital signal processing blocks etc. Analog-to-digital converter (ADC) provides interface to get feedback from the real world environment [6]. The PL architecture provides user configurable capabilities. ZYNQ PS-PL interface is supported by two AMBA switches which provide the connectivity among system components. One of the switch connects processor and standard interfaces like GigE, UART,SPI, etc. while another switch provides interfacing of PL and PS, DDR Memory for high speed data transfer. The switches implements AXI Bus Specification [8]. Data exchange between PL-PS, on FPGA in the PL, one can use AXI or EMIO interface signals. Figure 1 Shows block diagram of the Zynq-7000 SoC architecture [4, 7].

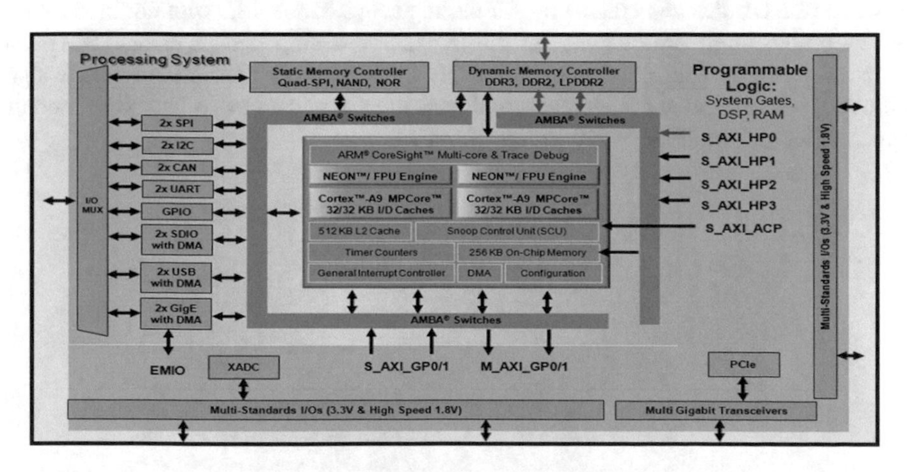

Fig. 1. ZYNQ SoC internal block diagram

For communication between PL-PS, three types of interfaces are there: High Performance port (HP Port), General Purpose port (GP Port) and Accelerator Coherence port (ACP). (1) There are 4 High Performance Ports (HP ports) with 64-bit width and can operate at more 150 MHz along with dedicated 1 KB FIFO (128 slots of 64-bit). HP ports provide bandwidth up to 1.2 GB/S. (2) 4 GP Ports provides low bandwidth, register based, memory mapped interface using 2 Master and 2 Slave ports. GP ports are utilized for register based control/status information exchange. (3) There is a Accelerated coherency port (ACP) is integrated in ZYNQ SoC. The ACP was developed by ARM for solving cache coherency issues using hardware when new accelerator blocks are integrated to a multi-core system. ACP port enables hardware accelerators to issue cache coherent memory requests to the CPU sub-system memory space [10, 11]. There exist other interfaces between PL and PS like PCAP, EMIO, Interrupts Lines, DMA flow control, Clock, and Debug. Table 1 shows summary of port types [8, 9].

Table 1. ZYNQ port comparison.

Port type	Type	Units (master +slave)	Cache coherent	Application
GP Port-32 Bit	AXI LITE	2M+2S	No	Control/status
HP Port-64Bit	AXI-FULL, AXI Stream	4S	No	High bandwidth
ACP Port-64Bit	AXI-FULL	1S	Yes	High bandwidth Cache coherent

Xilinx Vivado Design Suite is a software package which is used for synthesis and analysis of HDL designs created by Xilinx, replacing Xilinx ISE with additional tools for supporting high productivity on the system on a chip design using C/C++ and IP-based designs. Xilinx provides VIVADO HLS tools to translate advanced algorithms of wireless, medical, defense, and consumer applications to hardware module, which speed up the IP software process [3, 4].

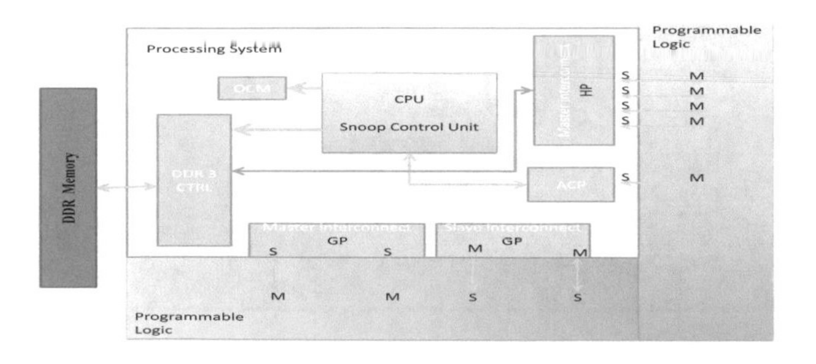

Fig. 2. ZYNQ SoC PL-PS interfaces

1.2 Data Transfer in ZYNQ SoC

For PS-PL communication, AXI master issues a AXI transaction along with valid 32 Bit address in system address space of 4 GB [12]. For high speed data transfer, two types of port are there in ZYNQ: (1) AXI HP port and (2) ACP port. The high performance port (HP) supports AXI master interface at PL, through which PL accesses DDR memory (Fig. 2). The PL master should be connected to one of these HP interfaces to get higher throughputs/bandwidth. The AXI Accelerator coherence slave Port (ACP) provides a similar user IP as the HP ports. The data is transferred by the PL master to the DDR using HP port is not coherent with the Cortex-A9 CPUs and it requires to program software instructions to make it coherent with the processor. As ACP port connected to CPU cache, allowing ACP transactions interacts with the CPU cache due to internal connectivity. This reduces collective latency for the data to be used by a CPU. These optional cache-coherent operations prevent the need of cache invalidation and flush of cache lines and thus improving overall application performance [12]. AXI4-LITE supports single beat data transfer and it is ideal for low speed communication like writing control information or reading status information between PL-PS. AXI-Full is the high performance bus supports burst data transfer up to 256 data transfer cycle for single address. For particular application it is required to choose proper interface for getting maximum throughput with very less power consumption.

2 Test Experiment: Data Transfer from PL to PS

The data generation is carried out at Virtex-5 FPGA in two DATA_LINE @64 Mbps along with DATA_VALID and CLOCK signals, which is sampled at the ZYNQ-PL module. Here, CLOCK is generated from Virtex-5 Board which realizes the source synchronous communication. For high speed, reliable data transfer between Virtex-5 and ZYNQ, LVDS (Low-voltage differential signaling) interface is utilized. At ZYNQ, input signal is sampled and digital data is received on the rising of the DATA_VALID signal. In the ZYNQ SoC, all the device controllers are interfaced with the system bus

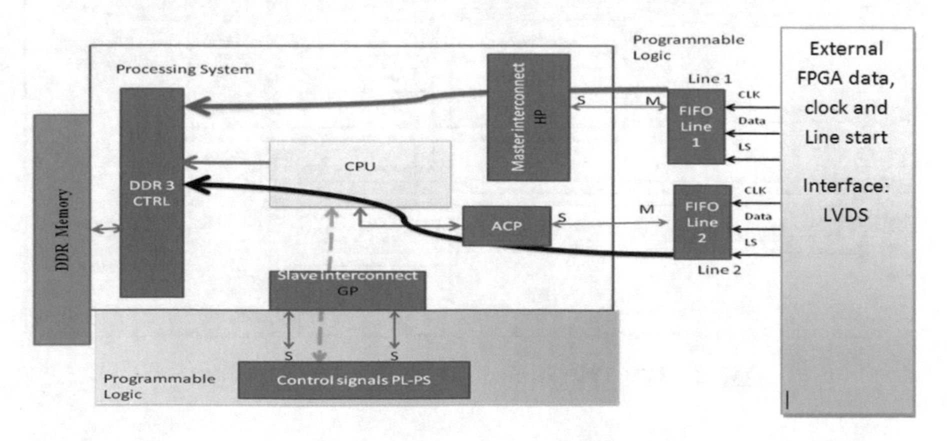

Fig. 3. ZYNQ SoC data interface system

using AXI Switches. PS is also connected with the DDR3 Memory using the AXI Dynamic Memory Controller on the AXI Bus. Figure 3 shows experimental system for transferring data from PL to PS using GP, HP and ACP ports.

For assessing the performance of HP and ACP ports, a PL module is designed with 2 HP Master PL System interfaces are connected with 1 HP Slave interface and 1 ACP slave interface of PS System respectively. AXI-LITE Slave Interface is connected with AXI-LITE Master Interface on GP Port, and Interrupt lines are connected with Generic Interrupt Controller using CONCAT IP. Figure 4 shows the VIVADO design for the above system. Steps for setting up the performance test experiment are summarized as follows: (1) Design of Hardware (2) Design of Device Driver and (3) Design of User Application.

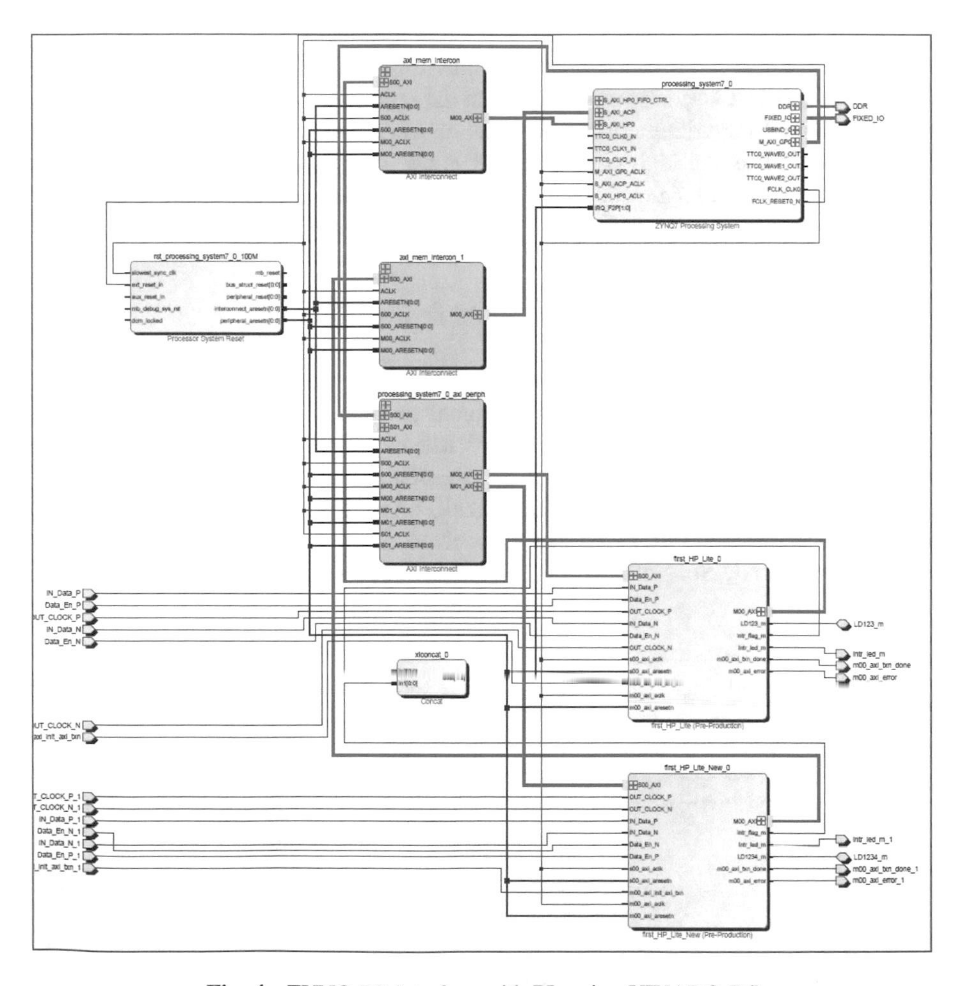

Fig. 4. ZYNQ PS interface with PL using VIVADO DS

2.1 Design of Hardware Programming: Design of Custom IP with Interfaces

Instantiation of a custom IP which consist of 1 GP port, 2 HP ports and Interrupt signals. In the design, GP port is a memory mapped AXI slave interface while HP and ACP ports are AXI master interfaces which communicate with mapped memory regions from DDR3, OCM etc. PS controls custom IP registers using AXI GP ports. Here HP/ACP ports acts as a bus master at PL for data transfer. Asynchronous data Acquisition is controlled using the DATA_VALID signal from external source along with CLOCK and DATA_LINE signals. As the digital signal is sampled, the data is temporary buffered and packetized in to 32 $\times$ 1 K FIFO in PL. As FIFO memory gets filled with 60% of FIFO depth, FIFO_FULL signal is raised and FIFO readout operation is initiated. AXI master writes data to pre mapped memory region at DDR3 RAM. After writing predefined size of memory Blocks, interrupt signal is raised (HIGH), which is served by PS and Interrupt Service Routine (ISR) is executed. Pre mapped memory address region set of DDR memory for interfaces is as shown in the Fig. 5. Resource utilization report for the Avnet ZED board is as shown in below Fig. 6.

Diagram × | Address Editor ×

Cell	Interface Pin	Base Name	Offset Address	Range	High Address
processing_system7_0					
Data (32 address bits : 4G)					
first_HP_Lite_0	S00_AXI	S00_AXI_reg	0x43C00000	4K	0x43C00FFF
first_HP_Lite_New_0	S00_AXI	S00_AXI_reg	0x43C01000	4K	0x43C01FFF
first_HP_Lite_0					
M00_AXI (32 address bits : 4G)					
processing_system7_0	S_AXI_HP0	HP0_DDR_LOWOCM	0x00000000	512M	0x1FFFFFFF
first_HP_Lite_New_0					
M00_AXI (32 address bits : 4G)					
processing_system7_0	S_AXI_ACP	ACP_IOP	0xE0000000	4M	0xE03FFFFF
processing_system7_0	S_AXI_ACP	ACP_DDR_LOWOCM	0x00000000	512M	0x1FFFFFFF
processing_system7_0	S_AXI_ACP	ACP_QSPI_LINEAR	0xFC000000	16M	0xFCFFFFFF

Fig. 5. ZYNQ memory region mapping for the custom design interfaces in VIVADO.

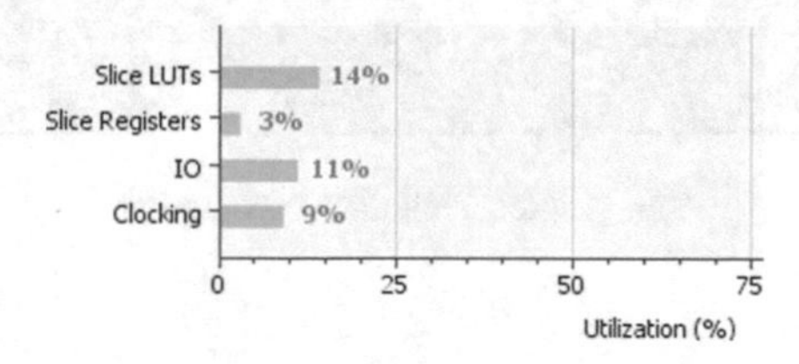

Fig. 6. ZYNQ PL (FPGA) resource utilization report in VIVADO DS.

2.2 Design of Device Driver and Application: Software Programming

Here, for experiment software is implemented on LINUX 4.4 Kernel System, so it requires to include the custom device driver module and user application for controlling the device as per the standard OS interfaces. This software is implemented as KERNEL Module and User Application code written in C language which includes:

Table 2. Data volume vs. time measured for 64Mbps date rate using HP port and ACP port.

Data volume	Time measured in NS (datarate:64 Mbps)	
	HP port	ACP port
4 MB	524288347	524285038
8 MB	1048583847	1048581558
12 MB	1572874840	1572878450
16 MB	2097172161	2097168633

1. Memory Allocation and Initialization.
2. Generic Interrupt Controller Initialization, Interrupt registration, Interrupt service routine.
3. Device Register Initialization to default values.
4. Application program waits for event of data arrival on LVDS input interface into sleep mode.
5. On event of interrupt, Interrupt service routine modifies device registers and perform the data transfer. In the ISR, low priority tasks like data processing and other can be scheduled based on critical time constraints.
6. When predefined volume of sampled data is written in to buffered memory, the data verification is carried out for every memory location.
7. After data verification, data from DDR memory is send to UART for display at slow speed using user applications (Fig. 7).

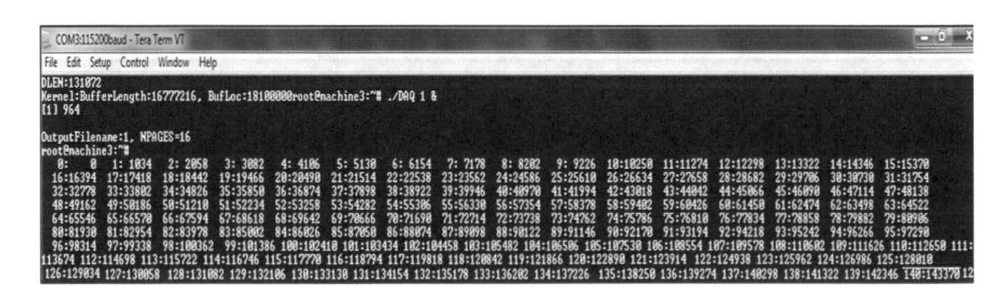

Fig. 7. Received data verification

3 Results

Figure 8 shows registered interrupt sequence. Here 4096 Pages (1Page = 4096B) data has been transferred, where for each 16 pages, PL interrupts the PS. In verified results, each location displayed starting data of each 1 page with 1024 locations. Here, initial data was started with 10, so for page data will be in multiple of 1024. E.g. for location 323, data will be (323*1024) + 10 = 330762. Total transferred data is 16 MB. Table 2 shows time required to write different volume of data for HP and ACP port.

```
andom: nonblocking pool is initialized
at /proc/interrupts
           CPU0       CPU1
27:          0          0       GIC  27  gt
29:        690        660       GIC  29  twd
35:          0          0       GIC  35  f800c000.ocmc
39:         43          0       GIC  39  f8007100.adc
40:        988          0       GIC  40  f8007000.devcfg
41:          0          0       GIC  41  f8005000.watchdog
43:        453          0       GIC  43  ttc_clockevent
45:          0          0       GIC  45  f8003000.dmac
46:        256          0       GIC  46  f8003000.dmac
47:          0          0       GIC  47  f8003000.dmac
48:          0          0       GIC  48  f8003000.dmac
49:          0          0       GIC  49  f8003000.dmac
51:          6          0       GIC  51  e000d000.spi
53:          1          0       GIC  53  zynq-otg, zynq-udc
54:        337          0       GIC  54  eth0
56:        176          0       GIC  56  mmc0
57:          0          0       GIC  57  cdns-i2c
72:          0          0       GIC  72  f8003000.dmac
73:          0          0       GIC  73  f8003000.dmac
74:          0          0       GIC  74  f8003000.dmac
75:          0          0       GIC  75  f8003000.dmac
82:        180          0       GIC  82  xuartps
91:        257          0       GIC  91  ATSA_DAQ
PI1:          0        300  Timer broadcast interrupts
```

Fig. 8. Registering interrupt with interrupt service routine in Linux v4.4 for HP0 Port

4 Conclusion

Based on above test design, implementation and achieved results, it can be concluded that, selection of appropriate ports should be carried based on the requirements of application. ZYNQ PS-PL interfacing provides rich set of interfaces where low, moderate and high data rates are supported as well as memory mapped interface with and without addressing modes are also supported. Data rate and bandwidth requirement justifies the need of application. For low bandwidth applications GP port is found to be suitable, but for high speed applications High Performance (HP) port and Accelerator Coherence Port (ACP) performs better. However, both port have advantages as well as disadvantages. ACP port has advantages of coherency for cache memory. For certain data processing application due to coherency processor will consume less power and can have high performance compare to HP port. However if any other computing task have higher priority, performance may degraded due to such coherent entry in L2 cache. In such case HP port is better since it provide same performance for data transfer in DDR memory as ACP port supports.

References

1. Powell, A., Silage, D.: Statistical performance of the ARM cortex A9 accelerator coherency port in the Xilinx ZYNQ SoC for real-time applications. In: International Conference on Reconfigurable Computing and FPGAs (ReConFig), pp. 1–6 (2015)
2. Xue, T., Pan, W., Gong, G., Zeng, M., Gong, H., Li, J.: Design of Giga bit Ethernet readout module based on ZYNQ for HPGe. In: 19th IEEE-NPSS Real Time Conference, pp. 1–4 (2014)
3. Xilinx Inc. UG702, Partial Reconfiguration user guide (2012)
4. Han, T., Liu, G.W., Cai, H., Wang, B.: The face detection and location system based on ZYNQ. In: FSKD 2014, pp. 835–839 (2014)
5. Xilinx Inc. UG925, Zynq-7000 EPP ZC702 Base Targeted Reference Design (2012)
6. Xilinx Inc. DS190, Zynq-7000 All Programmable SoC Overview (2016)

7. Xilinx Inc. UG585, Zynq-7000 Extensible Processing Platform Technical Reference Manual (2012)
8. Xilinx Inc. DS768, LogicCORE AXI Interconnect IP (v1.0.01a) (2012)
9. Xilinx Inc. UG1037, VIVADO design suite: AXI Reference Guide (2015)
10. ARM. Inc. Cortex-A9 MPCore Technical Reference Manual (2012)
11. Sadri, M., Weis, C., When, N., Benini, L.: Energy and performance exploration of accelerator coherency port using Xilinx ZYNQ. In: 10th FPGAWorld Conference (2013)
12. Erusalagandi, S.: Leveraging Data-Mover IPs for Data Movement in Zynq-7000 AP SoC Systems, WP459 (v1.0)

Matrix Factorization and Regression-Based Approach for Multi-Criteria Recommender System

Gouri Sankar Majumder[1(✉)], Pragya Dwivedi[1], and Vibhor Kant[2]

[1] Motilal Nehru National Institute of Technology Allahbad, Allahbad 211004, India
majumder.gs30@gmail.com, pragyadwi86@mnnit.ac.in
[2] The LNM Institute of Information Technology, Jaipur 302031, India
vibhor.kant@gmail.com

Abstract. Recommender systems (RS) try to solve information overload problem by providing the most relevant items to users from a large set of items. Collaborative filtering (CF), a popular approach in building RS, generates recommendations to users based on explicit ratings provided by the community of users. Currently many online platforms allow users to evaluate items based on multiple criteria along with an overall rating instead of single overall rating. Previous research work has shown that considering these multiple criteria ratings for recommendations improved the predictive accuracy of recommender systems.

In this paper, we propose a novel approach to increase predictive accuracy of multi-criteria recommender systems (MCRS). Firstly, we use matrix factorization to predict individual criteria ratings and then compute weights of individual criteria ratings through linear regression. Finally we predict overall rating using a weighted function of multiple criteria ratings. Through experiments on Yahoo! Movies dataset, we compare our proposed approach to baseline approaches and demonstrate its effectiveness in terms of predictive accuracy measures.

Keywords: Recommender systems · Multi-criteria recommender systems · Collaborative filtering · Matrix factorization · Linear regression

1 Introduction

In last 10–15 years, we witnessed exponential increase of information on Web. Recommender systems [4,6] are developed to help users to deal with this information overload problem. Recommender systems are systems that help users to find relevant items from large set of available items. Various approaches have been explored to build recommender systems effectively. Broadly these approaches can be classified into four categories like collaborative filtering based approach, content-based filtering based approach, knowledge-based filtering approach and hybrid approach. Adomivicius and Tuzhilin [3] provided a detailed review on recommender systems and approaches to build them.

S.C. Satapathy and A. Joshi (eds.), *Information and Communication Technology for Intelligent Systems (ICTIS 2017) - Volume 1*, Smart Innovation, Systems and Technologies 83, DOI 10.1007/978-3-319-63673-3_13

Traditional CF based recommender systems consider only overall rating of an item provided by user and generate recommendations based on only overall rating. But nowadays online platforms allow users to give rating for an item on multiple criteria along with an overall rating. For example a in movie domain a movie can be evaluated in several dimensions like acting, direction, visual effects, story and an overall rating. Multi-criteria recommender systems deal with such multi-dimensional rating data to provide more accurate recommendations to users. In this paper we propose a new approach to improve predictive accuracy of multi-criteria recommender systems.

Among all the techniques to build CF based recommender systems, matrix factorization performs better than other techniques. We propose to use matrix factorization to predict individual criteria ratings. We propose to learn regression model for each user to predict overall rating from these individual criteria ratings. Experimental results show that our approach performs better than traditional approaches of multi-criteria recommender systems.

This paper organised in following way. In Sect. 2, we discuss about related work and traditional approaches to build multi-criteria recommender systems. In Sect. 3, we present the matrix factorization and regression based approach proposed in this paper. Section 4 shows experimental evaluation of our proposed approach. Finally, we conclude our work in Sect. 5.

2 Background and Literature Review

In this section, we briefly discuss background and related work of multi-criteria recommender systems. Traditional CF based recommender systems are systems that try to find a rating function $R : Users \times Items \rightarrow R_0$. R_0 contains predicted overall rating of user-item pair. This type of recommender system considers only overall rating of an item for recommendation.

Multi-criteria recommender systems [2] try to predict overall rating of an item considering different criteria ratings. These criteria ratings contain some valuable information and considering these criteria ratings for recommendations lead to better performance of recommender systems. General form of rating function in multi-criteria recommender system is as follows:

$$R : User \times Items \rightarrow R_0 \times R_1 \times R_2 \times ... \times R_k \tag{1}$$

Here we need to predict overall rating R_0 along with other criteria ratings. Table 1. Shows a sample multi-criteria dataset from movie domain. Adomavicius and Kwon [1] proposed two basic approaches to build multi-criteria recommender systems. These are similarity based approach and aggregation function based approach. Basic ideas of these approaches are discussed below.

2.1 Similarity Based Approach

In similarity-based approach, we predict rating of a user u for an item i by considering the ratings of item i provided by users similar to u. To find similarity

Table 1. Sample multi-criteria rating data from movie domain.

User	Movie	Acting	Direction	Story	Visual effects	Overall
u_1	i_1	2	4	1	3	4
u_1	i_2	3	2	4	2	2
u_1	i_3	2	4	2	1	4
u_2	i_1	2	2	4	4	3
u_2	i_2	1	2	4	3	4

between two users for each individual criteria, we can use Pearson correlation coefficient or Cosine similarity. We aggregate these similarities of individual criteria by taking average or minimum to get overall similarity. Another way to find similarity between two users is using multi-dimensional distance metrics such as Euclidian or Chebyshev distance. Finally, we can predict the rating using weighted sum approach.

2.2 Aggregation Function Based Approach

In this approach we assume that overall rating of an item is dependent on individual criteria ratings. Overall rating (r_0) can be calculated using a function of other criteria ratings shown in Eq. 2. To approximate function f we can use linear regression techniques. To predict each individual criteria ratings we can use any single criteria collaborative filtering technique. Our proposed approach follows this aggregation function based approach.

$$r_0 = f(r_1, r_2, ..., r_k) = w_0 + w_1.r_1 + w_2.r_2 + ... + w_k.r_k \tag{2}$$

2.3 Literature Review

Jannach et al. [5,7] proposed to use support vector regression (SVR) instead of linear regression in aggregation function based approach. They also suggested to learn user-based as well as item-based regression model and use weighted combination of both to find overall rating. Jhalani et al. [8] proposed to use linear regression to find overall similarity as well as overall rating. Nilashi et al. [10] proposed an approach to identify similar users using clustering and then learn regression model for each cluster using SVR. Koren et al. [9] proposed different techniques of matrix factorization to solve single criteria collaborative filtering problem. We have employed matrix factorization in the area of MCRS through linear regression in our proposed approach.

3 Proposed Work

In this section, we describe our proposed approach to build multi-criteria recommender systems using matrix factorization and linear regression technique

(MCRS_MF_LR). Matrix factorization is used to predict individual criteria rating and linear regression is used to find weight of each criteria. We predict overall rating based on the individual criteria ratings using a weighted function of individual criteria ratings. In multi-ctiteria recommender systems we have a set of users $U = \{u_1, u_2, u_3, ..., u_n\}$, a set of items $I = \{i_1, i_2, i_3, ..., i_m\}$ and a set of criteria on which each item can be rated, $C = \{c_1, c_2, c_3, ..., c_k\}$. Each rating of an item by a user is represented as a rating vector $R(u, i) = (r_0, r_1, r_2, r_3, ..., r_k)$, where r_0 represents overall rating. Our aim is to predict $R(u, i)$ for any user-item pair where $R(u, i)$ does not present. Our proposed approach has three steps:

1. **Step 1:** Factorize each user-item rating matrix of each criteria to predict individual criteria rating.
2. **Step 2:** Learn user-based regression model to predict overall rating from individual criteria ratings.
3. **Step 3:** Predict overall rating for a user-item pair for which rating is not available.

Overview of the proposed approach is shown in Fig. 1. The details of these steps are as follows.

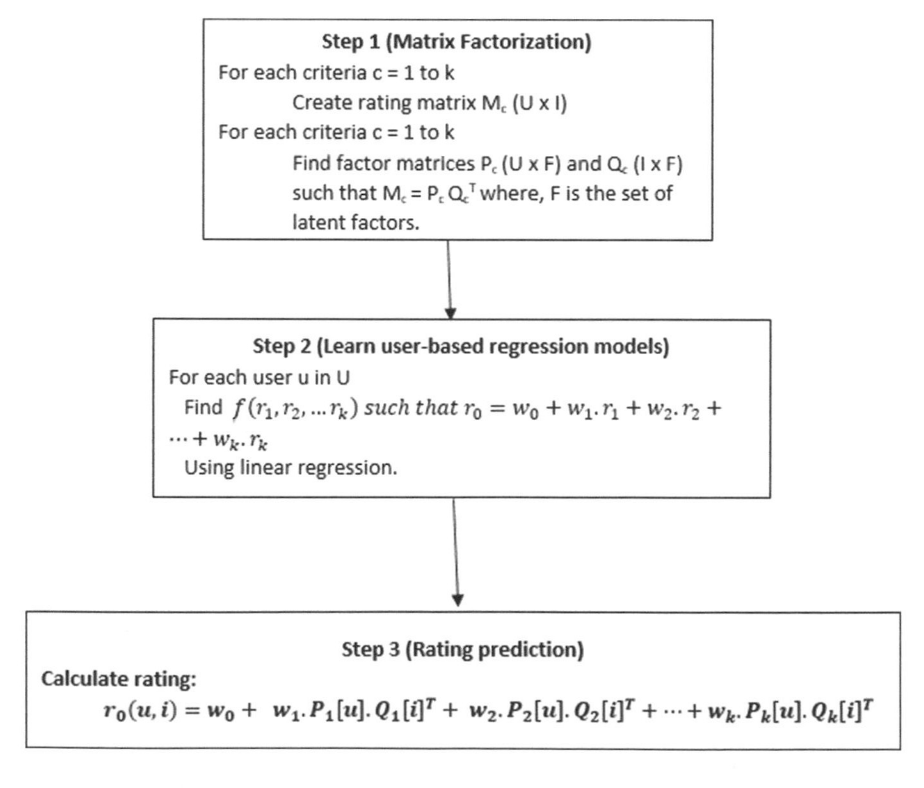

Fig. 1. Overview of matrix factorization and regression based approach.

3.1 Step 1: Matrix Factorization

Matrix factorization [9] is a latent factor based technique to build single criteria CF based recommender systems. In this approach, we find two factor matrices of user-item rating matrix; one related to users and one related to items. This factoring technique is based on identifying k latent factors that allows to predict the ratings. After matrix factorization we can represent each user as a k dimensional vector of latent factors and each item as a k dimensional vector of latent factors. Dot product of user vector and item vector is the predicted rating of the user-item pair. If R is the rating matrix, P is user factor matrix and Q is item factor matrix we can find these two matrix by solving the optimization problem [9] in Eq. 3

$$\min_{P^*,Q^*} \sum (R_{ui} - Q_i^T.P_u)^2 + \lambda(\|Q_i\|^2 + \|P_i\|^2) \quad (3)$$

Where λ is regularization constant.

Now in our approach we first create separate rating matrix of user and item for each criterion. Now we will apply matrix factorization to each individual criteria rating matrix and find their factor matrices. We use these factor matrices to predict individual criteria rating.

3.2 Step 2: Learn Regression Model

Overall rating of an item is dependent on other individual criteria ratings. We can represent this using Eq. 4.

$$r_0 = f(r_1, r_2, ..., r_k) = w_0 + w_1.r_1 + w_2.r_2 + ... + w_k.r_k \quad (4)$$

Where w_i is weights for each criteria for $i = 0, 1, ..., k$.

We find weights of each criteria using linear regression method. As each user has different priority for different criteria to give their overall rating, instead of learning a global regression model we propose to use user-based regression model to predict overall rating.

3.3 Step 3: Rating Prediction

In this step, we predict unknown rating of any user-item pair. First we calculate individual criteria ratings from factored matrices calculated in Step 1. Further, we combine these individual criteria ratings with their criteria weights to get overall ratings. Equation 5 is used to predict overall rating of user u for item i.

$$r_0(u,i) = w_0 + w_1.P_1[u].Q_1[i]^T + w_2.P_2[u].Q_2[i]^T + \cdots + w_k.P_k[u].Q_k[i]^T \quad (5)$$

Where w_i is weight of each criteria calculated from Step 2.

4 Experimental Analysis

We have performed several experiments on Yahoo! Movies dataset to demonstrate the effectiveness of proposed approach "Multi-criteria Recommender System based on Matrix Factorization & Linear Regression (MCRS_MF_LR)" over following existing baseline approaches:

- Single criterion recommender systems (SCRS) [1]
- Multi-criteria recommender systems using minimum similarity (MCRS_MS) [1]
- Multi criteria recommender systems using average similarity (MCRS_AS) [1]

4.1 Yahoo! Movies Dataset

Yahoo! movies dataset contains 6078 users and 976 movies. Each movie is expressed in terms of its four criteria namely, story, acting, direction, visual effects and one overall rating. Each user rates each movie based on four various criteria and also provides one overall rating on a 13 point rating scale. Since the dataset is sparse, therefore, we have created a sub-dataset namely YM_10_10 that contains those users who have rated more than 10 movies and those movies which have been rated by more than 10 users. We have created 4 samples of datasets for our experiments by selecting 100 random users from YM_10_10 dataset. Further, we divide the ratings of each user into two parts namely training movies (80%) and testing movies (20%).

4.2 Performance Measures

In order to evaluate the performance of our proposed approach, we have used following performance measures namely mean absolute error (MAE) & root mean squared error (RMSE) which are described in Eqs. 6 and 7 respectively.

$$MAE = \frac{\sum_{i=1}^{n} |\hat{x_i} - x_i|}{n} \tag{6}$$

$$RMSE = \sqrt{\frac{\sum_{i=1}^{n} (\hat{x_i} - x_i)^2}{n}} \tag{7}$$

Where $\hat{x_i}$ is predicted rating and x_i is actual rating and n is number of rating data present in testing movies dataset.

4.3 Results

Results presented in Table 2, depict the relative performances of schemes namely, SCRS, MCRS_MS, MCRS_AS and our proposed MCRS_MF_LR in terms of MAE & RMSE over various samples. Lower values of these measures indicate the better performance of the proposed approach. Based on the results expressed in Table 2, we can conclude that proposed approach outperform other baseline approaches in terms of MAE & RMSE.

Table 2. Performance comparison of the proposed approach with other approaches.

	Performance measures	Sample 1	Sample 2	Sample 3	Sample 4
SCRS	MAE	2.4591	2.3717	2.4026	2.3407
	RMSE	2.7332	2.7101	2.7258	2.6920
MCRS_MS	MAE	2.3618	2.2937	2.3178	2.2610
	RMSE	2.6732	2.6201	2.6571	2.6095
MCRS_AS	MAE	2.4390	2.3543	2.4098	2.3752
	RMSE	2.7205	2.7091	2.6857	2.6759
MCRS_MF_LR	MAE	**2.0742**	**1.9503**	**2.0188**	**1.9427**
	RMSE	**2.5204**	**2.3562**	**2.4532**	**2.3255**

5 Conclusion and Future Work

Previous works have shown that considering multi-criteria ratings in collaborative filtering leads to improved predictive accuracy of recommender systems. In this work we propose a new approach for multi-criteria recommender systems using matrix factorization and linear regression.

Matrix factorization is a well known technique to build single criteria recommender systems. We propose to use matrix factorization to predict individual criteria rating. As different user give priority to different criteria to give their overall rating, we propose to learn user-based regression model to predict overall rating based on individual criteria ratings. Experimental results show that our approach gives better performance compare to other traditional approaches.

In future we will try to evaluate our approach on dataset from other domain like tourism domain where multi-criteria ratings are available. Also we will try to explore other machine learning technique to predict overall rating from individual criteria ratings.

References

1. Adomavicius, G., Kwon, Y.: New recommendation techniques for multicriteria rating systems. IEEE Intell. Syst. **22**(3), 48–55 (2007)
2. Adomavicius, G., Kwon, Y.: Multi-criteria recommender systems. In: Recommender Systems Handbook, pp. 847–880. Springer (2015)
3. Adomavicius, G., Tuzhilin, A.: Toward the next generation of recommender systems: a survey of the state-of-the-art and possible extensions. IEEE Trans. Knowl. Data Eng. **17**(6), 734–749 (2005)
4. Felfernig, A., Friedrich, G., Schmidt-Thieme, L.: Guest editors' introduction: Recommender systems. IEEE Intell. Syst. **22**(3), 18–21 (2007)
5. Jannach, D., Karakaya, Z., Gedikli, F.: Accuracy improvements for multi-criteria recommender systems. In: Proceedings of the 13th ACM Conference on Electronic Commerce, pp. 674–689. ACM (2012)
6. Jannach, D., Zanker, M., Felfernig, A., Friedrich, G.: Recommender Systems: An Introduction. Cambridge University Press, Cambridge (2010)

7. Jannach, D., Zanker, M., Fuchs, M.: Leveraging multi-criteria customer feedback for satisfaction analysis and improved recommendations. Inf. Technol. Tourism **14**(2), 119–149 (2014)
8. Jhalani, T., Kant, V., Dwivedi, P.: A linear regression approach to multi-criteria recommender system. In: International Conference on Data Mining and Big Data, pp. 235–243. Springer (2016)
9. Koren, Y., Bell, R., Volinsky, C.: Matrix factorization techniques for recommender systems. Computer **42**(8), 30–37 (2009)
10. Nilashi, M., Jannach, D., bin Ibrahim, O., Ithnin, N.: Clustering-and regression-based multi-criteria collaborative filtering with incremental updates. Inf. Sci. **293**, 235–250 (2015)

Motivation to a Deadlock Detection in Mobile Agents with Pseudo-Code

Rashmi Priya[1(✉)] and R. Belwal[2]

[1] TMU, Moradabad, India
rashmi.slg@gmail.com
[2] AIT, Haldwani, India
r_belwal@rediffmail.com

Abstract. The solution presented locates locality of reference during the deadlock detection process by migrating detector agents to query multiple blocked agents. To message each blocked agent individually and gather their responses at the shadow agent itself is an alternative to this single migration. The pseudo code provides a context for the solution and insight into the responsibilities and activities performed by each entity.

Keywords: Deadlock · Agents · Pseudo code · Detector · Entity

1 Introduction

As presented in the previous section traditional distributed solutions commonly have fault and location assumptions that make them unsuitable for mobile agent systems. To solve this problem, mobile agent specific solutions are required. The properties of the presented deadlock detection algorithm illustrate how it is a fully adapted mobile agent solution. The presented technique is fault tolerant and robust. Lost agents or messages during the deadlock detection process do not represent a critical failure. This fault tolerance is due to three properties of the algorithm: the autonomous nature of the agents, the periodic nature of the detection process and the copying of deadlock information. Shadow deadlock detection and consumer agents execute asynchronously. They do not depend on continual communication during the deadlock detection process. The algorithm is designed around incremental construction of the global wait-for graph. Finally, therefore if a portion of the graph is lost, the next update will recover that information. Hence copying of the partial wait-for graph into deadlock detection agents make the loss or failure of a particular deadlock detection agent trivial and has no impact on the detection process, outside of slowing the process. Additional safeguards can be built into the agent hosts. such as agent crash detection, to improve fault tolerance.

1.1 Algorithm Motivation and Agent Properties

By limiting the number of messages that would be required in other solutions the detector migration reduces network load. It is difficult to compare the network load of

S.C. Satapathy and A. Joshi (eds.), *Information and Communication Technology for Intelligent Systems (ICTIS 2017) - Volume 1*, Smart Innovation, Systems and Technologies 83, DOI 10.1007/978-3-319-63673-3_14

this mobile agent solution to that generated in traditional distributed deadlock detection solutions due to the significantly different paradigm and properties of the environment. This is due to the parallel/distributed nature of the technique, which enforces the lack of a central point of messaging and coordination. This reduces the risk of flash congestion and allows the technique to handle deadlock involving many blocked agents.

The load is spread across many host environments, if the network load of the presented solution is considered as a whole. Additionally, network organization independence is guaranteed through a clear separation of mobile agents from the mechanics of routing and migration, the agents are not aware of the number of hosts in the mobile agent system and do not have explicit knowledge of resource locations. It should be noted that even though the solution is network independent, the topology is static once the algorithm begins. If the topology is allowed to change, a dynamic topology update protocol must execute in the background to provide new routes to the hosts.

A common use of mobile agents is to encapsulate complex protocols and interactions [6]. This technique uses the combination of shadow agents and deadlock detection agents to encapsulate a complex series of probes, interactions and acknowledgments. Additionally, these protocols are isolated from the consumer agent; therefore, can be easily modified and upgraded. The deadlock detection phase could be implemented as remote procedure calls or another form of distributed programming, but would require network organization assumptions and the continual exchange of messages. Detector and shadow agents carry out their deadlock detection tasks in an asynchronous manner. They coordinate their efforts in defined ways, but are able to keep working without regular contact and do not require constant supervision while carrying out tasks. This asynchronous and autonomous operation contributes to the previously discussed fault tolerance. For example, the combination of consumer, shadow and detector agents adapt to their environment to solve deadlock situations, shadow agents react independently to changing network conditions and the state of their target consumer agent to initiate the deadlock detection processing. Similarly, the separation of the implementation from facilities specific to a particular mobile agent system or operating system detector agents can react to network failures or the requests of other agents while gathering global wait-for graph information allows the solution to execute in a heterogeneous environment. Moreover, the separation of replica and detector agents from the consuming agents they monitor, allows them to be adapted to many different environments without (or with minor) modifications to the entities performing the work.

2 Deadlock Detection Pseudocode

This pseudo code provides a context for the solution and insight into the responsibilities and activities performed by each entity. This section presents pseudo-code of each element that plays a significant role in the presented solution. First, pseudo-code for the consumer, shadow and detection agent is presented. Finally, code for the mobile agent environment is presented.

2.1 Agent A

```
public class AgentA extends MobileAgent
{
public AgentA( String int heartbeat )
{
state = IDLE;
}
public void run()
{
while( true )
{
messages = getMessagesFromBlackboard
(agentId);
processMessages( messages );
switch ( state )
{
case IDLE:
case WAITING:
break;
case MOVING:
if( currentHost is not targetEnvironment )
{
postRouteRequest( targetEnvironment ) ;
{
else
{
state = IDLE;
{
break;
{
AND state
private void lockResource( String locktype,
String resourceName )
{
if( lockType equals "exclusive" )
{
get resource manager;
if resourceManager.lockResource(
resourceName,
lockType ) succeeds
{
if (resourceName equals
blockedResourceName )
state = IDLE;
{
sleep ( heartbeatDelay ) ;
{
private Vector processMessages( messages )
{
while ( more Messages )
if message was accepted remove from list;
return unprocessed messages;
{
private boolean processMessages
BlackboardEntry msg )
if ( message equals amove" AND state is IDLE
)
{
targetEnvironment = get target from message;
state = MOVING;
postBlackboardMsg( "route",
targetEnvironment ) ;
{
else if( message eqyals "lockw AND
state is IDLE OR WAITING )
{
extract lock type and resource from message ;
lockResource( lockType, resource );
if( message equals
{
extract resource from message;
unlockResource( resource );
{
{
else
state = WAITING;
blockedResourceName = resourceName;
postBlackboardMessage("agentBlock",
resourceName );
private void unlockResource( String
resourceName )
{
get Resource Manager
resourceManager.unlockResource(resourceNa
me);
}}
```

2.2 Agent B

```
public class ReplicaAgent extends
MobileAgent
(
public ReplicaAgent ( String id, String
targetAgent, int heartbeat )
{
state = IDLE;
targetAgentName = targetAgent;
reset locked resource list
reset detection info table
reset detector agent
{
public void run()
while( true )
{
if ( state is not MOVING )
{
messages = getMessagesFrom BlackBoard();
messages =processMessages(messages);
{
switch ( state)
{
case IDLE:
break;
get~messagesFromBlackboard();
processMessages( messages );
case MOVING:
if( currentHost is not targetEnviroment ) )
{
routeRequest( targetEnvironment );
{
else
{
state = IDLE;
}
break;
else if( message equals ablockedm ) )
{
blockedTarget( attachment #1 ),
retVal = true;
{
else if( message equals wblocked ) )
{
unblockedTarget ( ) ;
retVal = true;
{
}
sleep (heartbeatDelay) ;
}
}
private Vector processMessages( messages )
{
while( more Messages )
{
processMessage( currentMessage );
if message processed remove from list;
{
return unprocessed messages;
{
private boolean processMessage(
BlackboardEntry msg )
{
Vector attachments = msg.getAttachments();
if( message equals "move" AND state is IDLE)
{
String target = extract target from message;
targetEnvironment = target;
state = MOVING;
{
else if( message equals "addLock" )
{
addlock( attachment #l,
attachent #2,
attachment #3,
attachment #4 );
retVal = true;
else if( message equals "removeLock" )
{
removelock ( attachment #I, attachment #2 ) ;
state = IDLE;
retVal = true;
{
private void blockedTarget( Agent
blockedAgent,
String (resourceName )
{
Statel = TARGETBLOCKED;
owner = query host environment for owner of
resource;
blockedResourceName = resourceName;
localAgents = query host environment about
local agents;
```

```
else if( message equals "deadlockReport" ) )
{
processReturnOfDetector( attachment#1,
attachment #2, attachment #3 ) ;
retVal = true;
{
else if( message equals "deadlockInfoRequest"
) 1
retVal = true;
else
{
retVal = super.processMessage( msg );
{
{
public void exit ()
{
if( detector )
{
Remove ( detector from host environment);
{
super. exit () ;
private void addlock( String
environment,String resource,String owner,int
priority1 )
if( resource not already locked )
{
new r e s o u r c e I n f o ( env, res, owner, p r
i o r i t y1 ) ;
s t o r e resourceInfo i n locked resource list
}
}
Private void removeLock( S t r i n g envl, S t
r i n g resourceName1 )
{
i f ( resourceName1 is i n locked resource list )
{
remove resource from locked resource list
{
{
private void unblockedTarget1( )
{
state = IDLE;
{
if ( owner in localAgents )
Table .put ( targetAgentName, new
DetectionInfo( .. ) );
Detector1 = new DetectorAgentO;
postBlackboardMsg( d e t e c t o r );
postBlacKboardMsg( buildDetectorLocks () ) ;
numOfDetectionStarts++;
1astDetectionStartTirne = current time ;
{
{
p r i v a t e void processReturnOfDetector(
DetectorAgent agent )
switch( state )
{
case TARGET-BLOCKED :
if ( checkForDeadock(
agent.getDetectionTables()
reset detectionInfoTable;
{
case IDLE:
removeDetector () ;
(
switch ( state )
(
case TARGET-BLOCKED:
postBlackboardMsg( " s t a r t " ,
buildDetectorLocks() );
numOfDetectionStarts++; )
1astDetectionStartTime = currenttime ;
break;
case WAITING,FOR UNLOCK:
break;
{
{
p r i v a t e boolean checkForDeadlock( Vector
detectionTableList )
{
while ( detect Table List has more entries )
{
detectionTable = current detection table ;
agent List = get agent list from det. ;
while( agent List has more entries )
{
detection Info = detection info ;
```

```
if ( current agent name equals
targetAgentName )
{
deadlockFound = True;
{
{
{
return deadlockFound;
{
p r i v a t e void resolveDeadlock(
DetectorAgent agent )
{
if( state is TARGETBLOCKED )
{
cycleList = findElementsInCycle(
detectionInfoTable ) ;
while( cycleList has more elements )
{
1ockToBreak = lowest priority resource;
{
i f ( lockToBreak equals resource we are
blocked on )
{
postBlackboardMsg( "unlock",LockToBreak );
state = WAITING-FOR-UNLOCK;
{
{
private Vector findcycle ( Hashtable
detectionInfoTable )
{
Vcctor cycleVector = new Vector ( ) ;
cyclevector add( resource we are blocked on ) ;
info = find entry in detectionInfoTable
while( current entry agent name not equal to
our target agent )
{
info = find entry in detectionInfoTable ;
cyclevector add( info );
{
return cyclevector;
{
private void checkForDetectorDeath()
{
Date currentTime = current time;
BlackboardEntry msg;
if( num0fDetectionStarts > 0)
postBlackboardMsg ( "inject", detector ) ;
if( state is TARGET-BLOCKED )
{
postBlackboardMsg( "start",
buildDetectorLocks() ;
{
else if( state is WAITING-FOR-UNLOCK )
resolveDeadlock( detector.getIdentifier());
detector.getToken() );
{
1astDetectionStartTime = current time;
}}}}
```

2.3 Agent C

```
public class AgentC extends MobileAgent
{
public AgentC ( String id, int heartbeat,
ShadowAgent parent )
{
reset detection Table List;
reset resources To Vist;
reset targetEnvironment;
reset targetResource;
state = IDLE;
set parent = parent;
{
public void run ()
{
while ( true )
{
if ( state is not MOVING )
{
messages = getMessagesfromBlackboard () ;
messages = processMessages( messages );

}
switch ( state)
getMessagesfromBlackboard () ;
processMessages( messages );
(
case IDLE:
break;
case MOVING:
if( currentHost is not targetEnvironment ) )
{
else
{
state = CHECKING-LOCKS;
{
break;
 if( current Host is not targetEnvironment )
{
```

```
if( host.unlockResource( targetResource,
agentToNotify )
{
( state = RETURN-FROM-UNLOCK);
{
else
state = IDLE;
{
case RETURNRNFROM, the LOCK:
if( currentHost is not startingEnvironment )
(
Shadow,removeLock( targetEnvironment,
targetResource );
state = IDLE;
{
case DONE:
if( currenthost is not startingEnvironment )

(
state = REPORT-RESULTS;
)
case CHECKING-LOCKS:
checklocks ( ) ;
break;
case REPORT-RESULTS:
postMessageToBlackboard( shadowÀgent,
deadlockInfo );
break;
}
sleep( heartbeatDelay ) ;
private Vector processMessages( messages )
{
while ( more Messages )
{
processMessage( currentMessage );
if message processed remove from list;
{
return unprocessed messages;
{
private boolean processMessage(
BlackboardEntry rnsg )
{
attachments = msg,getAttachments();
if( message equals "startW ")
{
startDetection ( (Vector) attachment ;
retVal = true;
else if( message equals "unlock" ) {
startunlock ( attachment #1,attachment #2,
attachent #3 ) ;
retVal = true;
message ;
deadlockRequestResponse ( attachment #l ) ;
retVal = true;
{
else
{
super.processMessage( msg 1;
{
return retVal;
{
private void startDetection( resources )
( setVisitlist( resources ));
targetEnvironment( entry.getEnvlame() };
targetResource ( entry. getResName () ) ;
detectionTablelist( new Vector 1);
start ingEnvironment (getHost ().getName ());
state ( MOVING ) ;
{
private void checklocks ()
{
while( shadowlist has more elements )
count expected responses;
{
if( expected responses > 0)
(
state = WAITING-FOR-RESPONSE;
{
else
{
findNewTarget () ;
}
private void deadlockRequestResponse(
newTable )
{
shadowList = query current host for agents
blocked on
the resource we are visiting;
expectedResponses--;
detectionTablelist.add( newTable );
if( all expectertesponses received )
{
findNewTarget 0 ;
}
}
private void findNewTarget0
{
if( more resource to visit )
get next resource;
targetEnvironment = entry.getEnvName();
targetResource = entry.getResName();
{
else
{
targetEnvironment = startingEnvironment;
state = DONE;

}}}
```

2.4 Host Environment

```
public class AgentEnvironment extends
Thread
public AgentEnvironment( String name, int id,
int (1oggingLevel )
{
resourceManager = new ResourceManagerO;
topologyManager = new TopologyManager();
reset agentTable;
reset messageBoard;
reset blockedAgentTable;
globalIdentifier = id;
state = PROCESSING;
}
public void run ()
(
while( true )
{
checkErrorMessages () ;
updateRoutes () ;
sleep( 1000 );
}
)
public synchronized void agentEnter( Agent
newAgent )
{
if( state is PROCESSING )
{
agentTable.put( newAgent ) ;
newAgent.enter0;
}
}
private synchronized void agentExit( Agent
1eavlngAgent
{
if ( state is PROCESSING )
(
1eavingAgent. exit ( ) ;
)}
private void agentBlock( Agent blockedAgent,
String resourceName)
(
replica = find replica agent for blockedAgent;
if ( shadow found )
{
postBlackboardMsg( "blockedAgent",
resourceName ) ;
private void checkForMessages()
{
get messages from blackboard;
while ( more messages)
{
processMessage( current message );
}
if ( env is not null )
}
private void processMessage(
BlackboardEntry msg
attachments = rnsg.getAttachments0;
if( message equals "pause" ) )
state( PAUSED ) ;

else if( message equals wresumen ) )
(
state( PROCESSING ) ;
message equals
(
agentBlock ( msg . getAgent Id () , attachment
t1 ,attachment t2);
if( message equals 0)
{
routeRequest(msg.getAgentId(), attachment # 1
,attachment #2);
{
else if (( message equals *inject))
this.injectAgent( attachment)

else if( message equals "remove" ) )
{
removeAgent ( attachrnent #l ) ;
}
}
private boolean routeRequest( String
movingAgent, EnvironmentToken token,
String targetEnv )
{
if( state() is PROCESSING )
{
movingAgent = get moving agent from agent
tables;
(
return true;
)
if( check for shadow information in the token )
shadowAgentId = get shadow name from
token;
If ( check for shadow agent in agent tables )
{
shadow = get shadow agent from agent tables;
}
else
{
retVal = f alse;
}}
if ( retVal is true )
(
AgentEnvironment env = request route from
topologyManager;
(
```

3 Conclusion

The presented algorithm is designed with the unique properties and challenges of mobile agent systems as a motivating factor. As a result, the solution has some of the properties and features that are commonly found in mobile agent implementations. This section lists the properties of the proposed algorithm which make it a mobile agent solution. The solution is network organization independent. The algorithm makes no assumptions concerning network topology (i.e., ring). The number of hosts or node locations to support the solution. Resource-based routing and tracking of the nodes visited by a particular agent eliminate the need for explicit topology knowledge.

References

1. Walsh, T., Paciorek, N., Wong, D.: Security and reliability in Concordia. In: Mobility: Processes, Computer and Agents. Addison-Wesley, Reading (1999)
2. Mitsubishi Electric Information Technology Center: Concordia—Java Mobile Agent Technology. World Wide Web, January 2000. http://www.meitcacom/HSUProjects/Concordia/
3. University of Stuttgart: The Home of the Mole. World Wide Web, September 2000. http://mole.infonnatik.uni-stuttgart.de/
4. University of Tromse and Cornell University: TACOMA—Operating System Support for Mobile Agents. World Wide Web, August 1999. http://www.tacoma.cs.uit.no/
5. ObjectSpace, Inc.: ObjectSpace voyager core package technical overview. In: Mobility: Processes, Computer and Agents. Addison-Wesley, Reading (1999)
6. ObjectSpace, Inc.: ObjectSpace Product Information: Voyager. World Wide Web, September 2000. http://www.objectspace.com/products/voyager/
7. Fachbereich lnfonnatik and Johann-Wolfgang-Goethe-Universitaet Frankfurt: "ffMain". World Wide Web, February 2000. http://www.tm.informatik.uni.frankfurt.de/Projekte/MN
8. University of Geneva: The Messenger Project. World Wide Web, December 1997. http://cui.unige.ch/tios/msgr/
9. General Magic, Inc.: Odyssey. World Wide Web, November 2000. http://www.genmagic.com/
10. University of Modena: MARS (Mobile Agent Reactive Space). World Wide Web, October 2000. http://sirio.dsi.unimo.it/MOON/MARS/index.html

An Ontology Based Recommender System to Mitigate the Cold Start Problem in Personalized Web Search

Kamlesh Makwana(✉), Jay Patel, and Parth Shah

Charotar University of Science and Technology, Changa, Gujarat, India
{kamleshmakvana.it, jaypatel.it, parthshah.ce}@charusat.ac.in

Abstract. With the increase in the diversity of data available on the web, excellence of various searches and the need for personalizing the search results arises. The densely distributed web and heterogeneous information environment creates challenges for search engines such as Storage space, crawling speed, computational speed and retrieval of most relevant documents. It becomes difficult to identify the relevancy of the result due to instability in the search query context. In this paper, the framework to personalize web search through modeling user profile by content based analysis and recommendation model is proposed. The framework will use knowledgebase in form of query hierarchy which is specified for individual user to filter discovered results. The proposed approach is also used to discover current search context of particular user by alluding useful links through item-item collaborative filtering techniques. Due to integration of content based analysis and item to item collaborative filtering algorithm, the proposed framework will retrieved the results of user context on query and also suggest links that had been already clicked by the users within same context.

Keywords: Personalized web search · Recommendation · Ontology · Cold start problem · Information retrieval

1 Introduction

The Internet has become a huge communication mode through wide information access. This subsumes with informational, cultural and economical values to be specific. With the existence of Search Engines e.g. duck duck go, AOL, Bing and many more, the users utilize it for retrieving the relevant information. Although search engines can meet a user's general request, but lack to distinguish different users' specific need. For example if some user is a programmer and he/she searches for term as "bat". Result related to both bat as mammal and bat as category of sports are retrieved search engine retrieves. If any user who is interested in bat as category of mammal he/she would also get both the result on search term as *bat*. Thus, aim to personalize web search is to retrieve webpage for each user incorporating his/her interests. Web users issue queries to the search engines to fetch information on a number of topics. Due to diverse backgrounds and expectations for a specific query,

S.C. Satapathy and A. Joshi (eds.), *Information and Communication Technology for Intelligent Systems (ICTIS 2017) - Volume 1*, Smart Innovation, Systems and Technologies 83, DOI 10.1007/978-3-319-63673-3_15

existing search engines try to personalize results to better match the preferences of an individual user.

This exercise involves two major challenges. To build a user profile for every user in order to effectively identify the user interests and re-rank the results in a way that matches the interests of a given user. In short it requires to customize content of web sites and web search engine in such a way that particular user can find information that is more relevant to him/her. Process of customizing content of web sites or web search engine based on user preference is called web personalization [1].

Web Personalization is used to fetch the more relevant result based on user preferences. Through the personalized system user can retrieve result which she/he wants from very large amount of data. There is no. of research done to provide the personalized system in respective fields, but again size of web data is increasing roughly it becomes these techniques less efficient. Personalized Information Retrieval is an effective way that provides specific results to different users based on their query context [3]. The main issue here is how to obtain user's real-time information need. Again there is ample of research that has been done to provide personalized web results.

This problem can be evaded by using frequency of words in the model. One of the ways is to divide the number of times word occurs in a document by the total number of words in document and utilizing these frequencies. However, there is a pitfall with this approach: if the documents collection is associated with "java", the word "java" will have higher frequency in each document, and is therefore not suited to identify the content context precisely.

2 Related Work

This section describes existing work to personalized web search results. It first describes history of personalized web search and give the summary of various research have been done using different personalized web search techniques.

2.1 History of Personalized Web Search

A personalized Page-Rank was first proposed by Kenneth Wai-Ting Leung, Dik Lun Lee, Wilfred Ng and Hing Yuet Fung [14]. User's preference is discovered through click through data, the discovered preferences are stored in ontology that can be utilized to enhance search engine's ranking function. Fang Liu, Clement Yu and Weiyi Meng have proposed a personalized web search in 2002 [14]. PPR is modification of global Page-Rank algorithm that have compute page rank for large numbers of pages by analyzing link structure between different web page and user navigation. Various researches on personalized search have been carried out, also algorithms have been tested, it is still unpredictable. Personalization for different queries and under different search contexts is consistently less effective.

Personalized search techniques can be categories into three strategies [8].

- Based on content analysis
- Based on the hyperlink structure of the Web
- Based on user groups.

2.2 Personalized Search Based on Content Analysis

Personalized search aims to filter or re-rank web results by evaluating similarity in content between retrieved web search results and user profiles that are either specified by users or can be automatically learnt from a user's past activities which store approximations of user interests. Chen, Na, and Viktor K. Prasanna [13] proposed a Rank-Box, an adaptive ranking system to personalized web search. Based on user opinion on current query search it re-ranks the result. This ranking algorithm learns from user's feedback and replaces the current ranking algorithm with new machine learning based ranking technique. By analyzing user feedback Rank-Box learns to determine preference of user. In 2006 Agichtein, Eugene, and Zijian Zheng have proposed an effective approach of providing millions of user interactions with a engine to automatically inspect "best bet" top relevant results preferred by users [3].

2.3 Personalized Search Based on the Hyperlink Structure of the Web

Many strategies based on the hyperlink structure of the web have also been implemented. Personalized Page-Rank (PPR), which is a modification of the global page-rank algorithm. Hyperlink analysis improves the relevance of the web search results.

3 Proposed Work

This section draws an outline of the proposed approach for mitigating the cold-start problem in recommender systems. This approach is basically divided into three phases. In first phase web log file that has been generated in predefined format through user's clicks is analyzed. The knowledge base will be incorporated in form of ontology by analyzing web log file. To construct ontology we will use semi-automatic tool Protégé. At same time user will give some explicit information to the system. In second phase we create groups of similar users by using fuzzy C-means clustering technique [2]. Both the data of first and second phase is used as input of our system. In last phase ontology mapping is done based on item-to-item collaborative filtering technique.

Figure 1 shows architecture of proposed framework to filter and re-rank web search result with query ambiguities removal process and collaborative recommendation techniques. Initially when user enters into the system, he has given an unique-id number that he has to use for search any term in framework this technique is crucial to preserve privacy of user at client side as well to remove conflict for users who are using same machine to access this framework. The proposed framework is divided into two

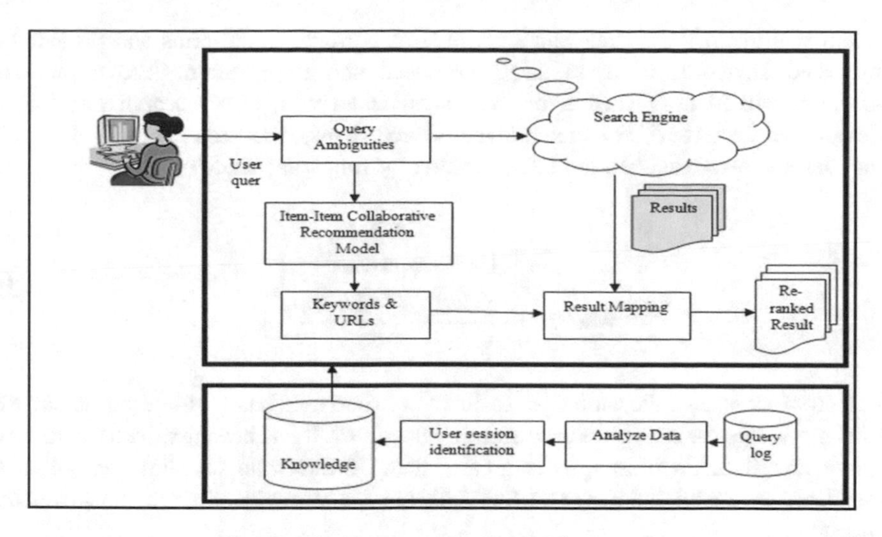

Fig. 1. Architecture of proposed framework

half offline phase and online phase. When user enters into the system query log file is analyzed at offline phase which contains the search history of particular user. One important point to consider is there is no need to preprocess the query log file because each entry in query log is generated dynamically and only contains information regarding user's search.

3.1 Query Ambiguities Removal

When any user will pass the query, this query will be fed to an interface query ambiguities removal process. This process will reformulate user's query based on some criteria that has been specified in next sections. After reformulation of user's query which is input to search engine to filter the retrieved results based on expanded query.

3.2 Knowledge Base Generation

As shown in Fig. 1 knowledge base is generated by analyzing query log data. Query log is generated from user's search and click information. Knowledge base represents taxonomy of queries searched by the all users. Root element of each tag represents the key of the individual users that distingue it from other user and also provides the privacy at client level.

3.3 Collaborative Recommendation Model

Proposed system will able to identify the current interest of users through item-item collaborative filtering. Users search is passed as input to the algorithms; proposed system will identify the irrelevant keywords of current search terms and provided as Suggested keywords to users using proposed above algorithm. Above proposed algorithm will identify each terms that is matched with user's search terms (both relevant and irrelevant). For each relevant search terms, it records the links clicked by that user and also calculate the interest value by following equation.

$$W = \begin{cases} 0 & if\, t > 1 \\ t & Otherwise \end{cases}$$
$$Where\, t = \left(\frac{End_{Time} - Start_{Time}}{\delta} \right) \quad (1)$$

Here, t represents the time span of links i.e. time user has spent on particular web links. δ presents the constant factor and its value is 10. It has been assumed in proposed system that if link's time span is greater than 10 min than this links is not to be considered as useful links. It this link information is excluded from the given link information.

Algorithm: Collaborative filtering to identify current interest and useful links.

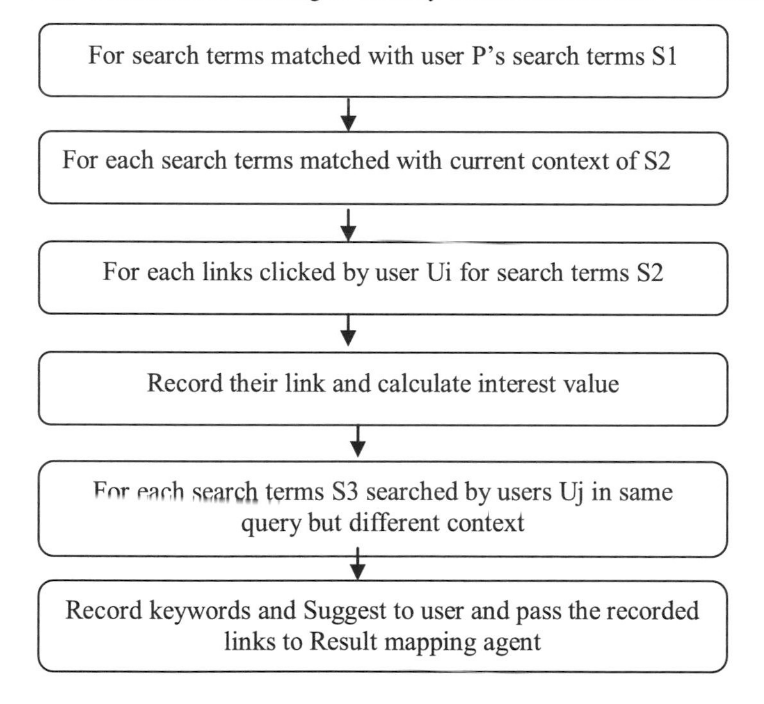

Table 1 shows the search histories of various users for different search terms. It contains the clicked links for query *apple* in different context such as apple in context of fruit as well as in context of computers/electronics for different users. It also shows the interest value of each link based on the Eq. 1 which shows the importance of each link for that particular user (Table 2).

Table 1. Search histories of various users for different search terms

User 1		User 2		User 3	
Search term: apple		Search term: apple computer		Search term: apple	
Link	Interest Value: t (seconds)	Link	Interest Value t (seconds)	Link	Interest Value t (seconds)
www.apple.com/in/	50	www.bestbuy.com/site/all-desktops/apple-imacs-minis…/pcmcat268200050003	250	www.nutrition-and-you.com/apple-fruit.html	240
www.bestbuy.com/site/all-desktops/apple-imacs-minis…/pcmcat268200050003	70	www.computershopper.com/brands/apple	300	www.besthealthmag.ca/best-eats/nutrition/15-health-benefits-of-eating-apples/	110
www.computershopper.com/brands/apple	500	http://www.webopedia.com/TERM/A/Apple_Computer.html	310	www.fruitsinfo.com/apples.php	480
www.harveynorman.com.au/computers-tablets/computers/apple-mac-computers	100	https://www.google.com/finance?q=NASDAQ:AAPL	100	www.healthy-aging-for-women-babyboomers.com/apple-fruit.html	350

Table 2. Recommended links to novice user for the query *apple*

User U_i, Query = apple
Recommended Links
www.computershopper.com/brands/apple
www.healthy-aging-for-women-babyboomers.com/apple-fruit.html
http://www.webopedia.com/TERM/A/Apple_Computer.html
www.computershopper.com/brands/apple
www.bestbuy.com/site/all-desktops/apple-imacs-minis…/pcmcat268200050003
https://www.google.com/finance?q=NASDAQ:AAPL
www.harveynorman.com.au/computers-tablets/computers/apple-mac-computers
www.bestbuy.com/site/all-desktops/apple-imacs-minis…/pcmcat268200050003
www.apple.com/in/
Suggested Keywords Apple fruit

4 Conclusion

The proposed framework presents an approach that personalize web search through modeling user profile by content based analysis and recommendation model. In this work user profile model is generated through click through data that stored in query logs. Query log is maintained for each user that is responsible to preserve user's privacy. Proposed framework will use knowledge base in form of query hierarchy which specified for individual user to filter discovered results. Knowledge base will be generated by analyzing query log file that is generated dynamically by framework and stored in web server. Re-ranking algorithm is use to re-rank retrieved result by search engine. Re-ranking is done through analysis the time spent by the user on particular web pages. The proposed approach is also used to identify search context of particular user by providing useful keywords through item-item collaborative filtering techniques. The proposed framework will also give some shorts of links that had been already clicked by the users within same context as current user has.

References

1. Makvana, K., Shah, P.: A novel approach to personalize web search through user profiling and query reformulation. In: 2014 International Conference on Data Mining and Intelligent Computing (ICDMIC). IEEE (2014)
2. Madia, N., Thakkar, A., Makvana, K.: Survey on recommendation system using semantic web mining. Int. J. Innov. Emerg. Res. Eng. **2**(2), 13–17 (2015)
3. Lu, Z., Zha, H., Yang, X., Lin, W., Zheng, Z.: A new algorithm for inferring user search goals with feedback sessions. IEEE Trans. Knowl. Data Eng. **25**(3), 502–513 (2013)
4. Jay, P., et al.: Review on web search personalization through semantic data. In: 2015 IEEE International Conference on Electrical, Computer and Communication Technologies (ICECCT). IEEE (2015)
5. Zhu, Z., Xu, J., Ren, X., Tian, Y., Li, L.: Query expansion based on a personalized web search model. In: Third International Conference on Semantics, Knowledge and Grid, pp. 128–133. IEEE (2007)
6. Kumar, R., Sharan, A.: Personalized web search using browsing history and domain knowledge. In: 2014 International Conference on Issues and Challenges in Intelligent Computing Techniques (ICICT). IEEE (2014)
7. Liu, F., Yu, C., Meng, W.: Personalized web search by mapping user queries to categories. In: Proceedings of the Eleventh International Conference on Information and Knowledge Management. ACM (2002)
8. Dou, Z., Song, R., Wen, J.R., Yuan, X.: Evaluating the effectiveness of personalized web search. IEEE Trans. Knowl. Data Eng. **21**(8), 1178–1190 (2009)
9. Shen, X., Tan, B., Zhai, C.: Implicit user modeling for personalized search. In: Proceedings of the 14th ACM International Conference on Information and Knowledge Management. ACM (2005)
10. Linden, G., Smith, B., York, J.: Amazon. com recommendations: item-to-item collaborative filtering. IEEE Internet Comput. **7**(1), 76–80 (2003)

11. Sieg, A., Mobasher, B., Burke, R.: Web search personalization with ontological user profiles. In: Proceedings of the sixteenth ACM Conference on Information and Knowledge Management. ACM (2007)
12. Yu, J., Liu, F.: Mining user context based on interactive computing for personalized Web search. In: 2010 2nd International Conference on Computer Engineering and Technology (ICCET), Vol. 2. IEEE (2010)
13. Chen, N., Prasanna, V.K.: Rankbox: an adaptive ranking system for mining complex semantic relationships using user feedback. In: 2012 IEEE 13th International Conference on Information Reuse and Integration (IRI). IEEE (2012)
14. Page, L., et al.: The PageRank Citation Ranking: Bringing Order to the Web. Stanford InfoLab, Stanford (1999)

A Study on e-Marketing and e-Commerce for Tourism Development in Hadoti Region of Rajasthan

Anukrati Sharma[1(✉)] and O.P. Rishi[2]

[1] Department of Commerce and Management, University of Kota, Kota, Rajasthan, India
dr.anukratishrama@gmail.com

[2] Department of Computer Science and Informatics, University of Kota, Kota, Rajasthan, India
dr.oprishi@uok.ac.in

Abstract. Travel, Tourism and Hospitality industry is growing day by day. It is providing immense opportunities of growth and development to a destination, region, state and Nation. The rapid changes in Information Communication Technologies generated the demand of e-marketing, e-commerce and content marketing in travel, tourism and hospitality industry. No doubts that ICT developments are impacting the demand and growth of tourism and its products phenomenally on the globe. On the other hand the region like Hadoti of Rajasthan state which is highly rich in tourism is lacking behind in the usages of e-commerce-marketing and content marketing in promotion of the attractions and other itineraries. The paper tries to explore the benefits of using e-commerce-marketing and e-business model for the destination image building and promotion of tourism of the region. The paper showcasing the benefits of shifting from the traditional promotional techniques to adapt e-commerce tools and techniques for travel, tourism and hospitality industry of Hadoti region.

Keywords: e-Marketing · e-Commerce · e Business model · Tourism · ICT

1 Introduction

e-Marketing and e-commerce are the buzz words for the promotion and marketing of travel, tourism and hospitality industry. An extensive approach could allow in enhancement of tourist visits for the Hadoti region-marketing, e-commerce and content marketing are bringing opportunities to the travel, tourism and hospitality industry at large. Information technology and communication creates essential modifications in the process, allocation and pattern of the tourism industry (Buhalis 2000). It is quite interesting and useful to find out where on the globe private organizations and government both recognized the need of e-commerce tools and techniques in travel, tourism and hospitality where we are lacking behind? In present era tourism and hospitality are the major sources of employment and income. It is a dynamic industry which needs sustainable information regarding its innovations, events and products. The recent demonetization in India also gives a push the usages of internet and ICT in

S.C. Satapathy and A. Joshi (eds.), *Information and Communication Technology for Intelligent Systems (ICTIS 2017) - Volume 1*, Smart Innovation, Systems and Technologies 83, DOI 10.1007/978-3-319-63673-3_16

tourism and hospitality industry. The changing needs and desires of the tourists in terms of satisfaction, safety and security also developed the urge of ICT usages in the said industry. At this stage no destination which would like to be recognized on the globe as Tourists Destination can ignore the usages of e-commerce-marketing and the recent trend of content marketing for tourism development.

2 Tourism Industry

Indian tourism industry is rising day by day. According to India Brand Equity Foundation the Indian tourism accounts for 7.5 percent of the Gross Domestic Product. Tourism industry is the third largest foreign exchange earner country. Not only this Indian tourism sector's total contribution to GDP is excepted to reach US $160.2 billion by 2026. The GDP contribution by the travel and tourism industry is suppose to grow to US$ 280.5 billion by 2025. On the another side the World Travel and Trade Council focused on improving infrastructure, introduction of e-visa services which have contributed significantly to the growth of Indian tourism industry. Indian tourism industry have lot to offer. It is remarkable that in 2014, the nation earned USD 19.7 billion earnings from foreign exchange from tourism-tourist visa scheme which is enabled by electronic travel authorization which has been launched by the Government of India in year 2014 for 43 countries proved as a boon for Indian Tourism.

Rajasthan which is one of the famous states for tourism and tourism products also witnessed a hike in tourists in 2015 to 2016 (up to December 2015) tourist arrivals in the state reached 36.66 million. On the other hand the Hadoti region of the state has witnessed a down fall in tourist arrivals both domestic as well as foreign tourists arrivals which has been shown in the table given below Table 1.

Table 1. Domestic and foreign tourist arrivals form 2012–2016 June

Year	Domestic tourist arrival	Foreign tourist arrival
2012	62029	1881
2013	63015	2889
2014	51467	3516
2015	90598	2574
2016 (June)	42876	1010

Source: Adapted from Hassan and Sharma 2017

There could be many reasons for the aforesaid down fall in the number of tourists at Hadoti region. The researchers in the present study would try to focus on one of a major reason i.e. the non or less usages of e-commerce-marketing and content marketing in the promotion and branding of the region as tourist destination

3 Tourism Industry

We cannot imagine our daily lives without internet. It's not surprising in this era if one tourist is excepting the he/she will get prior information before reaching the destination, while their visit and after their visit. In case of foreign tourists this exception goes high. Internet and mobile usages after the demonetization in India are coming more in demand. The travel, tourism and hospitality industry is an industry which needs constant communication with the potential tourists in every corner of the globe. Information technology and communication is probably the only tool and technique through which tourist destinations can attract and grab the higher number of tourists. Internet is not only a mean to provide information directly to the potential tourists but it also a better way to make a pre-satisfaction in the mind of the tourists, to identify the needs and desires of the tourists prior their visit. Technology advancements allow the pattern of exchange to be changed of clientele and businesses, which are the foundation of the entire marketing (Hanson 2000; Middleton 2001). The booking systems of tickets, hotels etc. made it easy for the travelers to approach the destinations. Worldwide the tourism industry has facilitated by ITC. (Buhalis 2003). ICTs in tourism industry include various components that consist computerized reservation systems, teleconferencing, video, video fliers, management information systems, airline electronic information systems, electronic money transfer, digital telephone networks, smart cards, mobile communication, e-mail, and Internet (Mansell and When 1998). Competitive advantage of organization can be achieved by tourism managers who clinch new information technology and vigorously contribute in the technology planning process to identify new users and manage their expansion (Moutinho 2000). The tourism industry process has been changed through E-innovation strategy (Hipp and Grupp 2005; Martin 2004).

3.1 e-Marketing

e-Marketing is an important technique to connect directly with the prospective customers. In the countries like India it is spreading its wings especially in metro cities. "e-Marketing describes about the company activities to inform, customers, communicate, promote and sell its aid through means of the Internet" (Kotler and Keller 2009:785). e-Marketing is tool by which the manufactures/sellers try to promote and market not only the gods but also the services. The shift in the service industry has been noticed because of the e-marketing strategies. Banking, insurance, tourism, travel and hospitality industries are using e-marketing tools like anything. Many times it has been observed that e-marketing is considered as a marketing campaign through the website. Actually e-marketing is vast and consist of several tools other than the company website. Although company website is a vital tool of e-marketing but considering it as the only tool of e-marketing is a wrong perception. Online distribution and payment processing is another important tool of e-marketing. The advancement in the technologies gives a boost in the usages of e-marketing specially in online distribution of the products. The recent demonetization in Indian economy and "go cash less" strategy has given a push to the usages of online payment processor. One of a new trend in e-marketing is collection of responses which can be very helpful in service industry. For grabbing new customers' online advertising methods has been used by many leading companies. The tourism, travel and hospitality industry understand the urge of the different tools of e–marketing. Although the website can be viewed as the basis for e-marketing activities in tourism (Andrlić 2007), there are numerous e-marketing tools that can be adopted by the hotel industry. While adopting the e-marketing strategies for the promotion of products and services one of the important factor these days are Content marketing. Content marketing works as back bone for attracting the tourist to a particular destination by reflecting the significant content which may include images of the destinations, map and details of the place likewise weather, transport facilities, accommodation, places to see, virtual tours and other special features of the particular destination.

3.2 e-Commerce and Tourism

Electronic commerce as business performance that first and foremost depends on web technology to function, and three different mediums are included in web technologies: the internet, the World Wide Web, and wireless transmissions on mobile telephone networks (Schneider 2011, p. 4). Marketers need to focus appropriately on the necessities of the e-Commerce and online tools and techniques to make wonders and increase sales. Brand imagining, Product Positioning, Product Awareness etc. all were supported by the help of online websites. E-commerce is also helpful to stop the undue advantages enjoyed by fake retail stores (Sharma 2013).

E-commerce in travel is a new method of business enterprises, which practically includes publishing, electronic information transmit, online ordering, electronic accounts and online payment services which are related to tourism organizations. This kind of ecommerce has brought remarkable changes to nations in present era.(Zhang 2011, p. 408.)

Because of the shortage of system understanding several of the conventional travel enterprises are till date choose to run their business with the practice of instruction booklet dimension, although many of companies are identified the significance of network usage. (Shen and Huang 2011, p. 179.)

E-commerce can be described as "different sort of profitable deals or functions between two or more persons, by the means of electronic medium and system, and targeting at the exchange of products and services. (Demetriades and Baltas 2003, p. 40). While adopting e-commerce tools and techniques the markets must use a content which is significant and boost up the moral of the target viewers. Thus, for creating a brand image the following points must be considered by the marketers:-

1. Be individual in the appeal
2. Be in conversation which is significant
3. Be a manufactured goods
4. Be community (Arora and Sharma 2013).

In present era the customers not only want the information but also they wish to feel delighted while visiting and watching the websites. Only getting information is not the

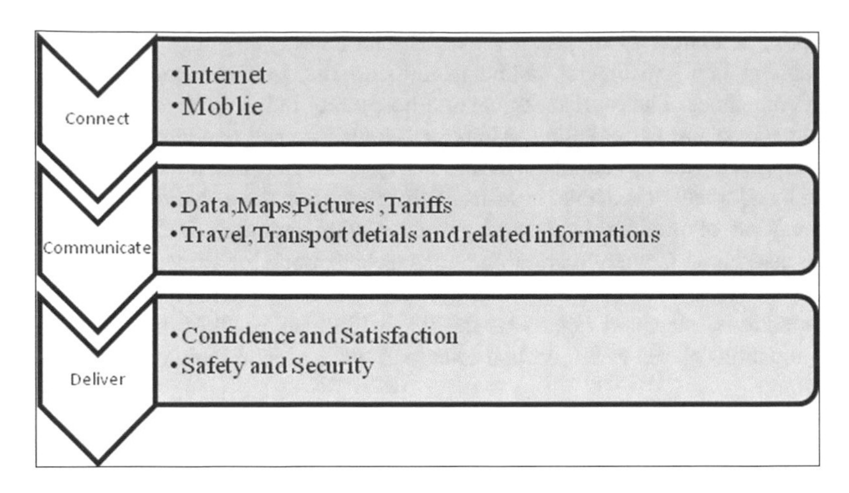

Fig. 1. CCD Model for e-Commerce and Tourism (Connect, Communicate and Deliver)

mere objective of the tourists. They would like to feel happy and relaxed while seeing the websites additionally they wish to feel safe and secure. The authors suggested a CCD model Connect, Communicate and Deliver for making potential tourist informed and delight Fig. 1.

3.2.1 Electronic Business Models

In present era the traditional business techniques and models has been replaced by internet and e-commerce. Timmers (1998) defined business business model as organized design for the good, service and information flows, an explanation of the a variety of business actors and of their roles, as well as a depiction of the possible benefits of these actors and finally a description of the sources of revenue. On the other hand Linder and Cantrell (2000) explained business model as the organisation's most important reason for creating significance". Hamel (2000) recognized four factors that determine a business model's prosperity latent-

- Effectiveness - Level to which the business conception is an well-organized method of providing benefits to customer
- Exclusivity- Level to which the business perception is distinctive.
- Balance - Level of balance among the fundamentals of the business theory.
- Revenue Generator - Level to which the business notion attain revenue generator which have the possible to create above average returns.

A business model defines how an organization creates funds by particularize where it is located in the value procession (Rappa 2001). Rappa's classification includes of nine generic e-business models, it categorize companies among the nature of their value plan or their approach of creating profits. Osteralder and Pigneur (2002) further endorsed a business model as a theoretical and architectural execution of a business strategy and as the base for the execution of business functions. The authors moreover formulated an e-business model that throw a light on the importance of e-business

Table 2. The pillars of an e-business model

Pillars	Content & components
Product innovation	The services and goods a firm offers, indicating a significant value to the customer and for which he/she is willing to pay. The essential factors are the value proposition a firm needs to present to special target client segments and the capabilities that a organization has to receive in an effort to give this worth to the customers
Infrastructure management	This element describes the value process composition that is quintessential to give the value proposition. This entails the events to create and supply value, the relationship between them, the in-residence resources and the company's partner network
Customer relationship	By using ICT firms can redefine the concept of seller -buyer relationship. The association capital the company creates and keeps with the customer, to be able to satisfy him and create sustainable income.
Financial aspect	It is transversal considering that all different pillars impact it. This element is composed of the earnings mannequin and its price constitution. The income model determines the firm's profit model and accordingly its ability to survive in competitor

Source: Adapted from Osterwalder and Pigneur 2002

issues and fundamentals companies have to think about in order to function productively. The model is based on four main pillars which are product innovation, infrastructure management, customer relationship, and financial aspects. Table 2 revels the interdependence of these factors with the detailed content and components.

The aforesaid table reveals that the four pillars are reciprocally supported and unified.

3.3 ICT Adoption for Tourism Development at Hadoti Region

Hadoti region of Rajasthan state is highly rich in its natural beauty, culture, heritage, water bodies and wildlife. The region consists of four districts namely Kota, Bundi, Baran and Jhalwar. Every corner and area of this region is full of surprises and differentiation in the tourists' products. Hadoti region have immense potential in Niche tourism likewise –Dark Tourism, Culinary Tourism, Agri-tourism, Sports and Adventure tourism. Despite of its rich cultural heritage, wildlife, water bodies and historical monuments this region is not yet gets fully exposed on the globe. There could be several reasons behind it but the researchers while the research found one of the most important reasons behind this are almost no usages e-marketing for the promotion and branding of the tourist destinations.

3.3.1 Wildlife at Hadoti and ICT

The region encompassing wide wildlife sanctuaries likewise Darrah Sanctuary, National Chambal Gharial Sanctuary, Bhensrodgarh Sanctuary, Ramgarh Sanctuary, Shergarh Sanctuary and Sorsan. Hadoti region is a paradise for the Wildlife and bird

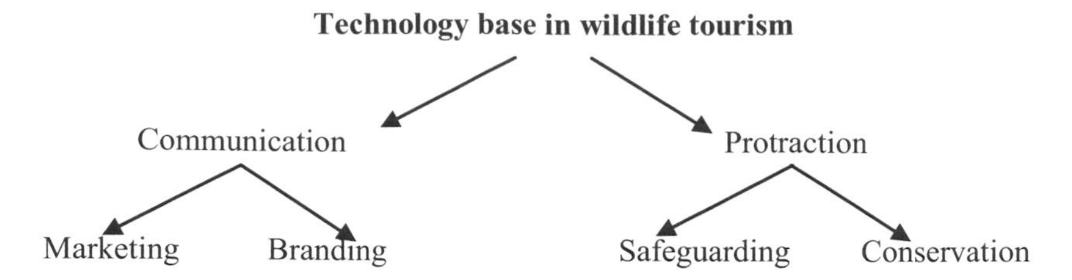

Fig. 2. Technology base in wildlife Tourism (Source: Adapted from Hassan and Sharma 2017)

watching. It gives ample opportunities to the visitors to see the natural habitat of animals. However, wildlife tourism is not getting that pace which is needed. The main reason behind it is lack of usages of information communication and technology usages. On the globe where different industries are recognizing the importance e-marketing on the other hand sanctuaries and bird watching destinations of this region don't have the websites. In present era we need ICT not only to circulate the information about the animals species, their natural habitat etc., but to give them protection and conservation. The below figure revels it Fig. 2.

The development of useful and workable technology experts still remains as one of the key challenges for marketing wildlife tourism. The availability of resources is also critical because technology use requires resource. Social networks mainly as Face book and Twitter are playing crucial roles in promoting wildlife tourism in many countries across the world. Technology becomes finely tuned with wildlife tourism marketing. Importance should be given to use and capitalize m-commerce, e-commerce, social media, blogging, and so on (Hassan and Sharma 2017a). It is really necessary to provide information through different media, including virtual tours, so visitors can get informed on the available wild animals and related visit sites, and by circulating such significant information, the wildlife attractions can be fairly set as part of tourism circuits and packages. Effective use of the Internet could include techniques like banner or video advertisements, pop ups etc. to showcase the pictures of animals, birds and sanctuaries. This helps visitors to plan a more encompassing and meaningful visits ahead of time while reducing possible disappointments and frustrations. By using proper marketing tools and means, planners, organizations and companies in the tourism sector can also propitiate the means for visitors to make their mind to visit the sanctuaries and zoo's (Hassan and Sharma 2017b). Technological gadgets are certainly contributing tourism promotion (Dadwal and Hassan 2015). In terms of strategic planning for wildlife tourism, it is relevant firstly "to connect the tourist desires and exceptions with the wildlife sanctuaries. Secondly it is important to write down their complaints regarding the wildlife visits experience for making the improvements" (Sharma 2014, p. 696). The information about the sanctuaries can work wonders for the potential tourists to attract towards wildlife sanctuaries. Lack of meticulous information about certain wildlife sanctuaries and parks is one of a major problem faced by wildlife tourism (Parveen and Sharma 2015).

3.3.2 Heritage and Cultural Tourism at Hadoti and ICT

The cultural heritage of Hadoti is very rich and diverse. The monuments, temples, forts, cuisine, fairs, festivals and crafts are giving the region a unique flavor to the region. The region covered many forts and palaces under its periphery. Taragarh fort at Bundi, Indergarh fort, Shergarh fort Baran, Nahargarh fort Baran, Gugor Fort Baran, Shahbad fort Baran and Gagron fort Jhalawar. A region with so many unique forts is difficult to find out in the state. While doing the research even the authors faced the problem to find out the details about the location, timing of the forts, and availability of facilities at nearby places of these forts. Only the details of Taragarh fort and Gagron fort are available on the website.

4 Conclusion

The research reveals that in spite of valuable tourism resources the Hadoti region is unable to attract tourists because of no usages of technology advancement tools. Lack of information related with the tourist destinations, missing road maps, directions, viability, timing, availability of facilities, transport to reach at the destination, images/pictures of the tourist destination etc. are few important aspects for any tourist which they desire to find out prior visit. While doing the research the authors themselves faced these problems. Although some of the travel agencies desired to use e-marketing and e-commerce for tourism promotion but they are still afraid. The local people are even not aware about few of the unique tourist destinations of the region. The paper stress upon the role of government in supporting tourism destinations through e-marketing and e-commerce. The present study draws the attention on the responses of some tourists who have shown their trust on the e-commerce transactions. The officials should also make sure that internet cost for users should be rational and reasonable as this will permit the growth of e-commerce to enlarge the horizon, which will bring positive results for the tourism sector. The state level and local level authorities should build a confidence in the mind of travel and tour operators, local people and tourists with an approach that the whole sector should have a cashless system and increase e-commerce methods for payments. The e-commerce transactions should be promoted such as movie e-tickets, e-entry tickets, transport e-ticket etc. Certainly to achieve these goals the government must create trust and improve on the promotional awareness. Education is also important in e- marketing and e-commerce some tourist destinations in the rural areas likewise Bilasgarh, Kanyadey are not known even by many of the locals. By the help of e-marketing and ecommerce people can get an outline of what is in the region to offer them. A foot forward by the government and private players can works wonder for the Hadoti tourism development.

References

Andrlić, B.: The use of e-marketing in tourism. Bus. Excell. **1**(2), 85–97 (2007)
Arora, S., Sharma, A.: Social media—a new marketing strategy. Int. J. Manag. 34 (2013)

Buhalis, D.: Tourism in an era of information technology. In: Faulkner, B., Moscardo, G., Laws, E. (eds.) Tourism in the Twenty-first Century: Lessons from Experience. Continuum, London (2000)

Buhalis, D.: e-Tourism. Information Technology for Strategic Tourism Management. Prentice Hall, New York (2003)

Demetriades, S., Baltas, G.: Electronic Commerce and Marketing. Rosili Publications, Athens (2003)

Dadwal, S., Hassan, A.: The augmented reality marketing: a merger of marketing and technology in tourism. In: Ray, N. (ed.) Emerging Innovative Marketing Strategies in the Tourism Industry, pp. 78–96. IGI Global, Hershey, PA (2015)

Hanson, W.: Principles of Internet Marketing. South Western College Publishing, Ohio, Cincinnati (2000)

Hamel, G.: Leading the Revolution. Harvard Business School Press, Boston (2000)

Hassan, A., Sharma, A.: Wildlife Tourism: Technology Adoption for Marketing and Conservation, pp. 74–79. Wilderness of Wildlife Tourism Apple, Academic Press, New York (2017a)

Hassan, A., Sharma, A.: Wildlife Tourism for Learning Experience: Some Anecdotal Evidences on the Royal Bengal Tiger and the Hadoti in India Wildlife Tourism Industry, pp. 17–20. Applied Ecology and Environmental Education Springer, New York (2017b)

Hipp, C., Grupp, H.: Innovation in the service sector: the demand for service specific innovation measurement concepts and typologies. Res. Policy **34**, 517–535 (2005)

Kotler, P., Keller, K.: Marketing Management, 13th edn. Prentice Hall, Upper Saddle River (2009)

Linder, J.C., Cantrell, S.: Changing Business Models: Surveying the Landscape. Institute for Strategic Change, Accenture, New York (2000)

Mansell, R., Wehn, U.: Knowledge Societies: Information Technology for Sustainable Development. Oxford :Oxford University Press (1998)

Middleton, V.T.C.: Marketing in Travel and Tourism (3rd edn.) Butterworth-Heinemann, Oxford (2001)

Martin, L.M.: E-innovation: internet impacts on small UK hospitality firms. Int. J. Contemp. Hosp. Manag. **16**(2), 82–90 (2004)

Moutinho, L.: Strategic Management in Tourism. CABI Publishing, UK (2000)

Osterwalder, A., Pigneur, Y.: An e-business model ontology for modelling e-business. In: 15th Bled Electronic Commerce, Bled, Slovenia, 17–19 June 2002

Parveen, W., Sharma, A.: Wildlife tourism: prominent panorama at Hadoti region of Rajasthan. Int. J. **3**(9), 1135–1149 (2015)

Rappa, M.: Business Models on the Web. (2001). http://digitalenterprise.org/models/models.html

Schneider, G.P.: Electronic Commerce. Cengage Learning Boston, Boston (2011)

Sharma, A.: A study on e-Commerce and online shopping: issues and influences. Int. J. Comput. Eng. Technol. **4**, 364–376 (2013)

Sharma, A.: A study on Rajasthan Wildlife Tourism: conservation and reformation. Elixir Market. Manag. **68**(2014), 91–97 (2014)

Shen, G., Huang, W.: Advance Research on Electronic Commerce, Web Application and Communication, Germany. Springer, New York (2011)

Timmers, P.: Business Models for electronic markets. J. Electron. Markets, 8(2), 3–8 (1998)

Zhang, T.: Future Computer, Communication, Control and Automation, Germany. Springer, New York (2011)

A Survey on Issues of Data Stream Mining in Classification

Ritika Jani(✉), Nirav Bhatt, and Chandni Shah

Charotar University of Science and Technology, Changa, Gujarat, India
ritikajani177@gmail.com,
{Niravbhatt.it, Chandnishah.it}@charusat.ac.in

Abstract. As Data Stream Mining is trending topic for Research nowadays and more users increases day by day with online stuff, the size of big data is also getting larger. In traditional data mining extracting knowledge is done mostly using offline phase. While in data stream, Extracting data is from the continuous arriving data or we can say from the online streams. Due to continuously arriving data, it cannot be stored in the memory for processing permanently. So examining of data as fast as possible is important. In this paper we would be interested to discuss about the data stream mining and the issues of stream classification, like Single scan, Load shedding, Memory Space, Class imbalance problem, Concept drift, and possible ways to solve those issues.

Keywords: Data stream classification · Imbalance class · Concept drift · Noise

1 Introduction

Data stream has the huge volume of continuous data. Imbalance class means, one class having more sample then another class, so we would not able to get the proper predication in case of data stream. Whereas the concept drift is the sudden change in the data. The algorithms for the data stream mining should be as fast and as light weighted as possible so that, Issues like limited Memory space, single scan of data and imbalance class can be solved. The existed learning algorithms are based on the fixed distribution of the data [1] Data sets under analysis are no longer only static databases, but also data streams in which concepts and data distributions may not be stable over time [1] As for an example weather prediction, fraud detection, energy demand and many other real-world applications [1]. There should be some novel approaches to handle these type of issues with data streams.

1.1 Major Issues in Data Stream Mining Classification

1.1.1 Single Scan

Because of the continuous arrival of real time data it is difficult to mine the useful information just in one pass. The problem with the single scan is that all the traditional FP algorithms typically require data/dataset to be stored and involved in two or more passes [2]. In case of using a single scan of data and somehow if data rate increases

S.C. Satapathy and A. Joshi (eds.), *Information and Communication Technology for Intelligent Systems (ICTIS 2017) - Volume 1*, Smart Innovation, Systems and Technologies 83, DOI 10.1007/978-3-319-63673-3_17

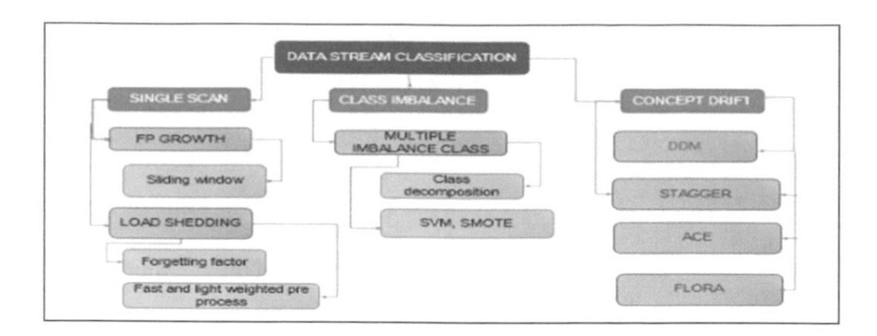

Fig. 1. Major issues in data stream classification.

then we would be forced to drop some amount of data which is not fitting in to the size of a stream, this will result in the problem of load shedding [2] (Fig. 1).

1.1.2 Memory Space

Stream mining algorithm with high memory cost is not applicable [3]. Design an algorithm which uses light weighted & preprocessing techniques. An algorithm that also learns how to compact data structure in order to store, update and retrieve info for improved and without structure efficiency of algorithm (being disturbed) [3].

1.1.3 Load Shedding

In Data Stream mining, when the new packet is arriving in the stream, and if the size of the stream is smaller than the size of the arrived packet, then The packets are going to be automatically dropped, because they did not fit in the small stream. It might possible that we lose the valuable information with that dropped packets, later which we might not even be able to retrieve again [2]. This problem is useful when it is difficult for the model to use the large data stream with high arriving ratio and very useful for querying data streams. To overcome this problem the concept of sliding window is used in which the defined size of the window is continued [4].

2 Qualities of the Good Classifier

2.1 Incremental Learning and Single Pass

Given algorithm read the block of data at the time of arrival, rather than reading it from beginning [5, 6] Which will ultimately save the processing time of a stream. The process to mine the data should be fast enough to extract the information without the loss of the data with single pass.

2.2 Limited Time and Memory

For treating high-speed continuous flow of data, streaming algorithms are online algorithms [7]. In streaming, examples are processed sequentially as well and can be examined in only a few passes (typically just one) [7]. These algorithms use limited memory and limited processing time per item [7].

3 Class Imbalance with Data Stream

The classifier with class imbalance data is where the number of the samples in each class is not equal. They classify most of the new samples in the majority class and they do not accurately predict the minority class [8]. The classifiers with imbalance class data are usually shifted to the majority class, so ultimately the accuracy to predict minority class is very poor and they are more likely interested to classify most new samples which are in the majority class [8].

Divide the imbalance data streams in number of chunks and according to novel majority class they adjust with previous data chunk. The best way is to assign the weight to each chunk and with the help of sliding window concept, if the previous chunk's weight is more than the current chunk's weight we need to replace the current one with the previous one [9, 10]. With the use of under sampling (usually removes the majority class) and over sampling (usually adding the minority class) we can balance the class. For changing the imbalance distribution of a class, balanced distribution is obtained by under sampling the majority class and over sampling the minority class. The problem of using under sampling method is, it could be possible that we might have to remove the majority classes which might be important for our examination. SMOTE (Synthetic Minority over Technique for high dimensional data) is one of the algorithm to solve the problem of imbalance class. So the oversampling technique is more preferable to obtain the balance class. Another algorithm is FSVM (Fussy Support Vector Machine) using dynamic sampling for multilayer perceptron network [11]. SVM on imbalance class produce suboptimal model which are biased towards the majority class [12]. SVM produce the effective solution for balanced data set so they are sensitive for imbalanced data set too [12]. Another method to deal with skewed is Cost sensitive learning. If the distribution of the class is getting changed then the cost of the class will also get changed [13]. So calculating cost at each distribution will improve the performance of the classifiers with the use of cost matrix [13].

4 Concept Drift in Data Stream

We know that, the distribution generating the items of a data stream can change over time because of the continuous flow of stream [5]. These changes depends on the analysis area. It is also denoted as temporal evolution, covariate shift, non- stationary, or concept drift [10]. As for an example – Consider that this is the season of the winter, we have trained our model in such a way that it can predict that after winter it will be the summer, but due to some climate change in the weather we face unexpected rains for some days, this kind of the change for the model was not predictable called drift, and the model gives us the wrong output.

Types of drifts: There are two main types of drift:

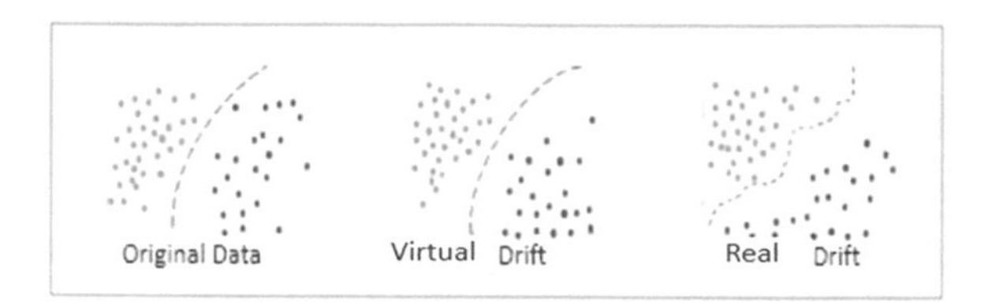

Fig. 2. Real v/s virtual drift [14]

Real drift – is the one where all the distribution of the class is not balanced. The instance of the Class distribution has been changed [6, 14] (Fig. 2).
Virtual drift – Distribution of the instance may change but the underlying concept does not, this may cause the problem for learner [6] (Fig. 3).

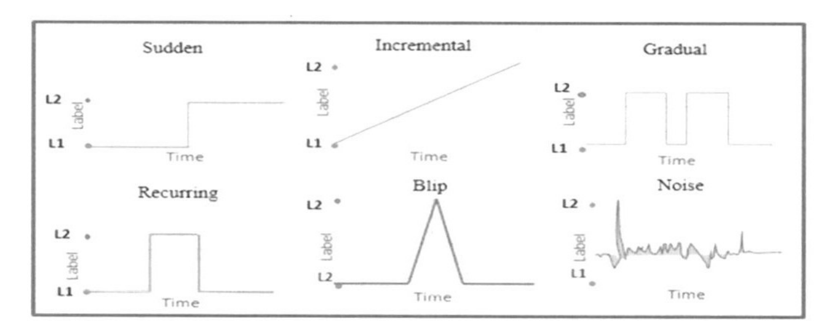

Fig. 3. Other drift types [5].

4.1 Approaches to Deal with the Concept Drift

4.1.1 Single Learner Approach

Traditional learning approaches like neural network, naïve bayse, nearest neighbor, decision rules falls in these category. The classifier incrementally update the single classifier and effectively respond to the Drift.

4.1.2 Ensemble Learner Approach

Combination of the classifiers which combine the series of the classifiers. Ensemble learning is the general way of improving classifier accuracy. This Approach has the ability to deal with the Repeated Drift, and also has the ability to deal with the traditional classifier Algorithms [6]. Boosting, Bagging and DWM (Dynamic Weighted Method) are the best popular methods in ensemble approach. In AdaBoost method the classifiers gets divided in to two sub classifiers in which half of them are correctly classified and half of them are wrongly classified (misclassify examples). They use the Position distribution for deciding the random probability, which is used for training. In Bagging the classifier train multiple methods on different samples, and average their prediction. By creating the multiple copies like this, we can improve the accuracy of model. When all the sample from same class are distributed the concept is stable, when the data distribution is different and the samples belongs to different class we say there is a concept drift [1]. While in DWM, rather than only updating the classifier it also maintains the weight of the ensemble classifier [15, 16]. There are also online approaches to solve the drift that are already existing like- STAGGER, FLORA (FLoting Rough Approximation), ACE (Adaptive Classifier Ensemble), SEA (Streaming Ensemble Algorithm) [4, 6]. A General Framework of concept drift [17] in data stream classification includes the continuously updated single classifier which is trained as the ensemble of classifier from the sequential data chunks. The prediction accuracy is given by combining number of classifiers weight [9, 17].

5 Drift Detection Methods

There are methods which pay attention on the drift. One of them is DDM and the other one is EDDM

5.1 DDM

The examples in Data stream are arriving one by one, if we consider they are arriving in bunch, then it would be easy for the model to make the next assumption and system can learn from the previous example [5]. Drift detection Method is one of the method which counts the number of errors produced by the learner during the prediction [5]. This method uses binomial distribution and that distribution gives the general form of the probability for the number of variables and number of errors in a sample of "n". For each ith point which is sampled in sequence (s_i), the rate of error is misclassifying probability (p_i) [5]. The standard deviation is given by the equation as shown below.

$$s_i = \sqrt{pi(1 - pi)/i} \tag{1}$$

5.2 EDDM

Early drift detection method, has been developed to detect the drift in the presence of the gradual drift [5]. The EDDM method works better than the DDM method and it is sensitive for the noise which has been present in Stream. This method considers the distance between two errors instead of considering only the number of errors in the classifier [5].

6 Decision Tree Methods in Data Stream Classification

When some attributes have calculated based on heuristic function, minor change in training data may change the result [19]. On other side, it is not a good way to train a minor change [18, 19]. But the only way for updating it is to replace it with a new one, which causes the decision-tree to drift concept and nonstop check the earlier selected split points to see whether they are still the best attributes or not [19], and if this is not the case, then to replace the current tree, new substitute sub-tree would be created [19]. Whenever the alternative one becomes more accurate on new data [18, 19]. Hoeffding bound is the well-known statistical criteria that is commonly used in decision trees which learns from data streams to decide when to split a node [18], i.e. hoeffding tree don't make any split until a statistical significance appears to be in favor of the current best attribute against the other ones [18]. For streams made up of discrete type of data, Hoeffding bounds assure that the output model of VFDT is asymptotically nearly matching to that of a batch learner [17]. VFDT produces decision-tree classifiers that are nearly similar to the tree produced by a conventional batch learner. Several variants of VFDT address the issue of concept drift. For example, the method concept-adapting very fast decision tree (CVFDT) works by maintaining a model with respect to sliding window of instance from the data stream. When the data is changed, the tree creates an alternate sub-tree of data with that changing node [9]. But this sub tree replaces the old tree only after it becomes accurate. The drawback of CVFDT is that it does not define any optimal window size. A small window size accurately predict current distribution whereas a large size accommodates several examples to work on and increases accuracy when the concept is stable and it will take the time [9]. This weakness can be overcome by a technique named hoeffding adaptive tree using Adaptive windowing (HAT-ADWIN) [9]. This method keeps a variable-length window of recently seen items which automatically grows when no change is apparent and shrinks it when data changes [9].

7 Concept Drift with Noise

The main problem with the stream mining is to identify the true drift and the noise. Whenever we find any kind of change in data we can not directly say whether it is by the true drift or by the noise itself. If the change is because of the noise then cleaning of the data is required, and if it is because of the drift then according to the drift the methods needs to be applied. If the data is changed by the noise then it might have been possible that some of the samples are misclassified, and because of this imbalanced class problem the results are not appropriate. The quality of the data is widely affected by noise or missing data. In information retrieval filed syntactic rules are usually defined to remove noise [16].

8 Future Work

The goal is to design and implement the stream algorithm which can distinguish between the true drift and the noise, and apply the methods according to the requirements. The final output would be the data stream of balance class without noise and drift.

9 Conclusion

One of the main issue with data stream mining is the Concept Drift, so identifying the drift and noise is the tough task, as soon as the noise gets recognized from the data and if there is any change in data we can clearly say that it is because of the true drift, so there is a need to implement such algorithms which can differentiate the noise and drift.

References

1. Breve, F., Zhao, L.: Practical competition and cooperation in networks for semi-supervised learning with concept drift. In: IEEE World Congress on Computational intelligence (2012)
2. Agrawal, L.S., Adane, D.S.: Models and issues in data stream mining. In: IJCSA (2016)
3. Kholghi, M.: An analytical framework for data stream mining techniques based on challenges and requirements. In: IJEST (2011)
4. Domingoes, P., Hulten, G.: Catching Up with the Data: Research Issues in Mining Data Stream. University of Washington, Seattle
5. Baena-Garcia, M., del Campo-Avila, J., Fidalgo, R., Bifet, A., Gavalda, R., Morales-Bueno, R.: Early Drift Detection Method, Spain (2006)
6. Hoens, T.R., Polikar, R., Chawla, N.V.: Learning from Streaming Data with Concept Drift and Imbalance: An Overwiew. Springer, Berlin (2012)
7. Gama, J., Zliobaite, I., Bifet, A., Pechenizky, M., Bouchachia, A.: A Survey on Concept Drift Adaption. ACM (2014)
8. Lusa, L., Blagus, R.: The class-imbalance problem for high-dimentional class prediction. In: 11th International Conference on Machine Learning and Application (2012)
9. Kotecha, R., Garg, S.: Data streams and privacy: two emerging issues in data classification. In: 5th Nirma University International Conference on Engineering (2015)
10. Godase, A.B.: Classification of Data Streams with Skewed Distribution, June 2012
11. Purwar, A., Singh, S.K.: Issuess in data mining: a comprehensive survey. In: International Conference on Computational Intelligence and Computing research. IEEE (2014)
12. Batuwita, R., Palade, V.: Class imbalance learning methods for support vector machines. In: Singapore-MIT Alliance for Research and Technology Centre, University of Oxford (2012)
13. Guo, X., Yin, Y., Dong, C., Yang, G., Zhou, G.: On the class imbalance problem. In: Fourth International Conference on Natural Computation. IEEE (2008)
14. Kadwe, Y., Suryawanshi, V.: A Review on Concept Drift, IOSR-JCE (2015)
15. Sindhu, P., Bhatia, M.P.S.: A Novel Online Ensemble Approach to Handle Concept Drifting Data Streams: Diversified Dynamic Weighted Majority. Springer, Berlin (2015)
16. Chu, F.: Mining Techniques for Data Streams and Sequences. University of California, Los Angeles (2005)
17. Wang, H., Fan, W., Yu, P.S., Han, J.: Mining Concept-Drifting Data Streams Using Ensemble Classifiers. ACM (2003)
18. Pears, R., Sakthithasn, S., Koh, Y.S.: Detecting Concept Change in Dynamic Data Streams: A Sequential Approach Based on Reservoir Sampling. Springer, Berlin (2014)
19. Mirzamomen, Z., Kangavari, M.R.: Evolving Fuzzy Min-Max Neural Network Based Decision Trees for Data Stream Classification. Springer, Berlin (2016)

An Algorithmic Approach for Recommendation of Movie Under a New User Cold Start Problem

Hemlata Katpara(✉) and Vimalkumar B. Vaghela

L.D. College of Engineering, Ahmedabad, Gujarat, India
hkatpara@gmail.com, vimalvaghela@gmail.com

Abstract. One of the most popular approach for personalized recommendations is the Collaborative filtering methods. The key point of this method is to find out similar users by calculating the similarities among them. The product is recommended to a user based on user-user similarity. For finding similarities, the measures like Cosine, Pearson correlation coefficient, Proximity-Impact-Popularity (PIP) measure, and Proximity-Significance-Singularity (PSS) measure can be used. The main issue with the recommendation system is the new user cold start problem where less ratings are available with the user. The user-user similarity matrix obtained with the help of above mentioned measures in terms of cold start problem is not more accurate. In this paper, we show that the hybrid measure is giving more accurate result then the other similarity measures. In this paper a detailed algorithm for the recommendation of a movie to a new cold start user for the hybrid measure is given. The experiments are done with Movie Lens dataset and the results are displayed in the form of user-user similarity metrics in order to obtain the user-user similarity matrix for the hybrid measure.

Keywords: Collaborative filtering · Similarity measures · Cold-start · Recommendation system

1 Introduction

Nowadays, we are living in the era of smart phones. We are spending lots of time in accessing social networking sites like Facebook, Twitter, WhatsApp and E-commerce sites through these smart phones. Users are not able to get satisfactory results for the information about the products. It is our good luck that on the social networking sites and on the e-commerce sites, the behavior of the user is recorded. So it is easy to retrieve the preference of the user. And in this context, the recommendation systems are used to recommend the product by analysing the user behavior [1, 5, 8].

Recommender system algorithms can be divided into following five different categories: collaborative, content-based, demographic, utility-based and knowledge-based [3]. With the help of the profile of a user or an item, the content based filtering is

S.C. Satapathy and A. Joshi (eds.), *Information and Communication Technology for Intelligent Systems (ICTIS 2017) - Volume 1*, Smart Innovation, Systems and Technologies 83, DOI 10.1007/978-3-319-63673-3_18

suggesting an item to an active user. We cannot check the quality of an item and cannot differentiate two items with content based filtering [5, 9].

Demographic recommendation systems use demographic data for similarity calculation among users. Utility based methods are using the utility function in which many factors of utility are used in making recommendations. Knowledge based methods introduces the knowledge models that are used for recommendation of complex products [3, 10].

The Collaborative filtering method is the most popular method to recommend items for users. It assigns the items to an active user with the help of similar users [1, 5, 9]. In this method, the user profile is the set of ratings which are given to items by other users and it is obtained by user-item rating matrix. The other way to obtain these ratings is by asking the user and capturing the behavior of a user with the system [2]. There are two methods under the collaborative filtering approach and they are: Memory based Collaborative filtering and Model based Collaborative filtering [1, 5, 8].

In memory based method, in order to calculate the similarities, the variety of similarity measures are used. In the second step, the method is finding the similar users of an active user using KNN algorithm. And in the last step, it is recommending the items to an active user [1, 5, 14]. Model based methods construct a model that shows the behavior of a user and then predict their ratings. The memory based methods are more accurate in recommending items to an active user but the computational time increases if the size of user-item rating matrix becomes large. The model based methods result in less accurate recommendation but they are faster in prediction time [1, 5, 8, 15, 16].

The problem that we discuss through the paper is the cold start problem. And as per the discussion, when a new user enter in the system, less ratings are with the user and with that much ratings system is not able to recommend items to that user because system is not able to guess the user's interests. In order to solve this problem, some systems are asking the user for giving ratings to different items. But it is not good to give a rating to an item without using that item. The new items are also affected by the cold start problem and those items are not suggested to an active user because they are not rated by enough number of users.

This paper focuses on variety of measures under the new user cold start problem and they are Pearson Correlation Coefficient (PCC), Proximity-Impact-Popularity (PIP) and Proximity-Significance-Singularity (PSS). In the paper, we introduce the new algorithm for the new hybrid measure that is a combination old existing measures. The algorithm results in recommendation of a movie to a new user who is a cold start user by using hybrid measure for calculation of user-user similarity matrix.

2 Related Work

To improve the accuracy of recommender system, many researchers have proposed variety of measures like Pearson Correlation Coefficient, Cosine measure, Proximity-Impact-Popularity (PIP) measure, Proximity-Significance-Singularity (PSS) measure. These measures consider the local information of ratings and not consider the global preference of the user ratings.

2.1 Existing Similarity Measures

Pearson Correlation Coefficient [1]

$$\mathrm{SIM}(u_i, u_j) = \frac{\sum_{p\in I}(r_{u,p} - \hat{r}_u)(r_{v,p} - \hat{r}_v)}{\sqrt{\sum_{p\in I}(r_{u,p} - \hat{r}_u)^2}\sqrt{\sum_{p\in I}(r_{v,p} - \hat{r}_v)^2}}. \tag{1}$$

where I is the set of common rated items by two different users. $r_{u,p}$ is the rating of u on item p and $r_{v,p}$ is the rating of item p by v.
$\hat{r}_u$ is the average rating of u and $\hat{r}_v$ is the average rating of v [1].

Proximity-Impact-Popularity (PIP) Measure [3–5]

This measure is comprising of factors: Proximity, Impact, and Popularity and hence the name is PIP. To find the similarity between two users using this measure is as follows:

$$\mathrm{SIM}(u_i, u_j) = \sum_{k\in ci,j} \mathrm{PIP}(rt_{ik}, rt_{jk}). \tag{2}$$

rtik and rtjk are the ratings of two different items. Ci,j is the set of co-rated items and PIP(rtik, rtjk) is the PIP score for the two ratings. For any two ratings, rt1 and rt2 [3, 4]

$$\mathrm{PIP}(rt_1, rt_2) = \mathrm{Proximity}(rt_1, rt_2) \times \mathrm{Impact}(rt_1, rt_2) \times \mathrm{Popularity}(rt_1, rt_2). \tag{3}$$

Two rating values are subtracted and the result gives the impact factor. It also check out that whether these ratings are following the rules or not. For the ratings which are in disagreement, penalty is given. An item can be preferred by or disliked by the user and it is represented by impact factor. For example, if two user rate 4 on an item, then it will show stronger preference than they rate 3 on that item [1, 4, 5]. If two ratings are not in agreement then the values of Proximity and Impact are calculated again and again. Popularity factor shows common things between two user's ratings. If two user's average ratings are far from the total user's average ratings then we can say that those two users are having higher degree of similarity [1, 4, 5].

$$\mathrm{Proximity}(\mathrm{rt}_1, \mathrm{rt}_2) = \{\{2.(\mathrm{R}_{\max} - \mathrm{R}_{\min}) + 1\} - \mathrm{D}(\mathrm{rt}_1, \mathrm{rt}_2)\}^2 \tag{4}$$

$$\mathrm{Impact}(\mathrm{rt}_1, \mathrm{rt}_2) = (|\mathrm{rt}_1 - \mathrm{R}_{\mathrm{med}}| + 1)(|\mathrm{rt}_2 - \mathrm{R}_{\mathrm{med}}| + 1) \text{ if Agreement}(\mathrm{rt}_1, \mathrm{rt}_2) \text{ is true} \tag{5}$$

$$\mathrm{Impact}(\mathrm{rt}_1, \mathrm{rt}_2) = \frac{1}{(|\mathrm{rt}_1 - \mathrm{R}_{\mathrm{med}}| + 1)(|\mathrm{rt}_2 - \mathrm{R}_{\mathrm{med}}| + 1)} \text{ if Agreement}(\mathrm{rt}_1, \mathrm{rt}_2) \text{ is false} \tag{6}$$

$$\mathrm{Popularity}(\mathrm{rt}_1, \mathrm{rt}_2) = 1 + \left(\frac{\mathrm{rt}_1 + \mathrm{rt}_2}{2} - \mu_k\right)^2 \text{ if } (\mathrm{rt}_1 > \mu_k \text{and } \mathrm{rt}_2 > \mu_k) \text{or } (\mathrm{rt}_1 < \mu_k \text{and } \mathrm{rt}_2 < \mu_k) \tag{7}$$

$$Popularity(rt_1, rt_2) = 1 \; otherwise \tag{8}$$

PSS (Proximity-Significance-Singularity) Measure [1]

$$\mathrm{sim}(\mathrm{u}, \mathrm{v})^{\mathrm{pss}} = \sum\nolimits_{\mathrm{x} \in \mathrm{I}} \mathrm{PSS}(\mathrm{r}_{\mathrm{u,x}}, \mathrm{r}_{\mathrm{v,x}}) \tag{9}$$

where:

$$\begin{aligned}\mathrm{PSS}(\mathrm{r}_{\mathrm{u,x}}, \mathrm{r}_{\mathrm{v,x}}) &= \mathrm{Proximity}(\mathrm{r}_{\mathrm{u,x}}, \mathrm{r}_{\mathrm{v,x}}) \times \mathrm{Significance}(\mathrm{r}_{\mathrm{u,x}}, \mathrm{r}_{\mathrm{v,x}}) \\ &\quad \times \mathrm{Singularity}(\mathrm{r}_{\mathrm{u,x}}, \mathrm{r}_{\mathrm{v,x}}).\end{aligned} \tag{10}$$

PSS measure is composed of three factors like Proximity, Significance and Singularity [1]. The Proximity factor of PSS measure is having same meaning as of Proximity factor of PIP measure. The name of second factor is significance and it shows that the ratings are more significant if they are more far from the median rating. The difference of two user's ratings with the other ratings are described by Singularity factor [1]. The Formulas for these three factors are as follows [1].

$$\text{Proximity } (r_{u,x}, r_{v,x}) = 1 - \frac{1}{1 + \exp(-|r_{u,x} - r_{v,x}|)}. \tag{11}$$

$$\text{Significance } (r_{u,x}, r_{v,x}) = \frac{1}{1 + \exp(-|r_{u,x} - r_{med}| \cdot |r_{v,x} - r_{med}|)} \tag{12}$$

$$\text{Singularity } (r_{u,x}, r_{v,x}) = 1 - \frac{1}{1 + \exp(-|\frac{r_{u,x} + r_{v,x}}{2} - \mu_x|)} \tag{13}$$

3 Proposed Workflow

3.1 Algorithm

```
Algorithm: RecSys(UId, MovId, Rating)
Input:
  UId    : User id of movie viewer who has rated the Movie.
  MovId  : Unique Identification of movie.
  Rating: Rating in the scale of [1-5] given by viewer.
 Output:
    Rated movie list according to user preference.

Procedure:
Step 0: [pre-processing]
      Convert xls-database to array User-MovieArray
Step 1:
      SimilarityMatrix=User-
      User_SimilarityMatrixGen(User-MovieArray);
 Step 2:
      KNNeighbour=KNNProc(MovId,UId);
 Step 3:
      Call Rating_Predict(User-MovieArray);
 Step 4:
      Recommend rated MovieList;
 Step 5:
      End

Proc User-User_SimilarityMatrixGen(User-MovieArray)
{
  For i=1 to User-MovieArray.rowlength
   For j=1 to User-MovieArray.columnlength
   Compute the similarity between User u=ui and User v=ui+1
      Compute:
```

$$sim(u,v)^{NHSM} = sim(u,v)^{JPSS} \cdot sim(u,v)^{URP}$$

where:

$$sim(u,v)^{URP} = 1 - \frac{1}{1+\exp(-|\mu_u - \mu_v| \cdot |\sigma_u - \sigma_v|)}$$

$$sim(u,v)^{JPSS} = sim(u,v)^{PSS} \cdot sim(u,v)^{Jaccard}$$

where:

$$sim(u,v)^{Jaccard} = \frac{|I_u \cap I_v|}{|I_u| \times |I_v|}$$

$$sim(u,v)^{pss} = \sum_{p \in I} PSS(r_{u,p}, r_{v,p})$$

```
   End
  End
    Return SimilarityMatrix()
}
```

```
Proc KNNProc(MovId,UId)
{
      Find the different UId for the given MovId from User-
      MovieArray
      Find the maximum similarity users from similarity
      matrix for different UId
      Return KNNeioghbour
}

Proc Rating_Predict(User-MovieArray)
{
      Make prediction of ratings for nearest neighbors of
      active user a
      For i=1 to User-MovieArray.rowlength;
            For j=1 to User-MovieArray.columnlength;

                  Compute
```

$$P_{a,i}=\frac{\acute{r}_a + \sum_{u=1}^{n}(r_{u,i}-\acute{r}_u)*w_{a,u}}{\sum_{u=1}^{n} w_{a,u}}$$

```
                  where:
```

$$w_{a,u}=\frac{\sum_{i=1}^{m}(r_{a,i}-\acute{r}_a)(r_{u,i}-\acute{r}_u)}{\sigma_a\sigma_u}$$

```
            End
      End
}
```

The above mentioned algorithm is for recommendation of a movie to a new user based on hybrid measure. The description formula for this hybrid measure is as follows:

3.2 Formalization of Hybrid Measure

$$sim(u,v)^{NHSM} = sim(u,v)^{JPSS} \cdot sim(u,v)^{URP}. \tag{14}$$

$$sim(u,v)^{URP} = 1 - \frac{1}{1 + \exp(-|\mu_u - \mu_v| . |\sigma_u - \sigma_v|)}. \tag{15}$$

$$sim(u,v)^{JPSS} = sim(u,v)^{PSS} \cdot sim(u,v)^{Jaccard}. \tag{16}$$

$$sim(u,v)^{Jaccard} = \frac{|I_u \cap I_v|}{|I_u| \times |I_v|}. \tag{17}$$

$$sim(u,v)^{pss} = \sum\nolimits_{p \in I} PSS(r_{u,p}, r_{v,p}). \tag{18}$$

4 Experiments

4.1 Datasets

We use the dataset Movie Lens for implementation purpose. The Movie Lens dataset ML-100K is taken for implementation. This dataset is having 100000 records for 943 users and 1682 movies. The dataset we collected from http://www.grouplens.org/. The dataset is having one data file from which we filtered the cold start data, the data in which users have rated only the 2% to 5% of total movies. All the experiments are performed on these filtered data.

For filtration of cold start data, we take randomly only 20 user id and for each user id we take records of 1 to 50 movie id. The dataset is having fields like user id, movie id, rating and timestamp value.

4.2 Results of Various Similarity Metrics in Order to Obtain Hybrid Measure

The following are some of the user-user similarity metrics generated with respect to below mentioned user-item rating matrix in order to get the matrix for hybrid measure (Figs. 1, 2, 3, 4, 6 and 7).

Table 1. User-Item rating matrix [1]

	Item 1	Item 2	Item 3	Item 4
User1	4	3	5	4
User2	5	3	–	–
User3	4	3	3	4
User4	2	1	–	–
User5	4	2	–	–

$$\begin{array}{c} \\ u1 \\ u2 \\ u3 \\ u4 \\ u5 \end{array} \begin{pmatrix} u1 & u2 & u3 & u4 & u5 \\ 0.47937 & 0.19013 & 0.42463 & 0.06007 & 0.19042 \\ 0.19013 & 0.23422 & 0.19013 & 0.05323 & 0.12449 \\ 0.42463 & 0.19013 & 0.47937 & 0.06007 & 0.19042 \\ 0.06007 & 0.05323 & 0.06007 & 0.13647 & 0.06397 \\ 0.19042 & 0.12449 & 0.19042 & 0.06397 & 0.22687 \end{pmatrix}$$

Fig. 1. PSS Similarity matrix

	u1	u2	u3	u4	u5
u1	0.25	0.25	0.25	0.25	0.25
u2	0.25	0.5	0.25	0.5	0.5
u3	0.25	0.25	0.25	0.25	0.25
u4	0.25	0.5	0.25	0.5	0.5
u5	0.25	0.5	0.25	0.5	0.5

Fig. 2. Modified Jaccard Similarity matrix

	u1	u2	u3	u4	u5
u1	0.11984	0.04753	0.10616	0.01502	0.04761
u2	0.04753	0.11711	0.04753	0.02662	0.06225
u3	0.10616	0.04753	0.11984	0.01502	0.04761
u4	0.01502	0.02662	0.01502	0.06824	0.03199
u5	0.04761	0.06225	0.04761	0.03199	0.11344

Fig. 3. JPSS similarity matrix

	u1	u2	u3	u4	u5
u1	0.5	0.5	0.47413	0.37337	0.4273
u2	0.5	0.5	0.43782	0.2227	0.5
u3	0.47413	0.43782	0.5	0.5	0.43782
u4	0.37337	0.2227	0.5	0.5	0.32082
u5	0.4273	0.5	0.43782	0.32082	0.5

Fig. 4. URP similarity matrix

	u1	u2	u3	u4	u5
u1	0.05992	0.02377	0.05033	0.00561	0.02034
u2	0.02377	0.05856	0.02081	0.00593	0.03113
u3	0.05033	0.02081	0.05992	0.00751	0.02084
u4	0.00561	0.00593	0.00751	0.03412	0.01026
u5	0.02034	0.03113	0.02084	0.01026	0.05672

Fig. 5. Hybrid similarity matrix

4.3 Advantages of Hybrid Measure with Respect to Old Similarity Metrics

We compare the hybrid similarity matrix mentioned in Fig. 5 with respect to the old similarity metrics as below.

$$\begin{array}{c} \\ u1 \\ u2 \\ u3 \\ u4 \end{array} \begin{pmatrix} u2 & u3 & u4 & u5 \\ 0.707 & 0.0 & 0.707 & 0.707 \\ & 1.0 & 1.0 & 1.0 \\ & & 1.0 & 1.0 \\ & & & 1.0 \end{pmatrix}$$

Fig. 6. PCC similarity matrix [1]

$$\begin{array}{c} \\ u1 \\ u2 \\ u3 \\ u4 \end{array} \begin{pmatrix} u2 & u3 & u4 & u5 \\ 0.743 & 1.0 & 0.167 & 0.506 \\ & 0.743 & 0.162 & 0.763 \\ & & 0.167 & 0.506 \\ & & & 0.767 \end{pmatrix}$$

Fig. 7. PIP similarity matrix [1]

1. If we compare these two metrics with the hybrid measure as shown in Fig. 5, we can say that in hybrid measure each and every similarity between two users is different.
2. Similarity between user2 and user5 should be higher than that of user2 and user4 according to Table 1 and it is not true for PCC but it is true for PIP and hybrid measure.
3. Similarity between user3 and user5 should be higher than that of user3 and user4 according to Table 1 and it is not true for PCC but it is true for PIP and hybrid measure.
4. Similarity between user3 and user5 should be higher than that of user3 and user4 according to Table 1 and it not true for PCC and PIP but it is true for hybrid measure.
5. Similarity between user1 and user4 and the similarity between user3 and user4 should not be same according to Table 1 but it is same in PCC and PIP, while it is different for hybrid measure.

5 Evaluation Methodology

To compare the quality of recommendation system, several types of evaluation metrics are available. They can be divided into following types.

Predictive Accuracy

The difference between predicted rating and the real rating is giving predictive accuracy. The most common metrics of this kind are as follows [2, 6, 7].

MAE (Mean Absolute Error)

It is the average of absolute errors over the all the predictions made by the collaborative filtering algorithms. It is calculated over all the ratings available in the evaluation set. For better accuracy the value of MAE should be small.

$$MAE = \frac{\sum_{i=1}^{MAX} |r_i - \hat{r}_i|}{MAX} \tag{19}$$

r_i and $\hat{r}_i$ are actual and predicted rating of an active user on an item. MAX shows the number of times prediction performed by the Collaborative filtering algorithm.

RMSE (Root Mean Squared Error)

The square root of square of difference between the predicted and the actual rating is known as Root mean squared error. Compared to MAE, RMSE punishes large errors [1, 6, 7].

$$\mathrm{RMSE} = \sqrt{\frac{1}{\mathrm{MAX}} \sum_{i=1}^{\mathrm{MAX}} \left(r_i - \hat{r}_i\right)^2} \tag{20}$$

Classification Accuracy

The qualitative performance of a recommendation system can be measured by classification accuracy. A list of recommendation items is given to an active user under the recommender system. Based on that, to evaluate the quality of a RS the following metrics are used [1, 6, 7].

Precision: It is the fraction of items in Lr that are relevant.

Recall: It is the fraction of total relevant items which are in the recommended list Lr.

$$\text{Precision} = \frac{|L_r \cap L_{rev}|}{|L_r|}. \tag{21}$$

$$\text{Recall} = \frac{|L_r \cap L_{rev}|}{|L_{rev}|}. \tag{22}$$

The values of precision and recall should be high for a system. The metrics precision and recall ae inversely related such that when precision increases, recall decreases.

6 Conclusion

This paper discuss about the cold-start problem in recommendation system. For various measures like PCC (Pearson Correlation Coefficient), PIP (Proximity-Impact-Popularity), user-user similarity metrics are derived and that metrics are compared with the user-user similarity matrix of hybrid measure. The comparison shows that the hybrid measure is more accurate in generating similarity metrics. In the paper an algorithm is proposed that is for recommendation of a movie to a new user who is a cold start user and the given hybrid measure is more accurate for finding user-user similarity.

References

1. Liu, H., Zheng, H., Mian, A., Tian, H., Zhu, X.: A new user similarity model to improve the accuracy of collaborative filtering. Knowl. Based Syst. **56**, 156–166 (2014)
2. Carneiro, V., Fernández, D., Formoso, V.: Comparison of collaborative filtering algorithms: limitations of current techniques and proposals for scalable, high-performance recommender system. ACM Trans. Web **5**(1), 1–33 (2011)

3. Ahn, H.J.: A new similarity measure for collaborative filtering to alleviate the new user cold-starting problem. Inf. Sci. **178**, 37–51 (2008)
4. Ahn, H.J.: A Hybrid Collaborative Filtering Recommender System using a new similarity measure, Hangzhou, China, 15–17 April 2007
5. Katpara, H., Vaghela, V.B.: Similarity measures for collaborative filtering to alleviate the new user cold start problem. In: 3rd International Conference on Multidisciplinary Research & Practice, vol. 4(1), pp. 233–238 (2016)
6. Resnick, P., Varian, H.R.: Recommender systems. Commun. ACM **40**(3), 56–58 (1997)
7. Miller, B.N., Albert, I., Lam, S.K., Konstan, J.A., Riedl, J.: Movie lens unplugged, experiences with an occasionally connected recommender system. In: Proceedings of the 8th International Conference on Intelligent User Interfaces, pp. 263–266 (2003)
8. Patra, B.K., Launonen, R., Nandi, V.O.S.: A new similarity measure using Bhattacharyya coefficient for collaborative filtering in sparse data. Knowl. Based Syst. **82**, 163–177 (2015)
9. Son, L.H.: Dealing with the new user cold-start problem in recommender systems: a comparative review. Inf. Syst. **58**, 87–104 (2014)
10. Safoury, L., Salah, A.: Exploiting user demographic attributes for solving cold-start problem in recommender system. Lect. Notes Softw. Eng. **1**(3), 303–307 (2013)
11. Bobadilla, J., Ortega, F., Hernando, A., Bernal, J.: A collaborative filtering approach to mitigate the new user cold start problem. Knowl. Based Syst. **26**, 225–238 (2011)
12. Bobadilla, J., Ortega, F., Hernando, A.: A collaborative filtering similarity measure based on singularities. Inf. Process. Manag. **48**, 204–217 (2012)
13. Bobadilla, J., Hernando, A., Orteqa, F., Gutirrez, A.: Collaborative filtering based on significances. Inf. Sci. **185**, 1–17 (2012)
14. Vaghela, V.B., Jadav, B.M.: Analysis of various sentiment classification techniques. Analysis **140**(3), 22–27 (2016)
15. Jadav, B.M., Vaghela, V.B.: Sentiment analysis using support vector machine based on feature selection and semantic analysis. Int. J. Comput. Appl. **146**(13), 26–30 (2016)
16. Pathak, H.H., Vaghela, V.B.: Partitioned RCF: an improved reversed collaborative filtering algorithm for maximizing recommendations. Int. J. Adv. Res. Sci. Eng. **5**(1), 1–7 (2016)

RED: Residual Energy and Distance Based Clustering to Avoid Energy Hole Problem in Self-organized Wireless Sensor Networks

Abhishek Chunawale(✉) and Sumedha Sirsikar(✉)

Department of Information Technology,
Maharashtra Institute of Technology, Pune, India
abhish.chunawale82@gmail.com,
sumedha.sirsikar@mitpune.edu.in

Abstract. Self-Organized Wireless Sensor Network (SOWSN) is a system of sensor nodes that takes global decisions through local interactions without involvement of any central entity. Wireless sensor nodes have constrained processing capability and energy. The key characteristic used to evaluate performance of Wireless Sensor Network (WSN) is its lifetime which depends on residual energy of nodes; hence the major challenge in WSN is the efficient use of available energy. Node clustering saves energy and also shows self-organization because global decision like Cluster Head (CH) selection is taken through mutual communication between nodes. In this paper, a new clustering method based on self-organization is implemented to boost lifetime of WSN. Sensor network is divided into regions. Cluster formation relies on Residual Energy (RE) and nearest Distance (D) from CH. Node with highest residual energy becomes CH. Rest nodes join the nearest CH. Clusters are broken when residual energy of CH falls below threshold energy; causing the sensor network to get self-organized into new clusters. RED also focuses to solve the energy hole problem caused due to higher energy consumption by CHs near Sink Node or Base Station (BS).

Keywords: Wireless Sensor Network · Self-organization · Residual Energy · Threshold energy · Node distance · Energy hole

1 Introduction

Wireless sensor nodes sense data from environment and transmit it directly to BS or Sink. But it is not appropriate for large network size because nodes at very large distance from Sink will deplete their energy soon. To solve this problem, sensor nodes are clustered. Instead of each node processing the data, a single node collects data, process it and transmit to sink node. Involvement of each and every sensor in data transmission increases congestion and data collisions in network. This will drain limited energy from network. Node clustering will address these issues. Clustering ensures efficient data transmission by reducing number of sensors trying to communicate data in the WSN. It also minimizes message overhead and number of dropped packets [1].

S.C. Satapathy and A. Joshi (eds.), *Information and Communication Technology for Intelligent Systems (ICTIS 2017) - Volume 1*, Smart Innovation, Systems and Technologies 83, DOI 10.1007/978-3-319-63673-3_19

1.1 Energy Hole Problem in WSN

In WSNs, all data is sent to sink, hence traffic close to the sink node is higher. The CHs and sensor nodes in this area will soon run out of energy. The sink will then be inaccessible and as a result, the residual energy of nodes will be wasted [2]. This problem is called as energy hole problem in WSN.

2 Literature Survey

Literature shows that a variety of clustering algorithms have been proposed till date. Also, there are different classifications of clustering algorithms based on attributes like number of nodes in a cluster, number of clusters, centralized or distributed clustering, CH selection criteria (e.g. based on weight calculation, residual energy, distance, probability, degree, location, etc.). One major category of clustering algorithms is distance based clustering algorithms. Table 1 throws light on such type of algorithms.

3 RED Algorithm

RED is refinement of our algorithm described in [3]. Here onwards the word 'PRE' is used to indicate previous algorithm. In PRE, a node joins CH if it is within the range of a CH; irrespective of distance from another CH. RED differs from PRE in following ways:

i. Single hop connectivity between nodes and multi-hop communication between CHs along with random deployment of nodes
ii. Problem of nodes coming under overlapped coverage areas of CH is solved as nodes join nearest CH
iii. Focus on energy hole problem
iv. Change in Threshold Energy (T) for reclustering

3.1 Selection of Cluster Head

The CH selection here is not centralized i.e. BS is not involved in selecting CH. It is purely on the basis of local information and communication among the nodes. It has two advantages: Problem of network bottleneck is removed and energy required for communication between BS and nodes far away from BS is saved since energy consumption and distance are directly proportional to each other. As there is no need to communicate with BS all the time for CH selection; energy is saved. Residual Energy (RE) is the prime factor in CH selection. The first step in RED is to find number of nodes falling in each nodes' coverage area and current residual energy of each sensor node. Nodes possessing highest residual energy will become Cluster Head provided that it has at least one neighbor.

Table 1. Comparison of distance based clustering algorithms to identify parameters for CH selection and cluster formation.

Algorithm & Year	Parameters for CH selection and cluster formation	Findings and future scope
Distance based clustering Routing scheme (2007) [4]	Probability based on: i. Distance between node and BS ii. Distance of farthest node from BS (dmax) iii. Distance from closest sensor node (dmin) iv. Ratio of Residual Energy to Initial Energy	• Energy-efficient load balanced clustering • More effective than LEACH (Low Energy Adaptive Clustering Hierarchy) & EECS (Energy Efficient Clustering Scheme) in prolonging the lifespan
New threshold assignment for LEACH and xLEACH (2010) [5]	i. Modification of threshold assignment in LEACH and xLEACH by introducing the distances of nodes from BS ii. Considered Median Distance as average of maximum and minimum distance of node from BS	• Longer lifetime and more uniform energy consumption in modified LEACH & xLEACH • **Future Scope:** To use "N" (Number of nodes) in threshold assignment
Unequal clustering scheme based LEACH (2010) [6]	i. Energy Ratio (of current energy to primary energy) and Competition Distance inserted in threshold assignment of LEACH ii. Adopted round-robin in CH election	• Better energy balancing • Prolonged network lifetime • Enhancement of network stability
LEACH-SC (LEACH-Selective Cluster) (2010) [7]	i. CH election is as per LEACH protocol ii. An ordinary node N will choose a CH which is nearest to the center point between N and the sink	• Reduces & balances overall energy consumption among sensors & extends lifetime of network as compared to LEACH
DECSA (Distance-Energy Cluster Structure Algorithm) (2012) [8]	i. A node becomes first round CH (False-Cluster-Head) if $r < T$. Where, r: random number r between 0 and 1 and T: Predefined Threshold ii. All nodes in cluster calculate threshold to elect CH i.e. k(i) [Node]. If k(i)[Node] > k(i) [False-Cluster-Head], then it becomes true CH, else False-Cluster-Head remains CH iii. k(i) is based on remaining energy and average distance of node i from all nodes in the same cluster	• The adverse effect on energy of CH caused because of non-uniform distribution of nodes in network is reduced • Direct communication between CH with low energy and far away from BS is avoided

(*continued*)

Table 1. (*continued*)

Algorithm & Year	Parameters for CH selection and cluster formation	Findings and future scope
LEACH-KED (Low-Energy Adaptive Clustering Hierarchy K-means Energy Distance) (2012) [9]	After clustering by K-means algorithm, CH is elected as per smallest cost calculated using: i. Weight values, α and β (weight of distance and surplus energy respectively) ii. Candidate CH position information iii. Distance of node from cluster's geometric center, maximum distance between any pair of nodes in current cluster iv. Distance between node and BS, and farthest distance of node to BS	• Selects the optimal node as CH • Uniform clustering • Extends the network life • Minimizes total network energy consumption
Distance Based Cluster Head Selection Algorithm (2012) [10]	i. CHs are selected as per net distance of each node from BS ii. Authors used matrix to store distance between two nodes	• Compared non-CH & CH based algorithm for energy & distance • **Future Scope:** To increase number of alive nodes
DBCP (Distance Based Cluster Protocol) (2013) [11]	i. Inserted initial energy and average distance between nodes and sink to calculate threshold	• Improves network lifetime and throughput • **Future Scope:** Introduce mobile nodes in the network
GEAR-CC (Centralized Clustering Geographic Energy Aware Routing) (2013) [12]	i. Clustering method is not fixed ii. Nodes don't know which cluster they belong to; they just pass on data to next node with specific ID without caring about routing protocols	• BS optimizes transmission mechanism for all nodes based on the global information of topology and energy • Avoids hot-spot phenomenon by balancing residual energy of each node
EAC (Energy Aware Clustering) (2013) [13]	i. Introduces the energy parameter for CH selection and distance parameter for non-CH to select CH	• Balances energy load & increases network lifetime • **Future Scope:** Compare EAC with other schemes & test it in suitable WSN test bed
DBCH (Distance Based Cluster Head Algorithm) (2015) [14]	i. Based on LEACH ii. New threshold calculation using: a. Node energy b. Distance of node from BS c. Distance of CH from BS	• Better node energy balance & enhanced network lifetime as compared to LEACH • **Future Scope:** Multi-hop routing

(*continued*)

Table 1. (*continued*)

Algorithm & Year	Parameters for CH selection and cluster formation	Findings and future scope
Routing algorithm based on Non Linear weight particle swarm optimization) (2015) [15]	i. BS selects CH with higher residual energy and better location according to location of BS ii. Cluster formation with well distributed nodes based on locations and residual energy	• Independent of node density • Uniform distribution of CH across the network
Clustering technique for WSN (2015) [16]	i. Node is selected as CH if its Residual Energy > Average Network Energy (Eligible CH List is prepared). Nodes join nearest CH	• Better network performance in terms of residual energy and number of dead nodes
MERA (Multi Clustered Energy Efficient Routing Algorithm) (2015) [17]	i. Network is partitioned into L clusters as per the distance from BS. Each cluster is further divided into L clusters ii. A node is nominated as Cluster Chain Leader (CCL) that communicates to its nearest CCL & finally to sink	• Better than PEGASIS (Power Efficient GAthering in Sensor Information Systems) • Achieves its maximum energy savings due to chain & communication between neighbors • **Future Scope:** Multiple mobile sensors and sink mobility
Manhattan distance approach (2015) [18]	i. Manhattan Distance (MD)	• Problem in data transmission due to obstacles is overcome
DACA (Distance Based Angular Clustering Algorithm) (2016) [19]	i. Each node builds its neighbor table with initial cost based on node distance, bandwidth & neighbor node ID ii. Random nodes are selected as the cluster centers iii. Each node joins the cluster according to smallest distance from centre node of cluster	• Removes problem of non-coverage area and path formation • **Future Scope:** Use swarm intelligence for the routing & aggregation & compare it against existing models
Energy resourceful distance based clustering & routing (2016) [20]	CH election probability based on: i. Average Energy of network ii. Residual Energy of node iii. Average Distance between nodes and BS iv. Position of a node from BS	• Better lifetime and scalability of the network

3.2 Clustering

Node distance plays a vital role in the cluster formation of RED. If transmission range is increased, then number of neighbors of a node will also increase. The cluster formation is as follows:

i. Each CH broadcasts join request along with its ID to nodes within its coverage area. Nodes within coverage area of CH will join that CH.
ii. If a node falls in coverage area of more than one CH, it calculates its distance from each one, and joins the CH with shortest distance. This is important because the more the distance, the more the energy consumption.
iii. Each node has 0 or 1 status. 0 means node is a part of some cluster. 1 indicates a free node. Only free nodes can participate in cluster formation.

Figure 1 shows cluster formation when coverage areas of CH overlap.

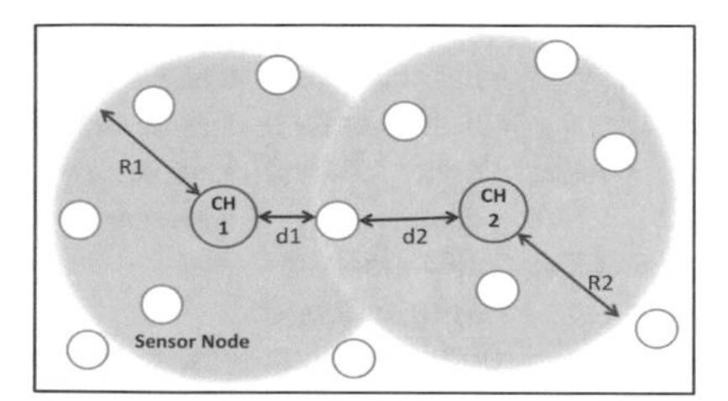

R1, R2 are the ranges of CH1 & CH2 and d1, d2 are distances of an overlapped node from CH1 & CH2. As d1<d2, it joins CH1.

Fig. 1. Cluster formation with minimum distance.

3.3 Reclustering with Self-organization

If a node remains CH for longer time till it becomes dead, the number of alive nodes in the network will fall drastically and lifetime of WSN will be reduced significantly. Changing the CHs near Sink region helps to solve energy hole problem. Once energy of CH E(CH) goes below Threshold Energy (T), that cluster is broken and nodes are made free (Node Status changes from 0 to 1). New CH is selected again on the basis of highest RE. Remaining nodes join new CH on the basis of their distances from new CH (Node Status again changes from 1 to 0). After reformation of cluster, it may happen that some nodes have d > R. But this problem is already solved in RED because of neighbor discovery.

4 Simulation and Results of RED

RED is simulated in NS2. Table 2 shows simulation parameters. Figure 2 shows the CH selection and cluster formation. Figure 3 shows the free nodes when cluster breaks.

Table 3 shows the CH and its member nodes before and after Reclustering. Figure 4 shows Reclustering.

Table 2. Simulation parameters for RED.

Parameters	Values
Area	2000 m × 1600 m
Channel type	Wireless
Number of nodes	40, 50,…110
Initial energy	100 J
Threshold energy (T)	85 J
Transmission range	350 m
Transmitting power	1 W
Receiving power	1 W
Simulation time	100 s

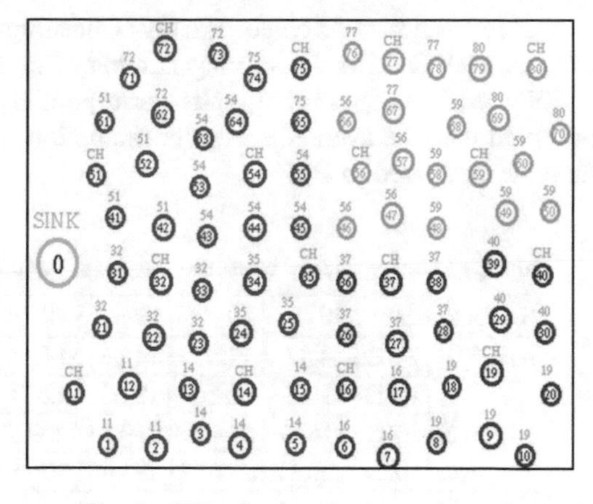

Fig. 2. CH selection & cluster formation.

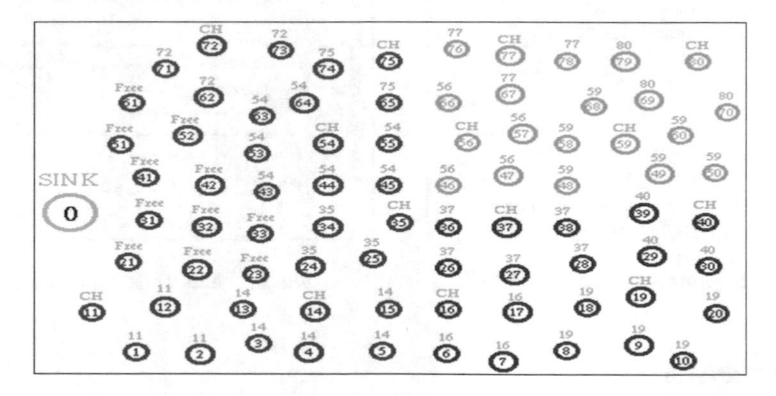

Fig. 3. Free nodes.

Table 3. Clustering and reclustering

	Before reclustering		After reclustering	
Cluster head ID	32	51	22	52
Cluster member node ID	21	41	21	41
	22	42	23	42
	23	52	31	51
	31	61	32	61
	33		33	

Fig. 4. Reclustering.

Table 4 shows Average Energy Consumption (AEC) and Network Lifetime (NL) of PRE & RED for varying number of nodes with transmission range 350 m.

Figures 5 and 6 show graphs corresponding to Table 4. From the graphs, it is observed that, the average energy consumption in RED is less and network lifetime is more as compared to PRE.

Table 4. Average energy consumption (AEC) and network lifetime (NL) (range: 350 m).

No. of nodes		40	50	60	70	80	90	100	110
AEC (J)	PRE	14.37	14.53	14.46	14.46	14.98	14.49	13.97	12.70
	RED	13.12	12.16	12.00	12.71	12.11	12.68	12.63	11.30
NL (Sec)	PRE	215.68	268.29	269.53	269.57	313.60	324.14	336.43	369.91
	RED	236.12	287.72	291.48	306.81	354.80	370.54	403.71	486.41

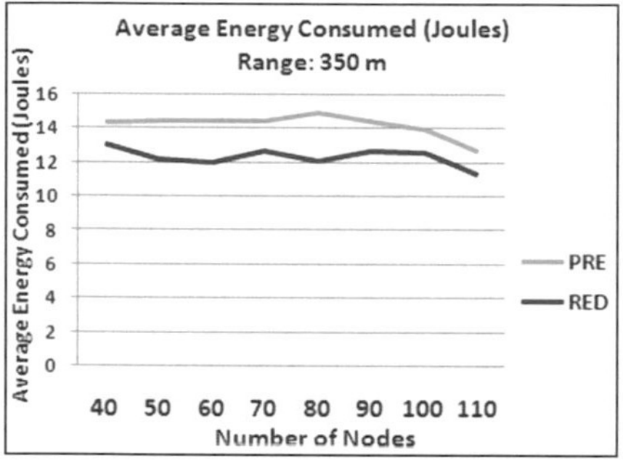

Fig. 5. Number of nodes v/s AEC

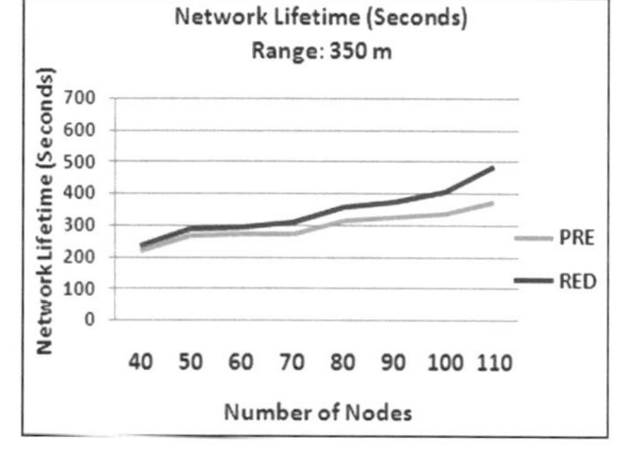

Fig. 6. Number of nodes v/s NL

5 Conclusion

The novelty of RED separates it from other state of the art approaches because most of them introduced distance factor into probability of CH selection. RED considers nearest distance for clustering. RED shows significant improvement in the results of PRE in terms of average energy consumption and network lifetime. RED further minimizes average energy consumption and extends network lifetime. In future, RED can be extended for implementation in real world applications.

References

1. Kumarawadu, P., Dechene, D.J., Luccini, M., Sauer, A.: Algorithms for node clustering in wireless sensor networks: a survey. In: 4th International Conference on Information and Automation for Sustainability, pp. 295–300 (2008)
2. Kumar, V., Jain, S., Tiwari, S.: Energy efficient clustering algorithms in wireless sensor networks: a survey. Int. J. Comput. Sci. Issues **8**(5), 1694–1814 (2011)

3. Chunawale, A., Sirsikar, S.: Minimization of average energy consumption to prolong lifetime of wireless sensor network. In: IEEE Global Conference on Wireless Computing and Networking (GCWCN), pp. 244–248 (2014)
4. Han, Y.J., Park, S.H., Eom, J.H., Chung, T.M.: Energy-efficient distance based clustering routing scheme for wireless sensor networks. In: International Conference on Computational Science and Its Applications, pp. 195–206. Springer, Berlin (2007)
5. Saadat, M., Saadat, R., Mirjalily, G.: Improving threshold assignment for cluster head selection in hierarchical wireless sensor networks. In: 5th International Symposium on Telecommunications (IST), pp. 409–414 (2010)
6. Ren, P., Qian, J., Li, L., Zhao, Z., Li, X.: Unequal clustering scheme based leach for wireless sensor networks. In: Fourth International Conference on Genetic and Evolutionary Computing (ICGEC), pp. 90–93 (2010)
7. Wang, J., Xin, Z., Junyuan, X., Zhengkun, M.: A distance-based clustering routing protocol in wireless sensor networks. In: 12th IEEE International Conference on Communication Technology (ICCT), pp. 648–651 (2010)
8. Yong, Z., Pei, Q.: A energy-efficient clustering routing algorithm based on distance and residual energy for wireless sensor networks, pp. 1882–1888. Elsevier, Amsterdam (2012)
9. Yunjie, J., Ming, L., Song, Z., Pengtao, D.: A clustering routing algorithm based on energy and distance in WSN. In: International Conference on Computer Distributed Control and Intelligent Environmental Monitoring (CDCIEM), pp. 9–12 (2012)
10. Kumar, B., Sharma, V.K.: Distance based cluster head selection algorithm for wireless sensor network. Int. J. Comput. Appl. **57**(9), 41–45 (2012)
11. Kumar, S., Prateek, M., Bhushan, B.: Distance based (DBCP) cluster protocol for heterogeneous wireless sensor network. Int. J. Comput. Appl. (0975–8887) **76**(9), 42–47 (2013)
12. Tang, B., Wang, D., Zhang, H.: A centralized clustering geographic energy aware routing for wireless sensor networks. In: 2013 IEEE International Conference on Systems, Man, and Cybernetics, p. 1 (2013)
13. Mohamed-Lamine, M.: New clustering scheme for wireless sensor networks. In: 8th International Workshop on Systems, Signal Processing & their Applications, pp. 487–491 (2013)
14. Sharma, R., Mishra, N., Srivastava, S.: A proposed energy efficient distance based cluster head (DBCH) algorithm: an improvement over LEACH. Procedia Comput. Sci. **57**, 807–814 (2015)
15. Wang, N., Zhou, Y., Liu, J.: An efficient routing algorithm to prolong network lifetime in wireless sensor networks. In: 10th International Conference on Communications and Networking in China (ChinaCom), pp. 322–325 (2015)
16. Desai, K., Rana, K.: Clustering technique for wireless sensor network. In: 1st International Conference on Next Generation Computing Technologies (NGCT), pp. 223–227 (2015)
17. Nayak, S.P., Rai, S.C., Pradhan, S.K.: MERA: a multi-clustered energy efficient routing algorithm in WSN. In: International Conference on Information Technology, pp. 37–42 (2015)
18. Gupta, S., Bhatia, V.: A Manhattan distance approach for energy optimization in wireless sensor network. In: 1st International Conference on Next Generation Computing Technologies (NGCT), pp. 203–206 (2015)
19. Kumar, N., Kaur, S.: Distance based angular clustering algorithm (DACA) for heterogeneous wireless sensor networks. In: Symposium on Colossal Data Analysis and Networking (CDAN), pp. 1–5 (2016)
20. Srividhya, V., Shankar, T., Karthikeyan, A., Gupta, P.: Energy resourceful distance based clustering and routing algorithm with competent channel allocation scheme for heterogeneous wireless sensor networks. Indian J. Sci. Technol. **9**(37) (2016)

An Automatic Segmentation Approach Towards the Objectification of Cyst Diagnosis in Periapical Dental Radiograph

Kavindra R. Jain[1,2](✉) and Narendra C. Chauhan[3]

[1] RKU, Rajkot, India
[2] Department of Electronics and Communication, G H Patel College of Engineering and Technology, V.V.Nagar, Anand, India
kavindrajain@gcet.ac.in
[3] Department of Information and Technology, A D Patel College of Engineering and Technology, New V.V. Nagar, Anand, India

Abstract. The crucial part of image segmentation is the proper selection of initial contour to start with the efficient process. The main reason for such kind of segmentation is to reduce the human interaction and moreover to have more accurate results. In this paper we trying to objectify the cyst diagnosis problem in periapical images with the help of automatic segmentation. We have utilized the internal and external energy of image forces that pull it toward features such as lines and edges, confining them precisely. Scale space continuation can be used to develop the catch area encompassing a component.

Keywords: Computer vision · Snakes model · Cyst · Dental radiograph · Internal and external energy of image

1 Introduction

The rapid advancement in Computer Vision have had a vast change on dental radiography. Digital intraoral Radiography when introduced to dentistry, various studies were being done for reliability, reproducibility and for the purpose of validation. Low level task of grey scale dental radiograph like edge and line detection have been utilized as autonomous process. Marr [1] in his view told that higher level information is not collected so far to undergo the process comprising of what is contained in the picture. In this paper author tries to reduce the energy as a tool for diagnosing the cyst development in decayed tooth based on periapical dental radiograph. Due to various shades of grey color available lots of noise information from the dental periapical radiograph creates problem in segmentation of cyst from the gum portion. So we need to improve the efficiency of the technique used to segment the periapical dental radiograph.

In the past few decades image fragmentation based on computer analysis has increasingly played a crucial role in the field of medical imaging. The medical images available for analysis are either corrupted by noise or sampling artifacts. To address

S.C. Satapathy and A. Joshi (eds.), *Information and Communication Technology for Intelligent Systems (ICTIS 2017) - Volume 1*, Smart Innovation, Systems and Technologies 83, DOI 10.1007/978-3-319-63673-3_20

these challenges deformable models been widely utilized as a part of therapeutic picture division. Deformable models are fundamentally bends or surfaces characterized inside a picture area that can move affected by inward powers, which are characterized inside the bend or surface itself, and outer strengths which are characterized from the picture information. The inward powers are intended to keep the model smooth amid mis-shapening. The outside powers are characterized to move the model toward a protest limit or other sought elements inside a picture. Obliging removed limits to be smooth and fusing other earlier data about the question shape. Deformable models offer heartiness to both picture clamor and limit crevices and permit incorporating limit components into a rational and steady numerical depiction. The initialization of segmentation starts with adding suitable energy terms to the process of minimization, it is possible for a user to push the model out of a local minimum toward the desired solution. The result is an active model that falls into the desired solution when placed near it. M. Kass [2] firstly introduced the basic model of snake or active contours in 1987, which are curves defined within an image domain that can move under the influence of internal forces coming from within the curve itself and external forces computed from the image data.

The next section discusses about the basic mathematical description of snakes. Section 3 describes the energy terms that can make a snake attracted to different types of important static, monocular features such as edges, lines and contours. Section 4 concentrates on the suggested approach for the cyst identification in periapical dental radiographs. Section 5 simulates the results and Sect. 6 concludes the paper.

2 Basic Snake Behavior

The crucial snake model is a controlled congruity spline influenced by picture qualities and outside impediment powers. The internal spline forces to constrain a piece adroit smoothness impediment. The image powers highlights like lines, edges, and subjective structures. The external basic qualities are accountable for putting the snake near the pined for neighborhood minimum. These forces can, start from a UI, modified attentional instruments. The shape is depicted in the (x, y) plane of a picture as a parametric bend.

$$V(s) = (x(s),\ y(s)) \tag{1}$$

Shape is said to take a vitality (Esnake) which is characterizes as the aggregate of three vitality terms.

$$E_{snake} = E_{\text{internal}} + E_{external} + E_{constraint} \tag{2}$$

The vitality terms are portrayed astutely in a way with the end goal that the last area of the form will have a base vitality (E min). Thusly our trouble of acquiring articles reductions to a vitality minimization trouble.

Internal Energy (E_{int})–Internal energy depends on the intrinsic property of the curve and addition of elastic energy and bending energy.

Elastic Energy ($E_{elastic}$)–The curve is care for as an elastic rubber band possessing elastic potential energy. It dispirit extending by introducing tension.

$$E_{elastic} = \frac{1}{2}\int_{s} \alpha(s)|v_s|^2 ds \quad (3)$$

where

$$v_s \equiv \frac{dv(s)}{ds} \quad (4)$$

Weight α(s) permits us to control elastic energy beside diverse parts of the contour. α basically utilized for reducing the size of region of interest.

Bending Energy ($E_{bending}$)–The snake is additionally considered to carry on like a thin metal strip offering ascend to bowing vitality. It is characterized as whole of squared bend of the shape.

$$E_{bending} = \frac{1}{2}\int_{s} \beta(s)|v_{ss}|^2 ds \quad (5)$$

Total internal energy of the snake can be defined as

$$\begin{aligned} E_{\text{int}} &= E_{elastic} + E_{bending} \\ &= \int_{s} \frac{1}{2}(\alpha|v_s|^2 + \beta|v_{ss}|^2)ds \end{aligned} \quad (6)$$

External energy of the contour (E_{ext})–It is obtained from the image. Characterize a capacity E image (x,y) with the goal that it acquire on its littler qualities at the components of intrigue, for example, limits

$$E_{ext} = \int_{s} E_{image}(v(s))ds \quad (7)$$

The problem at hand is to find a contour v(s) that reduce the energy functional

$$E_{snake} = \int_{s} \frac{1}{2}(\alpha|v_s|^2 + \beta|v_{ss}|^2)ds + \int_{s} E_{image}(v(s))ds \quad (8)$$

Using difference calculus and by concerning Euler-Lagrange differential equation we obtain following equation

$$\alpha v_s - \beta v_{ssss} - \nabla E_{image} = 0 \tag{9}$$

The contour deforms underneath the exploit of these forces. The improvement done in snake model for medical image segmentation is required for better objectification of cyst in intraoral dentistry. Hence the extraction of cyst in periapical images would be easier and much more accurate.

3 Active Contour Model

The basic theory behind active contour model lies on a curve whose behavior is going to be constrained by two aspects. The first and most important is due to the objective which is to perform a segmentation based on object and shape detection. So we need our contour to converge to the edges of the object we are interested in. The second constraint lies on the model of the contour we want to have. Due to the description we made before our contour function is going to be called $\gamma(s)$.

3.1 External Energy

The external energy is the component of our behavior function that describe how the deformable curve will match with objects of the image. To be attracted by a shape, we need to use a function that have the following properties: have at least one local minimum and be monotonic on areas around this point. To have such a function in an image we are going to use its gradient. We will enforce this behavior with a preprocessing consisting in a Gaussian Smoothing of the image to improve this aspect. So we can express the external energy function as:

$$E_{ext} = P(\gamma(s)) \tag{10}$$

Where P stands for a potential attraction field onto the edge of an object. So we have to consider this energy not only locally but for the whole contour C and to plug in the previous equation the properties we mentioned above. If the contour is closed we have:

$$E_{ext} = \oint_C \|\nabla I\|^2(\gamma(s))ds \tag{11}$$

Where I is representing the input image and ∇ is the spatial gradient function defined by:

$$\nabla I = \left(\frac{\partial I}{\partial x}, \frac{\partial I}{\partial y}\right) \tag{12}$$

Due to the fact that we want to minimize this energy, we are going to take the opposed value and introduce our Gaussian smoothing to enforce convergence to a local minimum. Using a weighting parameter on the external energy will allow us to increase the "visibility" of the gradient field by the snakes. So we can write it:

$$E_{ext} = -\delta \int_{A}^{B} \|\nabla(G_n * I\|^2(\gamma(s))ds \quad (13)$$

where δ is a real weighting value which for obvious reason would be positive and G_n is a Gaussian weighted kernel of dimension n.

3.2 Internal Energy

The internal energy is the component of the behavior function that describe the physical properties of our contour like smoothness or continuity and curvature. It is composed of two terms, the first one is describing the contour behavior regarding elasticity or smoothness. The second term is describing the curvature model of the curve and is function of the second derivative of our contour. If we put that into a mathematical form, we have:

$$E_{\text{int}} = f(\gamma'(s)) + g(\gamma''(s)) \quad (14)$$

The functions f and g are just going to be the Euclidean norm of the function. Then we need to specify that we want the energy for the whole curve C so not only for one spatial locations so we are going to sum this energy along the curve. Which will lead to two different cases. If our contour is closed:

$$E_{\text{int}} = \oint_{C} \|\gamma'(s)\|^2 + \|\gamma''(s)\|^2 ds \quad (15)$$

The previous definition is defining a model of smooth curve or function to describe the contour's behavior, which goes through the use of its derivative and in our case in their minimization. But, depending on the application and the choices of the user, the influence of these two aspects are not always equally important. So they need to be adjustable according to the situation.

3.3 Optimisation

Our goal in this section is to find the optimal parameters that will minimize the previous energy function we defined. This parameters are position vectors that will define the position of the snake which minimize this energy. Indeed we are looking for $\gamma(s) = (\gamma x, \gamma y)(s)$ such as E_{snake} is minimal. The optimization formulation of our problem is then:

$$s_{optimal} = \arg\min_{\gamma \in F} E(\gamma(s)) \quad (16)$$

Which means that we want to find the value of s which is the argument that makes the curve γ, of the set of possible curve, of minimal value regarding the energy function E.

3.4 Gaussian Smoothening

An important part of the set up to perform a segmentation using active contour is to preprocess the image using a Gaussian filtering. If our input image has a good quality, its edges are going to be extremely well defined, which means that there are going to be abrupt. In fact, we need to attract progressively the snake to the edge of the object, so we need to "attract" it. To do that we need to "spread" the edges around to create a more smoothed gradient around the edge. Another reason of applying a Gaussian filtering is regarding the noise issue.

3.5 Kernel Settings

The kernel size and type is really important as we mentioned before because it influences the way the edge is spread around. It also as a consequence of our simple convolution implementation reduce the size of the resulting image. This aspect can of course be noticed while looking at the intensity curve of the smoothed image where we can clearly see null areas on the sides of the intensity curve.

3.6 Contour Construction

As we said before, we are going to rely on the user to define an initial shape around the object that will serve as an initialization set up.

3.7 Stopping Criterion

In the original method presented by Kass et al. It seems to have no indications about a stopping criterion on the snake's evolution. When using the greedy method we have two stopping criterion. The first one is if the number of points that move at each iterations fall under a certain threshold value. The other one being that the number of iterations reaches another threshold value. Indeed, if we initialize the contour too far from the shape with a small kernel and a too low iteration threshold the algorithm will fail in reaching the shape. On the other hand if we have a too large number of iterations and a too close initialization we will spend time looking a contour that oscillate in the edge origin without significant improvements. This is maybe where the criterion can become useful.

4 Suggested Approach

The suggested approach is followed by below steps:

- Collection of periapical database images comprising of cyst diseases.
- Fix all the values of parameters so described in snakes model
- Initialize the contour by clicking at a sequence of points close to the object to be segmented, the contour is then defined as the set of consecutive points.
- Design a stopping criteria for the iterative process.
- Choose different initializations for starting contour, also with the different number of control points, and observe the evolution of the deformable contour.

After completing above procedure, we find out the edge of the affected part using canny afterward we measure the area of that part. For severity measurement, geometrical and textural features must be analyzed (Figs. 1, 2, 3 and 4).

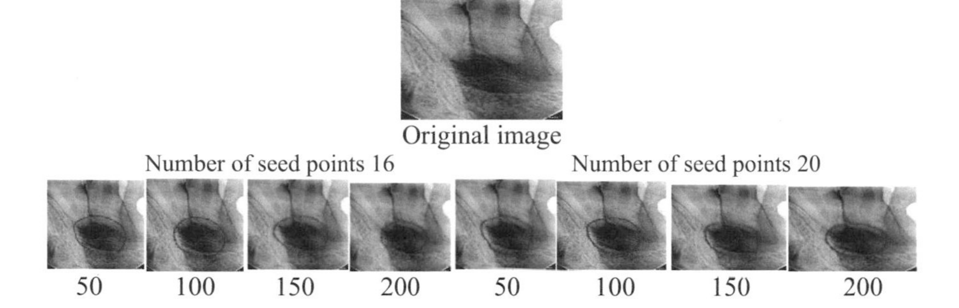

Fig. 1. Cyst identification of test 1 (patient 1) with various iterations and selected number of points

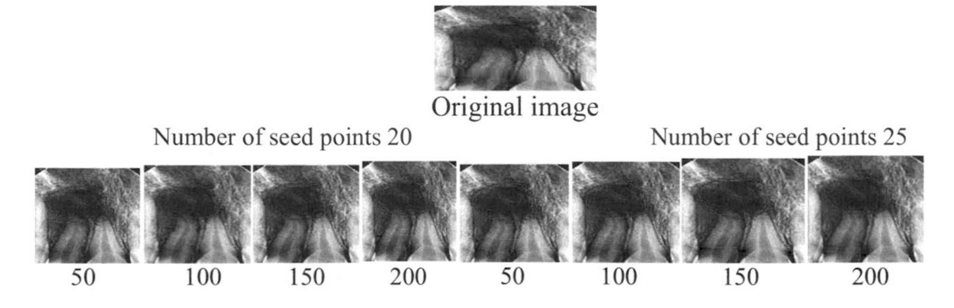

Fig. 2. Cyst identification of test 2 (patient 2) with various iterations and selected number of points

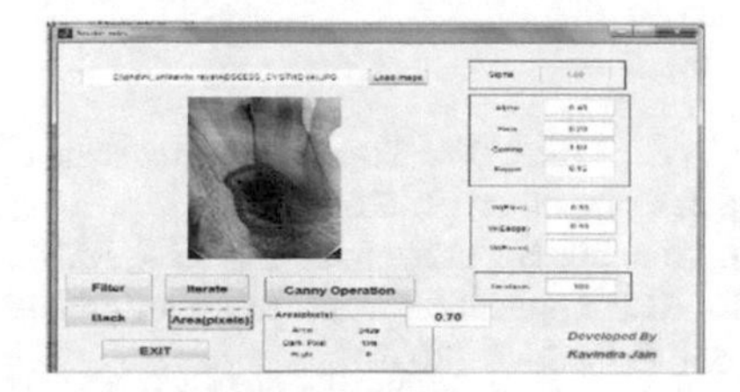

Fig. 3. Cyst identification of test 1 (patient 1) with various iterations and selected number of points (A GUI)

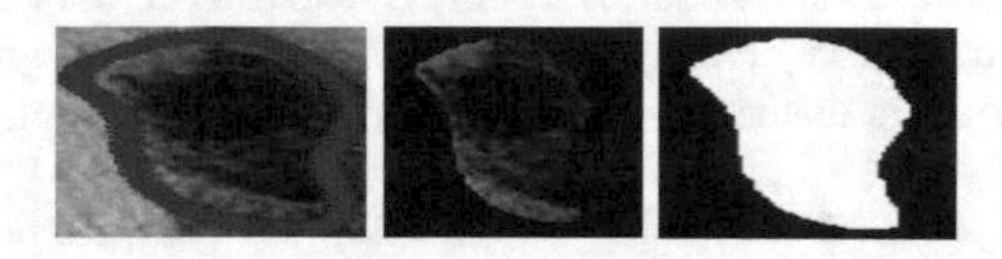

Fig. 4. Cyst identification of test 1 (patient 1) with filtering and final binaries detected output of resultant image

5 Result and Discussion

As described in Sect. 4 we have run the code on various database images and their results will be attached here. In this, various parameters α,β,γ, $W(E_{line})$, $W(E_{edge})$, $W(E_{term})$ are utilized for getting affected area of disease on various database images as shown in Table 1.

Table 1. Parameters and values for different patients and test images based on snakes model

Parameter	α (alpha)	β (beta)	γ (gamma)	k (kappa)	$W(E_{line})$
Values	0.4	0.2	1.0	0.15	0.3

6 Conclusion

We implemented GUI (Graphical User Interface) in MATLAB. In this GUI, we added some features like area, canny. This user friendly approach was quite satisfactory in providing a broad idea to doctor as well as patient for a precancerous treatment. As this tool has all the add-on to give a broad descriptive area of the affected patient with

The importance of the medical image in healthcare is constantly growing, making healthcare more effective and patient friendly. With innovative imaging technologies diseases can be detected earlier and more precisely: more useful for both patient and doctors.

Acknowledgment. I am highly indebted and thankful to Dr. Dhrumin Patel (G-3, swastik Apt., Opp Sub Jail, Near Bus Depo, Luncikui, Navsari-396445) for his massive support and suggestions for the purpose of my research work.

References

1. Kulkarni, G.A., Bhide, A.S., Patil, D.G.: Two degree greyscale differential method for teeth image recognition. Int. J. Comput. Appl. (2012)
2. Yousefi, B., Hakim, H., Motahir, N., Yousefi, P., Hosseini, M.M.: Visibility enhancement of digital dental X-ray for RCT application using bayesian classifier and two times wavelet image fusion. J. Am. Sci. **8**(1), 7–13 (2012)
3. Said, E.H., Fahmy, G., Nassar, D., Amar, H.: "Dental X-ray image segmentation" biometric technology for human identification. In: Proceedings of the SPIE, vol. 5404, pp. 409–417 (2004)
4. Dighe, S., Shriram, R.: Preprocessing, segmentation and matching of dental radiographs used in dental biometrics. Int. J. Sci. Appl. Inf. Technol. **1**(2), 52–56 (2012)
5. Said, E.H., Nassar, D.E.M., Fahmy, G., Ammar, H.H.: Teeth segmentation in digitized dental X-ray films using mathematical morphology. IEEE Trans. Inf. Forensic Secur. **1**(2), 178–189 (2006)
6. Sc, W., Pharoah, M.J.: Oral radiology—principles and interpretation, 5th edn. Selected illustration, O'Connor, D.: MOSBY (An affiliate of Elsevier) (2005). ISBN 0-323-02001
7. Rajendran, A., Sivapathasundharam, B.: Shafer's Tb. Textbook of Oral Pathology, 6th edn. Elsevier, India (2006)
8. Omanovic, M., Jeff, J.: Orchard exhaustive matching of dental X-rays for human forensic identification. J. Can. Soc. Forensic Sci. **41**(3), 1–11 (2008)
9. Jadhav, S., Shriram, R.: Dental biometrics used in forensic science. J. Eng. Res. Stud. **3**(1), 26–29 (2012)
10. Kass, M., Witkin, A., Terzopoulos, D.: Snakes: active contour models. Int. J. Comput. Vis. **1**, 321–331 (1988)
11. Michel, S., Koller, S.M., Ruh, M., Schwaninger, A: Do "image enhancement" functions really enhance X-ray image interpretation? In: Proceedings of the 29th Annual Cognitive Science Society, pp. 1301–1306 (2007)
12. Desai, N.P., Prajapati, D.B.: A simple and novel CBIR technique for features extraction using AM dental radiographs. In: CSNT (IEEE) 2013, Gwalior, pp. 198–202 (2012)

Real-Time Framework for Malware Detection Using Machine Learning Technique

Sharma Divya Mukesh[1(✉)], Jigar A. Raval[2], and Hardik Upadhyay[3]

[1] GTU-PG School, Ahmedabad, Gujarat, India
sharmadivya666@gmail.com
[2] Physical Research Laboratory, Ahmedabad, Gujarat, India
jigar@prl.res.in
[3] GPERI, Mehsana, Gujarat, India
hardik31385@gmail.com

Abstract. In this epoch, current web world where peoples groups are associated through correspondence channel and the majority of their information is facilitated on the web associated assets. Thusly the security is the significant concern of this internet community to protect the resources and to ensure the assets and the information facilitated on these networks. In current trends, the greater part of the end client are depending on the end security items, for example, Intrusion detection system, firewall, Anti-viruses etc. In this paper, we propose a machine learning based architecture to distinguish existing and recently developing malware by utilizing network and transport layer traffic features. This paper influences the precision of Semi-supervised learning in identifying new malware classes. We show the adequacy of the framework utilizing genuine network traces. Amid this research, we will execute and design the proactive network security mechanism which will gather the malware traces. Assist those gathered malware traces can be utilized to fortify the signature based discovery mechanism.

Keywords: Malware detection · Semi-supervised algorithm · ClamAV · Machine learning

1 Introduction

1.1 Malware Discovery from the Linux System

Malwares are a typical channel by which crackers encourage cybercrime. It has turned into a weapon whose utilization meets different dangers defied by security analysts over the globe. The crackers continue improving the intricacy of malcode to fulfill their notorious aims. Malware is a nonexclusive term to mean a wide range of undesirable programming (e.g., backdoors, virus, worms, spyware and Trojans) [5]. Various assaults made by the malware have represented a noteworthy security risk to PC clients. In this manner, malware identification is one of the computer security points that are of incredible intrigue.

As of now, the most huge line of guard against malware is hostile to infection programming items, for example, Kingsoft's Antivirus, Dr. Web and Norton.

S.C. Satapathy and A. Joshi (eds.), *Information and Communication Technology for Intelligent Systems (ICTIS 2017) - Volume 1*, Smart Innovation, Systems and Technologies 83, DOI 10.1007/978-3-319-63673-3_21

These generally utilized malware detection software apparatuses essentially utilize signature-based strategies to perceive dangers. Signature is basically a short series of bytes that are special for each existing malware so that future instances of it can be effectively ordered with a less error rate. In any case, this exemplary signature-based based strategy dependably neglects to identify variations of known malware or already obscure malware, in light of the fact that the malware essayists dependably embrace methods like obfuscation to bypass these signatures.

Various gadgets are running Linux because of its flexibility and open source nature. This has made Linux stage the objective for malware assaults, so it gets to be distinctly critical to identify and dissect the Linux malware. Today, there is have to break down Linux malwares in a mechanized approach to comprehend its capacities [3].

1.2 Commitment of Research Paper

With a specific end goal to manufacture compelling, automatic, and interpretable classifiers for malware recognition from the substantial, confused and unlabeled list, we need to address the accompanying difficulties:

- **How to develop a successful classifier to distinguish malware from the unlabeled dataset?** How to make the classifier less touchy to the lopsidedness and perform well for the vast and muddled dataset?
- **How to make the classifier interpretable?** The classifier ought to produce learning or patterns that are simple for the malware investigators to comprehend and interpret.
- **How to productively assemble the classifiers?** In our application, we have an aggregate of 5000 labeled file samples from ClamAV that are accessible for training: half of them are malware tests and the other half are amiable record tests. Testing strategies are expected to construct classifiers for such a vast information gathering to keep away from over-fitting and accomplish extraordinary viability and additionally high proficiency. In any case, the decisions of the class circulation and the extent of the training data in the inspecting technique for malware detection are not inconsequential and require careful examination.

In this paper, we depict our examination push to address the above difficulties. To address the interpretability issue, we fabricate various leveled classifiers since they can produce decides that are simple for malware investigators to comprehend and translate.

1.3 Content of the Paper

The straggling leftovers of this paper is dealt with as takes after. Sections 2 discuss about the machine learning information, Sect. 3 exhibits the outline of malware recognition framework architecture and Sect. 4 examines the related work. In Sect. 5, we portray the simulation study of malware detection framework. At long last, Sect. 6 concludes.

2 Machine Learning Information

Machine learning is a method for analysis of statistics that automates illustrative model constructing. Making use of calculations that iteratively benefit from facts, machine mastering allows systems to find concealed bits of statistics without being unequivocally tweaked wherein to look [10].

There are different usages of machine learning. It's very to recognize how much machine learning has finished in genuine applications. Machine learning is consistently associated in the disconnected training phase. Accordingly machine learning is used to improve the applications, for example, face recognition, face detection, speech recognition, genetics, image classification, weather forecast etc. Machine learning is associated in malware detection & classification to enhance the accuracy of the malware detection rate estimate. Machine learning makes it decently less requesting to make complex programming systems without much effort on the human side.

Underneath specified are essential six assignments required in Machine Learning process.

- **Classification** – It will categories new information in predefined trainingvia locating prescient getting to know characteristic.
- **Clustering** – Identify likeness in information and shape gatherings or groups
- **Summarization** – Locating a minimized illustration for a hard and fast or subset of records
- **Regression** – Discovering prescient learning capacity, which models information with the slightest mistake.
- **Dependency modeling** – Locating dating between elements or their features in a dataset.
- **Anomaly recognition** – Identify the most critical changes or mistakes in information set.

Data mining process utilizes framework acing calculations relying upon whether the class names are accommodated becoming more acquainted with, these calculations might be isolated into classifications supervised learning or unsupervised becoming acquainted with.

3 System Architecture

The inspiration to propose this approach is, about sixty percentage interruption and security infringement are inside the association. The security reports created analyser indicate obviously. If the traffic is separated into various classification astute, it will be more complex for further examination too. This building square would be same for open system traffic and authoritative. It is half breed behaviour and intrusion detection framework.

There are basic three phases of proposed architecture. This three phases are imagined as key segment of an expansive end security framework that screens the system stream and choosing whether it is malignant or amiable. Figure 1, demonstrate the proposed

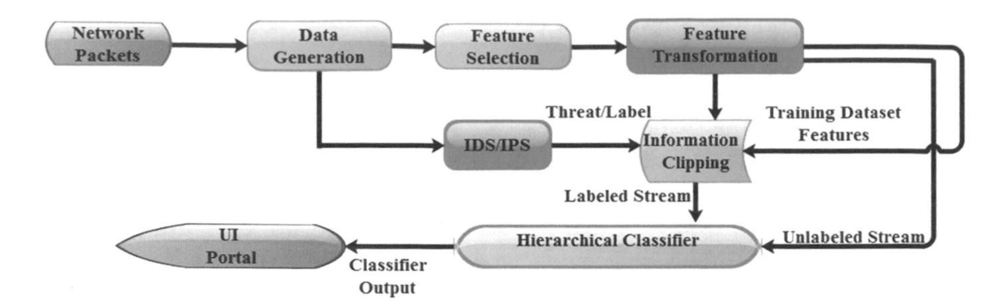

Fig. 1. System architecture

architecture, which consist three noteworthy segments: **Signature Based Malware Detection, Behaviour Based Malware Detection and Machine Learning Phase**

3.1 Phases of Proposed Architecture

First Phase: Signature Based Malware Detection
This module would accumulates every one of the packets from the network and match the approaching system activity with intrusion detection system signature for instance, approaching movement is tried on Clam-AV open source antivirus having different malware signatures (Fig. 2).

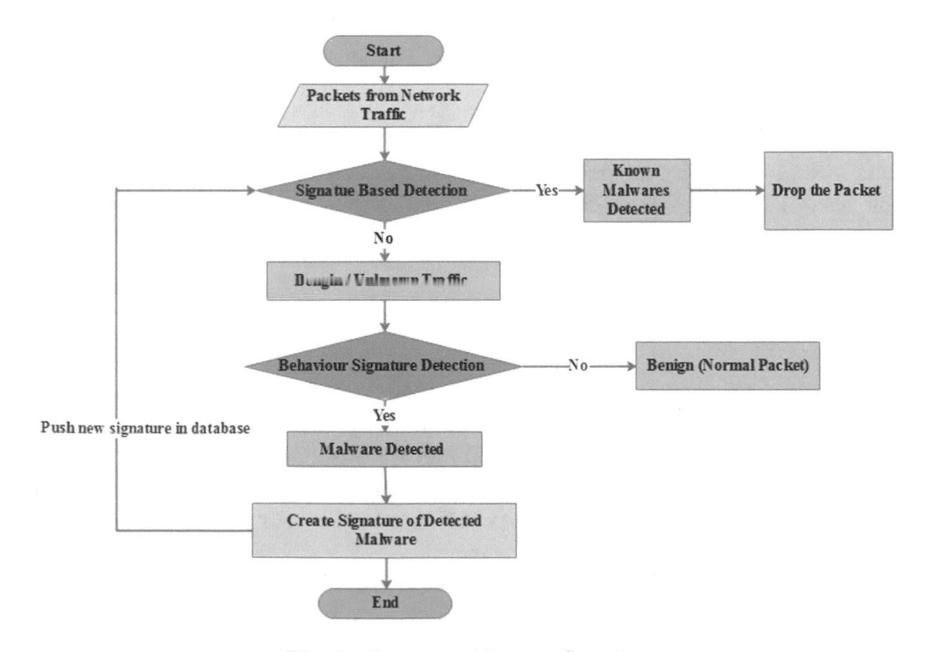

Fig. 2. System architecture flowchart

Second Phase: Behaviour Based Malware Detection
Intrusion detection and intrusion prevention system would perform DPI of packets after that it would names the stream conclude that it belongs to some existing threats On the off chance that the approaching traces will sidestep the Clam-AV then will consider that network traces for behavior based detection in which include feature selection is accomplished for incoming network traffic. That traffic would be consider as level to classifier in hierarchical class fifer phase of system architecture
Packet network packets and collect the value of network layer 3 and layer 4 features and gathering them in light of same stream. Features like packet per flow, bytes per second, packet inter arrival times and payload size.

Third Phase: Machine Learning Phase

Step 1: Input the information which is network packets.
Step 2: Parsing of network packet to collect feature selection.
Step 3: Construct hyper sphere based on view of classes saw in preparing set.
Step 4: Compute distance of test instance focal point of each hyper sphere.
Step 5: If the test instance lie outside the hyper sphere then class is anticipated as unknown malware sample.
Step 6: If the test instance is variant of existing malware then, it will live just inside a solitary hyper sphere.
Step 7: Compare the distance profile of the instances to all hyper sphere.

This structure utilizes Classifier work with Class-1 Support vector machine algorithm. This is semi supervised machine leaning algorithm, where each model invests huge energy in requesting cases from a specific malware class. This would have an ability to perceive new malware besides.

4 Related Work

An essential evaluation of the work has been completed as such far for malware recognition to show how the present review identified with what has as of now been finished. In light of the investigation of the security gadgets and their working capacity, it is important to put the proactive security instrument rather than the signature based techniques such as firewall, Network Intrusion Detection System (NIDS), etc. Since the identification algorithm of these network intrusion detection system depends on the how signature based antivirus apparatuses distinguish malevolent exercises. All these signature based gadgets are depend with respect to the pre-decided lead sets as assault database that as of now been distinguished and recorded. This leaves in the condition of known as obscure assaults. Thusly they won't have the capacity to distinguish the obscure assaults which bargain the PC framework. In view of the previously mentioned methods, there is a necessity to set up a proactive based security tried that would be capable review the whole system and in the meantime gather the assault follows which is known and obscure to them. The unknown attack traces can be utilized to fortify the network intrusion detection system, prevent intrusions from exposing and exploiting the vulnerabilities of the said network. One of such proactive system security instrument is as honeypots. Since honeypots are misleading frameworks, it will be exceptionally helpful secluded from everything the genuine estimation of the information that disregard the system.

In this section, we cover research efforts claim to detect malware variants. We group the research technique based into two groups'. Namely, they are signature-based, behavior-based.

4.1 Behavior Based Detection

API calls and in addition System calls are likewise utilized by for malware detection utilizing behaviour based malware detection approach (Mamoun Alazab et al. 2012, Hisham Shehata Galal, Yousef Bassyouni Mahdy et al. 2015, Stavros D. Nikolopoulos, Iosif Polenakis et al. 2016), perform well for classification by concept of dynamic analysis technique to extract API details of latest malware dataset inside controlled environment, API hooking technique is used means to trace the API calls invoked my OS. That traces will function as feature that is called as action. All this actions are then classified based on various algorithms. This model fails in front of malware samples that checks for the existence of virtual machine artifacts. For future improvement is needed the behaviour of malware actions, data flow dependence can be used which will give more insight for malware identification. While in the case of malware detection by considering system calls disadvantage is that grouping of system call to system calls groups leads to loss of information leading to higher rate false positive and Unknown malware to known malware failure rate is 83.42% i.e. classification rate [2, 8, 10].

4.2 Signature Based Detection

Keeping in mind the end goal to conquer the impediment of the generally applied used signature primarily based malware detection method, facts mining and system getting to know approaches are proposed for malware detection (Sachin Jain, Yogesh Kumar Meena et al. 2011; Han-Wei Hsiao, Deng-Neng Chen, TsungJu Wu et al. 2011), it is shown vindictive internet site detection, and it utilizes the spatial- temporal aggregating factors way to deal with detection system from Net Flow data For arrangement of packets it utilizes the induction learning technique on the way to differentiate between malicious and normal traffic pattern. Hence through the utilization of various variables it had been presumed that Net flow variable will produce more awful detection accuracy rate, combination on spatial and temporal variable may have greater accuracy rate [1, 7, 9]

In spite of the fact that Semi-supervised classification strategy performs well on extensive information gathering, it should be further explored for managing class imbalanced issue.

5 Simulation Study

The experiment of proposed system architecture has been portrayed in this section. Underneath Fig. 3, demonstrates the network architecture setup to implement propose malware detection architecture.

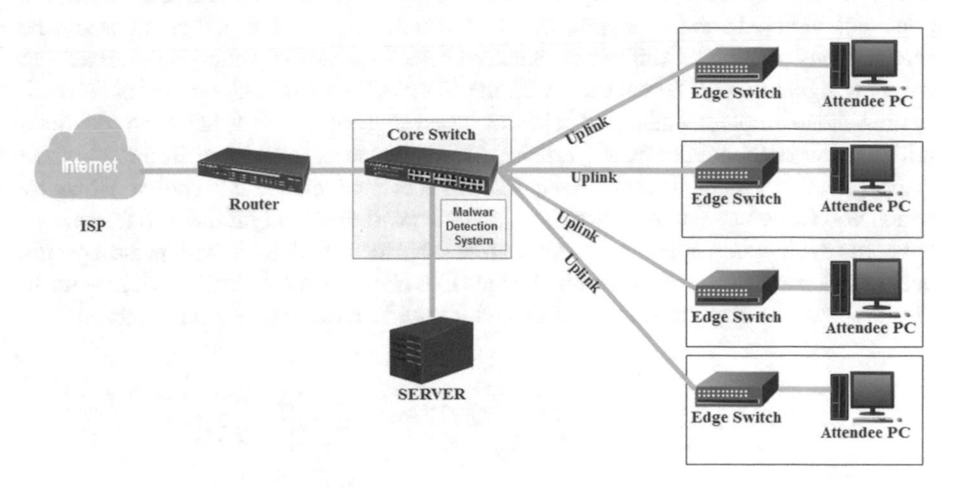

Fig. 3. Network architecture

We have utilized ClamAV for implementing signature based malware detection module which have essentially three database in which signature are stored in hex patterns, MD5 patterns and LDB signatures this are overhaul by utilizing freshClam command after establishment of ClamAV in the system.
The following are the steps which demonstrates to setup ClamAV:

Step 1: sudo apt-get install clamav
Step 2: sudo freshclam
Step 3: sudo nano freshclam.conf

Figure 4, demonstrates how existing signature are been utilized to coordinate the malware yet by utilizing this strategy zero-day malware whose signature doesn't not exist, can't been identified so its testing errand to beat this issue, by utilizing behavior based technique it would be supportively to detect new malware and variants of new malware.

```
root@kali:~/ForensicsTools/clamav/sigs/test# clamscan -d sig.ndb ../../malwarez/sploit.exe
../../malwarez/sploit.exe: Exploit.RingZ.UNOFFICIAL FOUND

----------- SCAN SUMMARY -----------
Known viruses: 1
Engine version: 0.97.8
Scanned directories: 0
Scanned files: 1
Infected files: 1
Data scanned: 0.05 MB
Data read: 0.05 MB (ratio 1.00:1)
Time: 0.014 sec (0 m 0 s)
```

Fig. 4. Scan Output from ClamAV

So as to overcome the challenge of signature based malware detection will use behavior based malware detection approach that is by considering feature selection and transformation of network traces which is unlabeled data set. This feature are consider as testing data. When the malicious traffic is encountered then it is being transferred to real database server then it will drop the request if signature matches with the existing signatures of malware. If the match is not found then the traffic is redirected towards the honeypot server. In the honeypot, the request will be executed and result would be screen, if any pernicious traffic is experienced here, then new signature is generated and that new signature will be interface with the Clam-AV or any IDS system. underneath figures shows implementation of behavior based malware. And It has been identified utilizingGlastopfhoneypot when remote file inclusion attack is performed on live website. In this we had upload lab.php file which is fundamentally backdoor Trojan by which we can bypass authorization. This malware will work only if the web framework have file inclusion vulnerability. Once malware identified then will parse specific network packet to identify behavior changes like client to server load or vice versa. In Fig. 5 shows the implementation of behavior based malware detection approach

Fig. 5. Remote file inclusion attack using malware

In Figs. 6, 7, graph shows the comparisons of existing and proposed system.

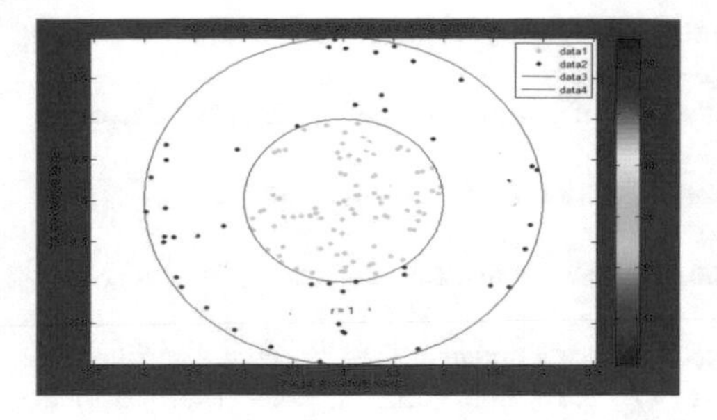

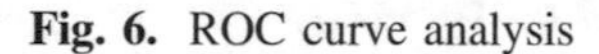
Fig. 6. ROC curve analysis

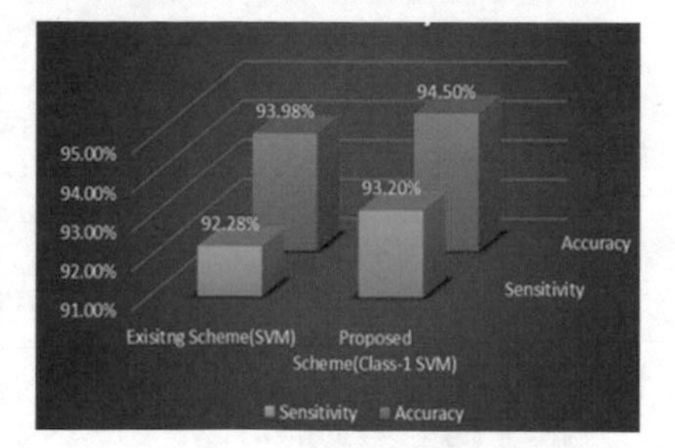

Fig. 7. Result analysis

6 Conclusion and Future Scope

To identify malicious packets, this paper present a new machine learning approach which depend on classification methodologies for detecting malware behavior. Features discover from the network and transport layer network stream attributes. In spite of the fact that the components are stronger for encrypted payload, it would have different concern which would hamper the appropriateness of extended complex learning algorithms. Besides, the supervised algorithm tough for distinguishing new variants of malware. Our proposed method addresses those difficulties and recognize streams of present and novel malwares with excessive accuracy. At last, we display a novel adjustment of class one Support vector machine to distinguish unheard-of malware.

Meanwhile, there are still many works to do in future. Currently, focuses on capturing only non-encrypted traffic. For destiny work, we strive to extend the method to SSL network traces and additionally do on line getting to know placing. Likewise, whilst the amount of training turns out to be significantly extensive, the forecast step is expensive due to the huge wide variety of hyper spheres that ought to be tried. To deal with this trouble, we plan to build up a revolutionary multiclass learning method.

References

1. Chuan, L.L., et al.: Design and development of a new scanning core engine for malware detection. In: 2012 18th Asia-Pacific Conference on Communications (APCC). IEEE (2012)
2. Hsiao, H.W., Chen, D.N., Wu, T.J.: Detecting hiding malicious website using network traffic mining approach. In: 2010 2nd International Conference on Education Technology and Computer, vol. 5. IEEE (2010)
3. Sochor, T., Zuzcak, M.: High-interaction linux honeypot architecture in recent perspective. In: International Conference on Computer Networks. Springer International Publishing (2016)
4. Bazrafshan, Z., et al.: A survey on heuristic malware detection techniques. In: 2013 5th Conference on Information and Knowledge Technology (IKT). IEEE (2013)
5. Galal, H.S., Mahdy, Y.B., Atiea, M.A.: Behavior-based features model for malware detection. J. Comput. Virol. Hacking Tech. **12**(2), 59–67 (2016)

6. Saeed, I.A., Selamat, A., Abuagoub, A.M.: A survey on malware and malware detection systems. Int. J. Comput. Appl. **67**(16), 25–31 (2013)
7. Ahmed, Irfan, Lhee, Kyung-suk: Classification of packet contents for malware detection. J. Comput. Virol. **7**(4), 279–295 (2011)
8. Anderson, B., et al.: Graph-based malware detection using dynamic analysis. J. Comput. Virol. **7**(4), 247–258 (2011)
9. Jain, S., Meena, Y.K.: Byte level n–gram analysis for malware detection. Comput. Netw. Intell. Comput. **157**, 51–59 (2011)
10. Alazab, M., et al.: Zero-day malware detection based on supervised learning algorithms of API call signatures. In: *Proceedings of the Ninth Australasian Data Mining Conference-Volume 121*. Australian Computer Society, Inc. (2011)
11. Nikolopoulos, S.D., Polenakis, I.: A graph-based model for malware detection and classification using system-call groups. J. Comput. Virol. Hacking Tech **13**(1), 1–18 (2016)

A Research Direction on Data Mining with IOT

Foram Chovatiya(✉), Purvi Prajapati, Jalpesh Vasa, and Jay Patel

Charotar University of Science and Technology, Anand, Gujarat, India
foramchovatiya@gmail.com, jalpeshvasa.it@gmail.com,
{purviprajapati.it, jaypatel.it}@charusat.ac.in

Abstract. The mission of connecting everything on the earth together via internet seems to be impossible. There will be the great effect on human life by Internet of Things (IOT), because with the help of IOT, many impossible things will become possible. IOT devices generates big data having useful, valuable and highly accurate data. It is difficult to extract the required information or data from the set of big data discovered by any device. For this purpose, data mining is used. Data mining will plays important role in constructing smart system that provides convenient services. It is required to extract data and knowledge from the connected things. For this purpose, various data mining techniques are used. Various algorithms such as classification, clustering, association rule mining etc. helps to mine data. This paper represents the different Data mining techniques, challenges, and Data mining issues with IOT.

Keywords: Data mining · Internet of things · Clustering · Classification · Frequent pattern

1 Introduction

IOT deals with connecting each and every things of the world via Internet [1–3]. The great progress on information technology and computer communication has made many application possible [4]. IOT is the next advanced generation of internet. It is believed that IOT will help in connecting trillion of nodes of various objects with the large web servers and cluster of a supercomputer. IOT also helps in integrating new computing technologies and communication [5]. Since, a decade, mobile devices and ubiquitous services help people to connect with anyone anywhere in the world. Nowadays, with the help of these devices, there is no limitation that corrupts the connection between people [6]. Many researchers working in different fields like academics, institutes and government departments have shown keen interest in modifying the internet by designing various systems like smart home, smart pen, intelligent transportation, global supply chain, healthcare [3, 7–9] etc.

S.C. Satapathy and A. Joshi (eds.), *Information and Communication Technology for Intelligent Systems (ICTIS 2017) - Volume 1*, Smart Innovation, Systems and Technologies 83, DOI 10.1007/978-3-319-63673-3_22

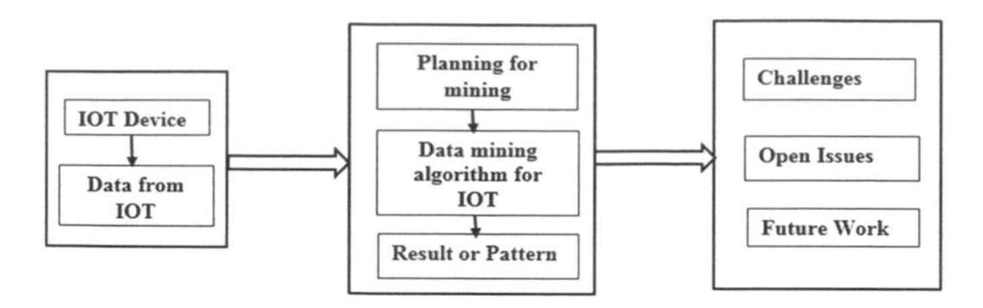

Fig. 1. Roadmap of data mining with IOT [4]

Above Figure Shows the cycle of the process required to design the survey of data mining with IOT. During the initial phase of the work, we need to review all the information about IOT device. We will extract the data stored in that device. Now our next step is to design such a type of data mining algorithm which will help in connecting data mining with IOT. While connecting data mining with IOT we may face some challenges and open issues that are described in the later section of this paper [4] (Fig. 1).

2 Data from IOT

It is obvious that IOT may create data consisting of much useful information. But, recently technical issues and challenges are major research topic along with different methods of handling these data. It is possible to solve the problem of [4] large IOT data is to design sensors capable of collecting only the interesting and useful data. To reduce the complexity of input data [4] is the recent research trend. Distributed computing, cloud computing and feature selection [4] are the popular data handling methods. It is assumed that big data will give born to more number of patterns from services and applications [4].

Moreover, how to handle large data obtained from IOT devices and finding hidden information from the data is important task. As described, data analysis for sensors and devices will be helpful in developing some useful system for the smart city or smart home. Many possible applications are possible to be developed from the large data analysis process. For fulfilling the task of finding hidden information from big data, Knowledge Discovery in Databases (KDD) successfully applied to different domains. Applied KDD has capability of finding "Something" or "interesting pattern" from IOT, with the help of following steps: data gathering, preprocessing, data mining and evaluation or decision making. Of these, data mining step play a main role in extracting interesting [4] pattern or knowledge from the big data. Data fusion, data transmission, big data and data decentralized issues has an impact on the system performance and IOT service quality [4].

3 Data Mining for IOT

The next section describes relationship between big data and data mining for IOT and detailed analysis and summarization of different data mining techniques for the Internet of Things [4].

3.1 Basics of Using Data Mining for IOT

It is easier [4] to use data mining for IOT for creating and analyzing data. Till the date, many studies have tried solving the problem of finding big data on IOT, without using effective and efficient analysis tools [4]. In recent times where big data is used widely, KDD systems and most traditional algorithms [4] are impossible to apply directly on the large amount of IOT data (Fig. 2).

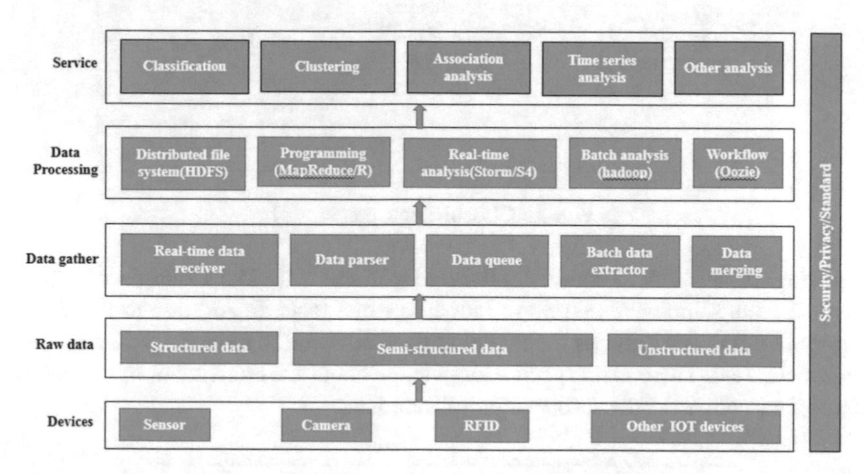

Fig. 2. Big data mining system for IOT [14]

For developing a high-performance data module of KDD for IOT [4], the applicable mining technologies is to be solved by KDD process: - These are Objective, Characteristics of data, and mining algorithm [4].

Objective: It is important to specify the limitations, assumptions, measurement and issues of the problem to get it solved [4]. The objective of the problem can be made clear with the help of above described parameters [4].

Data: Characteristics of data such as distribution, representation, size [4] etc. plays a vital role in data mining. Various data normally need to be processed differently. For example, R_i and R_j may be similar or not, but they need to be analyze differently if the semantics of the data is various [4].

Mining Algorithm: Data mining algorithm [4] can be designed easily with the specified needs of objective and data application. Here the question arises that, which data mining algorithm is suitable for [4] increasing high performance of the system or for the finding of a better service in [4] various IOT environments [4]. There are different techniques for extracting bigdata from IOT. These techniques are termed as Classification, Clustering and Association Rule Mining.

Classification: It is used to classify labeled as well as unlabeled patterns [4]. Labeled data basically considers the piece of unlabeled data with some information, class or tag related to it. For example, the picture of any animal is said to be the part of unlabeled data unless any information about it, its name, or tag like its sound is not clearly mentioned. Labeled data are achieved by making judgments about available piece of unlabeled data [4] (Fig. 3).

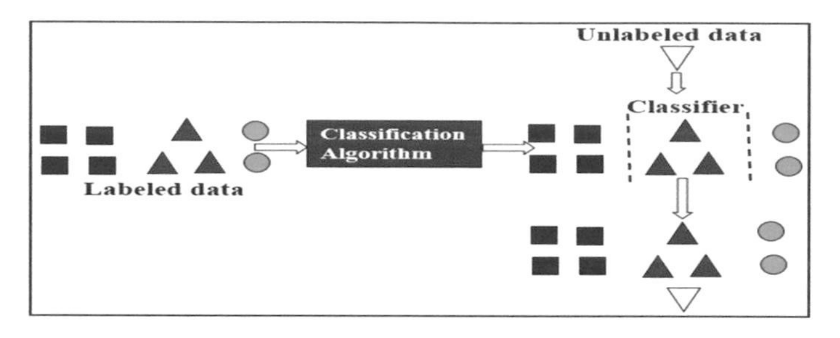

Fig. 3. Classification process [4]

Clustering: It is used to classify unlabeled patterns [4]. The example of unlabeled patterns are the samples of natural or human-created facts. In addition to this facts, the category of unlabeled patterns include photos, videos, audio recordings, x-rays, tweets, news articles, etc. There is no specific knowledge about the constituent of the unlabeled patterns, but contains only information of data (Fig. 4).

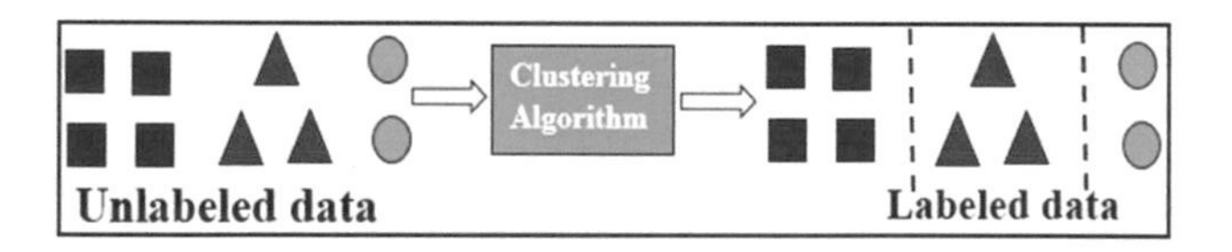

Fig. 4. Clustering process [4]

Association: It is used to find an event from the input pattern that doesn't occurs in particular order [4]. Sequential pattern is a part of association rule mining and is used to find an event from the input pattern that occur in particular order [4] (Fig. 5).

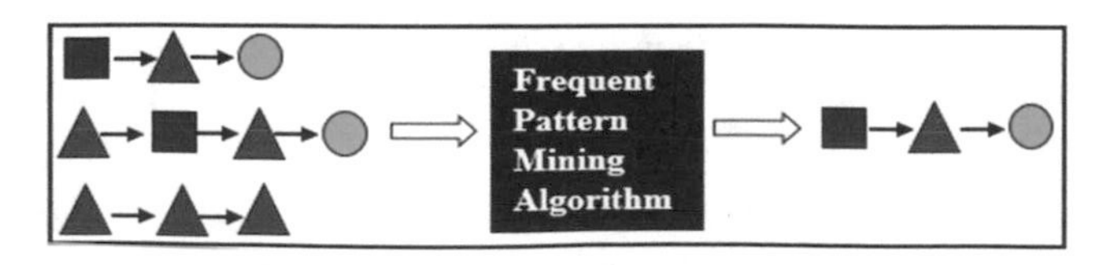

Fig. 5. Frequent pattern mining process [4]

Here we can apply total 12 numbers of possible combinations to extract bigdata from IOT as described in the following Table 1.

Table 1. Combination of algorithm

Algorithm	Possible	Combination
Clustering	Clustering → Classification Clustering → Frequent pattern	Clustering → Classification → Frequent pattern Clustering → Frequent pattern → Classification
Classification	Classification → Clustering Classification → Frequent pattern	Classification → Frequent pattern → Clustering Classification → Clustering → Frequent pattern
Frequency pattern	Frequency pattern → Clustering Frequency pattern → Classification	Frequency pattern → Clustering → Classification Frequency pattern → Classification → Clustering

The objective of finding an interesting hidden pattern may be different depending upon the goal. Many researchers try to provide better services by combining different mining techniques. Overall study and system designing is needed because a single technique or algorithm will not work for extracting [4] useful information and making a decision. For this reason, we describe the possible combination of mining technologies.

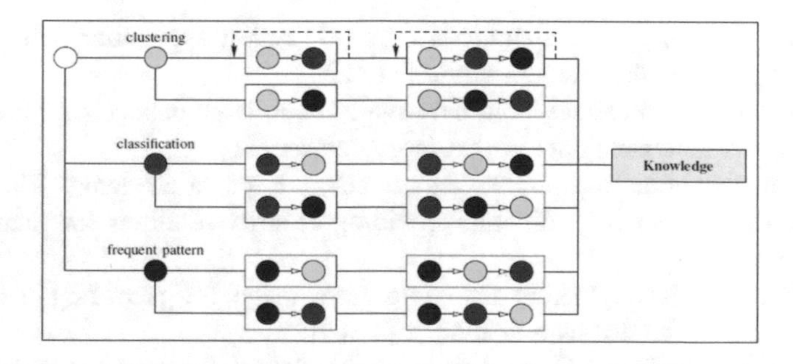

Fig. 6. Different combination of mining technologies for the IOT [4]

Figure 6 shows possible combination of mining technologies that could apply to extract hidden information. The first combination shows the use of clustering algorithm first and then classification algorithm. The second combination shows that classification algorithm to use first and after that clustering algorithm applied on the data. In the first combination, a set of classifiers is created without having any knowledge of input data or pattern. The second combination, classification is responsible for creating set and classifiers. On the set of the classifier, clustering algorithm used to add new classifier and create a new pattern. It is the combination of different mining technologies that enables the possibilities of handling patterns [4] or knowledge that enters the IOT incrementally.

For example, recognizing the human faces or behavior not previously in the knowledge database (the set of classifiers) [4]. The different combinations that can be used are, like

$$\text{Clustering} \rightarrow \text{classification} \rightarrow \text{frequent pattern}$$

The three techniques mentioned above consists of number of algorithms that are used to mine data from IOT devices. Some of them which are scalable and efficient for extracting data are decision tree algorithm (Classification technique), k mean algorithm (Clustering technique), and frequent pattern mining (Association rule mining).

It is mandatory to evaluate the quality of data that is obtained by applying one of the above mentioned algorithm. The best way to calculate and count the patterns in the right group is termed as "Accuracy Rate" (AR). It is defined by,

$$AR = \mathrm{Nc}/\mathrm{N_t} \tag{1}$$

where Nc is the total number of test patterns assigned correctly to the group they belong and N_t is the number of test patterns [10]. The other parameters which helps in measuring the quality of the accurate data are known as precision (P), and recall (R). The combination of P and R is termed as F-score which is defined as follows [10].

$$F = \frac{2\mathrm{PR}}{\mathrm{P} + \mathrm{R}} \tag{2}$$

4 Challenges in Data Mining With IOT

1. A major challenge is to upgrade a crime detecting application which includes advanced features that prevent crime [11, 12].
2. One challenge is to extract large data available in large data storage and to detect any noise or unreliable data in that large dataset [11].
3. Mining uncertain and incomplete data is also a big data challenge. The algorithm modification is also very difficult, providing security solutions for sharing data is not so easy [11].
4. Conversion of IOT generate data into knowledge for providing a convenient environment to the user is a challenge [4, 13].
5. Analysis and handling of large data is a difficult task for data mining purpose [4].
6. Selecting and implementing the better & efficient technology and mapping it with other technologies is difficult [4].
7. Building an intelligent system with the help of simple algorithm is not possible. For preparing an intelligent system, a number of algorithms must be fused to a single algorithm [4].
8. Internet connection may arise as a challenge when used in rapid speed device [4].
9. It is a challenge to build up an intelligent industrial IOT device, which includes smart city, green energy generation etc. [13].
10. Major challenges in security are widespread data collection, unexpected use of consumer data [13].
11. Sharing of standards and infrastructure gives rise to a security issue [3].

5 Major Issues in Data Mining with IOT

1. Parallel programming needs to be designed in such a way that every algorithm can be applied to it [1, 15].
2. The framework that considers security, privacy, data sharing, the growth of data size etc. is to be designed [11].
3. Infrastructure Perspective: IOT gives low computation and high throughput, but mining algorithm is designed for small size & low power consumption device. This creates an infrastructure issue [4, 16].
4. Data Perspective: Gathering data from different sources creates redundant data. The user needs to filter the redundant data for better system performance. Moreover, data generate from different from sources may become an obstacle [4, 17].
5. Algorithm Perspective: Some example of IOT needs adding classifiers dynamically and some needs adding classifier statically. So, we need some mining technologies to be fused that can classify the classifiers in a common way [4].
6. Privacy and Security: Privacy and Security remains an issue because every algorithm and technologies are not able to outperform privacy and security issue [4, 18]. For example, it is easy for companies to collect different customer data from various devices or sources and use data mining techniques to find the information that [2] helps in increasing sales volume but the issue is that many customers wouldn't like to disclose their privacy and security, such as retail, shopping behavior.
7. Massive Scaling: How to name, identify, authenticate, maintain, protect and use a large number of data is an open issue [18, 19].
8. Architecture and dependency: It is difficult to construct an architecture that connects a large number of things to the internet. Many things are dependent on each other, so removing any of them may generate an error [18].
9. Robustness: In robustness, devices location have to be known if clock drifts, the location of the device may not be accurate [18].

6 Conclusion

The concept of IOT has taken born from the requirement of managing, automating and exploring all devices, instruments and sources all mover the world. To deal simultaneously with people and IOT devices, there arises the need of data mining techniques. Various techniques of IOT supports in decision making and optimization of any application. When fused with IOT, data mining basically deals with discovering useful and interesting patterns from large data. Then number of algorithms are applied to extract important hidden information. Data mining techniques includes clustering, classification, pattern mining. Adding more to the introduction of IOT and data mining, this paper focuses on challenges and open issues of data mining. Characteristics of big data, data mining algorithms and analysis on challenges is also described. At this stage when the development of IOT is at initial stage, focus is more on developing efficient preprocessing mechanism and consequent developing effective mining technologies for describing rules of IOT data. In the addition to research trends going on nowadays in the world are discussed in this paper. The research trends are: (i) Data will be uploaded the internet once connected, but IOT faces come problem such as, data abstraction, data summarization and data fusion [2]. (ii) To mine data from the application controlling multi-media devices and controlling energy are also a topic of research.

References

1. Chen, F., Deng, P., Wan, J., Zhang, D., Vasilakos, A.V., Rong, X.: Data mining for the internet of things: literature review and challenges. Int. J. Distrib. Sens. Netw. (2015). Hindawi Publishing Corporation
2. Tsai, C.W., Lai, C.F., Chiang, M.C., Yang, L.T.: Data mining for internet of things: a survey. IEEE Commun. Surv. Tutor. **16**, 77–97 (2014)
3. Bhatia, S., Patel, S.: Analysis on different data mining techniques and algorithms used in IOT. Int. J. Eng. Res Appl. **2**(12), 611–615 (2015)
4. Stankovic, J.A.: Research directions for the internet of things. IEEE **1**, 3–9 (2014)
5. Bin, S., Yuan, L., Xiaoyi, W.: Research on data mining models for the internet of things. IEEE **9**, 127–132 (2010)
6. Chen, Y., Zhang, A.X.H.C.H: Research on data mining model in the internet of things. In: Proceedings of the International Conference on Automation, Mechanical Control and Computational Engineering (2015)
7. Chen, H., Chung, W., Xu, J.J., Wang, G., Qin, Y., Chau, M.: Crime data mining: a general framework and some examples. Computer **37**(4), 50–56 (2003)
8. Wu, X., Zhang, S.: Synthesizing high-frequency rules from different data sources. IEEE Trans. Knowl. Data Eng. **15**(2), 353–367 (2003)
9. Keller, T.,: Mining the internet of things: detection of false-positive RFID tag reads using low-level reader data. Ph.D. Dissertation, The University of St. Gallen, Germany (2011)
10. de Saint-Exupery, A.: Internet of things strategic research roadmap. European Research Cluster on the Internet of Things. Technical report (2009)
11. Baeza-Yates, R.A., Ribeiro-Neto, B.A.: Modern Information Retrieval. ACM Press Addison-Wesley, Boston (1999)
12. He, T., Stankovic, J., Lu, C., Abdelzaher, T.: A spatiotemporal communication protocol for wireless sensor networks. IEEE Transact. Parallel Distrib. Syst. **16**(10), 995–1006 (2005)
13. Liu, B.: Web Data Mining: Exploring Hyperlinks, Contents, and Usage Data. Springer, Berlin (2007)
14. Ashton, K.: That 'Internet of Things' things. RFID J. (2009). http://www.rfidjournal.com/article/print/4986
15. Auto-ID Labs: Masschusetts Institute of Technology (2012). http://www.autoidlabs.org/
16. Atzori, L., Iera, A., Morabito, G.: The internet of things: A survey. Comput. Netw. **54**(15), 2787–2805 (2010)
17. Miorandi, D., Sicari, S., DePellegrini, F., Chlamtac, I.: Internet of things: vision, applications and research challenges. Ad. Hoc. Netw. **10**(7), 1497–1516 (2012)
18. Bandyopadhyay, D., Sen, J.: Internet of things: applications and challenges in technology and standardization. Wirel. Pers. Commun. **58**(1), 49–69 (2011)
19. Domingo, M.C.: An overview of the internet for things for people with disabilities. J. Netw. Comput. Appl. **35**(2), 584–596 (2012)

Memory Optimization Paradigm for High Performance Energy Efficient GPU

Prashanth Voora(✉), Vipin Anand(✉), and Nilaykumar Patel(✉)

Intel Technology India Private Limited, Bangalore 560103, India
prashanthx85@gmail.com, anand.vipin@gmail.com,
nilay.nilpat@gmail.com

Abstract. The following paper propose a new memory management technique of handling the GState's in DDK and the same can be extended for sending command buffer which hold the job request to various USM/UVC blocks in GPU. We shall discuss the most generic design followed across various device driver and explains the drawback of using the existing design and introduce the new memory management technique along with some futuristic improvements. This design philosophy is more suited in GPU architecture's which have deferred multi-pass rendering for 3D graphics pipeline.

Keywords: GPU · Low-power · Memory optimization · Deferred rendering · 3G graphics pipeline · G-State · Command buffer

1 Introduction

1.1 Generic GPU S/W Driver

The S/W driver part of any GPU DDK is broadly classified into 4 major Layers that supports OpenGL and OpenCL backend. The four layers are GLES, OCL, CUMD and KMD. The GLES layer is responsible for implementing the OpenGL API's defined according to their specification supported. The GLES layer also take care of implementing EGL glue layer or implementing any other equivalent layer which specifies the interface with the native windowing system. The OCL layer implements the OpenCL specification API's. The CUMD acts as common user mode driver providing common functionalities to various upper layers like OpenGL, OpenCL, etc. The CUMD layer is responsible for maintaining active GState, generating Command buffer format to be understood by USM/UVC blocks of Graphics H/W, Handling job creation and various other innovation aspects handled in user mode space. The KMD driver is the kernel mode driver for the IP. Its main purpose is to manage the H/W and its main functionalities are Job Scheduling, Interrupt handling, etc., which interacts directly with the hardware (Fig. 1).

S.C. Satapathy and A. Joshi (eds.), *Information and Communication Technology for Intelligent Systems (ICTIS 2017) - Volume 1*, Smart Innovation, Systems and Technologies 83, DOI 10.1007/978-3-319-63673-3_23

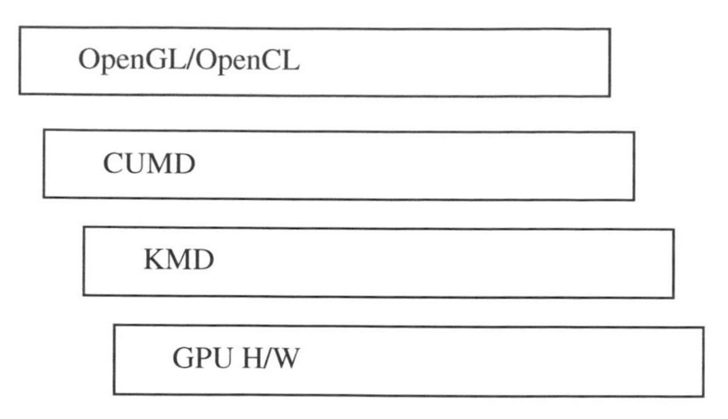

Fig. 1. GPU S/W layers

1.2 The Purpose of the Paper

The Purpose of the paper is to provide some design improvements in the existing CUMD driver related to memory management techniques used to maintain GSTATE across various draw calls and further extended to command buffer management. The CUMD driver in the current state uses three circular buffer for rendering a graphics frame. The three buffers Used for (a) GSTATE page/block pool memory for maintaining GState's. (b) Indirect command buffer pool for issuing indirect commands to the GPU. (c) Uniform buffer for maintaining Uniform variables between draw calls. This paper talks about the alternate usage of using the circular buffer in CUMD layer as it becomes very hard for re-use and deletion of individual units of a circular buffer.

1.3 Breakthrough Point of the Paper

Efficient way of handling GState and command buffer in CUMD and thus leading to improving the overall Performance and low power usage by the DDK in rendering a particular Frame.

2 Current Design

Considering the GState Circular pool is initialized with 128 bytes of block size as an example and further which it is referred to as GSTATE_PAGE_SIZE in further part of the document for a circular count of size 2048. The above numeric figures has been arrived based on the expected high draw call count when running various high computation graphics application like Manhattan benchmark across various key frames. The complex GState can be logically categorized into multiple of GSTATE_PAGE_SIZE grouped into different individual structures based on the various complex unit blocks involved in the pipeline. By proper padding in the grouped structures we could achieve the GState's to be present in multiples of GSTATE_PAGE_SIZE. The maximum number of GSTATE_PAGE_SIZE required for maintaining the complex GState is 64 units in size. Thus total memory pool of (64×128) bytes of memory required to store a complete GState. Also in some drivers they would also maintain a bit mask of 64 bits corresponding to 64 unit count of maintaining a complete GState. This bit mask helps in identifying the list of dirty GState pages between draw calls on state setting operation. Whenever the GLES/OCL driver tries to modify these GState page its corresponding bitmask would be set. Thus between successive draw calls a complete copy of GState of previous draw call is copied into new set of circular buffer and the new state setting will start appending to the newly created circular buffer. By this method it effectively maintains a GState for a draw call. This method of categorizing the complete GState structures to various GState pages are stored in various structure objects in CUMD for its internal implementation.

The GState circular buffer pool memory can be used to maintain two different types of circular buffer internally. The first type is known as shadow state GState buffer which is being exposed to upper layer of drivers like GLES, OCL etc., where whenever a request for GState modification is obtained then it modifies this type of circular buffer. The CUMD Layer is responsible for modifying this shadow state GState buffer. Also the CUMD Layer is also responsible for modifying the corresponding Dirty bit mask. The second type of circular buffer is where the actual GState association with the draw call. Every draw call should be associated with individual GState and this association is nothing but the reference to the current GState. Hence between draw calls a flush operation happens between shadow state GState and current GState. There is a direct one to one mapping between the shadow GState and current GState. All the memory required for shadow GState and current GState are taken from the GState circular block pool memory. This circular block pool memory is initialized during driver startup. On prior request it gives number of pages from the circular pool where each unit is of size GSTATE_PAGE_SIZE. This form of circular buffer becomes more inefficient when trying to re-use the same buffer and trying to delete some arbitrary GState pages for re-use from the circular block pool memory.

There is an indirect Command buffer pool of circular buffer for issuing indirect commands to the GPU. This indirect commands contains the various Jobs to be carried out by various USM/UVC in the GPU Hardware. The commands are directly issued to scheduling algorithm which in turn indirectly schedules across various cores.

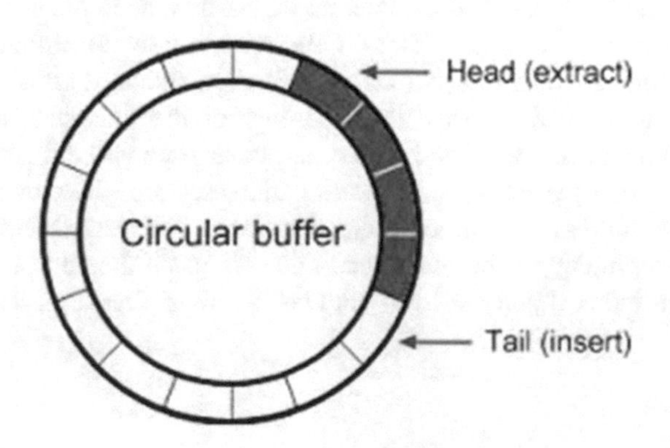

Fig. 2. Current design

The command buffer format is best understood by the protocol defined between S/W driver and Hardware. For each command buffer to be filled and send are directly loaded and filled from the indirect command circular buffer pool. The numeric equation of defining the circular buffer are best chosen based on the high number of draw calls together with various instructions which issues the command buffer. The various instruction that defines the command buffer various across different GPU and its protocol is best defined between the different GPU S/W part and the corresponding Hardware (Fig. 2).

3 New Design

In the new design the following changes are proposed to the Current design:

- Replace circular buffer for GState with unit state allocators corresponding to each individual's state.
- Remove dependency on GSTATE_PAGE_SIZE. Allocations to be done on actual sizes.
- There will be individual unit state allocators for each member representing the GSTATE. Hence for a particular draw call we have the indexes pointing from each of these unit state allocators to define that particular GSTATE.
- We all maintain the generation of unit state allocators. As sometime extreme cases we might see the unit state allocators has been fully utilized and need for further allocations, during that stage we shall initialize with unit state allocator with the next generator unit.
- A state vector would be generated for a particular state and the same vector will be used for further state matching occurrences to re-use the same state at different draw call.
- State matching will be done to prevent allocation for that state if an existing set of data is received. For this we calculate the hash value for a given set of unit state data & store it in unordered map with hash vale as the key and memory address where the data is stored as the value. When allocation needs to be done for a new set of unit state data we first calculate the hash value of new set of state data look up the hash value of new set of state data, look up the hash value in the unordered map & if there is a hit return the same address. Otherwise, a new allocation is done. Unordered map will be a data member of the unit state allocator class.
- Maintain an unordered set as a data member of the unit state allocation class. Initially, unordered set associated with a particular state will contain all the address available with that particular state. As the addresses are given our from the unordered set, the address will be re-arranged to move the least recently used address towards the beginning of the set. Hence once we reach the end of the set, we start re-using from index 0 and the LRU address is reused first (Fig. 3).

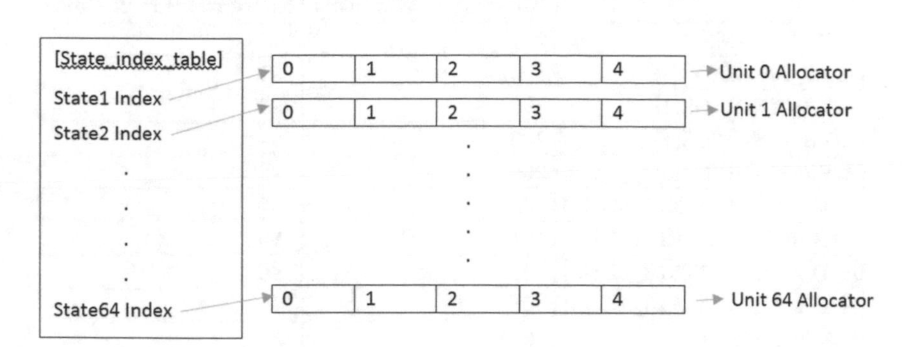

Fig. 3. New design representation

4 Improvements

The changes proposed in the new design when integrated is CUMD layer resulted in the following improvements:

- Memory allocations were less as there were sufficient sets of unit state data that were being repeated. In these cases memory was allocated only once and for rest of the cases same address as earlier is returned. The data below is collected after integrating the new design changes and tested with GLB2.7/f0017. The data below reflects <no of actual allocations>/<no of calls for allocation> for all 64 GSTATE pages.
- As we move from one circular buffer for all the states in the GState structure to individual unit state arrays, it becomes easier to match states to prevent duplicate allocations for the same set of data, removes dependency on GSTATE_PAGE_SIZE and it becomes easier to maintain indexes corresponding to different states and pass it on as State Index Table corresponding to a draw call.
- Significant Performance Improvement is also noticeable. Below data gives a comparison between current design & new design changes when tested with GLB2.7/f0017-trimmed & GLB21/f294-trimmed.
- With GLB2.7/f0017-trimmed, current design took 361.23 s to render 5 frames. With the new design changes it took 312.14 s to render same set of 5 frames.
- With GLB21/f294_trimmed, the current design took 118.48 s to render 2 frames. With the new design changes it takes 98.76 s to render 2 frames of the same trace (Table 1).

Table 1. Comparison of unit allocation between proposed and existing design

Unit state	No. of actual allocations	No. of calls for allocation
MISC_PARAMS	217	226
IA/VB_BINDINGS[0]	198	292
IA/VB_BINDINGS[1]	199	293
IA/VB_BINDINGS[2]	200	294
IA/VB_BINDINGS[3]	200	295
IA_DEFAULT_ATRIB_SET[0]	01	96
IA_DEFAULT_ATRIB_SET[1]	01	96
IA_DEFAULT_ATRIB_SET[2]	01	96
IA_DEFAULT_ATRIB_SET[3]	01	96
RASTER_STATE	05	52
VIEWPORT SET[0]	08	271
VIEWPORT SET[1]	08	271
VIEWPORT SET[2]	08	271
VIEWPORT SET[3]	08	271
SCISSOR STATE	07	38
CLIP STATE	01	24
POLY_STIPPLE	01	24
BLEND_STATE	09	43
DEPTH_STENCIL_STATE	07	114
FRAMEBUFFER_STATE	06	24
VS-BINDINGS[0]	306	710
VS-BINDINGS[1]	306	710
VS-BINDINGS[2]	306	710
VS-BINDINGS[3]	306	710
VS-BINDINGS[4]	306	710
VS-BINDINGS[5]	306	710
PS-BINDINGS[0]	01	144
PS-BINDINGS[0]	01	144
PS-BINDINGS[1]	01	144
PS-BINDINGS[2]	01	144
PS-BINDINGS[3]	01	144
PS-BINDINGS[4]	01	144
PS-BINDINGS[5]	01	144
HS-BINDINGS[0]	01	144
HS-BINDINGS[1]	01	144
HS-BINDINGS[2]	01	144
HS-BINDINGS[3]	01	144
HS-BINDINGS[4]	01	144
HS-BINDINGS[5]	01	144
GS-BINDINGS[0]	01	144

(*continued*)

Table 1. (*continued*)

Unit state	No. of actual allocations	No. of calls for allocation
GS-BINDINGS[1]	01	144
GS-BINDINGS[2]	01	144
GS-BINDINGS[3]	01	144
GS-BINDINGS[4]	01	144
GS-BINDINGS[5]	01	144
DS-BINDINGS[0]	01	144
DS-BINDINGS[1]	01	144
DS-BINDINGS[2]	01	144
DS-BINDINGS[3]	01	144
DS-BINDINGS[4]	01	144
DS-BINDINGS[5]	01	144
CS-BINDINGS[0]	01	144
CS-BINDINGS[1]	01	144
CS-BINDINGS[2]	01	144
CS-BINDINGS[3]	01	144
CS-BINDINGS[4]	01	144
CS-BINDINGS[5]	01	144
IPA_STATE	01	237

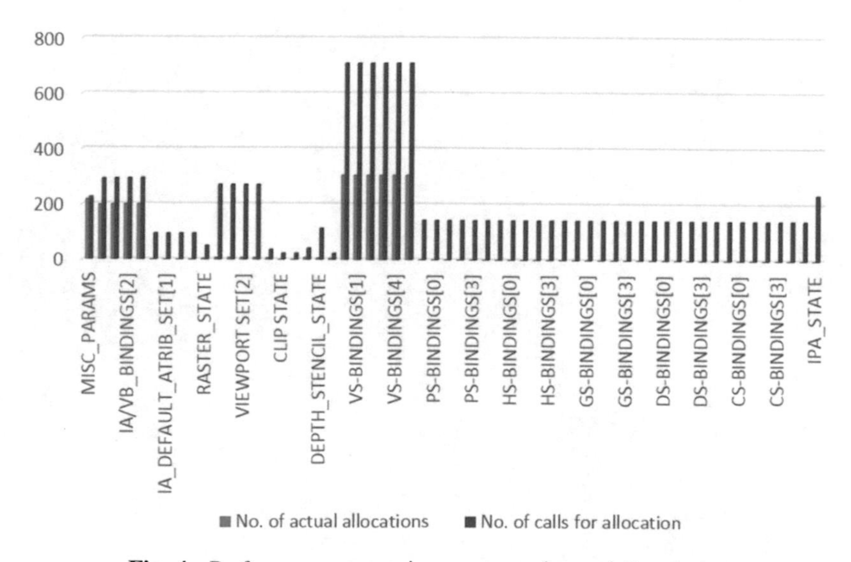

Fig. 4. Performance comparison: proposed vs existing design

5 Conclusion

This paper discussed about a new technique for supporting GState' s and command buffer usage in driver for increasing performance, reducing complexity & allow re-use of GState and command buffer. It also provides a means of significant improvement in usage of power. Thus resulting in a gradient improvement in PPA scores. Our Results demonstrate that using the new proposed design paradigm results in about 20–25% decrease in memory usage and less overhead in handling GState and Command buffer. These overheads are offset by the performance improvement from using GPUs and using the memory management techniques described in this paper improves the flexibility of the overall application (Fig. 4).

The above chart describes the performance comparison of actual allocations happened with new design paradigm depicted in blue color to the original allocation without any improvements depicted in red color. The following techniques presented in this paper will be necessary to ensure that their applicability is not limited to Gstate and command buffer allocation problem but also can be easily extended to other temporary internal memory allocators in the driver which uses ring memory and which requires re-using mechanism.

A Novel Approach for Predicting Ancillaries Ratings of Indian Low-Cost Airlines Using Clustering Techniques

Hari Bhaskar Sankaranarayanan(✉) and Viral Rathod

Amadeus Software Labs, Bangalore, India
{hari.sankaranarayanan,viral.rathod}@amadeus.com

Abstract. In this paper, we will present a novel approach for classifying and predicting airline passenger ratings for ancillaries using unsupervised learning techniques like K-Means and Expectation Maximization clustering. The datasets chosen for this study belong to Indian Low-Cost Airlines. The goal is to perform an empirical study and exploratory analysis for predicting the overall rating with respect to the individual ancillary services ratings. Our results suggest that while there is no clear pattern among the ratings that can lead to the overall rating from passengers, the factors like value for money can largely influence the overall rating. Low-cost airlines aggressively promote competitive fares and choice of ancillary services hence the passenger behavior towards the overall rating varies across the airline datasets.

Keywords: Clustering · India · LCC · Airlines · Reviews · Machine learning

1 Introduction

In the past decade, the rise of Low-Cost Carriers (LCC) provides a unique opportunity for air travelers to choose competitive fare options and ability to choose the ancillary services like baggage, meals on board, priority check-in, and seat selection. One of the primary reasons the passengers opt for LCC's is the affordability and value for money that can be obtained by choosing the services in à la carte manner. In this paper, we gathered a set of passenger reviews on LCCs to understand the pattern of ancillary services ratings and how they tend to influence the overall rating. Our analysis uses unsupervised learning techniques since the ratings are based on the subjective experience of the individual passenger and also the fuzziness associated with each rating. The passenger may or may not have opted for each service yet they rate the overall rating based on the services they have bought. We had used clustering approach since the customer segment and expectations vary based on the choice made by them. Our initial analysis shows that there is no clear pattern of influence among the ratings to influence overall rating with an exception that value for money emerges as a common theme among the airline reviews. The paper is organized into following sections. Section 2 discusses the related work in this area, Sect. 3 discusses the data set, Sect. 4 discusses the analysis approach, Sect. 5 presents the results of experiments and

S.C. Satapathy and A. Joshi (eds.), *Information and Communication Technology for Intelligent Systems (ICTIS 2017) - Volume 1*, Smart Innovation, Systems and Technologies 83, DOI 10.1007/978-3-319-63673-3_24

discussion around them, Sect. 6 highlights the limitations and future work and Sect. 7 provides the conclusion.

2 Related Work

The clustering approach is used widely in the topic of customer segmentation to understand the needs and expectations, especially for product marketing and Customer Relationship Management [1–3]. Data mining of online reviews is discussed across literature studies with a variety of techniques including fuzzy logic, clustering, and semi-supervised learning methods [4–8] and especially the literature work focuses mostly on text mining and sentiment analysis. Automated product marketing strategy is discussed in a paper based on the online reviews [9] and new avenues of opinion mining discussed with recent trends and technology advances [10] like context specific mining with respect to user specific needs and preferences.

3 Information About Datasets

The dataset collected from TripAdvisor website. Table 1 highlights the number of reviews.

Table 1. LCC review datasets

Airline	Number of reviews
Airline 1	1616
Airline 2	126
Airline 3	653

The next dataset contains information about tripadvisor information about review ratings. Table 2 highlights the rating characteristics entered by the passengers which are normalized to nominal values

Table 2. Airline rating characteristics

Rating attribute	Rating	Normalized nominal values
Seat comfort	1- Low to 5- High	Very Low, Low, Medium, High, Very High
Food	1- Low to 5- High	Very Low, Low, Medium, High, Very High
Legroom	1- Low to 5- High	Very Low, Low, Medium, High, Very High
Value for money	1- Low to 5- High	Very Low, Low, Medium, High, Very High
Check-in	1- Low to 5- High	Very Low, Low, Medium, High, Very High
Overall Rating	1- Low to 5- High	High, Medium, Low

4 Analysis Approach

For the purpose of our analysis, we did experimentation in two phase, the first phase is to study the observation using spreadsheets and also apply some supervised machine learning techniques to identify any correlation between individual ratings and overall ratings.

4.1 Phase 1

The empirical analysis of the datasets suggests the following:

1. Value for Money rating has a direct correlation with overall rating since 68.39% of medium and high rating lead to medium and high overall rating.
2. Lower rating of certain ancillary ratings doesn't necessarily lead to the lower overall rating. This is illustrated in Table 3 below.

Table 3. Analysis lower ancillary ratings vs high overall rating

	Airline 1	Airline 2	Airline 3
	Instances of overall rating 4 & 5	Instances of overall rating 4 & 5	Instance of overall rating 4 & 5
Seat comfort (1,2)	39/1616	1/126	11/653
Food (1,2)	67/1616	0/126	15/653
Value for money (1,2)	10/1616	0/126	2/653
Check-in (1,2)	12/1616	0/126	5/653
Legroom (1,2)	63/1616	1/126	22/653

3. Higher rating on certain ancillary rating doesn't necessarily lead to the higher overall rating. This is illustrated in Table 4 below.

Table 4. Analysis high ancillary ratings vs low overall rating

	Airline 1	Airline 2	Airline 3
	Instances of overall rating 1 & 2	Instances of overall rating 1 & 2	Instance of overall rating 1 & 2
Seat comfort (4,5)	13/1616	4/126	10/653
Food (4,5)	1/1616	1/126	3/653
Value for money (4,5)	7/1616	3/126	10/653
Check-in (4,5)	9/1616	1/126	8/653
Legroom (4,5)	13/1616	1/126	10/653

4. Table 5 lists missing values on each of ancillary ratings is quite significant as well which convey whether passengers didn't rate or not bought this service.

Table 5. Missing ancillary ratings

	Airline 1	Airline 2	Airline 3
Seat comfort	111/1616	9/126	46/653
Food	839/1616	68/126	345/653
Value for money	260/1616	21/126	98/653
Check-in	969/1616	79/126	379/653
Legroom	111/1616	10/126	45/653

5. We had also applied few supervised learning methods like Naïve Bayes, Logistic Regression and Logistic Model Trees for experimentation and the algorithms performed better with increasing number of training data using cross-fold validation technique. The comparison listed in Table 6.

Table 6. Comparison of prediction accuracy - supervised learning methods

	Airline 1	Airline 2	Airline 3
Logistic regression	80.53%	71.26%	70.99%
Logistic model trees	80.48%	72.27%	71.35%
Naïve Bayes	79.51%	75.10%	72.17%

Since we couldn't ascertain consistently on the patterns influence overall rating due to varied results on an inferring rating from one another and lot of missing values we experimented unsupervised learning methods as the second phase.

4.2 Phase 2

Clustering algorithms are used as part of unsupervised learning techniques to classify datasets. The goal is to identify the clusters based on the supplied values that are self-organized. We have used Expectation Maximization Clustering and K-Means Clustering for this purpose. EM performs better with missing values [11] and K-Means can be used with or without replacing missing values with mean/mode. We have used weka tool to perform the clustering experiments on the airline review datasets. In the result section, the clusters are evaluated to predict the classes (Ex: Overall Rating). The setup information for EM cluster for all the experiments is as follows in Table 7.

The setup information for K-Means cluster for all the experiments is as follows in Table 8.

Table 7. EM Setup Information

Parameter	Value
Max iterations	100 (Default)
Number of clusters	3
Number of folds	10 (Default)
Number of Kmean Runs	10 (Default)
Seed value	100 (Default)
Min std dev	1.0E−6 (Default)
Max number of clusters	−1 (Default)
Number of execution slots	1 (Default)
Min Log likelihood improvement CV	1.0E−6 (Default)
Min Log likelihood improvement iterating	1.0E−6 (Default

Table 8. Simple K-Means setup information

Parameter	Value
Max iterations	500 (Default)
Number of clusters	3
Don't replace missing values	True
Fast distance Calc	False (Default)
Seed value	10 (Default)
Initialization method	Random
Max number of clusters	−1 (Default)
Number of execution slots	1
Distance function	Euclidean (Default)
Preserve instance order	False
Canopy setup values	Default

5 Results and Discussion

In this section, we will tabulate the results and discuss the findings of the experiments. Results for Airline 1 on EM, KMeans clustering are listed in Tables 9 and 10.

5.1 Discussion on Airline 1 Results

The accuracy of the prediction for the EM is around 64% and K-Means around 59%. The cluster instances in EM show that Value for Money attribute is consistent across High and Medium predictions. The clusters in EM show the distribution for High, Very High values for all the attributes that lead to "High" overall rating cluster. For K-Means there is no clear pattern that can be inferred for High, Medium, and Low overall rating clusters. In reference to "Classes to clusters" the confusion matrix is better for EM algorithm.

Results for Airline 2 on EM, KMeans clustering are listed in Tables 11 and 12.

Table 9. Airline 1 - EM clusters for overall rating

Clustered instances		
0 966 (60%)		
1 426 (26%)		
2 224 (14%)		
Log likelihood: −5.10308		
Class attribute: overall rating		
Classes to clusters:		
0 1 2 <– assigned to cluster		
857 134 213 | High		
26 130 6 | Low		
83 162 5 | Medium		
Cluster 0 <– High		
Cluster 1 <– Medium		
Cluster 2 <– Low Incorrectly clustered instances:	591.0	36%

Table 10. Airline 1 – K-Means clusters for overall rating

Clustered Instances		
0 979 (61%)		
1 637 (39%)		
Class attribute: overall rating		
Classes to clusters:		
0 1 <– assigned to cluster		
784 420 | High		
101 61 | Low		
94 156 | Medium		
Cluster 0 <– High		
Cluster 1 <– Medium Incorrectly clustered instances:	676.0	41%

5.2 Discussion on Airline 2 Results

The accuracy of the prediction for the EM is around 60% and K-Means around 53%. The cluster instances in K-Means show that Value for Money attribute is inconsistent across High and Low predictions. For example Low cluster shows "Very High" overall rating value which is misleading in the experiment. The clusters in EM show the distribution for High, Very High values for all the attributes that lead to "High" overall rating cluster. There is no clear pattern that can be inferred for Medium and Low overall rating clusters. In reference to "Classes to clusters" the confusion matrix is poor for both the algorithms.

Results for Airline 3 on EM, KMeans clustering are listed in Tables 13 and 14.

Table 11. Airline 2 - EM clusters for overall rating

Clustered instances		
0 17 (13%)		
1 71 (56%)		
2 38 (30%)		
Log likelihood: −5.29259		
Class attribute: overall rating		
Classes to clusters:		
0 1 2 <– assigned to cluster		
15 55 3 | High		
2 8 16 | Medium		
0 8 19 | Low		
Cluster 0 <– Medium		
Cluster 1 <– High		
Cluster 2 <– Low		
Incorrectly clustered instances:	50	39%

Table 12. Airline 2 – K-Means clusters for overall rating

Clustered instances		
0 67 (53%)		
1 42 (33%)		
2 17 (13%)		
Class attribute: overall rating		
Classes to clusters:		
0 1 2 <– assigned to cluster		
48 25 0 | High		
5 10 11 | Medium		
14 7 6 | Low		
Cluster 0 <– High		
Cluster 1 <– Low		
Cluster 2 <– Medium		
Incorrectly clustered instances:	60	47%

5.3 Discussion on Airline 3 Results

The accuracy of the prediction for the EM is around 60% and K-Means around 62%. The cluster instances in K-Means show that Value for Money attribute matches the classification of High, Medium, Low predictions. The clusters in EM show the equal distribution for Value for Money in High and Medium predictions while Seat comfort, Leg room, and check-in distribution also influence Overall Rating to a greater extent.

Table 13. Airline 3 – EM clusters for overall rating

Clustered instances		
0 311 (48%)		
1 153 (23%)		
2 189 (29%)		
Log likelihood: -5.3872		
Class attribute: overall rating		
Classes to clusters:		
0 1 2 <– assigned to cluster		
120 46 19 | Medium		
144 8 167 | High		
47 99 3 | Low		
Cluster 0 <– Medium		
Cluster 1 <– Low		
Cluster 2 <– High		
Incorrectly clustered instances:	267	40%

Table 14. Airline 3 – K-Means clusters for overall rating

Clustered instances		
0 453 (69%)		
1 89 (14%)		
2 111 (17%)		
Class attribute: overall rating		
Classes to clusters:		
0 1 2 <– assigned to cluster		
83 34 68 | Medium		
285 7 27 | High		
85 48 16 | Low		
Cluster 0 <– High		
Cluster 1 <– Low		
Cluster 2 <– Medium		
Incorrectly clustered instances:	252	38%

In reference to "Classes to clusters" K-Means algorithm performs better even in terms of the confusion matrix.

A visualization of K-Means & EM cluster for Airline 3 is illustrated in Fig. 1 and 2 respectively.

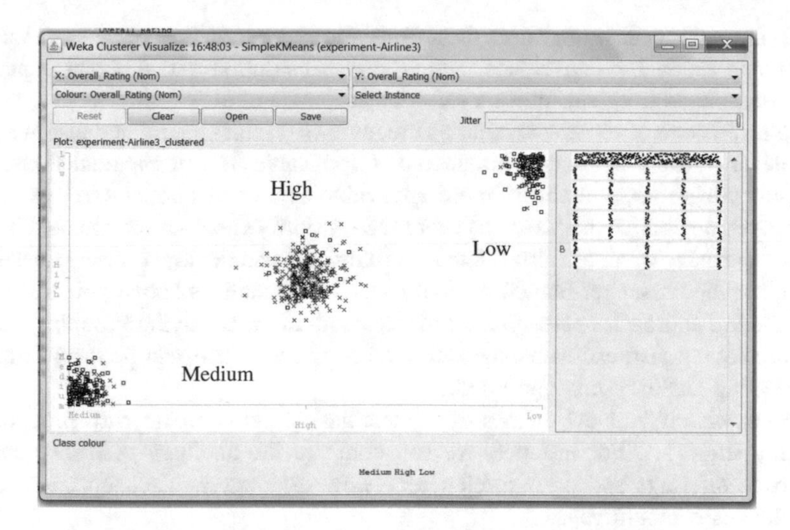

Fig. 1. Airline 3 – K-Means clusters for overall rating

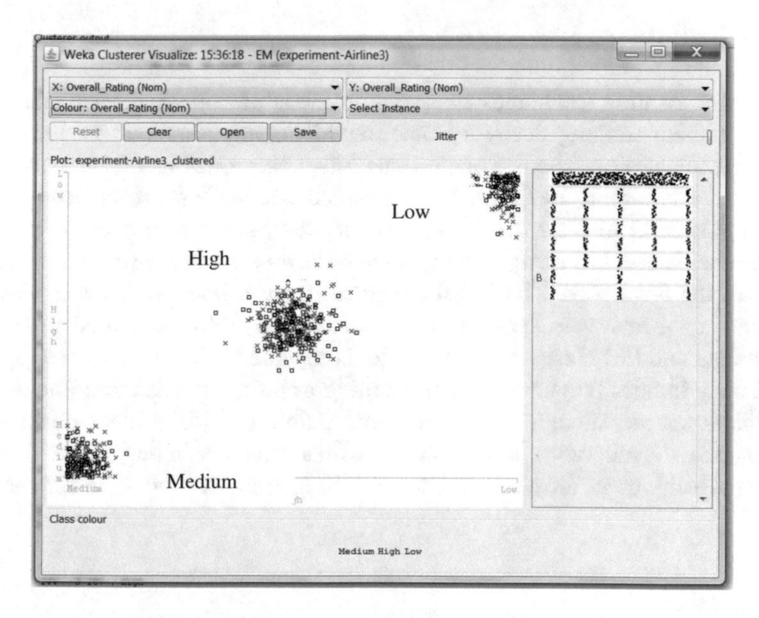

Fig. 2. Airline 3 – EM clusters for overall rating

6 Limitations and Future Work

The current research work has certain limitations and can be extended further in scope:

1. We didn't consider text mining using Natural Language Processing (NLP) to infer reasoning behind such rating patterns. In future, this can be done with more text mining techniques and augment the existing results.

2. Missing values mean that user didn't buy the service or do not wish to rate. We prefer to retain missing values without any substitution and this can experiment with missing values substitution using mean, mode et al.
3. Baggage is one of the key ancillaries the reviews do not capture and also we didn't consider In-flight entertainment since it is applicable for international flights. Seat comfort, legroom, and check-in are applicable as part of normal services as well, however, ancillaries like extra leg room seats, priority check-in service, and bundled plans inclusive of all are also offered. In future, we can deploy a survey question on whether the passenger bought ancillary services to address above points.
4. Clustering approach opens up a lot of perspectives on achieving grouping not only to predict but also estimate expectation on ancillaries. This can be used further for predicting ancillary mix and bundles.
5. We can classify review themes of the ancillaries and compare with product marketing strategies. For instance, we can compare the ancillary offerings, evaluate website offerings and draw inference on how the reviews stack up based on the ancillaries competitiveness.

7 Conclusion

Unsupervised learning techniques are useful in terms of achieving scale, and labeling attributes without training datasets hence experimenting passenger reviews datasets which are available in millions from various travel categories might be a good option than supervised learning methods. We attempted one such approach using EM and K-Means clustering to infer the relevance, accuracy, and consistency of predictions especially the dataset has many missing values since passengers may or may not have rated all of the ancillaries. The initial experiments of airline ancillaries review using clustering shown relatively lower accuracy results (10–20%) compared to the supervised learning and EM cluster performs better compared (60% and above) to K-Means. This logically implies that unsupervised learning technique needs more fine tuning for this problem space. Though the experiments didn't provide a clear conclusion on predicting the overall rating based on individual ancillary rating improvisation on clustering algorithms in form of semi supervised learning can be done further.

References

1. Namvar, M., Gholamian, M.R., KhakAbi, S.: A two phase clustering method for intelligent customer segmentation. In: 2010 International Conference on Intelligent Systems, Modelling and Simulation. IEEE (2010)
2. Kim, K., Ahn, H.: A recommender system using GA K-means clustering in an online shopping market. Expert Syst. Appl. **34**(2), 1200 (2008)
3. Hosseini, S.M.S., Maleki, A., Gholamian, M.R.: Cluster analysis using data mining approach to develop CRM methodology to assess the customer loyalty. Expert Syst. Appl. **37**(7), 5259 (2010)

4. Turney, P.D.: Thumbs up or thumbs down? Semantic orientation applied to unsupervised classification of reviews. In: Proceedings of the 40th Annual Meeting on Association for Computational Linguistics. Association for Computational Linguistics (2002)
5. Tsur, O., Rappoport, A.: RevRank: a fully unsupervised algorithm for selecting the most helpful book reviews. In: ICWSM (2009)
6. Zhai, Z., et al.: Clustering product features for opinion mining. In: Proceedings of the Fourth ACM International Conference on Web Search and Data Mining. ACM (2011)
7. Popescu, Ana-Maria, Etzioni, Orena: Extracting Product Features and Opinions from Reviews. Natural Language Processing and Text Mining. Springer, London (2007)
8. Chaovalit, P., Zhou, L.: Movie review mining: A comparison between supervised and unsupervised classification approaches. In: Proceedings of the 38th Annual Hawaii International Conference on System Sciences. IEEE (2005)
9. Lee, Thomas Y., Bradlow, Eric T.: Automated marketing research using online customer reviews. J. Mark. Res. **48**(5), 881 (2011)
10. Cambria, E., et al.: New avenues in opinion mining and sentiment analysis. IEEE Intell. Syst. **28**(2), 15 (2013)
11. Expectation maximization–to manage missing data, http://www.sicotests.com/psyarticle.asp?id=267

Secure Opportunistic Routing for Vehicular Ad Hoc Networks

Debasis Das(✉) and Harsha Vasudev

Department of CS&IS, BITS Pilani-K.K. Birla Goa Campus,
NH 17B Bypass Road, Zuarinagar, South Goa, Goa 403726, India
debasisd@goa.bits-pilani.ac.in, harshavasudev.dev@gmail.com

Abstract. Vehicular Ad-hoc Networks (VANETs) is a growing interest and research area over recent years for it offers enhanced safety and non-safety applications for transportation. Vehicular Ad-Hoc Networks (VANETs) is a developing technology that is yet unclear to many security issues. Security and routing concerns that are unique to VANET present great challenges. In this paper, we have proposed a scheme basically solve the black hole and gray hole attack in vehicular ad hoc networks. Black-hole and gray hole are the well-known routing attacks through which malicious nodes try to downgrade the communication performance of VANETs. In this paper, we focus on recent proposals that aim to enhance VANETs security and routing in systematic and architectural approaches.

Keywords: VANET · Security · Routing · Black-hole and gray hole

1 Introduction

A vehicular ad hoc network (VANET) uses cars as mobile nodes means dynamic nature of the node. Vehicular networks may require disseminating information to specific geographical areas. Generally two types of VANETs are Vehicle-to-Vehicle (V2V) and Vehicle-to-Road Side Units (V2R). Applications for vehicular networks may require disseminating information to specific geographical areas. Communication, Routing, and Security are the key research issues in VANETs. The research challenges for applications of VANETs remains to design of Intelligent algorithms is based on security and opportunistic routing [1,2] for the Intelligent transportation systems (ITS) to improve safety on the roads/highways/intersections [3]. The VANETs are a special case of MANETs due to the high mobility of nodes as well as a direction of the vehicular nodes. Managing this mobility is very important to ensure secure routing efficiency. A number of existing secure routing protocols are available, but none have been designed for vehicular networks and are not suitable for these networks mainly due to the constantly changing topology.

Opportunistic Routing (OR) [6] could be an extremely new routing technique that's an efficient, economical and heightens routing scheme for wireless networks

S.C. Satapathy and A. Joshi (eds.), *Information and Communication Technology for Intelligent Systems (ICTIS 2017) - Volume 1*, Smart Innovation, Systems and Technologies 83, DOI 10.1007/978-3-319-63673-3_25

(shown in Fig. 1) and the special case of a wireless network is VANETs [7]. In VANETs, the most challenge is that the extremely dynamic nature of vehicular nodes (i.e., vehicular node mobility) on roads or highways, so the extremely dynamic topology suggests that the dynamic topology changes over times mean frequently. The problem of OR is Black Hole Attack [4] at highway/intersection. Here the selected vehicle disseminates false information to neighbours' vehicles and sometimes the necessary information may be lost. The proposed Trust Value based OR model for black hole and gray hole attack [5,10] assumes that each vehicle which received the information is a source or starting the vehicle in a highway/intersection and participates in the forwarding of information. Actually, in the black hole and gray hole attack, the fraud vehicle receives the information from the source vehicle and all other immediate neighbors' vehicles drops the information, thinking that the selected vehicle will forward it. But in effect, the important information is dropped.

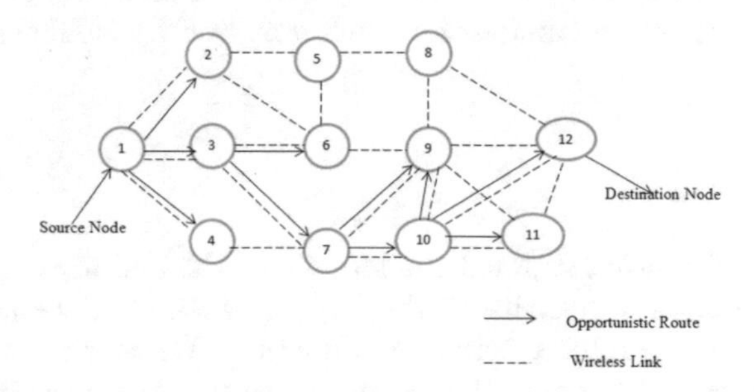

Fig. 1. Example of opportunistic routing in VANETs

Apart from security solutions such as cryptographic approaches [11–13] which consider the integrity of the data being transmitted, a category of attacks relates to the behavior of nodes in the routing. This may include injecting false information [10] in the network or preventing from collaboration with other ones when necessary. Our investigation and resolution development can detain mind the state of our transport material (city roads, highways, intersections, etc.) and track management policies to create our resolution deployable. This may offer the variety of different blessings to the social, material and therefore the setting like reduced hold up at town roads, cut back pollution, reduced track accidents, and then for the remainder of the paper is organized as follows. Section 2 describes the outline of the connected existing works. Section 3 explains the objective of the proposed works. Section 4 illustrates the planned algorithm for Secure Routing in VANETs. Section 5 describes the Implementation of secure routing in VANETs. Finally, we've conferred the conclusion in Sect. 6.

2 Literature Review

Geo-Opportunistic Routing is protocol planned by Kevin C Lee. Lee et al. [9] for conveyance Networks. The planned TOGO protocol for vehicular networks comes with topology assisted geographic routing with time serving to forward. Discrete-Time Markov Chains (DTMC) [5] is one of the important technique for addressing black-hole attack in Opportunistic Routing (OR). Cha [13] et al. Proposed a scheme, the Dual Weighted Trust for a stable Peer-to-Peer mechanism, by measuring the data transmission process based on data type and mobility. In that mechanism whenever a malicious node enters the network, it will be excluded by using reporting mechanism. Saheli et al. [10] introduced a new paradigm using Discrete Time Markov Chain (DMTC) for unicast opportunistic routing protocols in Wireless mesh networks. Here packet drop ratio calculation is also done for finding the uncooperative nodes. Mihaita et al. [12] presented the analysis of various cryptographic algorithms in the context of vehicular networks. Here, high levels of trust and security are guaranteed using various cryptographic algorithms. Vehicular Clouds (VC) [11] an extension of conventional cloud or in another way, VANETs to clouds. In VC, computing, vehicular resources, computing power, storage, internet connectivity can be shared between drivers. This paper mainly identifies and analyses a number of security challenges and threats in VCs.

3 Objective

Vehicular Networks have attracted the interest of the real life community, since many issues are open, especially in the research area of efficient routing techniques at highway and Intersection or Junction. While the most important objective has clearly been to enhance the general safety of conveyances traffic, trade labs, and the world are exploring novel conveyance applications like traffic management and on-board recreation. Rising conveyance applications typically necessitate wide-area coverage victimization multi-hop routing protocols for VANETs, that is that the major departure from safety applications that require solely native coverage. Underneath this circumstance, most impromptu routing protocols that discover and manage end-to-end methods is a smaller amount desirable as a result of high protocol overheads.

4 Proposed Model for Secure Routing in VANETs

The Opportunistic Routing (OR) based propose a scheme responsible for increasing the reliability of routing in Vehicular Ad-Hoc Networks (VANETs) [16]. Such type of routing protocol for multi-hop wireless networks [15] will be be less effective if some vehicular nodes in the wireless multihop environment act as a selfish or a malicious node at highway/interaction. We have proposed an intelligent algorithm for malicious node detection and solution is based on cryptographic algorithms. So our proposed secure routing algorithm based system designed to improve the security of data transmission over wireless networks.

4.1 Authentication

Authentication is very essential for effective Security. Authentication Problem is crucial in VANETs because if the vehicles failed to prove their authenticity (i.e., malicious) then they will not be allowed in the further communication process. The On-Board Unit (OBU) have proper mechanisms for checking the authenticity of both vehicle and driver. Below some authentication mechanisms are provided.

Existing Solutions. These are some existing mechanisms:

(1) Driver Ownership: A driver owns some unique identity (i.e., identity card, driving license etc.).
(2) User Knowledge: A user knows some unique things (i.e., passwords, human responses through secret questions etc.).
(3) Biometrics Solutions: These include the signature, thump expression, face and voice.

Proposed Scheme. All of these can be combined with the following new methods for establishing proper authentication.

1. Digital Licence Plates (DLPs) or Electronic Licence Plates: Each vehicle manufacturer should provide. Digital Number Plate mechanism for identification. This is the most powerful key for unique identification of a vehicle.
2. Driving Licence: This is a mandatory document that a driver should have.
3. A face recognition system is maintained to verify the driver's identity. This mechanism is added for high security, in the cases where the vehicle is legitimate and driver is a fraud. When the driver is fraud, he/she may have correct licence, but fails in the face detection system.

4.2 Trust Value

Trust plays a vital role in secure communication and Routing. Each vehicle should maintain a trust value. The possible values are 0 and 1. Trust value can be calculated using the above authentication techniques. By using these parameters such as Digital Licence Plate (DLP), Driving Licence and Face recognition system, trust value can be easily calculated. For example, if a vehicle have correct DLP, Driving Licence and Face recognition check then the trust value will be 1 and it is treated as a trusted vehicle. If a vehicle fails to prove any one of the parameters, then trust value will be 0, and the vehicle is treated as malicious. When a vehicle identified as malicious, then it will not allow to participate in the further secure communication process.

4.3 Secure Routing

It includes the following functionalities:

Select Shortest Route. Using OR mechanisms, vehicle coordination and actual forwarder (vehicle head) selection can be done. Not only other vehicle status, but also Environmental Conditions also checked, i.e., weather conditions, traffic jams, fog, road conditions etc.). This system will work for Multihop networks also. Whenever there is a large number of vehicles within the communication range of each other, then the information regarding next vehicles and environmental conditions are passed between the intermediate vehicles and at last the destined vehicle, through this secure routing method. OR algorithm finds a shortest path between legitimate vehicles only, because it avoids malicious vehicles through authentication and trust based calculation.

Malicious Vehicle Detection. Whenever a vehicle enters a communication range of other vehicles than its trust value is calculated based on the parameter value (i.e., Digital Licence Plate (DLP), Driving Licence and Face recognition system). If the calculated value is 0, then it is malicious and suddenly the actual forwarder vehicle broadcasts the information, that is, unique number of the vehicle to all vehicles within the range, and will not allow to participate in the routing process. Before sending sensitive information to each vehicle, the actual forwarder gets the knowledge of trustworthiness of each vehicle. Through that way the Malicious vehicle is detected and ensures that the message passes only through trusted vehicles.

5 Implementations

In our proposed scheme, we consider two different scenarios in the route discovery process: the first one only considers the path trust value factor, and the second one both considers the hop count factor and path trust value: (i) normal nodes, which the data packets are normally forwarded by those nodes; (ii) malicious nodes, which randomly drop the data packets when they receive the packets.

5.1 Simulation Environment

Our simulation model is built on the OMNET ++ and our model spans the area of $5 \times 5\,\text{km}^2$. We deliberate three different situations. In our simulation, the simulation parameters are as follows:

- The simulation time is 15 min.
- The node density of our proposed model is 25 nodes.
- The physical layer, and MAC layer support IEEE 802.11.

5.2 Results

Two metrics are evaluated, they are Network throughput and Average end-to-end (ETE) delay. In Fig. 2 represent the throughput comparison of Proposed Scheme, SD-TAODV and traditional AODV [14] in different data rates. In Fig. 2, we only consider the trust value factor. Through the Fig. 3, we can conclude that the performance of the proposed Proposed Scheme mechanism is better than the traditional AODV protocol and SD-TAODV. The average end-to-end delay of the proposed Improvised trust based ad hoc on-demand distance vector routing scheme is evaluated through the Fig. 3. In Fig. 3, we can see that the end-to-end delay of Proposed Scheme is lower than that of the traditional AODV and SD-TAODV [14] with different data rates.

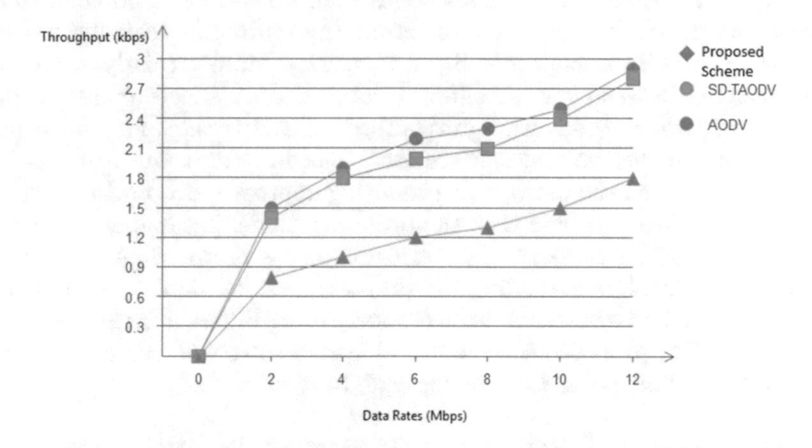

Fig. 2. Average throughput comparison

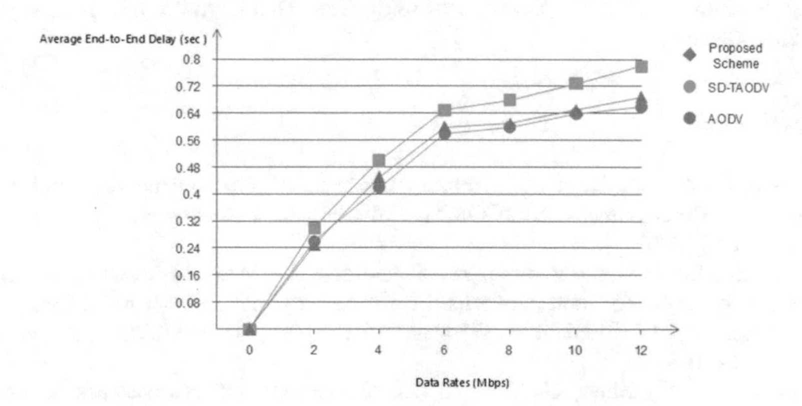

Fig. 3. Average end-to-end (ETE) delay comparison

5.3 Evaluation

In our proposed system, a trust based mechanism is defined for checking malicious vehicles [14]. The throughput of the proposed system is systematically relatively high because we are considering both forwarding and reversing node ratio. The delay of our system is very less. Authentication mechanism also used for checking the correctness of each vehicle. We compared our system with original AODV and TAODV [15,16] schemes, and it shows that our system is somewhat better than these existing systems.

6 Conclusion

In this paper, the effects of malicious nodes were studied for opportunistic routing protocols, which is one of the most promising routing in vehicular networks, where each vehicle have high mobility and density. More precisely, we assumed that the candidate set follows an effective coordination. A new analytical model designed and implemented using authentication and trust value which proves the existence of malicious nodes. More than that, in order to measure the effect of malicious nodes on the network, a promising approach, drop ratio calculation was introduced. We conclude that the proposed model is capable of validating the effects of malicious nodes on opportunistic routing protocols. A potential way of future works involves extending the proposed systematic model to include a defensive mechanism against malicious nodes, using different trust and reputation methods. Our proposed scheme introduced to solve the traffic jams, reduce the accidents and air pollution will continue to reduce.

Acknowledgement. This work is partially supported by Early Career Research Award from Science & Engineering Research Board (SERB), Department of Science & Technology (DST), Govt. of India, New Delhi, India (Project Number: ECR/2015/000256).

References

1. Biswas, S., Morris, R.: ExOR: opportunistic multi-hop routing for wireless networks. In: Proceedings of SIGCOMM, Philadelphia, Pennsylvania, USA, pp. 133–144. ACM (2005)
2. Tahooni, M., Darehshoorzadeh, A., Boukerche, A.: Mobility-based opportunistic routing for mobile ad-hoc networks. In: Proceedings of the 11th ACM Symposium on Performance Evaluation of Wireless Ad Hoc, Sensor, & Ubiquitous Networks, pp. 9–16 (2014)
3. Cai, X., He, Y., Zhao, C., Zhu, L., Li, C.: LSGO: link state aware geographic opportunistic routing protocol for VANETs. EURASIP J. Wirel. Commun. Netw. **2014**, 1–10 (2014)
4. Yuan, P., Fan, L., Liu, P., Tang, S.: Recent progress in routing protocols of mobile opportunistic networks: a clear taxonomy, analysis and evaluation. J. Netw. Comput. Appl. **62**, 163–170 (2016)

5. Salehi, M., Darehshoorzadeh, A., Boukerche, A.: On the effect of black-hole attack on opportunistic routing protocols. In: PE-WASUN-2015, pp. 93–100 (2015)
6. Boukerche, A., Darehshoorzadeh, A.: Opportunistic routing in wireless networks: models, algorithms, and classifications. ACM Comput. Surv. **47**(2), 1–26 (2015)
7. Castineira, F.G.: Opportunistic routing and delay tolerant networking in vehicular communication systems. In: Vehicular Communications and Networks, pp. 114–126. Elsevier (2015)
8. Lee, K.C., Gerla, M.: Opportunistic vehicular routing (invited paper). In: IEEE European Wireless Conference 2010, pp. 873–880 (2010)
9. Lee, K.C., Lee, U., Gerla, M.: Geo-opportunistic routing for vehicular networks. IEEE Commun. Mag. **48**, 164–170 (2010)
10. Salehi, M., Boukerche, A., Darehshoorzadeh, A.: Modeling and performance evaluation of security attacks on opportunistic routing protocols for multihop wireless networks. Ad Hoc Netw. **50**, 88–101 (2016)
11. Yan, G., Wen, D., Olariu, S., Weigle, M.C.: Security challenges in vehicular cloud computing. IEEE Trans. Intell. Transp. Syst. **14**(1), 284–294 (2013)
12. Mihaita, A., Dobrex, C., Mocanu, B., Popx, F., Cristea, V.: Analysis of security approaches for vehicular ad-hoc networks. In: 10th International Conference on P2P, Parallel, Grid, Cloud and Internet Computing, pp. 304–309 (2015)
13. Cha, H.J., Yang, H.K., Kim, Y.H.: A node management scheme for stable P2P service in mobile ad-hoc networks. Peer-to-Peer Netw. Appl. **9**, 558–565 (2016)
14. Zhang, D., Yu, F.R., Wei, Z., Boukerche, A.: Trust-based secure routing in software-defined vehicular ad hoc networks. ACM (2016)
15. Das, D., Misra, R., Raj, A.: Approximating geographic routing using coverage tree heuristics for wireless network. Wirel. Netw. (WINE) **21**(4), 1109–1118 (2015). Springer, US
16. Das, D., Misra, R.: Caching algorithm for fast hand off using AP graph with multiple vehicles for VANET. Int. J. Commun. Netw. Distrib. Syst. (IJCNDS) **14**(3), 219–236 (2015). Inderscience Publishers Switzerland

A Framework to Collect and Visualize User's Browser History for Better User Experience and Personalized Recommendations

Harish Kandala[1], B.K. Tripathy[1], and K. Manoj Kumar[2(✉)]

[1] School of Computing Science and Engineering, VIT University, Vellore, 632014, Tamil Nadu, India
{k.harish2013,tripathybk}@vit.ac.in

[2] Department of Computer Science and Engineering, Sri Venkateswara College of Engineering (SVCE), Tirupati, 517507, Andhra Pradesh, India
kandalamanojkumar@gmail.com

Abstract. In all the modern browsers, maintaining user's web history is one of the primary tasks. Browser history will help to summarize the activity of the user during a certain period. However, current browser history is not so efficient to visualize in a user-friendly manner and also doesn't provide enough information for personalized recommendations. One of the key reason is that browsers never maintain any inter-connection between history items. Overall history is maintained in a linear fashion with no information about how the user reached to a particular state. Another issue is that it is not possible to calculate how much time the user spent on any particular website using current history system. This paper provides a conceptual idea of solving these issues by providing a framework that solves this issue by introducing linked data and also describes how this can benefit in improving user experience and quality of recommendations.

Keywords: Browser history · User experience · Data extraction · Interactive visualization · Personalized recommendations

1 Introduction

Browser history plays a key role in understanding user behavior in the web. Using the history data one can provide recommendations to the user personalized to them which many search engines does this already. But, there are few issues with the current way of collecting browser history. The current system doesn't provide enough information to extract useful links between user activities. This disadvantage doesn't limit to information mining but also to the user experience. A linear list of history items is not an effective way to visualize any activity. Activity does have links between them. Activities are better to be visualized as tree or graph rather than lists. These links are the important part in which the current history system lacks. This paper provides few specifications to how to collect data from browser and also provides few UI prototypes of activity

S.C. Satapathy and A. Joshi (eds.), *Information and Communication Technology for Intelligent Systems (ICTIS 2017) - Volume 1*, Smart Innovation, Systems and Technologies 83, DOI 10.1007/978-3-319-63673-3_26

visualization. And this paper also discusses how the provided way of collecting history is better for recommender systems than the current way of doing it.

When it comes to user experience it depends on the user whether the proposed system actually makes any benefit to them. Though not all users care about their activity in the browsers but it is essential to have these specifics for professional users and researchers which generate a meaningful summary of their valuable activity. Consider a user who is researching on a particular topic and opened various links and from those links opened few more links and crawled deep into the internet. If the user wants to revise his research after a few days or weeks it will make no sense with all those lists of links. And also, if the user wants to share his research with his friend just by sending a list of links won't give a clear summary to the other person. The proposed system can be an extension of the current way of visualizing history.

And when it comes to recommendation engines having extra information definitely improve the quality of recommendations. There is a lot of research going on for providing recommendations using linked data. This framework solves the data extraction part of it for browsers. User privacy should also be taken into concern when designing such recommender systems [1]. Search engines like Google, Bing etc. already collect this way of linked data for better recommendations but that is confined only to that search engine [2]. Extending it to the higher level like the browser can have a huge impact on the user personalization.

The next sections of this paper discuss the issue with the current history item in detail by considering a real-life scenario and introduce a set of data properties that are preferred to be extracted from the browser. It also provides an UI prototype of how to visualize history timeline. And the final sections describe how this framework can improvise the personalized recommendations.

2 Background

Let's say the user has few questions or doubts about a certain topic. Like as expected any user will open his favorite search engine and enter this search query. But no search engine will give the clear answer to the user's query directly unless it is a general one like weather, time etc. As per user's query, search engine tries to get best possible links that can answer his question. All of those results have bits and pieces of information that are relevant to the user's query. So, the user opens various links and explore through the internet by opening lots of links. Once the user collected all the relevant information from those links he will close them.

Now the user wants to revisit those links or user got the same question after few weeks but don't have a clear idea what the solution is. With all the current features in a browser, the user will open his history but none of them will make any sense because there are hundreds of links in the history. Or in another way, the user will query it again in the search engine. In both ways, the user is wasting his time in going through the tedious repeating task.

Even though we are living in a data-driven society, as more and more databases are brought online [3], huge amounts of data are at fingertips and at the same time getting useful information from the available complex and enormous data is totally dependent on how well we can search and crawl through the internet. In technical terms, we can say that we have transitioned from the Classic Retrieval Model, to what is called, Berry-picking Search.

Any query won't be satisfied by the single final retrieved result but by a collection of various bits and pieces of information at each stage of the search. In other words, we do not usually query for something that can lead to a simple and single final result that clarifies our question [4], rather we search for terms and then explore the internet, connecting all the available bits and pieces of the answer as we crawl and read through the web of tabs.

Our search needs, including our browser history, cannot effectively handle our query with a single final source. We move through multiple sources with every new piece of information giving us new ideas and directions to follow. With the available pieces of information, we start searching on another topic and this search pattern will keep on fluctuating until we collected all the pieces of information.

And in another perspective, there is a lot of difference in opening the same domain from different parent sites. Like opening YouTube video from an educational site is a lot different from opening the same domain from entertainment sites. Without these parent links, recommender systems can misunderstand the user purpose and can give erroneous recommendations. The proposed specifications keep all these things into concern and provide an effective solution.

3 Extracting Essential Data from Browsers

In order to understand the current way, the browser history is collected it is good to look into the History API's of popular browsers. Here, Google Chrome History API is used. This API can provide all the history data that is being collected by the browser. By looking at the collected data, it shows that information available per history item isn't concise. And no relevant information about the parent item is available. All individual history items are independent of the other. A similar scenario is observed in other browsers too (Table 1).

Table 1. List of properties in the Chrome History API

Property	Description
id	The unique identifier for the item
url	The URL navigated to by a user
title	The title of the page when it was last loaded
lastVisitTime	When this page was last loaded, represented in milliseconds since the epoch
visitCount	The number of times the user has navigated to this page
typedCount	The number of times the user has navigated to this page by typing in the address

Our proposed framework suggests including the parameters mentioned in Table 2 to the History API. It is also observed that all the required data parameters included in the specifications are available to the browsers but they aren't stored by the History API.

And one should note that all these parameters may not be necessary for the certain task. Some of them might not be relevant for information mining like faviconUrl and windowType but they are relevant for improving the user experience of the history [5]. One can extract these parameters from either the provided API's of browser or it would be much better if browsers included these parameters into the History API by default.

Table 2. List of properties in the proposed history framework

Property	Description
id	The unique identifier for the item
parentId	Identifier of the referrer of the given item. 0 if none
transitionType	Transition type for this visit from its referrer. Following are the possible transition types: **linked:** If user reached this link by opening the link in a new tab from its referrer. The referrer is still available in this transition **overwrite:** This transition is same as the above except that link is opened in the same tab of the referrer. Referrer item will be overwritten with the current item **toplevel:** In this case, there is no transition and user reached this item either by manually entering the link or through bookmarks. parentId for this transition will be 0
url	The URL navigated to by a user
faviconUrl	URL for the favicon of the navigated item
title	The title of the page when it was last loaded
windowType	Describes the type of the window in which the current item resides **normal:** Regular window type where the user browse **popup:** Pop-up windows are usually used to display advertisements which are not relevant in history
startTime	Time when the user created this history item, represented in milliseconds since the epoch
endTime	Time when user closed this history item, represented in milliseconds since the epoch
activeTime	Contains intervals of time where the user was viewing this item during its life period, represented in milliseconds since the epoch
lastUpdated	Latest updated time of this item, represented in milliseconds since the epoch

Compared to the original History API, one can notice that few properties are added and few are removed in the proposed one. Few properties are removed because of their redundancy. Properties like visitCount and typedCount are derived attributes they can be derived from the collection of history items. Including them in every history item isn't necessary and they are redundant [6].

Properties that are added are parentId, transitionType, windowType, startTime, endTime, activeTime and lastUpdated. parentId and transitionType serve the purpose of defining interconnections between various history items and also describe the type of connection between those items. windowType is used to distinguish between relevant and irrelevant history items.

During most of the time even though multiple tabs were opened in a browser, at any given instant the user can handle only single tab. So just having startTime and endTime will give inaccurate estimations of user activity. Including activeTime can help to reduce this inaccuracy and give more detailed and accurate information.

4 Visualizing the Extracted Data

This UI prototype is a conceptual idea of visualizing the browser history by making use of the extracted data for better user experience. There are two components in the prototype. They are history item and the timeline.

Each history item can be visualized as rectangular bars where the length of the rectangle denotes the duration of the item [7]. These history items will have connections to its parent history item. And each history item can also visualize its active time by using different colors for an active and inactive time. In the Fig. 1 active time uses gray color and inactive time has a white color. Color schemes are up to the designer and developer. All the history items can be placed in a scrollable timeline where the user can freely navigate in the timeline. The user can also be able to scale the timeline (Fig. 2).

Fig. 1. UI component of the history item

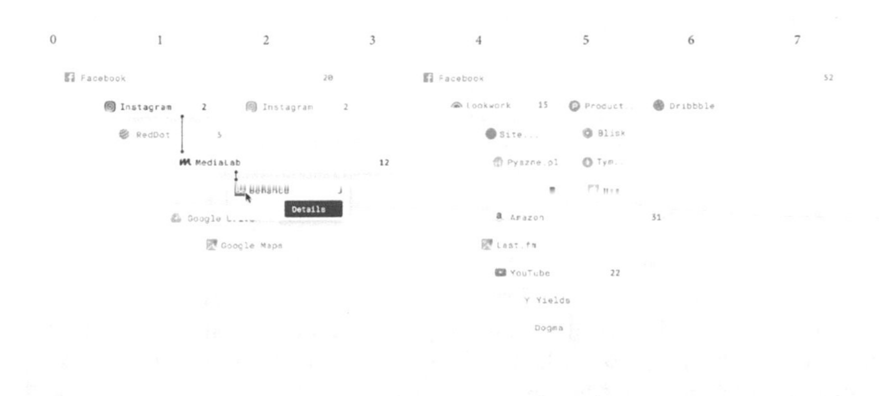

Fig. 2. Prototype of the history timeline (when the user hovered on a history item)

5 Improving the Quality of Recommendations

Personalized recommendations can be either helpful to the user or annoy the user. Lack of information can lead to faulty estimations of user behavior and annoys the user with poor recommendations [8]. Including additional data that give more connected information can most probably improve the quality of the recommendations. There are various methods and techniques to make use of the linked data for recommender systems which are out of the scope of this paper [9].

6 Future Work

Apart from the visualization of the history timeline and recommendations, whole history data can also be used to produce the statistics of the user activity and these statistics can be used to develop a model that understands the user and help the user to focus on his task when the user is being diverted too much from his work.

7 Conclusions

The web browser is one of the most used software for any user. Being one of the most used software, activity of browsers isn't given much importance. All the data that can be extracted from the browsers aren't being utilized for understanding user behavior. And they aren't being used for providing a better user experience too. This paper tried fixing the flaws in the current browser history system and provided a framework introducing a new way to collect and visualize history data. It also showed that linked data can substantially improve the quality of recommendations.

References

1. Jany Shabu, S.L., Manoj Kumar, K.: Preserving user's privacy in personalized search. Int. J. Appl. Eng. Res. (IJAER) **9**(22), 16269–16276 (2014). ISSN/E-ISSN: 0973-4562/1087-1090
2. Manoj Kumar, K., Vikram, M.: Disclosure of user's profile in personalized search for enhanced privacy. Int. J. Appl. Eng. Res. (IJAER) **10**(16), 37261–37266 (2015). ISSN/E-ISSN: 0973-4562/1087-1090
3. Manoj Kumar, K., Tejasree, S., Swarnalatha, S.: Effective implementation of data segregation & extraction using big data in E-health insurance as a service. In: 2016 3rd International Conference on Advanced Computing and Communication Systems (ICACCS), Coimbatore, pp. 1–5 (2016)
4. Praveen Kumar, R., Manoj Kumar, K., Tejasree, S., Aswini, R.: Review on cost effective and dynamic security provision strategy of staging data items in cloud. Res. J. Pharm. Biol. Chem. Sci. (RJPBCS) **7**(6), 1592–1597 (2016). ISSN: 0975-8585
5. Ter Louw, M., Lim, J.S., Venkatakrishnan, V.N.: Extensible web browser security. In: International Conference on Detection of Intrusions and Malware, and Vulnerability Assessment, pp. 1–19. Springer, Heidelberg (2007)
6. Jang-Jaccard, J., Nepal, S.: A survey of emerging threats in cybersecurity. J. Comput. Syst. Sci. **80**(5), 973–993 (2014)

7. Wang, L., Xiang, J., Jing, J., Zhang, L.: Towards fine-grained access control on browser extensions. In: International Conference on Information Security Practice and Experience, pp. 158–169. Springer, Heidelberg (2012)
8. Isinkaye, F.O., Folajimi, Y.O., Ojokoh, B.A.: Recommendation systems: principles, methods and evaluation. Egypt. Inf. J. **16**(3), 261–273 (2015)
9. Rouzbeh, M., Davis, J.G.: Recommendations using linked data. In: Proceedings of the 5th PhD Workshop on Information and Knowledge, 02 November (2012)

Big Data Analytics Towards a Framework for a Smart City

Devesh Kumar Srivastava(✉) and Ayush Singh

School of Computing and IT, Manipal University Jaipur, Jaipur, Rajasthan, India
devesh988@yahoo.com, mail.ayushsingh@gmail.com

Abstract. While collecting and gathering the data from various domains of a city, one never fail to criticize the city officials as to how they have improved the conditions of livability, economic development and sustainability. A city is always taking initiatives to augment the existing framework or to implement a new framework altogether for transportation system, solid waste management, water supply system and other mission specific domains by gathering relevant data. This process of endless growth of data marts and datasets and analyzing it to gain valuable insight into the domain is being known by the business society today as Big Data Analytics. In our research work, we discussed the challenges and hurdles that are faced during the development of a smart city. Along the way, we will also make use of some of the recent advances in cloud computing, sensing technologies and provide a framework for the smart city using internet of things which can effectively contribute in the enhancement of current models of smart city. We discussed the components of our framework which is proposed for the city, and analytics for large datasets with their applications in various domains.

Keywords: Internet of things (IOT) · Information and communications technology (ICT) · Integrated data centers (IDC) · BUCI · MUCI

1 Introduction

For the last 15 years, we are witnessing a worldwide population boom all around the globe, which is forcing people in small areas to move to big cities to survive in severe and extreme conditions. This move forces a drastic change in lifestyle related activities of people due to various environmental factors that we see in Metropolitan Areas. As these cities were originated and developed as a part of historical planning that was adequate for the population of that time, today these cities are facing several problems and crisis in various domains. The most visible form of mismanagement that can be seen in today's life is in the cities' Public and Private Transportation System, Water Sewage System, Communication Systems, Solid Waste Disposal System etc. Because of the rising populations in metropolitan areas and key hub areas, even the task forces are facing daily problems and are incapable of finding solutions to these issues [1]. The Environmental Factors are just acting like a cherry on top of a cake. People avoiding public transportation systems and still using their personal vehicles for transportation are the root cause for increasing Noise, Air and Traffic pollution. More than 59% of Air pollution in New Delhi is because of pollutants released by an average car in 2015. Rest

S.C. Satapathy and A. Joshi (eds.), *Information and Communication Technology for Intelligent Systems (ICTIS 2017) - Volume 1*, Smart Innovation, Systems and Technologies 83, DOI 10.1007/978-3-319-63673-3_27

is covered by Industrial emissions and CFCs from common home appliances. A recent study conducted by Delhi Flood Control Department in 2015 cited that more 20 mm of rainfall can put lower areas of the city at a heavy risk of Water Clogging which could result in heavy disruption of daily commute [2]. In August 2016, A rainfall of more than 100 mm virtually closed the National Highway No. 8 and I.G.I. Airport Terminal 1 and 3 which caused total meltdown of transportation system costing the State Government millions and resulted in Cancellation of more than 63 Domestic and 22 International Flights from Airport citing a revenue loss of more than 66 million rupees in a single day [3].

In order to propose ways to meet the challenges of developing a Smart City, this work is aimed towards a framework which lays out a heterogeneous taxonomy for development of a smart city. The study also aims at using basic and advanced ICT Technologies for the implementation practices and identifying risks and opportunities for major stakeholders in smart city. Objectives of this work are as follows:

1. To develop a basic layout efficiently and a conceptual framework of a Smart City Implementation Practice and to do an assessment of loopholes and how to effectively overcome them.
2. To strategically use IoT Technologies and equipment to make the implementation more time and energy efficient and gather large datasets and analyze them for development and regulatory purposes of the framework.

This research work would fill some conceptual and practical loopholes for a heterogeneous approach in the topic of developing a smart city and specializing in the implementation of services like telecom infrastructure and the fundamentals of governance.

2 Literature Background

The basic idea of developing a smart city originated due to the genesis of buzzwords like 'tech city', 'intelligent city', 'digital city' etc. All these buzzwords basically define one common ideal and a single approach. But, the fact worth noting is that the idea of building a smart city and its governance is itself very inconsistent and unclear. The definition of a 'hi-tech city', 'intelligent city', 'modern city' or an 'electronic city' are sending out a wave of ideas that perceive the city to be technically abled and data driven, ICT is considered to be the driving force behind the ideal. 'Intelligent City' collects data from different localities and sectors and provides it in form of information to the public via the biggest public resource, Internet. Other scholars explain a similar definition that 'a smart city is a community which is connected through an efficient infrastructure of communication which is flexible, service oriented and powerful and is based on open standards of industry; where innovative services completes and assists the key stakeholders in completing their tasks efficiently and timely. Further discussions have revealed that indeed, a 'tech city' is based all upon collection of data and using it to further close the loopholes in the framework and the infrastructure and help in the implementation practices by the assistance of IoT devices [1]. The 'digital city' defines a city which makes ubiquitous data available across an urban intelligence embedded infrastructure which uses intelligent equipment in the streets, bridges, roads like street lamp, traffic light etc. The term 'ubiquitous city' originates from South Korea where they define a smart city as a city which is managed by networked infrastructure and provides its citizens with much needed services through it with either a BUCI (fixed u-City infrastructure) or a MUCI (mobile u-City infrastructure) [6], providing and taking assistance of advanced sensors and actuators. The underlying idea of a BUCI city stresses the fact of converging of all IT services within a metropolitan city which must be accessible from every location and should be available anytime. That's where from, an idea of a 'digital city' originates [7]. However, its main goal is to emphasize on social development towards a human to human perspective (Fig. 1).

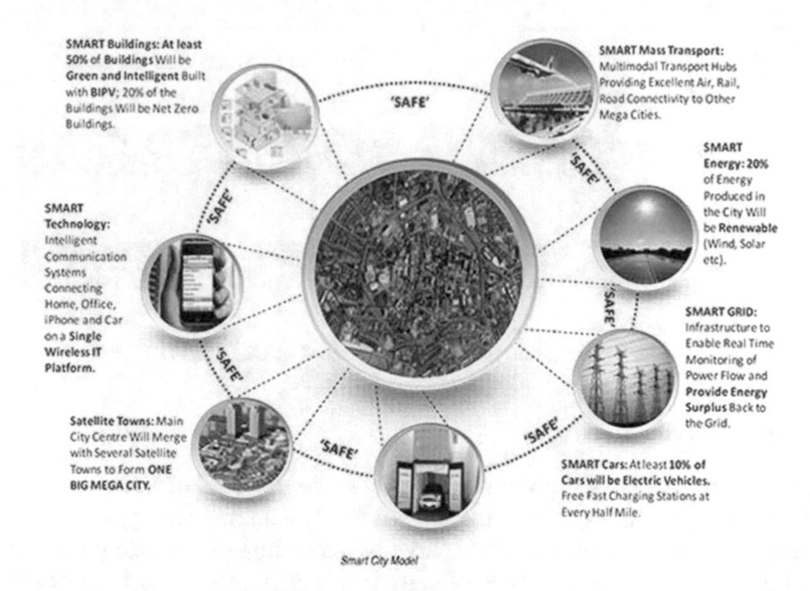

Fig. 1. View of a smart city

This multi directional flow of data is analyzed and is used by organizations (public or private) to provide us with more accurate services and simultaneously improving their own recommendation system by asking users their preferences in forms of feedback and surveys [7]. This data exchange and analysis is only possible after user consents to its data for sharing and analyzing processes.

3 Data Analysis

The following research study is based on multiple cases in order to address the ever growing, evolving field. Today, we are seeing different cities from around the world employing and adapting smart services and applications with respect to different strategies. While, some of the cities have already started to implement their long planned smart green framework, other are still at an earlier stage of planning or are beginning with one-step-at-a-time approach. As a number of cities prepare for smart services and green framework, for the sake of this study, we are considering two of the vital cities in northern corridor of India, namely New Delhi (DEL) and Jaipur (JAI). The smart city projects for these two cities were initially recommended by the Planning Commission of the 2 states Delhi and Rajasthan under the Smart City Project of the Prime Minister of India [2]. This study than took suggestions and recommendations from volunteers and participants that were already acquainted with the smart city development from all around the world and the city officials who were assigned the task of building a smart city. In strict terms of its dimensions and its sub-dimensions, we evaluated a count of 62 and 39 individual services and projects that are being provided by DEL and JAI respectively (Fig. 2).

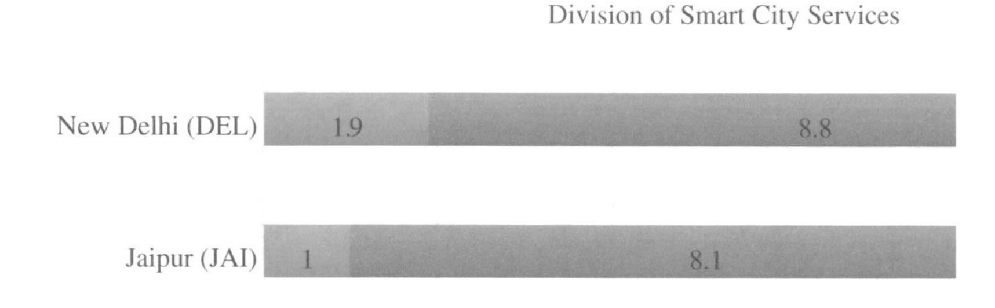

Fig. 2. Division of smart city services in DEL and JAI

3.1 Digital Freedom

Both, DEL and JAI has a defined understanding of the potential of an open source platform in boosting the transparency throughout the system and are, to their level best, exploiting those advantages as a vital source of innovation to generate public services to, in turn, can elicit a higher number of participants. It is a critical task for a smart city to maintain a balance between transparency of open data, encouraging developers and having constraints of availability of data. Figure below shows that DEL emerges with more open data platforms (APIs) then JAI and suggests that DEL is more active in terms of transparency of open data.

Both the cities, when it comes to public openness, take a distinguishable stand and either target a wide area approach throughout the field for public transparency or select a specified area for designing and development of app, e.g.- parks and recreation, sanitation. But polls from various studies show that a number of public departments are turning their heads away from the wide area approach because they are reluctant to open their data up due to its sensitivity or confidentiality. But rules and legislation highlighting the open data movement is stifling with the cross departmental laws and finally, results in a delay of data openness.

3.2 Dedicated Solutions

We will see in the following figure the division of approximately every service and solution that is being provided by DEL and JAI. It also depicts that, in opposition to the popular belief, JAI has scored higher percentage in terms of service diversity. The largest number of initiatives in JAI addresses the transportation sector (27%) and public facilities with their managements (20%). After the largest portion, we can acknowledge the tourism sector with a staggering (16%) followed by cultural recreation with (15%).

These services, smart or not, have been designed and developed by the ICT technologies and companies since 2013 and have been constantly supervised with smart city development officials. However, we recently found out that they are tilting to move towards a participatory service design and beginning to use an open data API platform (Fig. 3).

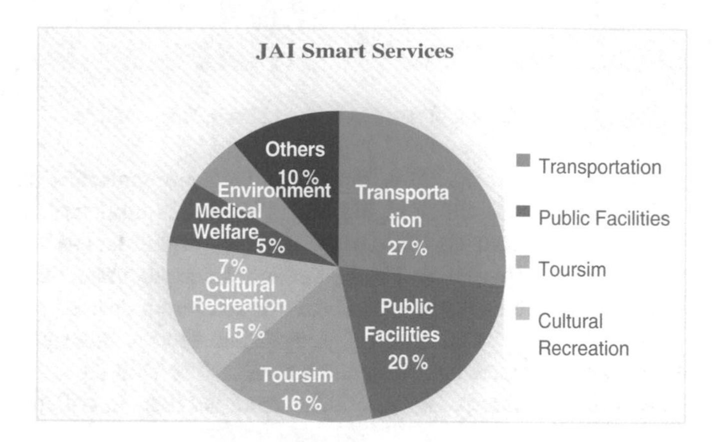

Fig. 3. Smart services in JAI

The outcomes of the study that were done on these services imply a weak correlation to the city's already functioning open data platform API, never mind the diversity of the services provided. Nevertheless, the open big data APIs in New Delhi has been very meticulously designed between the years of 2013 and 2014, strongly suggesting that DEL still retains the hopeful potential to rapidly evolve in the forthcoming years into a strong market oriented platform, but blame of the yesteryear's failure can be directed towards fickle government policies and population boom.

3.3 Creating Collaborations

As we depicted in the upcoming figure, our findings reveal that both DEL and JAI have created successful multitude of collaborations and partnerships for building a smart city. 91% of all the services in the JAI were created and developed by central government and either designed explicitly by local state agencies or was designed with contract to private firms and companies. Meanwhile, only a portion of 26% of the services has been outsourced under the state government in DEL. The government of Rajasthan has recently underwritten a good number of pilot and R&D projects in which a multitude of cities, like Mumbai, Pune, Bangalore and even New Delhi itself has shown interest and have bid for providing funds [4]. Most of these startup projects are initiated in a level of pilot projects or a test bed created for further development of services. However, most of the services provided by the JAI have been developed from a huge master plan or in some instances, have been completely left with unsuccessful IT divisions (Fig. 4).

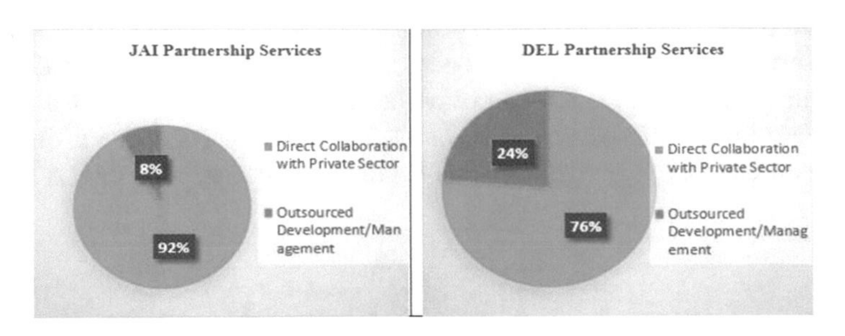

Fig. 4. Comparison of partnership services in DEL and JAI

Resulting features indicate that DEL and most metropolitan cities have already adopted a model of governance which highlights their private-public collaborations as a salient feature using an approach of bottom-up fashion, whereas Jai has been a leading pioneer and a strong initiator of the top-down approach. Being accurate, DEL is concentrated on building a structured eco system in order to deliver value added services and still being less involved than usual in redesigning urban spaces, like by creating geo tagging technology or geo fence. JAI has gone a step further with the development of its infrastructure while simultaneously providing more services in collaboration with the private sector.

3.4 Modern Sustenance

This will come in context with the collaborations developed with public and private sectors as a sustainable model is based on the systems we are using being intelligent by them and provide services as such. In figure below, most of the services in both the cities are based on the use of GPS technology. Nearly, 3 quarters of the services are developed and works by utilizing GPS. JAI is pioneering the field in the exploration of the wide range of intelligent approaches which features some 38% of its own services [2]. These services will, in the long run, add the value to the unintegrated urban spaces. In DEL, ignoring the use of GPS technology, some smart services use the artificially intelligent analytical processing tools for real time information domains like Air Traffic Control Systems, Transportation, Cab Services, and Metro Projects etc. Based on the case analysis of the services, it also appears that DEL smart services also has a less footprint and hence promotes environmental sustenance. More than 45% of the services of DEL directly or indirectly affect the environment in an adverse way, whereas the tally in JAI is above 62% easily. But state governments have taken necessary actions to ensure that every organization, small or large, charging big tax figures according to their electricity usage and carbon footprint determined by their appliances. Over 2 years, government has managed to reduce over 21,000 tons of local carbon emission and helped save those organizations tens of thousands to lakhs of rupees. DEL also integrates the all new GIS technology with its services to promote renewable sources of energy like wind and solar energy as a part of their 'GO GREEN' initiative (Fig. 5).

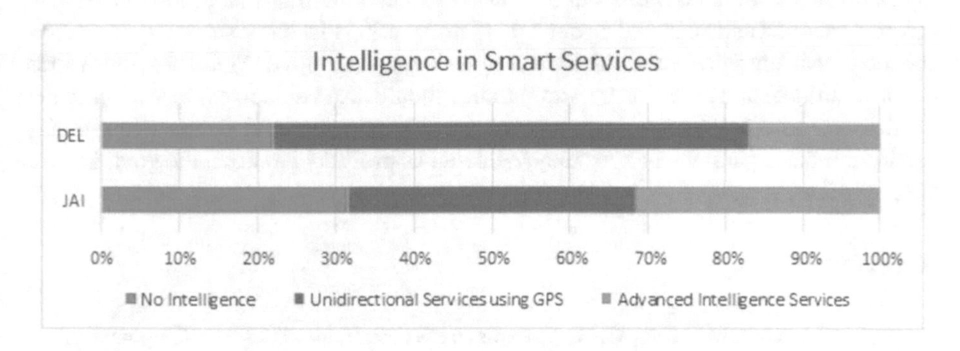

Fig. 5. Comparison of intelligence in smart services

3.5 Governance of a Smart City

DEL, over the past 6 years, had a plan in place. With its former chief minister and her cabinet, DEL was long planning to become a smart city and as the years concluded, it showed that its long term plan that it was designing since 2008, was more formidable and adaptable. It clearly had the better plan of governance, particularly the IT sector did significantly, partnering with different types of private organizations to design and develop services that are vital for citizens. Leadership is currently being shared with major corporations in order to avoid conflict and endorsing the approach of giving an equal opportunity to everyone to communicate in major decisions [8]. Even citizens were regularly polled and surveyed on public centers to share their view on major policy changes that were being considered by the municipal division. When compared to DEL, organizational infrastructure and ability of JAI is somewhat weaker basically due to its availability of more 'GREENER' services than DEL and has less direct or indirect impact. In 2013, the city's chief minister appointed the first chief innovation officer for its communication, transportation, IoT and other civic services. However, the efforts that were taken were overshadowed by limited internal resources as Rajasthan is a bigger state than DEL which is a Union Territory and a state [9].

4 Conclusions and Future Work

The findings of this research study have been difficult to visualize and depict because of the number of cities in India that have made their data publicly available for educational purposes and analysis. This study has been limited to only 2 cities, New Delhi (DEL) and Jaipur (JAI). To further confirm and refine this framework, it could be applied to development of other undermined and unfocused cities. If we compare India to other countries, US, for example, has San Francisco (SFO), New York City (NYC), Boston (BOS) etc. has the largest number of open APIs. More number of studies would strengthen our understanding of developing a smart city and services and help other scholars to further explore the area of development of a smart city. Other recommended topics for future work comprises of the questions like how smart cities and its services can be developed and analyzed through this framework. How other smart city initiatives and local small entrepreneurs with their companies and startups can reinforce innovation in this traditional and orthodox system. Studies should also begin to explore other scales of measurements for every single one of the 6 parameters and define evolutionary versions of these parameters after they validated as profitable to both, the govt. and the citizens of that city.

References

1. Lee, J.H., Hancock, M.G., Hu, M.-C.: Towards an effective framework for building smart cities: lessons from Seoul and San Francisco. Technol. Forecast. Social Change. **89**, 80–99 (2014)
2. Duggar, The Core of Smart City Must Be Smart Governance, Forrester Research, Inc., Cambridge, MA, 2011
3. van Ark, B., Broersma, L., den Hertog, P.: Service Innovation, Performance and Policy: A Review, Research Series 6. Ministry of Economic Affairs, The Hague (2003)
4. Kapoor, E., Kemerer, C.F.: Network externalities in microcomputer software: an econometric analysis of the spreadsheet market. Manag. Sci. **42**(12), 1627–1647 (1996)
5. Chourabi, H., Nam, T., Walker, S., Gil-García, J.R., Mellouli, S., Nahon, K., Pardo, T.A., Scholl, H.J.: Understanding smart cities: an integrative framework. In: Proceeding of HICSS, pp. 2289–2297 (2012)
6. Chesbrough, H.W.: Open Innovation. Harvard University Press, Boston (2003)
7. CISCO, Smart Cities Expose: 10 Cities in Transition 2012, Smart Connected Communities Institute (2012) http://www.smartconnectedcommunities.org
8. Dirks, S., Keeling, M.: A Vision of Smarter Cities: How Cities Can Lead the Way into a Prosperous and Sustainable Future. IBM Global Business Services, Somers, NY (2009)
9. Dirks, S., Gurudev, C., Keeling, M.: Smarter Cities for Smarter Growth: How Cities Can Optimize Their Systems for the Talent-Based Economy. IBM, Somers, NY (2010)

Multi-focus Image Fusion Using Deep Belief Network

Vaidehi Deshmukh(✉), Arti Khaparde, and Sana Shaikh

Department of Electronics and Telecommunication,
Maharashtra Institute of Technology, SPPU, Pune, India
{vaidehi.deshmukh,Arti.khaparde}@mitpune.edu.in,
sana01.1992@gmail.com

Abstract. Multi-focus images may be fused to get the relevant information of a particular scene. Due to the limited depth of field of a convex lens of a camera, some objects in the image may not be focused. These images are fused to get all-in-focus image. This paper proposes an innovative way to fuse multi-focus images. The proposed algorithm calculates weights indicating the sharp regions of input images with the help of Deep Belief Network (DBN) and then fuses input images using weighted superimposition fusion rule. The proposed algorithm is analyzed and examined using various parameters like entropy, mutual information, SSIM, IQI etc.

Keywords: Image fusion · Superimposed · Deep learning · Restricted boltzman machine

1 Introduction

Multi-view information of the scene obtained from different images captured using different angles and variable focal length of the lens of a camera. In practice, all cameras consist of convex lens and have limited depth of a field. Hence while capturing an image using such camera; objects appear either focused or blurred depending upon their distance from the camera [2].

For human perception or machine vision, a well-focused image is preferred. These images are also useful in biomedical imaging, microscopic imaging, object recognition, military operations, machine vision, and so on.

Image fusion algorithms work in both spatial domain and transform domain. Spatial domain methods deal with the pixels gray level value whereas transform domain methods decompose input image into constituent images of different resolution and then fuse together depending upon the fusion rule.

Spatial domain methods mainly work upon the image blocks. Sharper image blocks from multi-focus images of the same scene are selected based upon the sharpness measure such as spatial frequency [3], EoG (Energy of Gradient) [4], Sum-Modified Laplacian operator [5] and sometimes using focus detection algorithm [6–10]. The main difficulty of these methods is to select the block of optimum size, which decides the quality of fused image. In [10], authors have used genetic algorithm to optimize size of the block to get better fused image. Also [5] applies sharpness measure to each pixel

S.C. Satapathy and A. Joshi (eds.), *Information and Communication Technology for Intelligent Systems (ICTIS 2017) - Volume 1*, Smart Innovation, Systems and Technologies 83, DOI 10.1007/978-3-319-63673-3_28

instead of whole block and fused pixel is obtained using weighted addition of selected clear pixels from input images.

Generally, transform based methods consist of three steps viz. decomposition, coefficients fusion and reconstruction. These methods use DCT [3], DWT [11], Laplacian pyramid [12], Wavelet, Curvelet [13], Shearlet [14] and contourlet [15] etc. transforms for decomposition of images. The salient information of the images can be extracted using such multi-scale decomposition effectively; but it may produce halo artifacts near the edges.

Above methods are further extended with the use of neural networks [16] and pulse coupled neural networks (PCNN) [17, 18]. In these papers, neural networks provide sharper regions of input images when features extracted from the blocks of input images are applied to them. Finally, these sharper regions are combined based upon the fusion rule which gives the final enhanced fused image.

Deep neural networks or Deep Belief Networks (DBN) are composed of restricted Boltzman machines. This network is used for fusion of multi-focus images in [19] wherein multi-focus input images are decomposed into low and high frequency sub-bands and then fused image is obtained based on features extracted using auto encoder model of DBN and high frequency sub-band.

In this paper, authors propose the abstract algorithm using DBN which gives the fused image when input images are directly applied to it. There is no need for feature extraction or decomposition of input images. DBN accepts the pixel values of input images and returns values of weights corresponding to sharper regions in images. Then images are fused using weighted superimposition fusion rule.

The paper gives the brief idea about the structure of Deep Belief Network (DBN) and its use for the image fusion in Sect. 2. Section 3 describes various parameters which are used to examine the method of fusion described here. Section 4 shows the results and Sect. 5 concludes the paper.

2 Deep Belief Network

Deep learning, which is a branch of machine learning, uses a hierarchical representation to learn high-level abstraction in data. This new and emerging approach has been widely used in speech and audio processing, language processing and modeling and artificial intelligence domains such as computer vision etc. [20] presents a very deep review and analysis of state-of-the-art methods and DBN to various applications. Research in the field of artificial neural networks paves the way for Deep Learning Networks. Increasing number of hidden layers in Feed-forward MLPs (Multi-Layer Perceptron) gives Deep Neural Network (DNN). Deep architecture is required to process natural signals. Vision and audition systems are the two mechanisms in human body; which collect the information from outside world. This information is converted to language form or image form using layered hierarchical structures present in those systems. As mentioned earlier, deep learning algorithms use layered or hierarchical structure and extract the complex information from received inputs. Hence, deep architectures, with the help of deep learning algorithms can be used to process these natural signals efficiently [22].

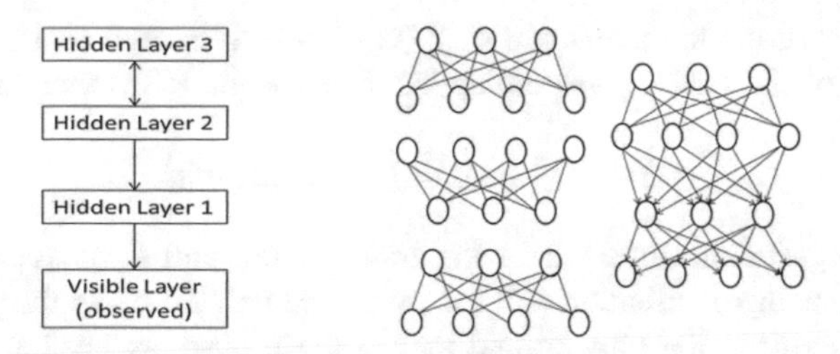

Fig. 1. (a) Schematic block diagram of DBN, (b) Left: RBM layers which use samples derived from lower layers to train the next Right: Deep Belief Network obtained by stacking trained RBM layers.

A Deep Belief Network (DBN) is basically Deep Neural Network which uses Bayesian probabilistic generative models to acquire correlation in input data for classification, pattern decomposition and reconstruction purpose. A DBN consists of many layers of restricted Boltzmann machines (RBMs) (Fig. 1). A layer-by-layer greedy learning algorithm optimizes DBN weights and forms the main component of DBN. Sometimes, hidden layers of DBN are trained using unsupervised learning and then fine-tuned using back propagation [21].

An RBM contains two layers viz. hidden layer and visible layer. Each layer consists of neuron-like units which are not connected to each other, whereas all visible units are connected to hidden units. Each node of the visible layer gets input which is multiplied by the weight in the hidden layer node and then bias value is added to it. This is the simple feed-forward movement of inputs through the network. This is decomposition used to calculate the probability of output given a weighted input (training the network). Outputs from hidden layer are passed to the visible layer with bias to reconstruct the input from given activations (testing the network). Two energy functions each corresponding to the data and reconstructed model are defined. Their difference is used to update the weights as given by (3) [18]. This energy can be given by:

$$E(v,\ h;\theta) = -\ v^{T}Wh - b^{T}v - a^{T}h\ = \Sigma_{i=1}^{D}\Sigma_{i=1}^{R}W_{ij}v_{i}h_{j} - \Sigma_{i=1}^{D}b_{i}v_{i} - \Sigma_{i=1}^{R}a_{j}h_{j} \quad (1)$$

In (1), v represents visible units, h hidden units and θ are model parameters. W represents symmetric interaction term between visible and hidden units and a, b are biasing terms. The probability of the visible layer in the RBM is:

$$\begin{aligned} P(v;\theta) &= [1/Z(\theta)]\Sigma_{h}\exp\left(v^{T}Wh + b^{T}v + a^{T}h\right) \\ &= [1/Z(\theta)]\exp\left(b^{T}v\right)\Pi_{j=1}^{R}\Sigma_{hj\in\{0,1\}}\exp\left(a_{j}h_{j} + \Sigma_{i=1}^{D}W_{ij}v_{i}h_{j}\right) \\ &= [1/Z(\theta)]\exp\left(b^{T}v\right)\Pi_{j=1}^{R}\left(1 + \exp\left(a_{j} + \Sigma_{i=1}^{D}W_{ij}v_{i}h_{j}\right)\right) \end{aligned} \quad (2)$$

In (2), Z represents normalization factor.

The gradient of the log likelihood of P (v; θ), which is used as a cost function for the fine tuning of the DBN, gives the update rule for the RBM weights as:

$$\Delta W_{ij} = E_{data}\left(v_i h_j\right) - E_{model}\left(v_i h_j\right) \tag{3}$$

$E_{data}(v_ih_j)$ is expectation observed in the training set, and $E_{model}(v_ih_j)$ is that same expectation under the distribution defined by the model. To make the performance of RBM better, energy in the training and that in the testing are need to be similar. So $E_{model}(v_ih_j)$ i.e. energy corresponding to the testing is approximated using contrastive divergence (CD) which uses Gibbs sampler. In this way, RBM can be used to portray the data using its probability distribution. When RBM is used for classifying the data, then energy function $E_{data}(v_ih_j)$ corresponding to data (training) is also approximated. This is "fine tuning" of RBM for classification tasks.

Partially trained RBMs are arranged layer by layer from bottom to make a DBN. Once the RBM is trained completely, activation probabilities of one RBM are used as training data set for the other [18].

Using this DBN, feature vectors of input images are obtained. Mean of feature vectors are calculated. These mean values W_1 and W_2 are multiplied with input images and these weighted images are added to obtain the fused image.

$$F_{final} = W_1 \times Img_1 + W_2 \times Img_2 \tag{4}$$

(4) states the superimposition fusion rule. The block diagram for the process is shown below:

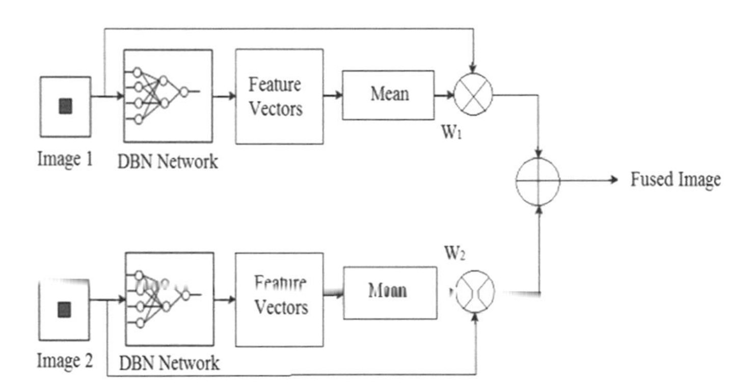

Fig. 2. Steps of Proposed method

3 Evaluation Parameters for Output Images

Quality of the fusion process can be examined using objective as well as subjective parameters. Subjective evaluation of fusion process includes human factor and hence it depends largely upon the individual examiner. Objective parameters are, therefore, preferred for the evaluation. These are statistical parameters which indicate the quality of the fused image. [15, 23–26].

Root Mean Squared Error: It directly calculates the variation of pixel values of two images and compares them. It is a good indicator of spectral quality of image. If its value is close to zero, then the combined image is close to a reference image.

$$\text{RMSE} = \{(1/mn)\,[\Sigma_{i=0}^{M-1}\Sigma_{j=0}^{N-1}\text{Img}_1(i,\ j) - \text{img}_2(i,\ j)]^2\}^{1/2} \quad (5)$$

m, n: size of the images Img1 and Img2.

Peak Signal to Noise Ratio: Peak signal to noise ratio gives the ratio of the maximum power of a signal and the power of corrupting noise. Higher value of PSNR indicates that fusion is superior.

$$\text{PSNR} = 20\log_{10}(\max/\text{MSE}) \quad (6)$$

Max: Maximum gray level value of an image.

Structural Similarity Index: SSIM compares the local patterns of pixel intensities between the images. Higher value of SSIM shows that the images are similar. Its maximum value is 1.

$$\text{SSIM}(x, y) = (2\mu_x\mu_y + c_1)(2\sigma_{xy} + c_2) / \left(\mu_x^2 + \mu_y^2 + c_1\right)(\sigma_x^2 + \sigma_y^2 + c_2) \quad (7)$$

Where μ_x = average of x; μ_y = average of y; σ_x^2 and σ_y^2 are variances of x and y respectively; σ_{xy} = covariance of x and y;

$$c_1 = (k_1L)^2,\ c_2 = (k_2L)^2 \quad (8)$$

L is the dynamic range of the pixel-values; $k_1 = 0.01$and $k_2 = 0.03$ by default.

Entropy: Entropy indicates the amount of information present in an image. Mathematically it is given as:

$$E = -\Sigma_{i=0}^{L-1}P_i\log_2 P_i \quad (9)$$

P_i is the probability of occurrence of a pixel value in the image, L is the number of grey levels in image. Increase in entropy value after fusion indicates that fused image contains more information than input images.

Mutual Information: It is an indicator of the information obtained from the source images and the quantity that is conveyed by the fused image. Higher value of MI indicates the better image quality.

$$\text{MI} = \text{MIF}, \text{I1} + \text{MIF}, \text{I2} \quad (10)$$

$$\text{MIF}, \text{I1} = \Sigma\ \text{PF}, \text{I1}(f, i1)\ \log2\ (\text{PF}, \text{I1}(f, i1)/\text{PF}(f)\ \text{PI1}(i1)) \quad (11)$$

$$\text{MIF}, \text{I2} = \Sigma\ \text{PF}, \text{I2}(f, i2)\ \log2\ (\text{PF}, \text{I2}(f, i2)/\text{PF}(f)\ \text{PI2}(i2)) \quad (12)$$

Where $MI_{F,I1}$ denotes the mutual information between the fused image and the first input image, $MI_{F,I2}$ denotes the mutual information between the second input image and the fused image, $P_{F,I1}(f,i1)$ and $P_{F,I2}(f,i2)$ are the joint histograms of fused image and input, $P_{I1}(i1))$, $P_{I2}(i2)$ and $P_F(f)$ are the histograms of two input images and the fused image respectively.

Run Time: Run time is the time in seconds consumed by the algorithm for its execution.

Image Quality Index [9]: Image quality index states the distortion in the fused image in terms of three factors viz. loss of correlation, luminance and contrast distortions. The range of IQI is −1 to 1.

Mathematically,

$$IQI = 4\sigma_{xy}\mathbf{xy} / \{(\sigma_x^2 + \sigma_y^2)\left[(\mathbf{x})^2 + (\mathbf{y})^2\right]\} \tag{13}$$

Where **x** and **y** indicate means of x and y respectively, σ_{xy} is the covariance of x and y; σ_x^2, σ_y^2 are the variances of x and y respectively.

4 Experiments

4.1 Experimental Settings

DBN, as discussed in Sect. 2 is coded using Python and implemented on Intel dual core, i5, 3.14 GHz processor. The algorithm is tested on different database images as well as real time images. Pixel values of input images are directly applied to the DBN. This DBN is composed of three hidden layers and an output layer containing two neurons. Different evaluation parameters mentioned previously are used for analyasis of results.

4.2 Results

The algorithm is quantitatively evaluated using the parameters mentioned previously. Table 1 compares results for two algorithms viz. edge-GA-superimposition implemented in [1] and DBN-superimposition implemented in this paper (Fig. 2). These results are compared based on different parameters for database images and real time images. Figure 3 shows one sample output image obtained after implementing the proposed method on database images and real time images. It was observed that proposed method works well to produce bright fused image. As seen from table I, PSNR, SSIM and MI values obtained from DBN-superimposition method are better than those obtained from Edge-GA-superimposition for database images. However, Edge-GA-superimposition method works better for real time images. DBN-superimposition method gives slightly higher values for MI indicating that fused images have higher information content. Also values of SSIM are close to 1 which indicates that fusion quality is superior.

(a)

(b)

(c)

Fig. 3. (a, b) Data base input images, (c) fused image

Table I Comparision of results for different image sets

Image set	Database image sets		Real-time image sets	
Method/parameter	*Edge-GA-superimposition*	*DBN-superimposition*	*Edge-GA-superimposition*	*DBN-superimposition*
PSNR	34.409	34.817	21.076	21.016
SSIM	0.936	0.943	0.806	0.805
ENTROPY	7.359	7.101	7.336	7.116
MI	7.212	7.262	5.192	5.218
IQI	0.891	0.861	0.601	0.557
TIME(S)	0.229	1.104	0.241	1.214

5 Conclusion

Multi-focus images (database and real-time) are fused using DBN-superimposition method. The results obtained from proposed technique show that the algorithm works well for image fusion. The added advantage is its simplicity as raw pixel values are applied to DBN which itself extracts the features of image and calculates the weights for fusion. This reduces the computation load required for pre-processing of input images and increases the abstraction level of the algorithm.

Further time required for fusion of images can be reduced if optimisation of number of layers and number of neurons in each layer is done. Also effect of this optimisation on performance of the algorithm can be studied. Presently, DBN runs for the fixed number of epochs. Convergence criterion can be applied in order to reduce number of epochs and hence time consumed.

References

1. Khaparde, A., Deshmukh, V.: Optimized multi-focus image fusion using genetic algorithm. Adv. Sci. Technol. Eng. Syst. J. **2**(1), 51–56 (2017)
2. Aslantas, V., Kurban, R.: Fusion of multi-focus images using differential evolution algorithm. Expert Syst. Appl. **37**, 8861–8870 (2010)

3. Cao, L., Jin, L., Tao, H., Li, G., Zhuang, Z., Zhang, Y.: Multi-focus image fusion based on spatial frequency in discrete cosine transform domain. IEEE Signal Process. Lett. **22**(2), 220–224 (2015)
4. Zhang, Y., et al.: Multi-focus image fusion based on cartoon-texture image decomposition, Optik - Int. J. Light Electron Opt. (2015). doi:10.1016/j.ijleo.2015.10.098
5. Qu, X., Li, L., Guo D.: Multi-focus image fusion with structure driven adaptive regions. In: Proceedings of International Conference on Internet Multimedia Computing and Service, Xiamen, China — 10–12 July 2014). doi:10.1145/2632856.2632914
6. Zhang, X., Li, X., Li, Z., Feng, Y.: Multi-focus image fusion using image-partition-based focus detection. Sig. Process. **102**, 64–76 (2014)
7. Pertuz, S., Puig, D., Garcia, M.A., Fusiello, A.: Generation of all-in-focus images by noise-robust selective fusion of limited depth-of-field images. IEEE Trans. Image Process. **22**(3), 1242–1251 (2013)
8. Zhang, B., Lu, X., Pei, H., Liu, H., Zhao, Y., Zhou, W.: Multi-focus image fusion algorithm based on focused region extraction. Neurocomputing (2016). doi:10.1016/j.neucom.2015.09.092
9. Zhang, X., et al.: A new multi focus image fusion based on spectrum comparison. Sig. Process. (2016). doi:10.1016/j.sigpro.2016.01.006i
10. Xiao, J., Liu, T., Zhang, Y., Zou, B., Lei, J., Li, Q.: Multi-focus image fusion based on depth extraction with inhomogeneous diffusion equation. Signal Process. (2016). doi:10.1016/j.sigpro.2016.01.014
11. Lacewell, C.W., Gebril, M., Buaba, R., Homaifarn, A.: Optimization of image fusion using genetic algorithms and discrete wavelet transform. Radar Signal Image Process. IEEE (2010)
12. Liao, C., Liu, Y., Jiang, M.: Multi-focus image fusion using Laplacian Pyramid and Gabor filters. Appl. Mech. Mater. **373–375**, 530–535 (2013)
13. Li, S., Yang, B.: Multifocus image fusion by combining curvelet and wavelet transform. Pattern Recogn. Lett. **29**, 1295–1301 (2008)
14. Guorong, G., Luping, X., Dongzhu, F.: Multi-focus image fusion based on non-subsampled shearlet transform. IET Image Process. **7**(6), 633–639 (2013)
15. Yang, Y., Tong, S., Huang, S., Lin, P.: Multifocus image fusion based on NSCT and focused area detection. IEEE Sens. J. **15**(5), 2824–2838 (2015)
16. Li, S., Kwok, J.T., Wang, Y.: Multi-focus image fusion using artificial neural networks. Elsevier Pattern Recognit. Lett. **23**, 985–997 (2002)
17. Arawal, D., Singhai, J.: Multi-focus image fusion using modified pulse coupled neural network for improved image quality. IET Image Process. **4**(6), 443–451 (2010). doi:10.1049/iet-ipr.2009.0194
18. Wang, Z., Ma, Y., Jason, G.: Multi-focus image fusion using PCNN. Elsevier Pattern Recognit. Lett. **43**, 2003–2016 (2010)
19. Fan, L., Zehua, C., Jing, C.: A new multi-focus image fusion method based on deep neural network model. J. Shandong Univ. (Eng. Sci.) **46**(3), 7–13 (2016). doi:10.6040/j.issn.1672-3961.2.2015.106
20. Guo, Y., Oerlemans, A., Lao, S., Wu, S., Lew, M.S.: Deep learning for visual understanding: a review. Neurocomputing **187**, 27–48 (2016)
21. Deng, L., Yu, D.: Deep learning for signal and information processing. In: ICASSP (2012)
22. Deng, L., Yu, D.: Deep learning methods and applications. Found. Trends Signal Process. **7** (3–4), 197–387 (2013)
23. Wang, W. W, Shui, P. L., Song, G. X.: Multifocus image fusion in wavelet domain, Proceedings of the Second International Conference on Machine Learning and Cybernetics (2003)

24. Zaveri, T., Zaveri, M.: A novel region based image fusion method using high-boost filtering, Proceedings of the 2009 IEEE International Conference on Systems, Man and Cybernetics, San Antonio, TX, USA (2009)
25. Hugo, R., Albuquerque, T. I.R., Cavalcanti, G.D.C.: Image fusion combining frequency domain techniques based on focus. In: IEEE 24th International Conference on Tools with Artificial Intelligence (2012)
26. Wang, W., Chang, F.: A multi-focus image fusion method based on laplacian pyramid. J. Comput. **6**(12), 2559–2566 (2011)

Implementation of K-mean Algorithm Using Big Data in Health Informatics

V. Kakulapati[1(✉)], V.K. Pentapati[2], and S.R. Kattamuri[1]

[1] Sreenidhi Institute of Science and Technology, Yamnampet, Gatakeswar, Hyderabad, Telangana, India
vldms@yahoo.com

[2] MEC, Hyderabad, Telangana, India
vijay_p51@yahoo.com

Abstract. Big information could be a new technology to spot the dataset's giant in size and complication. The tremendous growth-rate of huge information with appearance contemporary scientific techniques for informative assortment, large amount of medic's specialty and Health scientific discipline. Giant amounts of heterogeneous medic's information became on the market in varied aid organizations. This Medical information may be associate sanctionative resource for account insights for up concern delivery and reducing misuse. The immenseness and complication of that dataset's yield challenges in analyses and succeeding applications to sensible medical surrounding. There is a sensible problem to judge and infer the info victimization inevitable ways. For extracting helpful data in effective and economical information investigation ways are necessary. Data processing bunch methodology is one that helps in characteristic attention-grabbing patterns from huge information. Several smart applications wide used formula is k-mean. This k-means formula is computationally valuable and conjointly the following cluster quality is heavily depending upon the selection of the initial centurions. This paper proposes a k-means formula is with refined initial centurions. To figure out the initial centurions associated with nursing improved methodology to various clusters the data points is also an assignment. The final results show that the projected formula produces clusters with higher precision in less than working out time.

Keywords: Big information · Health · K-mean · Datasets · Algorithms

1 Introduction

Clinical informatics is on the cusp of it's nearly all fun amount thus far, stepping into a replacement era wherever technology is getting all the manner right down to handle Brobdingnagian knowledge, transfer relating to unlimited potential for data growth. The process and large knowledge analytics square measure serving to understand the objectives of identification, delighting, assisting, and healing all the long-sufferings in want of health care, among the best purpose of this sphere being developed, the Output of Health Care or the standard of concern that, health care will give to complete customers i.e. tolerant.

S.C. Satapathy and A. Joshi (eds.), *Information and Communication Technology for Intelligent Systems (ICTIS 2017) - Volume 1*, Smart Innovation, Systems and Technologies 83, DOI 10.1007/978-3-319-63673-3_29

Clinical informatics is additionally a grouping of Data science and technology inside the sphere of health care. A square measure varied current areas of analysis inside the sphere of Health informatics, to boot as Bio-informatics, Image informatics (e.g. human informatics, medical informatics, Public Health issues and put together travel Bioinformatics (TBI). Analysis tired that clinical informatics as altogether its subsets will vary from knowledge attainment, recovery, storage, and analytics victimization process methods. The possibility of this work goes to be analysis with the intention of utilizes the process thus on answer queries all over the varied levels of health.

All of these works tired with a specific subset of medical informatics uses knowledge from the specific intensity of individual subsistence [1]: Bio-informatics make use of tissue level knowledge, Nero-informatics employs tissue level knowledge, Health informatics pertains the patient intensity information and Public fitness informatics uses population knowledge. Every one of these subsets do commonly cover (for instance, one review would perhaps think about learning from an attempt among contiguous levels), in any case, inside the enthusiasm of minimizing perplexity, we relate degree slant to all through this work can arrange a review bolstered the premier successful information level that is utilized as this work is getting the chance to be organized per learning utilization. Additionally, inside a given knowledge level that we've bent to the square measure getting to split down knowledge supported the kind of questioning learning tries for response, wherever every problem level is of a relatively equivalent scope to a minimum 1 all told info intensity. The twill level is of equivalent capacity to individual-scale natural science queries, the extent of long-suffering knowledge is expounded to health queries and therefore the extent of community knowledge is simply like the outbreak-level queries.

An extent for information employed by subset of travel Bioinformatics, on the choice entrust, exploits the knowledge from all of that intensity, from the tissue level for the entire population. Especially, Travel Bio-informatics is exclusively targeted on integration knowledge from the Bio-informatics intensity to the superior intensity, as results of history, this intensity has been inaccessible within the experiment and take apart from numerous patient-facing intensity (Neuro-informatics, Health informatics etc.). Travel Bio-Informatics, in this manner, prepare of blending learning from all intensity of human presence is also an additionally innovative trend in clinical informatics. The principal level of queries that Travel Bio-Informatics at last tries to reply on the clinical level, all by itself, answers will encourage enhancing HCO for patients. Analysis all through all levels of openness information, the victimization varied process and analytical techniques, is additionally accustomed facilitate the health care system, built alternatives quicker, numerous accurately, and much of with efficiency, tired a more cost-effective manner than whereas not victimization such ways that within which.

2 Related Work

Clustering As against the classification task, once we have a tendency to tend to perform the cluster analysis, category labels of knowledge objects/instances don't seem to be the gift among the work set, associated thence kind the kind of learning that possesses to occur the associate unattended type. The foremost aim of bunch ways in which during which is to cluster the info objects in step with some live of similarity, that the info objects in one cluster or cluster unit terribly quite like each other whereas being terribly totally fully completely different from the info object happiness to the residual cluster. In various statements, endeavor is to augment the intra-class closeness and minimize the between class likeness [2]. Among the health care profession, this kind of disadvantage is fascinating to urge knowledge some drug, treatment or pathological state [3].

Ubeyli et al. [4] conferred associate approach supported the achievement of the k-means clustering to automatic exposure of erythematosquamous diseases. Clustering algorithm K-Means [5, 10] is one that in a position to be the accustomed partition varies of knowledge dots into the given form of clusters. It anticipates that quantity of clusters is given as input to rule. "K" among the name "k-means" stands for the quantity of clusters. K-means rule assumes that every knowledge incorporates one equal numeric price. If not, when information focuses have multi-attribute values, distances between the points calculated observe geometrician distance. At some point of this review, the algorithm's task is to assign the info points to a minimum of one between the five clusters. They tested with 5 categories, so as that they removed the info for the patients World Health Organization has the pathological state of the sixth category and therefore the rule was accustomed notice 5 health problem indications were used. They obtained ninety four percentage prediction accuracy of the K-means bunch. With the goal that they completed that the arranged K-mean clusters is utilized in investigating erythematosquamous ailments by taking into consideration the mis-classification rates. Tsakoumis et al. [6] planned a unique cluster method for description and tagging of pigmented skin lesion among the medication picture. To the segment tissue picture, the skin injury to be assessment got the opportunity to be isolated from the solid skin. Mostly, locale based division routes in which amid which unit being utilized for this reason [7, 8]. The arranged technique utilized the principal half Analysis. They attempted to premium retrieval of cancer illustrations in skin lesion photographs. The outcome has been trying by the predominance of this method against ancient ones.

On the quantity of the mining platform sector, at present, parallel programming models [9] like Map prune unit being used for the aim of study and mining of knowledge. Map prune is additionally a set-oriented parallel computing replica. Still there is a particular gap in performance with relative databases. Rising the performance of Map prune and enhancing quantity the number of time nature of large-scale method have received a large amount of attention, with Map prune parallel programming being applied to numerous machine learning and process algorithms. Process algorithms typically got to be compelled to scan through the work data for getting the statistics to resolve or optimized model.

In case of favor of knowledge mining algorithms, data evolution is additionally a regular development in planet systems. However, as a result of the drawback statement differs, consequently the information can dissent. As associate example, once we have a tendency to tend to go to the doctor for the treatment, that doctor's treatment the program endlessly adjusts with the patient conditions.

Developments in Big Data [11, 13, 14] have associated with the open provide package revolution. Enormous companies as Facebook, Yahoo!, Twitter, and joined in profit and contribute functioning on open provide comes. Brobdingnagian data, communications deals with Hadoop and numerous connected packages as follows:

Hadoop [12]: package for information concentrated disseminated applications, principally based is totally among the Map prune programming model and a distributed arrangement established as HDFS. Hadoop grants composing applications that quickly strategy a lot of information in parallel to an enormous bunch of figure nodes.

A Map prune work partitions the data dataset into free subsets those are taking care of by guide assignments in parallel. This progression of mapping is then trailed by a stage of lessening assignments. These prune assignments utilize the yield of maps to encourage the last word for the consequences of the task.

3 Design of Mining Algorithms

Big data challenges are classified into three sectors. Those are:

- Mining the platform
- Protection
- Mining algorithms

Fundamentally, large information is hold on at totally completely totally dissimilar places and along the information may be increased as results of knowledge stays on increasing endlessly. In this way, to accumulate all the data hung on at very surprising spots is that seriously expensive. Suppose on the off chance that we have a twisted to tend to utilize these typical process ways that the unit that is utilized for mining the little scale data in our notes or a personal computer for mining of big information, then it'd turned out to relate hindrance in it. As the consequences of the regular way's unit needed information to be loaded in main memory, however, we've huge main memory.

To keep up protection, is one in each preeminent point of data mining calculations. In a matter of seconds, to mine data from huge data, parallel registering based for the most part total calculations like the Map prune unit that is utilized. In such algorithms, vast data sets the unit isolated into the kind of subsets then, algorithm part connected to individual subsets. Finally, the summing up algorithm units applied to the results of mining algorithms to fulfill the objective of the big process. All through whole strategy, security statements plainly break as we have a bowed to tend to partition one broad information into the kind of smaller datasets.

While developing with such algorithms, the varied challenge's unit occurred. As associate example, there are unit blind men observant the large elephant. Most are the unit making on the attempt to predict their conclusion on what the difficulty is confessedly. Someone is the expression that the difficulty is additionally a hose; somebody says it's a tree or pipe, etc. the terribly most area unit just observant an area of that big elephant and not the total, therefore the results of every blind person's prediction area unit, some things totally different than terribly what it's.

Similarly, once we've got a bent to divide the large information in to kind of subsets, and apply the mining algorithms on those subsets, the results of these mining algorithms won't often propose the North yank country to the particular result as we've got a bent to would really like once we've got a bent to gather the results on. Throughout this paper we've got a bent to tend to applied k-mean agglomeration rule on large information to the sort of potential clusters.

3.1 Clustering Algorithm K-means

Input:
D = d1, d2, ….., dn/set of n information items. K // set of chosen clusters.
Output: K clusters. Steps:

1. Subjectively pick K information things from D as beginning centroids;
2. do again
   Distribute everything di to the group which has the nearest Centroid;
   Ascertain the new mean for each cluster;
3. Until the union model is met.
   Despite the fact that algorithm is effective in producing clusters with some useful applications, some disadvantages exist. The unpredictability of unique K-Means algorithm is very high, especially in big data. The intricacy is found from the equation O (nK l) where n is the number of data points, K the number of clusters and l the number of iterations. Kind of clusters in view of the underlying centurion's determination. The final cluster accuracy relies on upon the underlying centurions chosen.

4 Experimental Result

The following methods were implemented and tested. We executed the experiments in Java environment. The information sets accessible in the UCI information storehouse was utilized for testing. The information sets accessible in the UCI information was

Name	Age	Sex	Month	Place	Disease
AbdullahAlmulla	22	female	March	Ooty	gestational diabetes
AbhasSarkar	23	male	February	Chennai	Down syndrome
AbidiasCarneirodaSilva	21	female	October	Bangalore	Edward syndrome
AbrahamGarcia	32	male	March	Mumbai	trisomies
AceepYaamy	24	female	March	Delhi	Patau syndrome
AdamAngel	35	male	April	Chennai	preeclampsia
AdanMencariBakat	21	female	October	Delhi	microcephaly
AdelaHuerta	32	male	March	Mumbai	spina bifida
AdhamMrasm	24	female	September	Mumbai	bradycardia
AdinaniNkinda	26	male	April	Madurai	hydronephrosis
AdodoAdoxWarndia	28	female	October	Bangalore	preeclampsia
AdrianVazquez	28	male	March	Coimbatore	massive fetomaternal h...
AdrianoAssis	32	female	September	Delhi	Hepatosplenomegaly
AfandiMustaqim	27	male	September	Bangalore	hemiplegia
AgdaLima	31	female	January	Bangalore	omphalocele
AgungMahardhika	45	male	March	Ooty	Meningomyelocele
AhTingBee	24	female	March	Trichy	Hydrops fetalis
AhmadRoshan	20	male	October	Chennai	Ascites
AhmedFoysal	28	female	September	Coimbatore	hydrothorax
AhmedWhite	22	male	March	Chennai	pericardial effusion
AidaZamir	29	female	January	Delhi	endometriosis

Fig. 1. Data set of health records

Cluster ID	Name	Age	Sex	Month	Place	Disease
0	AbdullahAlmulla	22	female	March	Ooty	gestational diabetes
0	AhmedWhite	22	male	March	Chennai	pericardial effusion
0	AjuloTosin	32	male	March	Madurai	gestational diabetes
0	AliOmar	22	female	September	Trichy	pericardial effusion
0	AllanaHearn	24	female	January	Bangalore	gestational diabetes
0	AndreMeireles	32	male	June	Mumbai	pericardial effusion
0	AndrewHaldeman	44	male	January	Coimbatore	gestational diabetes
0	AntoninaKorneeva	24	female	July	Madurai	pericardial effusion
0	AprilGriffith	24	female	July	Pondy	gestational diabetes
0	AvinashKeshari	44	male	September	Bangalore	pericardial effusion
0	BAAbdulMuthalib	43	female	May	Trichy	pericardial effusion
1	AbhasSarkar	23	male	February	Chennai	Down syndrome
1	AidaZamir	29	female	January	Delhi	endometriosis
1	AkhyasArief	30	female	September	Mumbai	Down syndrome
1	AliceSmith	23	male	November	Delhi	endometriosis
1	AlmaSumbalAlah	35	male	December	Chennai	Down syndrome
1	AndreaKrieger	30	male	August	Bangalore	endometriosis
1	AndriEnggi	36	male	August	Ooty	Down syndrome
1	AntonioNolasco	35	female	July	Ooty	endometriosis
1	AraleKoto	26	female	February	Hyderabad	Down syndrome
1	AyNoppharat	36	female	October	Madurai	endometriosis

Fig. 2. Data set after applying the k-mean algorithm with instances whenever patient

utilized for testing. We utilized a medical database for diverse diseases. On these databases we create clusters by using the K-mean clustering algorithm. Subtle elements of the information sets utilized are abridged in the Figs. 1 and 2.

5 Conclusion

Big data keeps on growing throughout consecutive years, and each data individual needs to deal with the substantially more measure of information consistently. This information goes to be loads of varying, bigger, and quicker. We tend to say some insights regarding the topics, and what we tend to require under the consideration unit of measurement the foremost problems and conjointly the most challenges for the long haul. The tremendous information is flying into the innovative closing Frontier for the logical information investigation and for trade applications. We tend to tend to face live at the beginning of a replacement era where immense processing to seek out knowledge.

6 Future Enhancement

There square measure many be necessary for the future difficulties in the Brobdingnagian information administration and investigation that emerge from the character of data. It isn't clear, however, but Associate in Nursing optimum style of analytics systems have to be compelled to be to have the impact on the historic information and from the period data at an analogous point. Practice sampling, we've got an inclination to the square measure losing data, yet additions in space may even be all together of greatness. Practice may be blending lessen the inadequate sets which can then be utilized for determination difficult the machine learning issues in parallel. A principle undertaking of expansive information investigation could be expressed gratitude toward to picture the outcomes. Therefore of learning is the Brobdingnagian, it's truly the strong to chase out the basic representation.

References

1. Chen, J., Qian, F., Yan, W., Shen, B.: Translational biomedical informatics in the cloud: present and future. BioMed. Res. Int. **2013**, 8 (2013). doi:10.1155/2013/658925
2. Han, J., Kamber, M.: Data Mining Concepts and Techniques. Elsevier Inc., San Francisco (2006)
3. Houston, A.L., Chen, H., Hubbard, S.M., Schatz, B.R., Ng, T.D., Sewell, R.R., Tolle, K.M.: Medical data mining on the internet: research on a cancer information system. Artif. Intell. Rev. **13**(5–6), 437–466 (2000)
4. Ubeyli, E.D., Dogdu, E.: Automatic detection of erythematosquamous diseases using k-means clustering. J. Med. Syst. **34**, 179–184 (2010)
5. Margaret, H.D.: Data Mining: Introductory and Advanced Topics. Prentice-Hall, Englewood Cliffs (2003)
6. Tasoulis, S.K., Doukas, C.N.: Classification of dermatological images using advanced clustering techniques. In: 32nd Annual International Conference of the IEEE EMBS, Buenos Aires, Argentina, pp. 6721–6742 (2010)
7. Mohammadi, E., Duzgun, S.: Statistical Data Analysis and Modeling of Skin Diseases (2008)
8. Storm. http://storm-project.net
9. Zikopoulos, P., Eaton, C., deRoos, D., Deutsch, T., Lapis, G.: IBM Understanding Big Data: Analytics for Enterprise Class Hadoop and Streaming Data. McGraw-Hill Companies, Incorporated, New York (2011)
10. Feldman, D., Schmidt, M., Sohler, C.: Turning big data into tiny data: constant-size corsets for k-means, PCA and projective clustering. In: SODA (2013)
11. Gantz, J., Reinsel, D.: IDC: The Digital Universe in 2020: Big Data, Bigger Digital Shadows, and Biggest Growth in the Far East, December 2012
12. Apache Hadoop. http://hadoop.apache.org
13. Gartner. http://www.gartner.com/it-glossary/bigdata
14. Gopalkrishnan, V., Steiner, D., Lewis, H., Guszcza, J.: Big data, big business: bridging the gap. In: Proceedings of the 1st International Workshop on Big Data, Streams and Heterogeneous Source Mining: Algorithms, Systems, Programming Models and Applications, Big Mine 2012, pp. 7–11 (2012)

SSD Implementation and Spark Integration

K. Soumya(✉) and M. Arunkumar

Computer Science Department, ICET, Mulavoor, India
soumyakonline@gmail.com, arunpvmn@gmail.com

Abstract. One of the main challenges in Big data is the processing speed and scalability. Solid State Drive (SSD) helps for faster processing than HDD. Here along with SSD, Spark is also accompanied with hadoop framework for more scalability and fast processing. Apache Spark is a general-purpose engine for large-scale data processing on any cluster. It is a framework which can afford more than 8000 nodes in a cluster Spark allows for code reuse across batch, interactive, and streaming applications. Spark is much faster than MapReduce. It was generally coded from Java; Spark supports not only Java, but also Python and Scala, which is a newer language that contains some attractive properties for manipulating data. Spark runs up to 100 times faster than Hadoop MapReduce in memory and 10 times faster on disk. This paper tries to integrate spark with Hadoop ecosystem along the SSD. It increases the processing speed.

Keywords: MapReduce · Spark · Hadoop · SSD

1 Introduction

Big data can be described in terms of volume, velocity and variety. Big data cannot be processed by traditional data management system. Examples of big data are Black box data, Social media data, Stock exchange data, Search engine data etc.

Traditional approaches fails to handle the big data. These system works well with less voluminous data. So the Hadoop is introduced [1].

Hadoop is a software framework for supporting large data processing in a distributed computing setup. It has a master-slave architecture. It contains two layers—HDFS and MapReduce. HDFS stands for the Hadoop Distributed File System. It is the storage layer. It maintains high fault tolerance capability. It contains Name node and data node which act as master and slave respectively. It maintains the fault tolerance by replicating files in to different data nodes. So if one fails, that work will be taken by other data node.

A MapReduce framework is another highlight of hadoop. It splits *the* input data-set into independent chunks and they are processed parallely by the *map tasks*. The *reduce* task combines all the intermediate results with respect to the key [2].

Studies shows that SSD plays an important role in the performance of the hadoop cluster. We always look forward to process large amount of data with in short time. After the map phase, n MapReduce reduce process takes place. It must undergo sort-merge processes. SSD store the data using a semiconductor. So it is much faster than HDD (Hard Disk Drive). SSD performs faster for random reads than HDD.

S.C. Satapathy and A. Joshi (eds.), *Information and Communication Technology for Intelligent Systems (ICTIS 2017) - Volume 1*, Smart Innovation, Systems and Technologies 83, DOI 10.1007/978-3-319-63673-3_30

This paper tries to distinguish the HDD and SSD in terms of performance. Another concept proposed by this paper is the integration of spark with hadoop.

Spark is not an alternative for MapReduce. It add extensibility to MapReduce in interactive and iterative jobs. As stated in [7] it is much faster than MapReduce since it uses in memory computing be processed by traditional data management system. Examples of big data are Black box data, Social media data, Stock exchange data, Search engine data etc.

2 Back Ground

2.1 HDFS

Hadoop MapReduce frame work will work better with large data. HDFS is meant for storing large data. It has a master-slave architecture (Fig. 1).

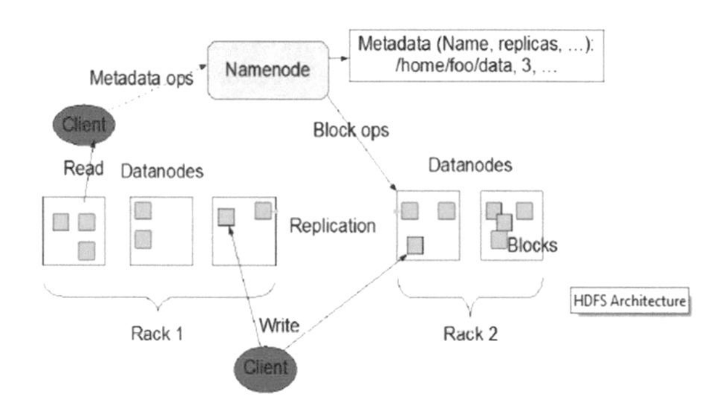

Fig. 1. HDFS architecture

Name Node: Files and directories are placed in the Name Node. The file is divided in to large blocks. Each block is replicated at multiple data nodes. The client will first contact the name node for the data blocks. The client then writes the data in pipeline mode.

Name node contains two other features-image and journal. Image contains metadata and journal is the log file. It contains the modification log of image file [4].

Data Node: Each block is replicated in the data node. While the connection is established the data node and the name node will perform a handshake. During the handshaking the namespace id of name node and the software version of the data node is cross checked. If they are not matched the connection will not be established [4].

After handshake, the data node is registered. Each data node will have unique storage IDs. The data node will send a block report to the name node. It includes block ids, Block length etc. When the client writes an application, a data node will be leased by the name node. After three seconds the lease should be renewed. It is done by sending a heartbeat, which is an indication of the liveliness of the data node [4].

Checkpoint Node: Name node also play the role of checkpoint node and back up node. It returns a modified checkpoint node after downloading and merging the existing checkpoint node to the name node [4].

Backup Node: Backup node is similar to the checkpoint node. But it does not need to down load the existing one. It is more efficient than checkpoint node [4].

Block placement: Blocks are placed in racks. Rack awareness is another feature of HDFS. Racks are communicated using the switches. HDFS replica management is also interesting. Not more than two copies of replica is stored in same rack. No data node contain more than a replica [4].

2.2 MapReduce

MapReduce is a framework or programming model for processing and generating large datasets [5]. MapReduce parallelize the execution in cluster of commodity machines. It process the input data set and emits key value pairs. MapReduce comprises of two jobs, map job and reduce job. The programming model is like the following.

Map function: Takes the input key value pair and produce intermediate key value pairs [5].

Reduce function: It takes the intermediate result, merges to produce smaller results. The reduce function allows to hold large files in memory [5]. MapReduce computation can be used in Distributed Grep, Count of URL Access Frequency, Reverse Web-Link Graph, Term-Vector per Host, Inverted Index, Distributed Sort etc. [5] (Fig. 2).

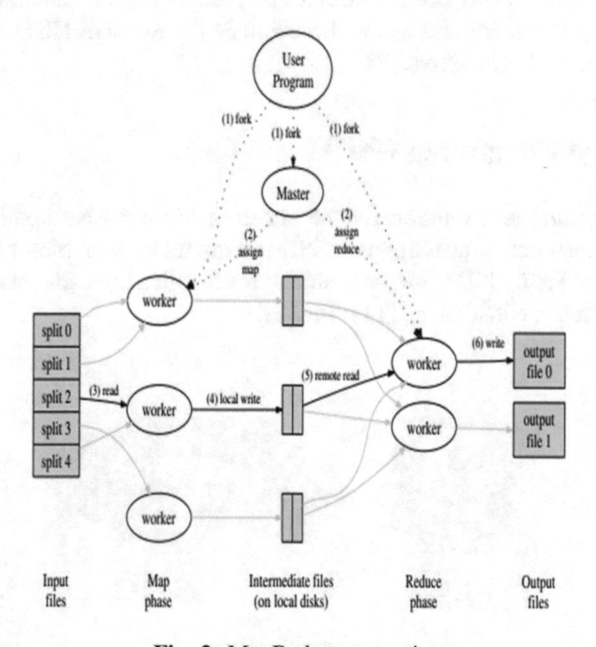

Fig. 2 MapReduce execution

The input data is splitted in to several chunks in different machines (16 megabytes to 64 megabytes (MB)).

It has a master-slave architecture. So one copy will act as the master. The master will assign workers. The map worker will read the input and process it and emits intermediate key value pairs. It is then stored in a memory. The buffered pairs are then moved to local disk. The location of these pairs are passed to the master. The master then assign the reducer workers to reduce task. The shuffle and sort operations are carried out here. After the completion of the MapReduce program the output will be available in the output files [5].

2.2.1 Limitation of MapReduce

The MapReduce is undoubtedly an efficient framework for processing large data over commodity cluster. But it is having some limitations as mentioned in [6]. It will not work properly for small files. MapReduce will not perform better for iterative and interactive jobs. It access sequentially.

2.3 SSD And Hadoop

SSD is Solid State Drive, used for faster processing. It tackle the draw backs of the conventional magnetic disks (HDD), which consumes more energy and takes time to process data. SSD is used mainly due to their low latency and high throughput. It achieves high IOPS because it avoids the physical disk rotation [8]. The experiments held in [8] shows that SSDs were capable of ~ 1.3 Gbps sequential read or write. SSD in hadoop plays a vital role in improving the processing speed.

MapReduce job has two kinds of data access, one is large sequential access and the other is small random reads and writes. By using SSD instead of HDD can increase the processing speed and throughput [8].

3 In-Storage Computing (ISC)

In-storage computing is a solution for the boosting I/O intensive application. It share the meaning of near data processing too. SSD is a computer with high performance low power processor [10]. SSDs are best suited for applications that use random data accesses and small payload sizes [11] (Fig. 3).

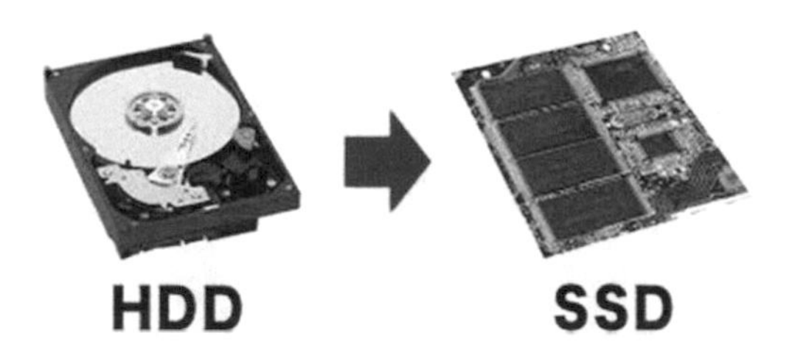

Fig. 3 SSD and HDD

Table 1 SSD versus HDD

	SSD	HDD
Battery life	Less power draw	More power draw
Cost	Expensive	Very cheap
Capacity	1 TB–4 TB	500 GB–10 TB
OS boot time	10–13 s	30–40 s
Copy/write speed	200–500 MB/s	50–120 MB/s
File open speed	30% faster than HDD	Slow

3.1 SSD Hardware Architecture

A SSD comprises of a NAND flash memory array, a SSD controller, and DRAM. The SSD controller contains 4 parts. One is a host interface controller, embedded processors, a DRAM controller and a flash media controller. The commands are processed by the host controller interface and distribute them to the embedded processors. They pass it to the flash controller. Embedded processors are having a tightly coupled memory (SRAM) for storage. DRAM is used for buffering data [13] (Fig. 4).

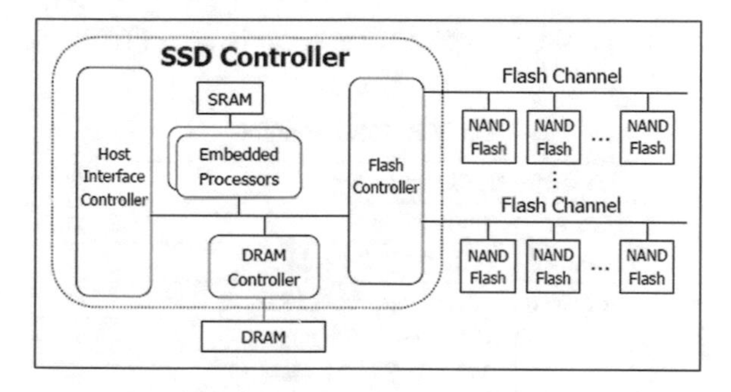

Fig. 4 SSD hardware architecture

4 Proposed System

In the existing system, the MapReduce framework requires more time for processing the data. A solid-state drive (SSD, also known as a solid-state disk) is a solid-state storage device that uses integrated circuit assemblies as memory to store data persistently. SSD technology primarily uses electronic interfaces compatible with traditional block input/output (I/O) hard disk drives (HDDs), which permit simple replacements in common applications [30].

Here, this paper tries to implement SSD on Virtual machine using th software VMware Workstation. Unlike the regular hard disks that are electromechanical devices containing moving parts, SSDs use semiconductors as their storage medium and have no moving parts. It enables usage of SSD as swap space for improved system performance. It increases virtual machine consolidation ratio as SSDs can provide very high I/O throughput [31].

4.1 Apache Spark Integration With Hadoop

Industries are using Hadoop widely to analyze and process their data sets. The reason is that Hadoop framework is based on a simple programming model (MapReduce) MapReduce enables a computing solution that is scalable, flexible, fault-tolerant and cost effective. We always prefer vast data preceding in least amount of time. So as we are more concerned about speed instead of MapReduce we prefer Spark.

Spark was introduced by Apache Software Foundation for enhance the speed of the Hadoop computational computing software process [17].

Spark is not going to replace Hadoop and is not, dependent on Hadoop because it has its own cluster management. Hadoop is just one of the ways to implement Spark.

Spark uses Hadoop in two ways—one is **storage** and second is **processing**. Apache Spark is a lightning-fast cluster computing technology. It is designed for fast computation. MapReduce will not work efficiently for iterative and interactive process. Spark efficiently work on these jobs. The main feature of Spark is its in-memory cluster computing that increases the processing speed of an application [18].

4.2 Process Model

In this proposed system a single node cluster setup is used (Table 2).

Table 2 Configuration

Property	Configuration
Cluster	Single node
Memory	4 GB Intel Core i3
Software	Native hadoop 0.18.0 Windows 8, 64 bits Jdk1.6.0_27, 64 bits, Cygwin, WampServer 2.0, 64 bits
IDE	Eclipse 3.3.1

Here a dataset is loaded in the hadoop system. The dataset is preprocessed. Here the word count dataset is used. After the dataset is preprocessed, the stop words are removed. Then the task is splitted. The categorization will be take place. Here single node cluster is used. So the same system will be act as master and slave. Task is then allocated. Task allocation is nothing but counting the words. It is then passed to MapReduce. It will emit the dataset as key value pair. Finally the result is uploaded. We can retrieve the result in the same system (Fig. 5).

SSD is having influence in the speed of processing. It is an in storage computing technology. Virtual SSD can be also implemented. Table 1 shows the comparison of SSD and HDD.

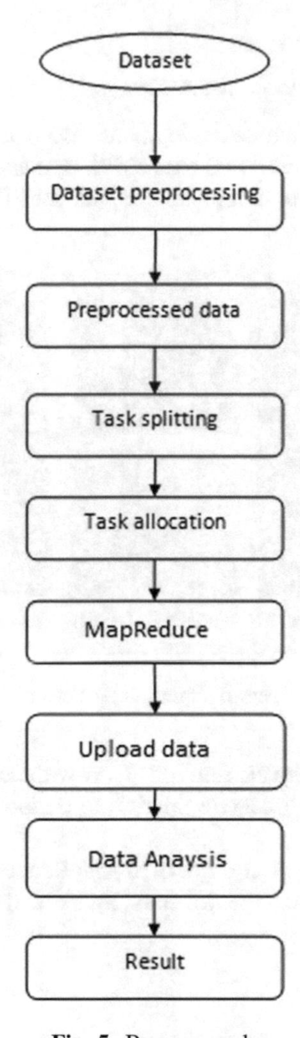

Fig. 5 Process mode

4.3 Features of Apache Spark

- ***Speed*** – Spark run an application in Hadoop cluster, up to 100 times faster in memory, and 10 times faster when running on disk.
- ***Supports multiple languages*** – It provides built-in APIs in Java, Scala, or Python. So we are able to write applications in different languages.
- ***Advanced Analytics*** – Spark supports 'Map' and 'reduce'. It also provide support for SQL queries, Streaming data, Machine learning (ML), and Graph algorithms [7].

4.4 Spark Built On Hadoop

Spark can be deployed in hadoop in three ways.

- **Standalone** – Spark Standalone deployment says that Spark occupies the place on top of HDFS and space is allocated for HDFS. In this model Spark and MapReduce will run by hand in hand to complete all spark jobs [7] (Fig. 6).

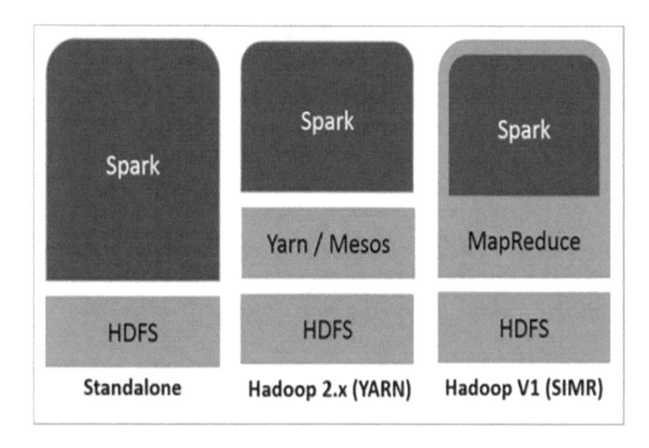

Fig. 6 Spark deployment

- **Hadoop Yarn** – In this, spark runs on Yarn without any pre-installation or root access required. It helps to integrate Spark into Hadoop ecosystem or Hadoop stack [7].
- **Spark in MapReduce (SIMR)** – Spark in MapReduce is used to launch spark job. With SIMR, user will be able to start Spark and uses its shell without any administrative access [7].

4.5 Components of Spark

These are the different components of Spark (Fig. 7).

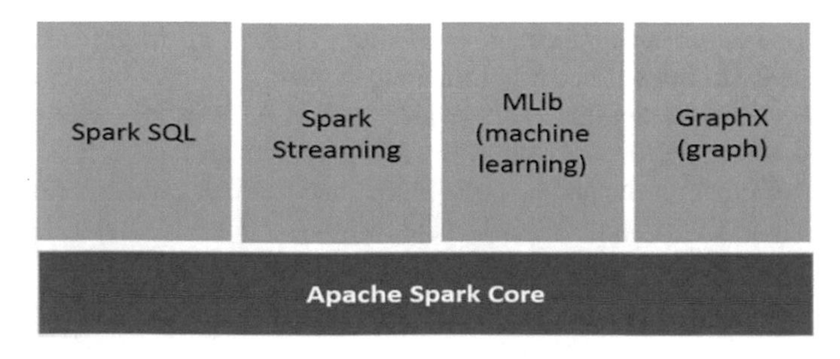

Fig. 7 Spark architecture

Apache Spark Core: It is the general execution engine for spark platform that all other functionality is built upon. It provides In-Memory computing and referencing datasets in external storage systems.

Spark SQL: Spark SQL is a component on top of Spark Core that introduces a new data structure for spark called RDD, which provides support for structured and semi-structured data.

Spark Streaming: Spark streaming is an extension of the core Spark API. It split data in mini-batches and performs RDD (Resilient Distributed Datasets) transformations on those mini-batches of data.RDD has two main operations: Action and Transformations.

MLlib (Machine Learning Library): MLlib is a distributed learning framework above Spark because of the distributed memory-based Spark architecture. Spark MLlib is nine times as fast as the Hadoop disk-based version of **Apache Mahout.**

GraphX: It is a distributed graph-processing framework on top of Spark. It provides an optimized runtime for this abstraction [7].

4.6 RDD-Resilient Distributed Datasets

Resilient Distributed Datasets (RDD) is a fundamental data structure of Spark. It is an immutable distributed collection of objects. RDD is an abstraction of the data that's held in memory, which is a lot faster than storing and fetching things from disk. This already gives you a significant speed boost over other systems [29].

RDDs are also safer to use because the transformations keep the original data in lineages, returning new RDDs with the transformations applied. This allows Spark to reconstruct the data with all the changes should something go wrong with one of the nodes in the cluster, such as a power failure.

DStream takes the concept of RDDs and applies it to streams. A DStream is simply a stream of RDDs, giving all of the advantages of speed and safety in near real time [29].

In this paper, idea of spark integration is proposed. It is in order to make the performance more better than native hadoop. Studies shows that spark has improved the performance of hadoop applications. So if we integrate SSD and spark together we will get more better results (Fig. 8).

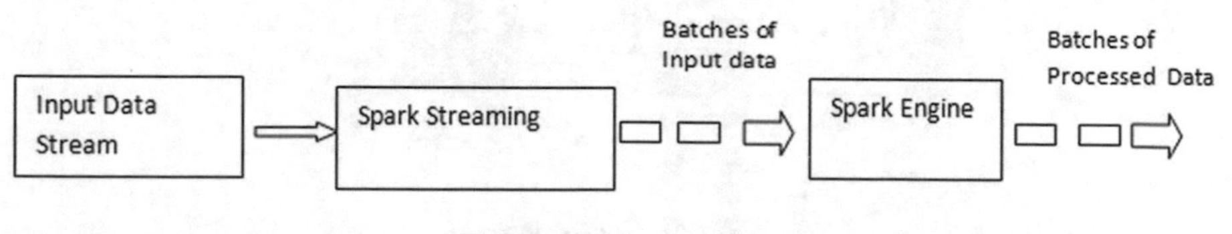

Fig. 8 Spark streaming

Input data can be received from different data sources, like kafka, flume etc. These data will be processed in streams and output is given after functions like map, reduce, join and window. The processed data will be given to the databases or file systems.

If we are integrating Spark with SSD we will get a better result. Studies are showing that spark integration.

In this paper we tried a small application using Spark. A simple word count has been tested using the spark.

Prerequesties: We need to install JDK 1.7. Install Spark from spark website.

After spark has installed, the Wordcount application has run. The speciality of Spark is already mentioned, RDD. Here cache function is called. So spark does not want to compute all the time. Spark will not store data in memory. It will store only when we call an action.

5 Performance Analysis

Here is the analysis of the SSD with hadoop and hadoop without SSD is given. From this graph we can understand that usage of SSD increases the performance of Hadoop. Hadoop without SSD is taking more time than hadoop with SSD. Virtualization of SSD is also possible.

Here, three cases are taken, Wordcount, Shuffle and HDFS read and write are taken. In the above three cases the performance SSD is much more better than hadoop without using SSD (Figs. 9 and 10; Tables 3 and 4).

Table 3 Performance of SSD

Category	HDD	SDD
Wordcount	1	0.6
Shuffle	1.5	0.8
HDFS read and write	0.8	0.5

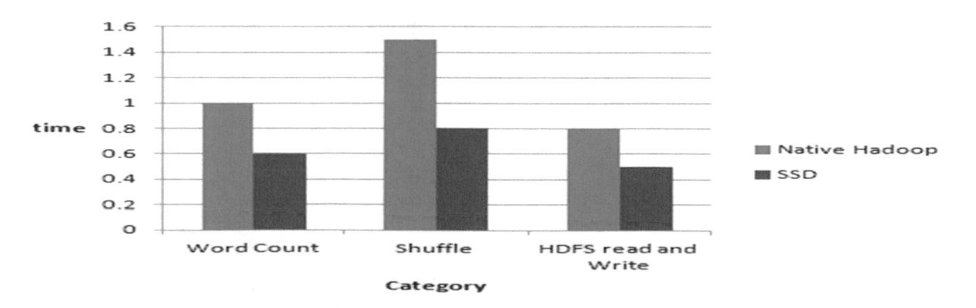

Fig. 9 Comparison

Table 4 Throughput comparison

Record size (byte)	Hadoop	Spark
1	8	15
2	15	20

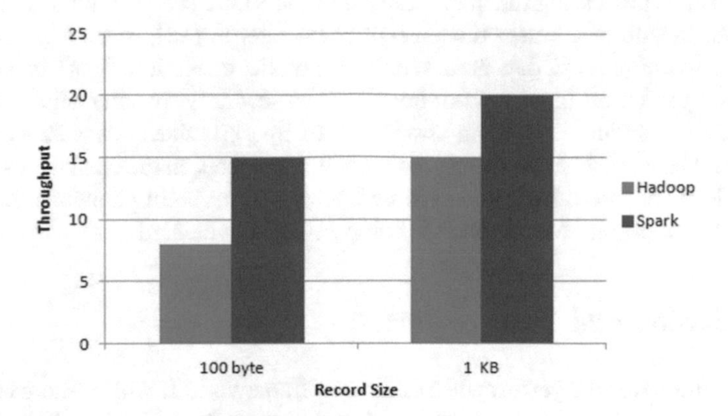

Fig. 10 Throughput comparison

6 Related Works

So many related works and discussions have been evolved about Hadoop and its enhancements. This section covers a literature survey done on Hadoop. In [6] the disadvantage of existing HDFS is scalability of Namenode. When the usage of memory meets its maximum, the Namenode will not respond. So this paper suggest usage of multiple name nodes. So if one name node fails we can continue the processing with other namenodes. MapReduce is having limitations as stated in [9]. They were not good for interactive and iterative programs. This paper illustrate about the concept called MRlite. It consist of master slave architecture. Master controls parallel execution. It has NFS server. Master accepts the data from client and distribute to slaves. Hadoop enhancement can be done using a cache, describe in [10]. The proposed system uses Locality Sensitive Hashing (LSH). Where a cached data is stored in the distributed file system. Map phase will execute the file system. Then the result is stored in the cache. It contains five modules like: Preprocessing files, Creating file vector, Create signature, then using LSH to find the nearest neighbor. Preprocessing will remove all the stop words and the signature will be generated based on TFID (Term Frequency Inverse Document Frequencies) Then by using LSH they will find the related files. It take less time to the processing the files. In [14–16] SSD and HDD are compared and they have mentioned that SSD increases the throughput and consume less power. They have done experiments with DFSIO benchmark, Join, Random Writer etc. They conclude SSD perform better than HDD. Hibench suit is also used. In [9] the possibilities and challenges of implementing SSD in Hadoop frame work is depicted. While SSD is implemented so many challenges are emerged. Data representation discrepancy, Data Split problem, since SSD will not move data, It is performing in-memory computing. There occurs system interface discrepancy. SSD cannot directly contact with the host Hadoop system. They introduced File to LBA converter to solve this issue. In [17] the concept of Apache Spark is introduced. It is a general purpose cluster computing engine provides application programs interfaces in various programming languages like Java, Scala, Python etc. Spark supports in-memory computing. It makes processing much faster. Spark uses RDD. It does not need to store any intermediary result. Spark uses the concept of lineage. In [18], focus is given to those that reuse a working set of data across multiple parallel execution. Spark can be mainly used for two use cases: first one is or iterative jobs secondly for interactive jobs, where MapReduce is a failure. The main abstraction of Spark is that it uses RDD (Resilient Distributed Dataset). It is read only collection of objects partitioned across a set of machines.RDD achieves fault tolerance by lineage. It can be implemented in Scala. In [19] Apache Spark and MapReduce is compared and evaluated.

7 Conclusion and Future Work

SSD usage improves the performance of hadoop framework. It will be more convenient for the Big data industry. Spark can be integrated with the SSD. It will improves the processing time because of the feature RDD.

Spark is fast in memory data processing. It track lineage of operations used to derive data. During failure, it use lineage to recompute data. For distributed storage, the read throughput can be improved using caching, however, the write throughput is limited by both disk and network bandwidth due to data replication for fault-tolerance.

Acknowledgements. We would like to thank Dr. Muhammad Sitheeq and Prof. Rosna P. Haroon of Computer Science Department at Ilahia College of Engineering for their valuable feed back.

References

1. www.tutorialspoint.com/hadoop/hadoop_tutorial_in_pdf.htm
2. Bhosale, H.S., Gadekar, D.P.: A review paper on Big Data and Hadoop. Int. J. Innov. Res. Sci. Eng. Technol. **4**(10), 1–3 (2014)
3. Kang, S.-H., Koo, D.-H., Kang, W.-H., Lee, S.-W.: A case for flash memory SSD in hadoop applications. Int. J. Control Autom. **6**(1), 201–210 (2013)
4. Shvachko, K., Kuang, H., Radia, S., Chansler, R.: The Hadoop Distributed File System. In: Proceedings of MSST2010, May 2010, IEEE
5. Dean, J., Ghemawa, S.: MapReduce: simplified data processing on large clusters. In: OSDI 2004
6. Ma, Z., Gu, L.: The limitation of MapReduce: a probing case and a lightweight solution. In: Proceedings of 1st International Conference on Cloud Computing, GRIDs, and Virtualization (Cloud Computing 2010)
7. Srinivas Jonnalagadda, V., Srikanth, P., Thumati, K., Nallamala, S.H.: A review study of Apache Spark in Big data processing. Int. J. Comput. Sci. Trends Technol. (IJCST) **4**(3), 93–98 (2016)

8. Kambatla, K., Chen, Y.: The truth about MapReduce performance on SSDs. In: Proceedings of the 28th Large Installation System Administration Conference (LISA14). https://www.usenix.org/conference/lisa14/conferenceprogram/presentation/kambatla
9. Park, D., Kee, Y.-S.: In-storage computing for Hadoop MapReduce framework: challenges and possibilities. IEEE Trans. Comput. doi:10.1109/TC.2016.2595566
10. http://www.flashmemorysummit.com/English/Collaterals/Proceedings/2015/20150813_S301D_Ki.pdf
11. http://www.dell.com/downloads/global/products/pvaul/en/ssd_vs_hdd_price_and_performance_study.pdf
12. http://www.storagereview.com/ssd_vs_hdd
13. Kang, S.-H., Koo, D.-H., Kang, W.-H., Lee, S.-W.: A case for flash memory SSD in hadoop applications. Int. J. Control Autom. **6**(1), 201–210 (2013)
14. Do, J., Kee, Y.-S., Patel, J.M., Park, C., Park, K., DeWitt, D.J.: The truth about MapReduce performance on SSDs. In: Large Installation System Administration Conference (LISA14) (2014)
15. https://www.infoq.com/articles/apache-spark-introduction
16. Saxena, P., Chou, J.: How much solid state drive can improve the performance of hadoop cluster? Performance evaluation of Hadoop on SSD and HDD. Int. J. Mod. Commun. Technol. Res. (IJMCTR) **4**(6), 72–78 (2016)
17. Shoro, A.G., Soomro, T.R.: Big data analysis: Ap spark perspective. Glob. J. Comput. Sci. Technol. C Softw. Data Eng. (2015)
18. Zaharia, M., Chowdhury, M., Franklin, M.J., Shenker, S., Stoica, I.: Spark: Cluster Computing with Working Sets. University of California, Berkeley (2015)
19. Gopalani, S., Arora, R.: Comparing Apache Spark and Map Reduce with performance analysis using K-means. Int. J. Comput. Appl. (2015). doi:10.5120/19788-0531
20. Capriolo, E., Wampler, D., Rutherglen, J.: Programming Hive. O'Reilly Media, Sebastopol (2012)
21. Karau, H., Konwinski, A., Wendell, P., Zaharia, M.: Learning Spark. O'Reilly Media, Sebastopol (2015)
22. Thomas, L., Syama, R.: Survey on MapReduce scheduling algorithms. Int. J. Eng. Trends Technol. (IJETT) (2014)
23. White, T.: Hadoop: The Definitive Guide, 4th edn. O'Reilly Media, Sebastopol
24. Machine Learning: Wikipedia (2014). http://en.wikipedia.org/wiki/Machine_learning
25. SparkJobFlow – Databricks. https://databrickstraining.s3.amazonaws.com/slides/advancedspark-training.pdf
26. Zaharia, M., Konwinski, A., Joseph, A.D., Katz, R., Stoica, I. Improving MapReduce Performance in Heterogeneous Environments. In: 8th USENIX Symposium on Operating Systems Design and Implementation, University of California, Berkeley
27. Khanam, Z., Agarwal, S.: Map-reduce implementations: survey and performance comparison. Int. J. Comput. Sci. Inf. Technol. (IJCSIT) (2015). doi:10.5121/ijcsit.2015.7410.119
28. Armbrust, M., Das, T., Davidson, A., Ghodsi, A., Or, A., Rosen, J., Stoica, I., Wendell, P., Xin, R., Zaharia, M.: Scaling spark in the real world: performance and usability. Proc. VLDB Endow. **8**(12), 1840–1843 (2015)
29. www.mapr.com
30. Wikipedia
31. https://pubs.vmware.com
32. Garion, S.: Big Data Analytics Hadoop and Spark. PhD, IBM Research, Haifa

Invisible Color Image Authentication in SVD Domain

Canavoy Narahari Sujatha(✉)

Department of Electronics and Communications, SNIST, Ghatkesar
Hyderabad 501301, Telangana, India
sujathareddy2014@gmail.com

Abstract. In this paper, singular value decomposition based color image authentication scheme is proposed. Singular values of the host image are modified with that of a secret image. Modifications are optimized to obtain the maximum robustness without losing visual quality. Experiments are done using color images as host and watermark. Simulation results show that the present algorithm is highly resistant to various image processing attacks with significant improvement in perceptibility. The performance is measured objectively in terms of Mean Square Error, Peak Signal to Noise Ratio and Correlation Factor.

Keywords: Watermark · SVD · Embedding · Extraction · PSNR · CF

1 Introduction

In recent years, illegal distribution of multimedia over the internet has been grown. Digital media can be transmitted efficiently through communication networks without losing the quality. Copied version of the digital media is similar to original multimedia. This causes serious concern about manipulation of digital content and illegal usage [1]. Earlier, protection of digital media has been done with the help of encryption algorithms. But using encryption alone is not sufficient for protecting digital media, because in this process digital media has to be decrypted to identify the hidden information. Then the decrypted data can be easily distributed and manipulated. Thus it leads to introduce a new hiding scheme known as watermarking. Watermarking is a process to embed copyright information into original multimedia for authentication process. The embedded data should be invisible and robust against various possible attacks.

Various watermarking schemes have been proposed. These schemes are categorized as spatial domain and frequency domain schemes [2]. In spatial domain scheme, watermarking has been done by directly manipulating the pixel values of cover image. In the frequency domain scheme, data can be embedded by modifying coefficients of the selected transform. According to transparency, embedding can be done either in visible way or invisible way [3]. In visible watermarking, embedding has been done in such a way that the watermark is visible on host image. Though it has been used for longtime, it is not secured. Therefore visible watermarking is used for owner identification but for all other applications invisible watermarking is used. In invisible watermarking, the embedded data is perceptually unseen and it holds the quality of host image. Invisible

S.C. Satapathy and A. Joshi (eds.), *Information and Communication Technology for Intelligent Systems (ICTIS 2017) - Volume 1*, Smart Innovation, Systems and Technologies 83, DOI 10.1007/978-3-319-63673-3_31

watermarking can be implemented using number of techniques. Generally, the simplest technique used for invisible watermarking is Least Significant Bit method in which embedding has done in LSBs of host image. In the proposed method, Singular values of host image are used to hide the singular values of watermark image invisibly. Before going into details of developed watermarking scheme, the brief about SVD is discussed in the following section.

2 Singular Value Decomposition (SVD)

Nowadays, SVD is of great significance both in mathematics theory and applications. SVD is an important factorization of a real or complex matrix in signal processing. According to spectral theorem, normal matrices can be diagonalized unitarily with a basis of eigen vectors. Thus the SVD can be treated as a generalization of the spectral theorem to arbitrary matrices.

Suppose A is a $m \times n$ image matrix, then the factorization of A is given by

$$\mathrm{SVD}\,(\mathrm{A}) = \mathrm{USV}^{\mathrm{T}} \tag{1}$$

Where, U is a $m \times m$ unitary matrix

S is a $m \times n$ matrix with square root of eigen values as singular values

V is a $n \times n$ unitary matrix

In singular value decomposition, the matrix V contains a set of orthonormal input vector directions for the matrix A. the matrix S contains non-negative components on the diagonal and zeros off the diagonal [4, 5]. These non-negative components represent the intensity value of the image. Slight change in these singular values does not affect the quality of the image and also don't change much even after attacks. The matrix U contains a set of orthonormal output basis vector directions for the matrix A. The singular vectors U and V specify the algebraic properties of an image. SVD uses non-fixed orthogonal bases unlike other transforms which uses fixed orthogonal bases [6, 7]. Watermarking schemes make use of these properties to give good accuracy, imperceptibility and robustness in proving the authentication of watermarked image.

3 Proposed Schemes

There are two steps in proposed watermarking method like watermark embedding procedure and watermark extraction procedure.

3.1 Embedding Scheme

Let the host color image and watermark image of size $n \times n$. Three color planes R, G and B are isolated from the host image. Each plane is decomposed into three matrices using SVD [8, 9]. R, G and B planes are separated from the watermark image. Each plane of watermark image is also decomposed with SVD. Then singular values of R plane of host image are embedded with those of R plane of watermark image at a particular scaling factor [10, 11]. Similarly, singular values of all planes of host image are modified with the corresponding singular values of respective planes of watermark image at chosen scaling factor. At the last stage inverse SVD is calculated for individual planes using the respective singular vectors of the planes in host image. Thus the watermarked color image is produced by concatenating the modified R, G and B planes. PSNR is calculated between host and watermarked images for objective measure of quality [12]. Block diagram for embedding process is shown in Fig. 1.

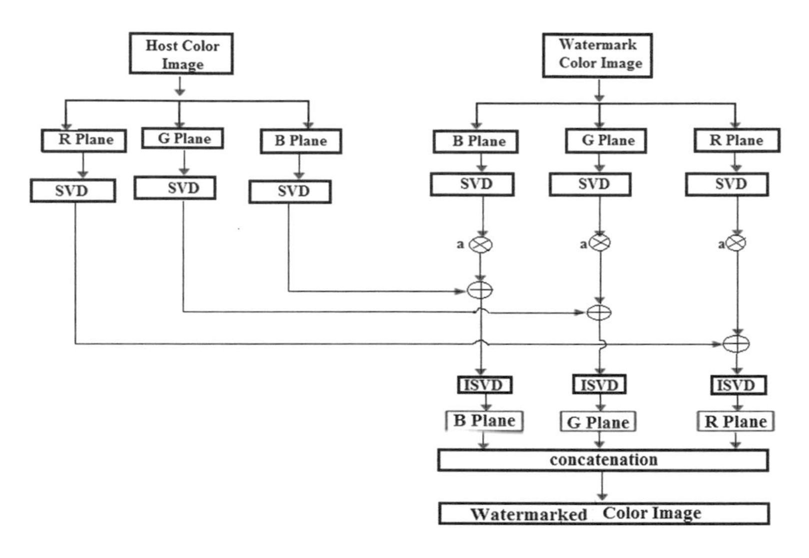

Fig. 1. Block diagram for watermark embedding

3.2 Extracting Scheme

In the extraction process, watermarked image is isolated into three color planes. Each plane is factorized using singular value decomposition to get modified singular values. Then the inverse SVD is applied on the recovered singular values of individual planes using singular vectors of original watermark to identify the hidden data [13, 14]. Therefore the retrieved R, G, B planes are combined to form color watermark image. After extracting the watermark image, correlation factor is calculated between original and extracted watermark images to justify the similarity between them. To test the robustness, the algorithm is subjected to various kinds of attacks and the above steps are followed to extract the watermark under attacks. Again MSE, PSNR and CF are calculated for various watermark images to judge the efficiency of the algorithm. Figure 2 shows the block diagram for watermark extraction.

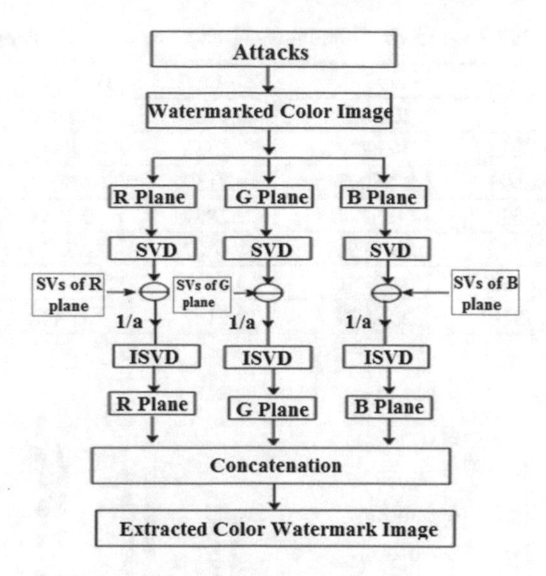

Fig. 2. Block diagram for watermark extraction

4 Simulation Results & Analysis

All experiments have been carried out using Matlab as computing environment and programming language. The proposed algorithm is tested on variety of images. Results are attained by using 256 × 256 pixels size host and watermark color images. The obtained MSE and PSNR values at different scaling factors are displayed in Tables 1 and 2 respectively. The same variations for various host images with SVU logo are graphically shown in Figs. 3 and 4. Without distortions, the correlation factor is found to be unity which says that the extracted logo is exactly similar to embedded logo without any degradation.

Table 1. MSE values for different host images at various scaling factors.

Scaling factor	Peppers	Lena	Sunset	Balloon	Autumn
0.01	0.3656	1.5303	1.4568	1.5845	0.2976
0.03	2.8884	13.3596	11.4213	13.3252	2.7028
0.05	7.6587	35.6538	27.6385	36.5308	7.2028
0.07	13.0692	56.1486	38.4372	56.8384	12.1140
0.1	19.7188	69.4264	46.4526	69.3078	15.0861
0.5	42.0270	84.2894	68.4815	81.9974	23.5666
1.0	43.4433	84.3896	70.9763	82.9200	24.7623

Table 2. PSNR values for different host images at various Scaling factors.

Scaling factor	Peppers	Lena	Sunset	Balloon	Autumn
0.01	52.5012	46.2829	46.4968	46.1320	53.3944
0.03	43.5242	36.8729	37.5536	36.8841	43.8126
0.05	39.2893	32.6097	33.7157	32.5042	39.5558
0.07	36.9683	30.6374	32.2833	30.5844	37.2979
0.1	35.1820	29.7156	31.4607	29.7230	36.3450
0.5	31.8955	28.8731	29.7751	28.9928	34.4078
1.0	31.7516	28.8679	29.6197	28.9442	34.1929

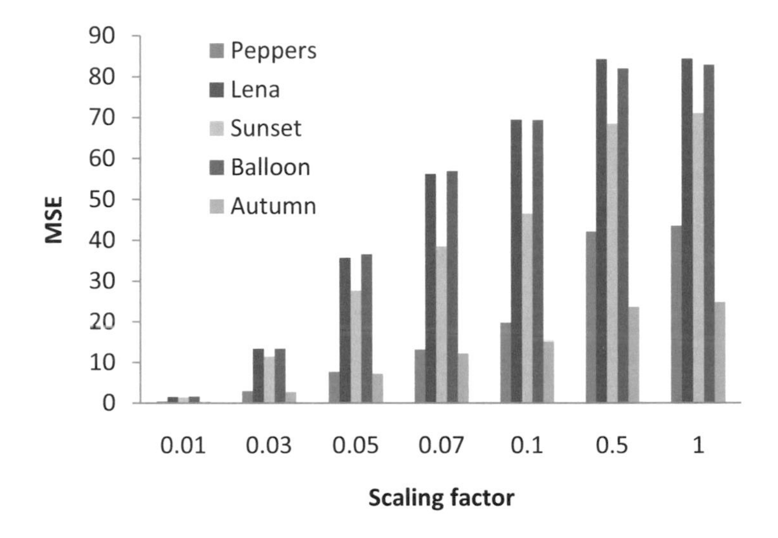

Fig. 3. Change in MSE values with various scaling factors for different host images.

The robustness of the present scheme is tested under six practical conditions such as adding Salt & Pepper noise, Gaussian noise, Median, Average filtering, Sharpening and Rotation. To examine the performance of the algorithm, Lena image is taken as a host image with different logos and the scaling factor is set at 0.1.

If embedding is done at small value of scaling factor, then the image quality is higher, but the robustness of algorithm is weak under attacks. Otherwise if embedding takes place at higher scaling factor, then the image quality is reduced and gives more robustness.

The algorithm is tested for variety of logo images by embedding them in standard Lena image, but for the sake of space, here the results are given against SVU and SNIST logos as shown in Tables 3 and 4. Graphically results for all logos at the specified scaling factor of 0.1 are shown in Fig. 5.

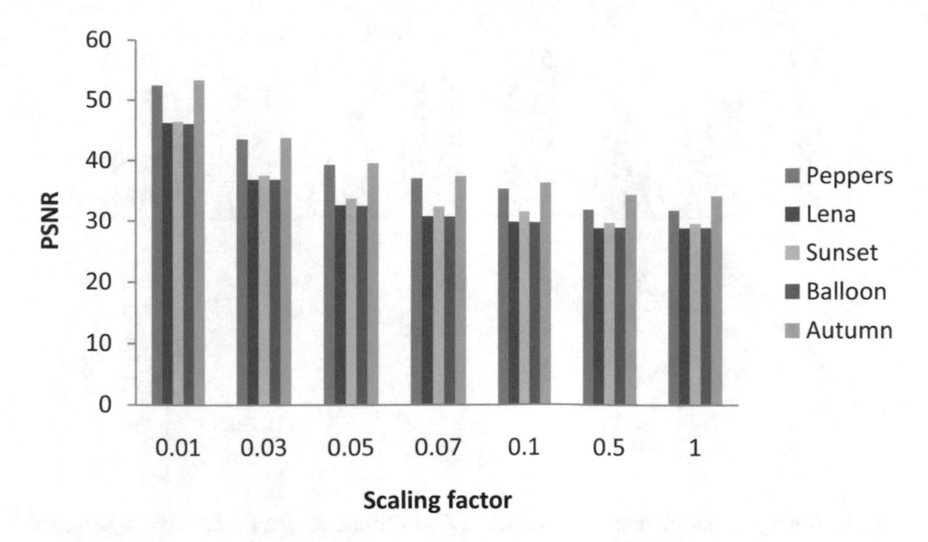

Fig. 4. Change in PSNR values with various scaling factors for different host images.

Table 3. Performance metrics for SVU logo against various attacks.

Attacks	MSE	PSNR	CF
Salt & Pepper noise (0.02)	267.6518	23.8551	0.6573
Gaussian (0.001)	323.6742	23.0297	0.5767
Median filtering (3 × 3)	155.0846	26.2251	0.9443
Average filtering (3 × 3)	158.4984	26.1306	0.7667
Sharpening	514.6877	21.0154	0.6072
Rotation	2.6628e + 03	13.8774	0.7481

Table 4. Performance metrics for SNIST logo against various attacks.

Attacks	MSE	PSNR	CF
Salt & Pepper noise (0.02)	243.1714	24.2717	0.6709
Gaussian (0.001)	300.4615	23.3529	0.5922
Median filtering (3 × 3)	125.3651	27.1490	0.9472
Average filtering (3 × 3)	128.7011	27.0350	0.8060
Sharpening	454.7729	21.5529	0.5914
Rotation	2.6435e + 03	13.9090	0.5423

From the Fig. 5, it has been observed that the Lena host image with different logos is more robust to Median filtering among six mentioned distortions. Marked Lena image with SNIST logo is more robust to Salt & Pepper noise, Gaussian noise, Median and Average filtering attacks. With SVU logo, it is highly robust to Sharpening and rotation attacks. Among all the distortions, SNIST logo is more robust to median filtering attack as JNTU logo is least resistant to rotation.

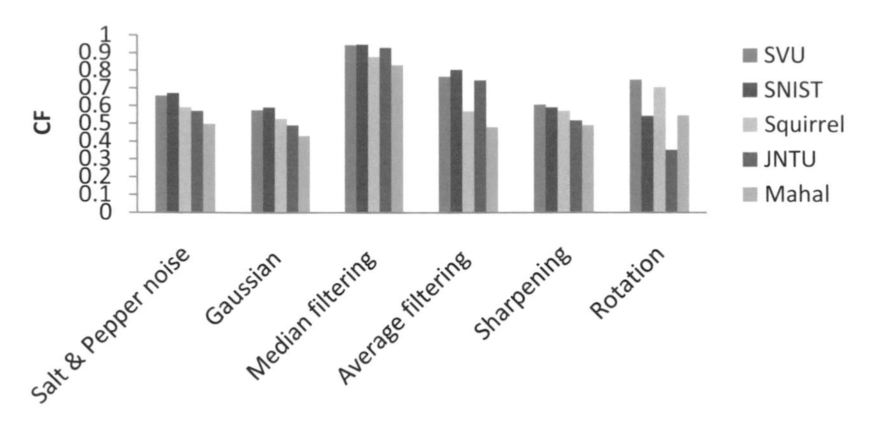

Fig. 5. Change in correlation factor values against various attacks for different logo mages.

Original images shown in Fig. 6 (i) to (v) are embedded with logos (vi) to (x) respectively, measured CF and PSNR values also displayed against median filtering. From the results, it has been analyzed that the proposed scheme can withstand certain attacks at specified scaling factor of 0.1. But this algorithm is not resistant to distortions like cropping, compression and histogram equalization etc. And also understood that the efficiency of the present scheme changes for various logo images embedded in different chosen host images. Thus, based on the luminance values of host and watermark images,

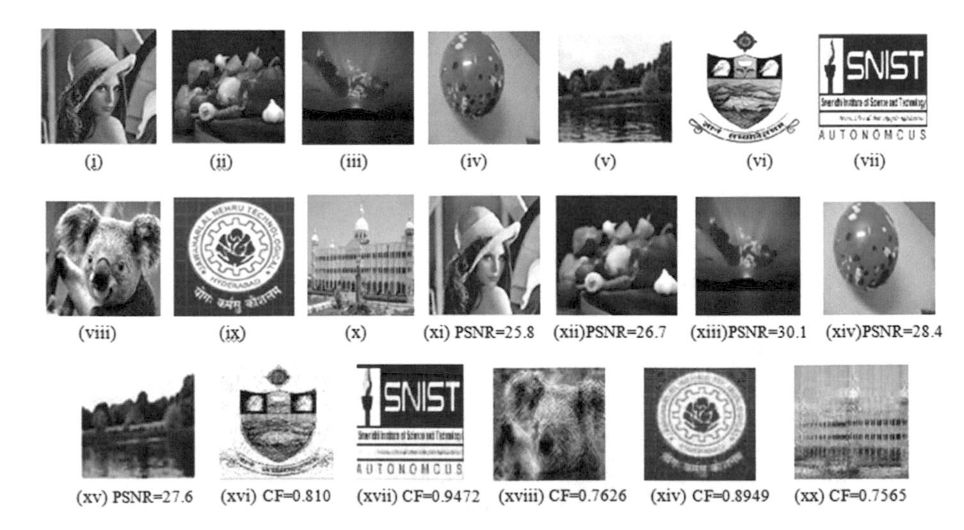

Fig. 6. (i) to (v) are the original color images, (vi) to (x) are the watermark color images, (xi) to (xv) are the watermarked images attacked by median filtering with 3×3 mask, (xvi) to (xx) are the recovered watermarks from median filtered watermarked images at scaling factor of 0.1

the performance of the proposed scheme is affected under attacks. Under attacks, extraction of logos is accepted with minimal distortions.

5 Conclusion

This paper presents an invisible watermarking scheme to insert a color logo in color image using SVD. Simulation results explore the resistance of the proposed scheme to different image processing attacks such as noise addition, median filtering, average filtering, sharpening and rotation. In this method, original image information is required in extracting the hidden data which is known as non-blind watermarking. An additional advantage of this scheme is that the color logo itself serves as a hidden key for logo extraction. This algorithm ensures the rightful ownership of watermarked image to prove the authentication.

References

1. Potdar V.M., Han S., Chang E.: A survey of digital image watermarking techniques. In: Proceedings of IEEE International Conference on Industrial Informatics, pp. 709–716 (2005)
2. Nikolaidis, N., Pitas, I.: Robust image watermarking in the spatial domain. Sig. Process. **66**(3), 385–403 (1998)
3. Chandra, D.V.S.: Digital image watermarking using singular value decomposition. Circuits and Systems. MWSCAS-2002, vol. 3, pp. 264–267 (2002)
4. Liu, R., Tan, T.: A SVD - based watermarking scheme for protecting rightful ownership. IEEE Trans. Multimed. **4**(1), 121–128 (2002)
5. Zhang, X.P., Li, K.: Comments on: an SVD-Based watermarking scheme for protecting rightful ownership. IEEE Trans. Multimed **7**(2), 593–594 (2005)
6. Wu, Y.D.: On the security of an SVD-based ownership watermarking. IEEE Trans. Multimed. **7**(4), 624–627 (2005)
7. Andrews, H., Patterson, C.: Singular value decomposition (SVD) image coding. IEEE Trans. Commun. **24**(4), 425–432 (1976)
8. Sun, X., Liu, J., Sun, J., Zhang, Q., Ji, W.: A robust image watermarking scheme based on the relationship of SVD. In: Proceedings of IIH-MSP, IEEE, pp. 731–734 (2008)
9. Basso, A., Bergadano, F., Cavagnino, D., Pomponiu, V., Vernone, A.: A novel block based watermarking scheme using the SVD transform. J. Algorithms **2**, 46–75 (2009)
10. Qing, X., Jianquan, X., Yunhua, X.: Performance test and analysis of image watermarking algorithm based on singular value decomposition. In: International Conference on Computer Engineering and Technology, IEEE, pp. 282–286 (2010)
11. Benhocine, A., Laouamer, L.: New images watermarking scheme based on singular value decomposition. J. Inf. Hiding Multimed. Signal Process. **4**(1), 9–18 (2013)
12. Sujatha, C.N., Sathyanarayana, P.: Non blind color image watermarking scheme using SVD. In: IEEE Conference on Electrical & Electronics Engineering. Bangalore (2014)
13. Huang, H., Chen, D.-F., Lin, C.-C., Chen, S.-T., Hsu, W.-C.: Improving SVD-based image watermarking via block-by-block optimization on singular values. EURASIP J. Image Video Process. **2015**(1), 25 (2015)
14. Yao, L., Yuan, C., Qiang, J., Feng, S., Nie, S.: Asymmetric color image encryption based on singular value decomposition. Opt. Lasers Eng. **89**, 80–87 (2016)

Cost Efficient Intelligent Vehicle Surveillance System

S.L. Kiran(✉) and M. Supriya

Department of Computer Science and Engineering, Amrita School of Engineering, Amrita Vishwa Vidyapeetham, Amrita University, Bengaluru, India
kiranattingal11@gmail.com, m_supriya@blr.amrita.edu

Abstract. The rapid development in the field of electronics, provide a secured environment for the human to live in. This paper introduces a model, "Intelligent Vehicle Surveillance System", which is designed to reduce the risk involved in losing the vehicles and it also provides notification of occurrence of any accidents, which will reduce the rate of deaths in a very cost efficient manner. This paper introduces a tracking system which passes alert to the owner of the vehicle immediately regarding a theft or an accident of the vehicle with the precise location of the vehicle. There are different methods to identify and continuously track the location of remote vehicle. This proposed system has a single board GPS equipped with GSM and Arduino microcontroller attached in the vehicle. As the vehicle moves its location gets updated via SMS. User can provide real-time control, by sending messages, controlling the vehicle, changing direction as well as on and off. A Software attached would help to read, analyze, process and store the incoming SMS. This system finds its application in real time traffic monitoring. The current system will provide monitoring information from anywhere.

Keywords: GSM module · GPS module · Vibration module · Arduino · SIM900 · L293D · Surveillance

1 Introduction

Diurnally the vehicle theft incident is increasing. According to the survey, more than over 5000 vehicles are purloined in our country each year as per the reports by National Crime Records Bureau, Ministry of Home Affairs, Government of India, New Delhi. The number of accidents is also growing slowly however steady, year by year. Trucks and two-wheelers are alone answerable for an additional four-hundred deaths than by any other vehicles. The afternoon rush on the roads constructs graveyard for another hundreds each day. Also, variety of individuals abraded in road accidents succumb to their injuries a couple of days once the accident, and such deaths don't seem to be classified as results of road accidents. In rural areas, motor vehicle accidents involving animals are quite common. Near farms, accidents with tractors and alternative farm vehicles on the aspect of the

S.C. Satapathy and A. Joshi (eds.), *Information and Communication Technology for Intelligent Systems (ICTIS 2017) - Volume 1*, Smart Innovation, Systems and Technologies 83, DOI 10.1007/978-3-319-63673-3_32

road may occur. Rural roads are usually the smallest amount well lit, still, that create a lot of accidents during night. No matter where it occured but the victims should be provided best treatment once it occurs. The person should be given great care and for that the people around must be able to spot the accident and rush for help [1]. Though the victim do not acknowledge any injuries promptly, there could also be some lurking below the surface. A professional will examine his injuries and find him back on the trail to a full recovery. Night accidents are a lot more dangerous than those that occur at daytime. Therefore we need a quick responsive system that determine and track the vehicle [2]. There are other cases where the vehicle at the front to be tracked [3].

2 Related Works

There are many works that tracks a vehicle using the GPS system but paper [4] talks about GSM enabled vehicle tracking system. This was the cheapest anti-theft and vehicular tracking system. This has a small kit with many components and a GSM module in it. The owner can send SMS to switch the system ON and OFF. The message is actually an instruction to the microcontroller which actually turns it on and off. The microcontroller gets the location of the vehicle and it stimulates the GSM module to send the location to the owners mobile as a short message. The delivery message is sent back to the microcontroller with the help of same GSM module. The advantage of this system is that it helps the owner in tracking the vehicle very quickly, and reduces the complexities compared to other systems.

The frequency of road accidents is increasing significantly because of the rise in use of cellular phone while driving. In order to avoid this situation, the work in [5] proposes a reference in which a mobile stand is provided for the driver to fix his phone. Otherwise, the micro controller forces the driver to stop the vehicle in order to continue the conversation on cell-phone. This leads to minimum chance of accidents. Furthermore if road mishaps occur, this system allows to send an emergency message to the rescue teams.

Aarthi et al. [6] proposed a technique of extracting the registration details of a defaulting vehicle that exceeds the speed limit.

A mechanism proposed in [7] senses any accident by the vehicle and intimates to the preprogrammed mobile numbers like the owner of the vehicle, ambulance, police for immediate remedy. GSM technology is used to send the position of the vehicle as a SMS to those numbers. Using these messages the position of the vehicle can be obtained by the owner of the vehicle and this process make sure that the human life is safe though the vehicle in damaged. This proposed method uses small switching function (reset) for disconnecting the signal and when a car is met with an accident, it waits for some time and immediately the car and the GPS coordinate of the location are messaged to the nearby hospitals, thereby ensuring timely help to the needy. A smart anti-theft system which not only tracks the exact location of the vehicle but also tries to prevent the theft is discussed in [8]. GPS will provide the location information via satellites.

If the vehicle is met with some accident, then a message is automatically sent for help. The user can control other technologies installed in the system as well with his mobile phone. It includes the engine ignition cutoff, fuel supply cut-off, electric shock system (installed on steering wheel) and paint spray. This complete system is designed taking in consideration the low range vehicles to provide them extreme security.

2.1 GSM

Global System for Mobile communications (GSM) is a technology used for communicating over the mobile network. We can track the vehicle continuously and also inform to the Local ambulance if the vehicle is met with any accident using GSM technology. This is an affordable device that reduces the matter related to accident notification and antitheft management. If the user is somewhere off from the vehicle and he needs to understand wherever his vehicle is correct from the place he's standing, he has got to send a predefined message to the modem [9].

Features of GSM are

Value added features: GSM provides value added features. Because of this more than 450 million people all over the world is currently using GSM technology.

No Additional Charges: No additional charges are involved. As GSM is used in more than 200 countries all over the world, we can simply use our GSM phones when we are going through these countries.

Secure Data Transmission: GSM is a completely dependable device. There is no down time until an excessive electric typhoon damages the transceiver or cellular at someplace.

Variety of Service: Variety of service providers and handsets are available in the market which helps the customer to choose from a variety of options.

Less Consumption of Power: Less consumption of power and extensive coverage. So the user can use GSM spectrum to transmit any amount of data. Cheaper call rates compared to others moreover the messaging options are also free. The quality of the calls compared to GSM is much better than CDMA.

2.2 GPS Module

There is a family of stand-alone GPS receivers that offers high performance u-blox 6 positioning engines. This is the NEO-6 collection. The receivers are very value effective that they provide several connectivity options in a miniature $16 \times 12.2 \times 2.4$ mm package. Their architecture is very compact and they have high energy and reminiscence option which makes these NEO-6 modules perfect for cellular gadgets that operates on battery.

2.3 Microcontroller Arduino

Arduino board designs use a diversity of microprocessors and controllers. The Mega 2560 microcontroller board is one of the variety from arduino which is based on the ATmega2560 has 54 digital input/output pins, 16 analog inputs, 4 UARTs (hardware serial ports), a USB connection, a 16 MHz crystal oscillator, an ICSP header, a power jack and a reset button. It supports a microcontroller. It can be connected to the computer with a USB or it can be powered to an adapter or a battery and it will get started. The Mega 2560 board is compatible with most shields designed for the Uno and the former boards. Figure 1 shows the pin diagram of Arduino mega 2560.

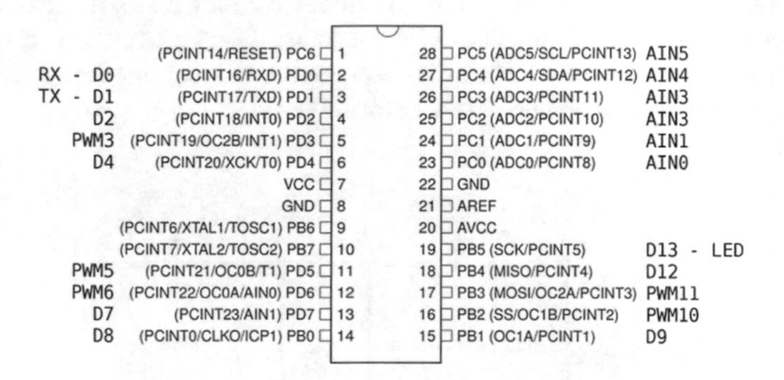

Fig. 1. Pin diagram of Arduino Mega 2560

3 Design Methodology

There are a great volume of areas where a need exists for some gadgets that identifies and tracks the geographic area of a vehicle and does a steady monitoring of the same. Passing of alert messages to the owner or any person directed owner about the theft or accident that happens in every much necessary and this should happen without any delay. The location of the vehicle is actually shared. This paper introduces a tracking gadget which includes worldwide Positioning device (GPS) and GSM. GPS and GSM together determine the vicinity of the vehicle. A microcontroller equipped with GPS and GSM modems on its sides and is organized as a single board embedded machine. This is attached to the vehicle [10]. Throughout the car movement, its area can be suggested by the use of SMS. A software is created to read, technique, analyses and save the incoming SMS messages. Using of the GSM and GPS technologies, permits the device to track the car and gives the maximum up to date statistics approximately about the vehicle. If a pre-defined message is dispatched by the proprietor, it routinely stops the car and it can offer actual time control. The motive of this system is to layout and combine a new mechanism that's incorporated with GPS-GSM to offer features like the location information and real time tracking using the SMS and also to control the speed.

Block Diagram. This project is implemented by building a protocol known as robo vehicle. This robo vehicle consists of two dc motors, a voltage amplifier L293D, microcontroller ARDUINO MEGA 2560 and switches. DC motors are used to move the vehicle in forward, backward, right and left. But the current supplied by the controller is not sufficient for rotating the dc motor. Hence, an amplifier L293D which amplifies the current is used to produce the sufficient current. For the notification of thefts and accidents, we are interfacing GSM and GPS modules on robo vehicle through serial communication. Figure 2 shows the block diagram of proposed system. GSM and GPS are both CMOS devices and the controller is a TTL device. The final output for all CMOS devices is RS232. The internal voltage levels for CMOS devices are 18-23 v. As the controller is a TTL device it operates at 5 v. In order to convert the voltage levels from 18-23 to 5 v we used MAX232 which acts as amplifier [11].

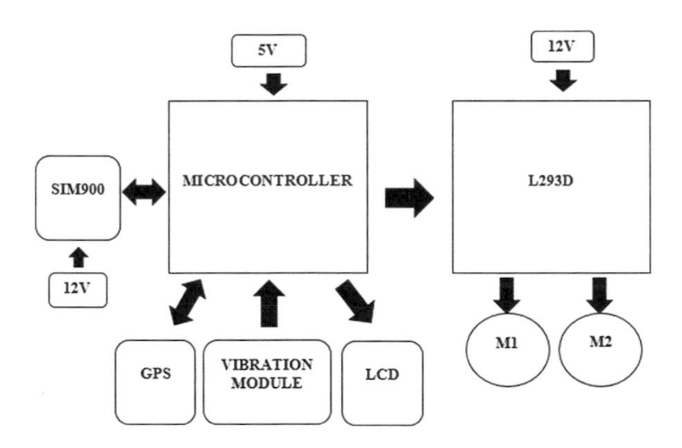

Fig. 2. Block diagram

In this proposed system which is an inexpensive tool, the troubles associated with coincidence notification and antitheft manage have been reduced. The controlling unit will be constant to the automobile. The controlling unit consists of the microcontroller and the GSM and GPS are interfaced to microcontroller [12]. The microcontroller constantly checks whether it has acquired any message from the modem. Figure 3 shows the geographical picture of the area in which the project was tested.

Fig. 3. Geographical image

4 Results

When the vehicle is stolen or met with an accident, first predefined text 'WRU' (that means Where Are You) is send to the SIM card inserted in the GSM modem. By using the AT commands the messages are send or received.

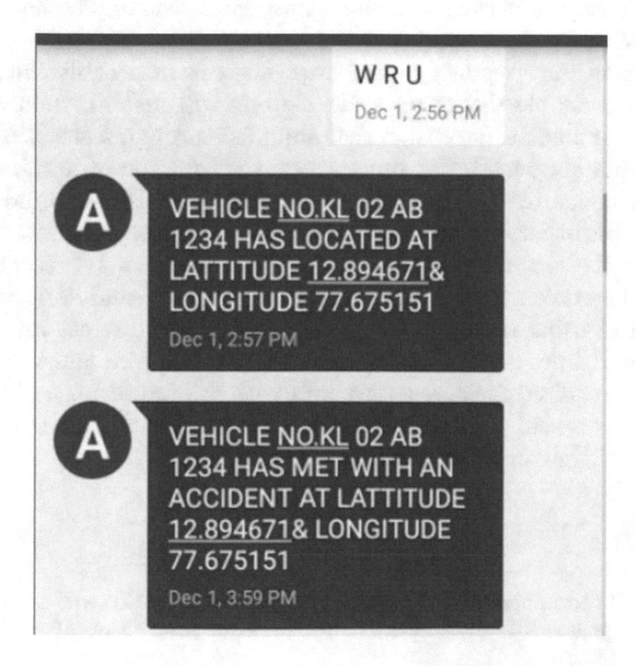

Fig. 4. W R U message and its response and scenario after accident occurs

Then the GPS locates the position of the vehicle and the GSM receives the latitude and longitude values of that particular place where the vehicle is located. Now through AT commands GSM sends the information to user's mobile immediately. It can be any type of mobile. Thus the stolen vehicle can be easily identified within no time. If we want to stop the vehicle the user send the message 'STOP' to GSM. Whenever the vehicle is met with an accident, vibration module gets activated and sends a high signal to the controller which activates the GSM modem. Now the GSM sends a reply to owner that Your vehicle is met with an accident at LT: 12.892115 LG: 77.679150, that is latitude and longitude of the position. By clicking the link which is send through GSM module to the specified number, it will automatically redirect to the map. So we can spot the vehicle. Figure 4 shows the message sent by the system after sending W R U message and the response after accident occurred.

5 Conclusion

The intention of the venture is to decrease the street accidents which causes the loss of worthwhile human life and other valuable items. Besides, the mechanism for the protection of the vehicle is also furnished to keep away from the robbery action. On this rapid moving global, new technologies have been evolved every second for our human existence style development. There have been considerable development in car technology already and nonetheless to come. Due to these technologies, now we are enjoying the vital consolations and safety. But there are lot of accidents occurring nowadays. Its miles because of multiplied automobile density, violating guidelines and carelessness. The embedded technology is used to prevent accidents, the use of cellular phones even as driving and many others. As a result by using imposing this noticeably cheap and without problems available device on a vehicle one will make certain a great deal extra security and exclusivity than that supplied through a traditional lock and key. If accidents happens in far flung areas, the function of auto-providing the coincidence vicinity to the emergency facilities for assist and guide is likewise provided. On the other hand, the safety for the car is also superior. This is made viable due to the fact the theft automobile location can be recognized to the person and the vehicle gasoline can be reduce off and middle lock is enabled. With the aid of using these concepts, we hope that the street injuries because of violating guidelines and carelessness could be minimized and this may be one of the venture required for now-a-days and with the importance of low cost. We can enforce this machine for real time programs. This will be improved in the future to actual time applications.

References

1. Iyyappan, S., Nandagopal, V.: Automatic accident detection and ambulance rescue with intelligent traffic light system. Int. J. Adv. Res. Electr. Electron. Instrum. Eng. **2**(4) (2014)

2. Goud, V., Padmaja, V.: Vehicle accident automatic detection and remote alarm device. Int. J. Reconfigurable Embedded Syst. (IJRES) **1**(2), 49–54 (2012). ISSN: 2089-4864
3. Chen, H., Chiang, Y., Chang, F., Wang, H.: Toward real-time precise point positioning: differential GPS based on IGS ultra rapid product. In: SICE Annual Conference, The Grand Hotel, Taipei, Taiwan, 18–21 August 2010
4. Prasanth Ganesh, G.S., Balaji, B., Srinivasa Varadhan, T.A.: Improving encryption performance using anti-theft tracking system for automobiles. In: IEEE Industrial Technology (ICIT), September 2014
5. Bagade, A.P.: Cell phone usage while driving avoidance with GSM-RF based accident emergency alert system. Int. J. Adv. Res. Comput. Commun. Eng. **2**(5) (2014)
6. Aarthi, R., Arunkumar, C., Padmavathi, S.: A survey of different stages for monitoring traffic rule violation. In: 4th International Conference on Global Trends in Information Systems and Software Applications, ObCom 2011, Vellore, India. Code 92172, 9–11 December 2011
7. Boopathi, S., Govindaraju, K., Sangeetha, M., Jagadeeshraja, M., Dhanasu, M.: Real time based smart vehicle monitoring and alert using GSM. Int. J. Adv. Res. Comput. Commun. Eng. **3**(11) (2014)
8. Divya, G., Sabitha, A., Sai Sudha, D., Spandana, K., Swapna, N., Hepsiba, J.: Advanced vehicle security system with theft control and accident notification using GSM and GPS module. Int. J. Innov. Res. Electr. Electron. Instrum. Control Eng. (2016)
9. Nair, B.B., Keerthana, T., Barani, P.R., Kaushik, A., Sathees, A., Nair, A.S.: A GSM-based versatile unmanned ground vehicle. In: International Conference on Emerging Trends in Robotics and Communication Technologies, INTERACT-2010, Chennai, India, 3–5 December 2010
10. Joshi, M.S., Mahajan, D.V.: Arm 7 based theft control, accident detection and vehicle positioning system. Int. J. Innov. Technol. Explor. Eng. (IJITEE), **4**(2), (2014). ISSN: 2278-3075
11. SIMCom Wireless Solutions Co. SIM548C Hardware Design-V1.01 Manual. http://wm.sim.com/upfile/200949111924.pdf. Accessed 9 Jan 2014
12. Kochar, P.P., Supriya, M.: Vehicle speed control using Zigbee and GPS. In: Smart Trends in Information Technology and Computer Communications, SmartCom 2016. Communications in Computer and Information Science, CCIS, vol. 628. Springer, Singapore (2016)

An Approach to Analyze Data Corruption and Identify Misbehaving Server

Nishi Patel and Gaurang Panchal(✉)

U & P U. Patel Department of Computer Engineering,
Chandubhai S Patel Institute of Technology, Changa, India
nishipatel5293@gmail.com, gaurangpanchal.ce@charusat.ac.in

Abstract. Many studies have derived multiple ways to achieve security in the server and integrating the data in multiple servers by detecting the misbehavior in the server. The data is secured on server using encryption techniques before dividing into fragments before storing on virtual cloud. This study focuses different perspective of storing data on virtual cloud to maintain integrity by storing the fragments of address of data. Hence, the data remains secure and only the address of the data is transmitted when divided in fragments and data is secured with encryption, so it would be difficult for third party to decrypt and access on server. Thus, security level has been increased on cloud platform though data stored on server is more secure and integrity is maintained throughout the cloud platform.

Keywords: Cloud data storage · Data corruption · Misbehaving server · Data security · AES encryption

1 Introduction

In today's era Cloud computing is a new and Internet based technology which provides a services like infrastructure as a service, Platform as a Service and Software as a Service [1–5]. Cloud Computing provides a facility to end user for on demand accessing resources which is virtualized on the cloud by Anytime Anywhere [6]. In this technology Cloud data storage is used for storage of data which offers an on demand data outsourcing as a service [7–10]. The main use of Cloud data storage is data written once by the user and rarely read by the user so it is necessary to ensure its integrity of data. So, the cloud computing is beneficial due to low cost and high efficiency [10–15]. But the Security is a major issue of Cloud data storage which contains the some sensitive data. In this paper, the research work goal is to design a single approach which uses, an encryption algorithm, Using standard encryption algorithm where we encrypt users data and convert into cipher text so no other malicious user can decrypt it, a fragmentation which is used for data fragmentation, an addressing scheme to provide a higher security of data and allocate an address to each fragmented

S.C. Satapathy and A. Joshi (eds.), *Information and Communication Technology for Intelligent Systems (ICTIS 2017) - Volume 1*, Smart Innovation, Systems and Technologies 83, DOI 10.1007/978-3-319-63673-3_33

data which preserve the data integrity and prevent the data corruption. So, using this approach we can ensure the data integrity and identify misbehaving server [15–34].

2 Related Work

C. Wang *et al.* [2] proposed a concept of Ensuring Cloud Data Storage correctness and identify misbehaving server. So they proposed a system which include the file distribution using Reed-Solomon code, Challenge Token Pre-computation, Correctness verification for identify external or internal threats and error localization. Due to burden of user they provide the auditing system to audit the data and identify misbehaving server. In addition to it, also provides the dynamic data operation like Update, Delete, Append, Insert operation.

G. Rathanam *et al.* [3] proposed a system which is dynamic secure storage system (DSSS). In this system they provide the Dynamic remote data integrity checking method and also how to detect the threats and preserve the data Integrity. For security purpose they used the Improved RSA Double encryption algorithm to pencrypt the data.

S. Shen *et al.* [4] proposed an integrity check scheme. Using this scheme their storage system can handle storage server failure as well as storage server corruption. In this paper they compare the previous work, so in previous years for Data Security they used single key server. Some authors describe the use of coding system likes Network Coding, Integrity Protected Error Correcting and Luby transform code. For Integrity tag they used the Aggregative Verification instead of Aggregative Verification Homomorphism. But this authors have described the use multiple key servers for Key management, Secure Decentralized Erasure Code for code, Homomorphic Ciphertexts for data security and Aggregative Verification Homomorphism for Integrity tag.

D. Ruiying *et al.* [1] proposed the framework PoOR (Proof of Ownership and proof of Retrivability) for mutual validitation. In past only one way validation is used, but in this framework Clients prove to the server their ownership of files and verify the retrievability of the files without uploading and downloading them.

M. Derfouf *et al.* [5] proposed the system for the solution of secure data storage. This Proposed System provides the encryption algorithm for secure data storage, fragmentation technique for dividing data into fragment, and duplication of fragmented data.

3 Problem Formulation

In this section, we discuss about the problem formulation. The basic flow of the system model is shown in Fig. 1.

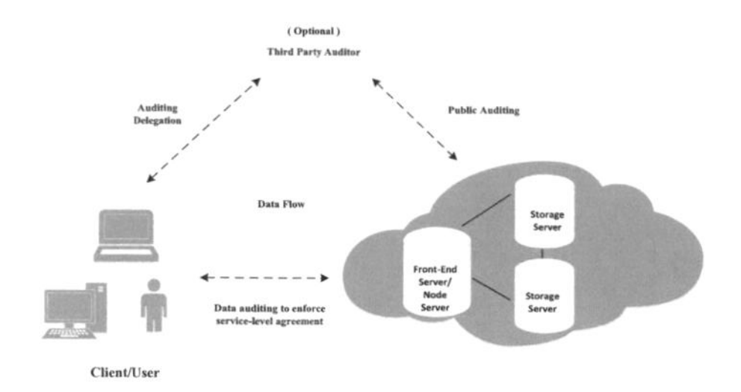

Fig. 1. System model.

This Service Model provides the following Entities:

1. User: Enables the access of data which is already stored on the cloud data storage. Also user can enable storing of data on the cloud.
2. Cloud Data Storage: This service is used for storing data.
3. Data Integrity Checking: This method will detect the threats which corrupt the data of storage.

So In this cloud data storage user store their data in encrypted format. In this model user can encrypt their data using encryption algorithm. And then this encrypted data are fragmented automatically using fragmentation which is provided by the specific cloud. This data are stored on the different cloud storage server. In this model, restitution program is used with which data can be encrypted.

Design Goals; To Ensure the Data security and identify misbehaving server, we aim to design single approach for verification and achieve the following Goals; "Time Complexity: Decrease the time complexity to retrieve data which is requested by the users." "Storage Complexity: Increase the storage space to reduce the size using address of encrypted data instead of encrypted data." "Security: To achieve security, single addressing technique is used to allocate the address of each fragmented data."

4 Proposed Approach

This proposed model first encrypt the whole data before it fragment into the small part. So that's why in this model Advanced Standard Encryption (AES) algorithm (see Fig. 2) is used which is a symmetric key algorithm and most standard algorithm.

This Proposed model provides the data integrity checking scheme to enhance the data security and it also provides the mechanism to identify misbehaving server against the different types of attacks and unauthorised access of data. So this model contains the encryption algorithm, namely AES which will used to generate the cipher text of original data. This encrypted data is fragmented using the fragmentation technique. This fragmentation is done automatically by

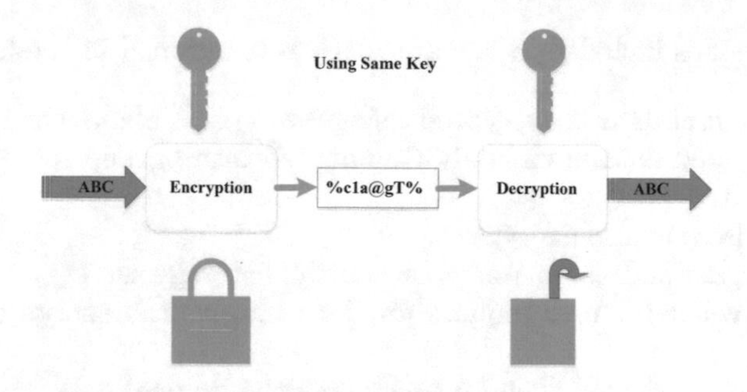

Fig. 2. Basic flow of AES encryption mechanism.

the cloud. This fragmented data is not only stored directly on the cloud storage server, but in this approach single centralized server is used, which contains the fragmented data. This server allocates single address to each fragmented data. After this allocation once again encryption algorithm is used to encrypt the address of data so approach is made to use double encryption algorithm. This encrypted address are stored on the different cloud storage server. So whenever user will request for the data, the user gets the address of encrypted data and using this address the user request to the data and gets the original data if and only if the user is an authenticated user. A pictorial representations of the proposed approach is shown in Fig. 3

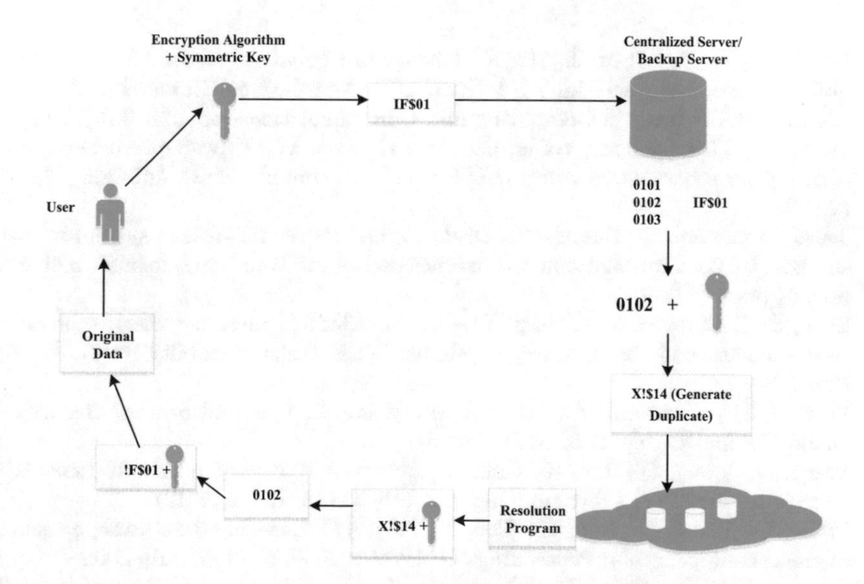

Fig. 3. Proposed approach.

The different steps included in our approach is mentioned in the following.

1. Client's original data is encrypted using encryption algorithm
2. This Encrypted data is virtually fragmented into backup server and gives a address to it.
3. So this address is also encrypted.
4. This encrypted address is stored on multiple storage servers on the cloud.
5. After this whenever user request for data he gets the encrypted address of that data.
6. If user is authenticated then he easily gets the original data.
7. Original data (i.e. Plaintext)

5 Conclusion

In this paper, I have proposed a new approach to analyze data corruption and identify misbehaving server. This approach contains the encryption algorithm and fragmentation. Moreover, to provide higher level security I have added new mechanism, that is addressing scheme to allocate the address for each fragmented data and this address is also encrypted by the encryption algorithm and then stored on cloud data storage server which achieves the time complexity, storage complexity and security.

References

1. Du, R., Deng, L., Chen, J., He, K., Zheng, M.: Proofs of ownership and retrievability in cloud storage. In: 2014 IEEE 13th International Conference on Trust, Security and Privacy in Computing and Communications, pp. 328–331 (2014)
2. Wang, C., Chow, S.S.M., Wang, Q., Ren, K., Lou, W.: Privacy-preserving public auditing for secure cloud storage. IEEE Trans. Comput. **62**(2), 362–366, 372–373 (2013)
3. Jeeva Rathanam, G., Sumalatha, M.R.: Dynamics secure storage system in cloud services. In: 2014 International Conference on Recent Trends in Informational Technology, pp. 1–5 (2014)
4. Shen, S.-T., Lin, H.-Y., Tzeng, W.-G.: An effective integrity check scheme for secure erasure code-based storage systems. IEEE Trans. Reliabilty **64**(3), 841–844 (2015)
5. Derfouf, M., Mimoumi, A., Eleuldj, M.: Vulnerabilities and Storage Security in Cloud Computing, pp. 2–5. IEEE (2015)
6. Wang, C., Wang, Q., Ren, K., Cao, N., Lou, W.: Towards secure and dependable storage services in cloud computing, pp. 1–6, 11–13. IEEE (2012)
7. Wang, C., Wang, Q., Ren, K., Cao, N., Lou, W.: Ensuring data storage security in cloud computing. In: Proceedings Of IWQos 2009, pp. 1–9, July 2009
8. Feng, B., Ma, X., Guo, C., Shi, H., Fu, Z., Qiu, T.: An efficient protocol with bidirectional verification for storage security in cloud computing, pp. 2–6. IEEE Access (2016)
9. Li, J., Zhou, K., Ren, J.: Security and efficiency trade-offs for cloud computing and storage, 15, pp. 148–150. IEEE (2015). ISBN 978-1-4799-8594-4

10. Ashalatha, R., Agarkhed, J., Patil, S.: Data storage security algorithms for multi cloud environment. In: International Conference on Advances in Electrical, Information, Communication and Bio-Informatics (AEEICB 2016), pp. 697–699 (2016)
11. http://www.infoworld.com/
12. http://vmartinezdelacruz.com/in-a-nutshell-how-openstack-works
13. http://www.ankitjain.info/articles/Cryptography_ankit4.htms
14. Amazon.com, Amazon Web Services (AWS) (2009), http://aws.amazon.com/
15. Sun Microsyatems, Inc., Building customer trust in cloud computing with transparent security, November 2009, http://www.sun.com/
16. A. Juels, Kaliski Jr., B.S.: Pors: proofs of retrievaility for large files. In: Proceedings Of CCS 2007, Alexabdria, VA, pp. 584–597, October 2007
17. Ganatra, A., Panchal, G., Kosta, Y., Gajjar, C.: Initial classification through back propagation in a neural network following optimization through GA to evaluate the fitness of an algorithm. Int. J. Comput. Sci. Inf. Technol. **3**(1), 98–116 (2011)
18. Panchal, G., Ganatra, A., Kosta, Y., Panchal, D.: Forecasting employee retention probability using back propagation neural network algorithm. IEEE 2010 Second International Conference on Machine Learning and Computing (ICMLC), Bangalore, India, pp. 248–251 (2010)
19. Panchal, G., Ganatra, A., Shah, P., Panchal, D.: Determination of over-learning and over-fitting problem in back propagation neural network. Int. J. Soft Comput. **2**(2), 40–51 (2011)
20. Panchal, G., Ganatra, A., Kosta, Y., Panchal, D.: Behaviour analysis of multilayer perceptrons with multiple hidden neurons and hidden layers. Int. J. Comput. Theory Eng. **3**(2), 332–337 (2011)
21. Panchal, G., Panchal, D.: Solving NP hard problems using genetic algorithm. Int. J. Comput. Sci. Inf. Technol. **6**(2), 1824–1827 (2015)
22. Panchal, G., Panchal, D.: Efficient attribute evaluation, extraction and selection techniques for data classification. Int. J. Comput. Sci. Inf. Technol. **6**(2), 1828–1831 (2015)
23. Panchal, G., Panchal, D.: Forecasting electrical load for home appliances using genetic algorithm based back propagation neural network. Int. J. Adv. Res. Comput. Eng. Technol. (IJARCET) **4**(4), 1503–1506 (2015)
24. Panchal, G., Panchal, D.: Hybridization of genetic algorithm and neural network for optimization problem. Int. J. Adv. Res. Comput. Eng. Technol. (IJARCET) **4**(4), 1507–1511 (2015)
25. Panchal, G., Samanta, D.: Comparable features and same cryptography key generation using biometric fingerprint image. In: 2nd IEEE International Conference on Advances in Electrical, Electronics, Information, Communication and Bio-Informatics (AEEICB), pp. 1–6 (2016)
26. Panchal, G., Samanta, D.: Directional area based minutiae selection and cryptographic key generation using biometric fingerprint. In: 1st International Conference on Computational Intelligence and Informatics, pp. 1–8. Springer (2016)
27. Panchal, G., Samanta, D., Barman, S.: Biometric-based cryptography for digital content protection without any key storage, pp. 1–18. Springer (Multimedia Tools and Application) (2017)
28. Panchal, G., Kosta, Y., Ganatra, A., Panchal, D.: Electrical load forecasting using genetic algorithm based back propagation network. In: 1st International Conference on Data Management, IMT Ghaziabad. MacMillan Publication (2009)
29. Patel, G., Panchal, G.: A chaff-point based approach for cancelable template generation of fingerprint data. In: International Conference on ICT for Intelligent Systems (ICTIS 2017), p. 6 (2017)

30. Patel, J., Panchal, G.: An IOT based portable smart meeting space with real-time room occupancy. In: International Conference on Internet of Things for Technological Development (IoT4TD-2017), pp. 1–6 (2017)
31. Soni, K., Panchal, G.: Data security in recommendation system using homomorphic encryption. In: International Conference on ICT for Intelligent Systems (ICTIS 2017), pp. 1–6 (2017)
32. Bhimani, P., Panchal, G.: Message delivery guarantee and status up- date of clients based on IOT-AMQP. In: International Conference on Internet of Things for Technological Development (IoT4TD-2017), pp. 1–6 (2017)
33. Mehta, S., Panchal, G.: File distribution preparation with file retrieval and error recovery in cloud environment. In: International Conference on ICT for Intelligent Systems (ICTIS 2017), p. 6 (2017)
34. Kosta, Y., Panchal, D., Panchal, G., Ganatra, A.: Searching most efficient neural network architecture using Akaikes information criterion (AIC). Int. J. Comput. Appl. **1**(5), 41–44 (2010)

Open Issues in Named Data Networking – A Survey

Prajapati Zalak, Kalaria Aemi, Gaurang Raval(✉), Vijay Ukani, and Sharada Valiveti

Institute of Technology, Nirma University, Ahmedabad, India
{15mcen21,15mcen09,gaurang.raval,
vijay.ukani,sharada.valiveti}@nirmauni.ac.in

Abstract. Internet is now being used as content distribution network also. Internet users are interested in specific contents rather than host machines where the content is located. Named Data Networking (NDN) is a step towards future Internet architecture that would be based on named data rather than numerically identified hosts. Many projects are in progress to architect the structure of the future Internet. It is envisaged that NDN would provide functional efficiency with named content applied as the core concept. NDN offers lot of research opportunities to contribute for design of future Internet architecture. In this paper we analyze the modern Internet Protocol (IP) based host- centric architecture and explore the newer scalable and more efficient architecture, based on content-centric approach. Open research issues in various area of NDN are also addressed in the paper.

Keywords: Named Data Networking · IoT · Information centric networks · In-network cache control · Content delivery networks

1 Introduction

Use of Internet is gradually increasing in every facet of life. The existing TCP/IP based Internet model does not cater to holistic Internet requirement, especially in the domain of faster data dissemination. The current Internet architecture is glued with presence of TCP and IP protocols. The IP protocol connects two hosts which are identified by a numerical identifier. NDN aka Information-Centric or Content-centric networking is a more recent concept to efficiently control transmission of videos and on demand required data on the Internet. NDN focuses on what the data is, rather than where the data is. Traditional TCP/IP protocol works on host-based content delivery whereas NDN mainly focuses on content based packet routing. IP has finite namespace as compared to NDN. NDN uses approximate state and IP uses exact state. NDN is application friendly and gives faster response too. Though the data is highly distributed, performance of NDN

S.C. Satapathy and A. Joshi (eds.), *Information and Communication Technology for Intelligent Systems (ICTIS 2017) - Volume 1*, Smart Innovation, Systems and Technologies 83, DOI 10.1007/978-3-319-63673-3_34

is said to be effective. NDN can be reasonable solution for repeated and duplicate delivery of large sized media contents. It reduces bandwidth consumption with reduced response time.

In WSN, sensors do the job of collecting data from the filed. In IoT, every sensing device connects with other such devices globally. So there is definitely a higher requirement of the heterogeneous and potent network to support such applications. NDN can be applied to collect sensed data from the node. WSN or IoT devices require in-network caching and content based security, which is already supported by NDN. For IoT systems, open challenges like network configuration, scalability, robustness are supported in NDN by directly managing server functionality at the network layer.

Algorithm 1. Interest packet processing

```
1  if Content in CS then
2  |  Return content to consumer ;
3  else
4  |  Forward to PIT ;
5  |  if Content entry in PIT then
6  |  |  Update PIT;
7  |  else
8  |  |  Forward to FIB ;
9  |  |  if Content found in FIB
   |  |   then
10 |  |  |  Forward interest packet;
11 |  |  |  Insert Entry in PIT;
12 |  |  else
13 |  |  |  Discard Interest packet;
```

Algorithm 2. Data packet processing

```
1  if Data in PIT then
2  |  Delete entry from PIT;
3  |  Forward to CS;
4  |  Cache in CS;
5  |  Serve Consumer;
6  else
7  |  Discard Data packet ;
```

In NDN, the major design parameters to be considered are scalability, memory, self-regulating network traffic, efficient multicast, load balancing, content replica, security etc. It also contains NDN forwarding fabric enhanced with packet overhearing to reduce collision and duplication over shared medium. It uses a top-down method to develop high-level NDN architecture for IoT system specific content [3]. Figure 1 shows the dissimilarities between traditional packet forwarding model and the NDN model [1]. In the domain of ICN, since 2004 many researchers have contributed with projects like Combined Broadcast and Content Based (CBCB)-2004, Data Oriented Network Architecture (DONA)-2007, Network of Information (NetInf)-2009, Named Data Networking (NDN)-2009, and Publish Subscribe Internet Technology (PURSUIT)-2010 [4,5,15]. This paper explores the benefits of NDN for modern networks with variety of content needs and compares it with traditional Data-Centric routing.

2 NDN Functioning

Named Data Networking follows content-based forwarding process instead of IP based forwarding. It contains two types of packets, interest packet and data packet. Data request sent by consumer is processed as interest packet. Response to that particular interest packet is known as data packet. NDN routers consist of Forwarding Information Base (FIB), Content Store (CS) and Pending Interest Table (PIT). PIT (Pending Interest Table) table of NDN contains the name of interest, nonce (random number), the list of the interfaces and the list of outgoing interfaces [17]. FIB (Forward Interest Base) table has features like Routing information to contain name prefix and stale time of interest packet. Algorithm 1 shows the processing of interest packet arrival at the NDN router whereas Algorithm 2 shows the same for data packet.

FIB works with different strategies like BestRouteStrategy, BroadcastStrategy, ClientControlStrategy, etc. The FIB contains outgoing interface (routing process) and name related information. The CS works as a buffer and uses different replacement policies. Least Recently used (LRU), Least Frequently Used (LFU) and First in First out (FIFO) are the names of different CS strategies. PIT has a table that manages unserved interest packets.

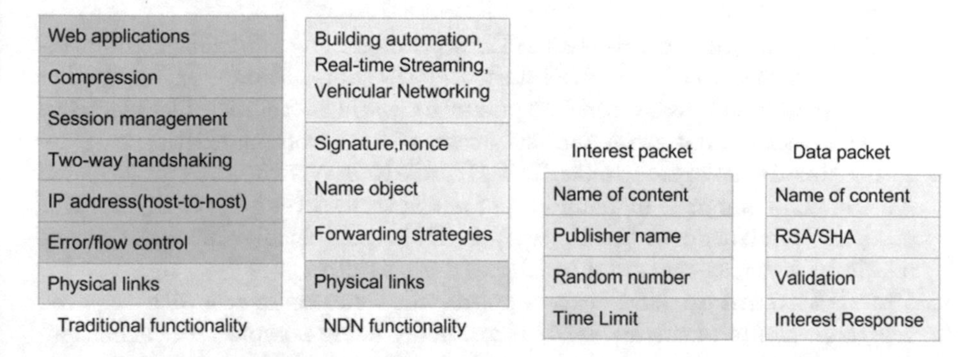

Fig. 1. Comparison of traditional and NDN model

Fig. 2. Format of interest and data packets

Figure 1 shows comparison between layered architecture of OSI and NDN with different functionalities at different layers. In NDN architecture, first layer deals with physical links like wired or wireless transmissions. Second layer works on name object forwarding strategies, where it finds network problems and sends data to alternative paths. The third layer focuses on content rather than address. The fourth layer deals with security. It applies digital signature and nonce on consumer and producer sides. Also, public key encryption is performed on the name object. The fifth layer interacts with different web applications.

Figure 2 shows the format of interest and data packet with basic changes. User put their request in interest packet and get response by content provider in data packet. NDN uses hierarchical naming structure which can be easily understood by human e.g. video requested by user on Youtube regarding networking will be routed in NDN like Youtube/video/networking. Interest packet contains name of requested data, publisher name and nonce which contains random number as well as time limit as shown in Fig. 2. As NDN follows receiver driven approach, corresponding data packet is also associated with interest packet. Mainly four fields are included in data packet like name of requested content, authentication, security and requested data.

3 Applications of NDN

NDN applies the concept of in-network storage for faster data delivery. Robust and scalable data dissemination are supported by NDN. Amount of traffic on the networks have increased due to increased use of multimedia, voice, data etc. Following are the possible applications of NDN:

1. **Real-time Conferencing:** Real-time conferencing is finding many takers day-by-day. The Chrono Chat is an application which supports multi-user chat and supports chat service that works on peer-to-peer connectivity. NDN video uses NDN specific approaches like rate adaptation and congestion control.
2. **Building Automation Systems:** Enterprise building automation and management systems can be NDN based. NDN application can be further extended with necessary sensing points for various facilities like electrical demand monitoring system and fast access of data from an existing system.
3. **Vehicular Networks:** Unlike TCP/IP, NDN is not the location-oriented system. NDN supports high amount of media traffic over 3G, Wi-Fi, etc. Due to location independent functionality, NDN based vehicular networks can be relatively faster as compared to IP based architecture.
4. **Video Streaming:** NDN can support video transmissions with required degree of quality of service. NDN is gradually being accepted for such applications. NDN can support prerecorded as well as live video transmissions.
5. **ChronoShare:** NDN provides decentralized file sharing system like ChronoShare. Functionality of ChronoShare is similar to Dropbox. NDN also supports multicast method to share decentralized files.

4 Open Issues in NDN

Researchers have been working since years on various issues related to technologies involved in networks to cater to all classes of consumers. There are many open research areas in NDN, like applicability of NDN in IoT, Cache control, Routing and Forwarding, Naming, Security, Congestion control etc. Each of these are discussed in detail in the following subsections.

4.1 IoT Adaptation

NDN can be working solution for repeated and duplicate delivery of large multimedia content which is located at independent places. NDN aims to reduce bandwidth consumption by retrieving data from any nearby router which holds valid data in its cache. Researchers have proposed the Diffserv based model for NDN which differs from IPdiffer model [13]. In this paper, the researchers proposed NDN architecture embedded with existing CDN to improve the efficiency in delivering data. CDN improves delivery of a large amount of data but it leads to problems like complexity and inefficiency. nCDN [12] is CDN based pure NDN overlay network. Erich et. al. used NLSR protocol for creating routing table with focus on request routing and content routing. Authors claim in the paper that nCDN provides better reliability, QoS and Scalability compared to traditional CDN. NDN provides commutation infrastructure to wireless networks to handle a large amount of traffic in the network. Authors [2] proposed push based data delivery using NDN. Push based traffic can handle periodic and event triggered data, which focuses on the reliability of data. Authors have also proposed three schemes to evaluate the efficiency of network, energy efficiency etc.

4.2 Cache Control

NDN applied in-network caching which supports efficient data retrieval. In IP based architecture cache is maintained at server. Router cache functionality is not integral part of IP based approach. Data retrieval process in NDN is faster as compared to traditional IP based model with less bandwidth consumption. When content provider is not reachable, data can be accessed quickly if it is cached at router [18]. NDN caches provide some features like cache transparency, cache ubiquity and fine granularity [16,17]. Cache transparency enables sharing of cache content independently. Cache ubiquity provides content availability in caches. In fine granularity, large content is divided into small chunks and the operations are performed on smaller chunks. This seems to be more effective then operating on large file. Cache performance may be affected due to pollution attacks, which increases traffic on the network by requesting useless data. Caching algorithm for IOT system contains parameters like freshness and presence of constrained resource. pCasting parameter is the probability of caching strategy for IoT system. pCasting algorithm depends on the freshness of data, energy level and storage capability. Using ndnSIM, it is evaluated that as pCasting increases and the delay reduces in retrieval of data [8].

4.3 Routing and Forwarding

The foundation of Named Data Networking lies in the routing process. Interest packets are compared with cache content during routing process and that describes the importance of cache strategy. In NDN, announcement forwarding, scalability, and multipath support are considered as desirable parameters. Many research challenges exist in the routing field like routing based on naming data, scalability, aggregation of naming data, look-up through name routing, hybrid routing, etc. Name aggregation, active publishing based on popularity and passive serving are scalability related solutions as explored in this paper [6]. NLSR (Named-data Link State Routing) is mainly designed to fulfill the requirements like naming, trust management, multipath management and information dissemination [10]. NDN router forwards the packet and maintains the state of the packet which resulted into efficient packet delivery [19]. Traditional TCP/IP has robust routing and inefficient forwarding as it strictly follows the routing table. So, it has no adaptability in today's architecture. NDN provides adaptability in forwarding process which maintains the state of the packet, detects the network problem and sends the packet to the alternative path or finds multiple paths. In NDN, data or interest packet is cached. So it has fast data retrieval than IP Based forwarding.

In NDN, each router maintains all information about producers and consumer in pending interest table. With adaptive forwarding NDN applies best effort data delivery or attempts to get data via the effective path. It has inherent mechanisms like interest NACK, retransmitted interest and interface probing [17]. Data is forwarded in NDN using PIT and FIB table. In FIB table, forwarding performance information marks interface using color scheme (Green, Yellow, Red). It shows which interface uses higher rate limit compared to other interfaces for effective traffic control [16,17] Issues in NDN forwarding techniques like;

1. As network size increases router gets more interfaces in PIT table, size of the PIT table needs to be controlled
2. At every router, when interest or data packet arrives, how to perform lookup efficiently on PIT table.

4.4 Naming

Naming is the core part of NDN. It has aggregate and hierarchical naming strategies which are easily understood by the user. Content names have the parameters to route data in the network like version, segment number and digest algorithm used for providing security to end users. NDN hierarchical naming is the same as URL in IP networks. User-friendly naming in NDN removes scalability issues of traditional architecture. So we can retrieve data dynamically because it is location independent [11]. Attribute naming can be useful for improving scalability in NDN. In name based routing user request is forwarded based on name in interest packet [5]. Some issues in naming are still unexplored like how to aggregate different naming in globally acceptable way.

4.5 Security

In NDN, content is rendered through the network. So security is the most important challenge for NDN. Different types of Distributed Denial of Service (DDoS) attacks are possible in the network. NDN mainly has two types of attacks like Interest Flooding Attack and Content/Cache Poisoning attack. Future NDN architecture must have provisions to tackle different DDoS attacks. Some possible changes in existing NDN architecture can be like interest packet could have enroll ID field, validation request packet and validation response packet. NDN routing table has additional fields for validation purpose [9]. Access table maintains allowable ID and Deniable ID as a field. Pending validation table is required in existing NDN architecture [7]. Different types of security attacks possible in NDN are like reflection attack, Prefix Hijacking, Black-Holing and DNS Cache Poisoning [7]. NDN security research challenges still remains open like data integrity, original packet authentication, named data object binding with real-world identities, access control, encryption, traffic filtering, state overloading etc.

4.6 Congestion Control

When compared to traditional TCP/IP concept, NDN is characterized based on the following significant features:

1. Content based routing instead of location based routing
2. In-Network Caching, routing and forwarding are separate in NDN
3. Packet forwarding according to receiver feedback
4. Works on receiver driven multisourcc and multipath transport

In traditional TCP/IP, congestion control is based on RTO (Re-transmission Time-Out), duplicate ACK and congestion control window. In NDN, these parameters are not considered for congestion control. NDN congestion control has 3 mechanism. They are the receiver based Control, hop-by-hop control and hybrid method. Cache strategies need to be modified according to varying transport layer parameters. The cost of congestion control mechanism needs to be analyzed in detail with different set of requirements. Content mobility, multipath congestion management, application specific congestion control mechanism and interest packet aggregation are the fields which require further improvement [14].

5 Conclusion

Compared to traditional TCP/IP-based architectures, NDN has many new characteristics. NDN provides faster, efficient, scalable and cost effective services. Some open issues like naming, congestion control, caching, security, routing, forwarding etc. still remains to be explored. This paper explored various researchers work pointing out the open challenges in NDN. Applicability of NDN may be further examined for data request and retrieval service from IoT point of view.

References

1. Amadeo, M., Campolo, C., Iera, A., Molinaro, A.: Named data networking for IOT: an architectural perspective. In: 2014 European Conference on Networks and Communications (EuCNC), pp. 1–5. IEEE (2014)

2. Amadeo, M., Campolo, C., Molinaro, A.: Internet of things via named data networking: the support of push traffic. In: 2014 International Conference and Workshop on the Network of the Future (NOF), pp. 1–5. IEEE (2014)
3. Amadeo, M., Campolo, C., Molinaro, A., Mitton, N.: Named data networking: a natural design for data collection in wireless sensor networks. In: 2013 IFIP Wireless Days (WD), pp. 1–6 (2013)
4. Barakabitze, A.A., Xiaoheng, T., Tan, G.: A survey on naming, name resolution and data routing. In: Information Centric Networking (ICN), vol. 3, pp. 8322–8330 (2010)
5. Bari, M.F., Chowdhury, S.R., Ahmed, R., Boutaba, R., Mathieu, B.: A survey of naming and routing in information-centric networks. IEEE Commun. Mag. **50**(12), 44–53 (2012)
6. Dai, H., Lu, J., Wang, Y., Liu, B.: A two-layer intra-domain routing scheme for named data networking. In: 2012 IEEE Global Communications Conference (GLOBECOM), pp. 2815–2820. IEEE (2012)
7. Gasti, P., Tsudik, G., Uzun, E., Zhang, L.: DoS and DDoS in named data networking. In: 2013 22nd International Conference on Computer Communication and Networks (ICCCN), pp. 1–7. IEEE (2013)
8. Hail, M.A., Amadeo, M., Molinaro, A., Fischer, S.: Caching in named data networking for the wireless internet of things. In: 2015 International Conference on Recent Advances in Internet of Things (RIoT), pp. 1–6. IEEE (2015)
9. Hemanathan, V., Anusha, N.: Role based content access control in NDN. J. Innov. Technol. Educ. **2**(1), 65–73 (2015)
10. Hoque, A., Amin, S.O., Alyyan, A., Zhang, B., Zhang, L., Wang, L.: Nlsr: named-data link state routing protocol. In: Proceedings of the 3rd ACM SIGCOMM workshop on Information-centric networking, pp. 15–20. ACM (2013)
11. Jacobson, V., Smetters, D.K., Thornton, J.D., Plass, M.F., Briggs, N.H., Braynard, R.L.: Networking named content. In: Proceedings of the 5th International Conference on Emerging Networking Experiments and Technologies, pp. 1–12. ACM (2009)
12. Jiang, X., Bi, J.: nCDN: CDN enhanced with NDN. In: 2014 IEEE Conference on Computer Communications Workshops (INFOCOM WKSHPS), pp. 440–445. IEEE (2014)
13. Kim, Y., Kim, Y., Yeom, I.: Differentiated services in named-data networking. In: 2014 IEEE Conference on Computer Communications Workshops (INFOCOM WKSHPS), pp. 452–457. IEEE (2014)
14. Ren, Y., Li, J., Shi, S., Li, L., Wang, G., Zhang, B.: Congestion control in named data networking-a survey. Comput. Commun. **86**, 1–11 (2016)
15. Tyson, G., Sastry, N., Rimac, I., Cuevas, R., Mauthe, A.: A survey of mobility in information-centric networks: challenges and research directions. In: Proceedings of the 1st ACM workshop on Emerging Name-Oriented Mobile Networking Design-Architecture, Algorithms, and Applications, pp. 1–6. ACM (2012)
16. Yi, C., Afanasyev, A., Moiseenko, I., Wang, L., Zhang, B., Zhang, L.: A case for stateful forwarding plane. Comput. Commun. **36**(7), 779–791 (2013)
17. Yi, C., Afanasyev, A., Wang, L., Zhang, B., Zhang, L.: Adaptive forwarding in named data networking. ACM SIGCOMM Comput. Commun. Rev. **42**(3), 62–67 (2012)
18. Zhang, G., Li, Y., Lin, T.: Caching in information centric networking: a survey. Comput. Netw. **57**(16), 3128–3141 (2013)
19. Zhang, L., Afanasyev, A., Burke, J., Jacobson, V., Crowley, P., Papadopoulos, C., Wang, L., Zhang, B., et al.: Named data networking. ACM SIGCOMM Comput. Commun. Rev. **44**(3), 66–73 (2014)

Employee Attrition Analysis Using Predictive Techniques

Devesh Kumar Srivastava and Priyanka Nair(✉)

Department of CSE and IT, Manipal University Jaipur, Jaipur, India
devesh988@yahoo.com, priyankanair@live.com

Abstract. Employee churn is an unsolicited aftermath of our blooming economy. Attrition may be defined as voluntary or involuntary resignation of a serving employee from an organization. Employee churn can incur a colossal cost to the firm. However, furtherance to prediction and control over attrition can give quality results. Earmarking the *risk of attrition,* the management can take required steps to retain the high valued talent. Workforce Analytics can be applied to reduce the overall business risk by predicting the employee churn. Predictive Analytics is the field of study that employs statistical analysis, data mining techniques and machine learning to predict the future events with accuracy based on past and current situation. The paper presents a framework for predicting the employee attrition with respect to voluntary termination employing predictive analytics.

Keywords: Turnover prediction · Predictive analytics · Data mining · Employee attrition · Predictive algorithms

1 Introduction

When an employee leaves an organization, the critical information of the organization is carried along. Considering the inimical repercussion on the work place productivity and massive cost borne to the firm on account of replacing the talent cluster, there is a dire need of administering attrition rate. A certain amount of turnover is unavoidable, but too much can ruin a company. The current scope of the work is limited to predicting the probability of voluntary resignation of a serving employee. Attracting and retaining great talent is a major challenge. According to research it costs 30 to 400% of an employee's annual salary to replace them [1].

Different organizations acquire several strategies to retain the employees. The strategies may include increase in pay, opportunities for travel, frequent job rotations, etc. It is important to retain the employees since they are essential asset to the organization. Predictive Analytics exploits the features of statistical analysis and data mining to predict the future events based on historical data. Various predictive techniques can be employed on the same data set with varied rate of accuracy considering the various data points. Literature exhibits that attrition indicates that the reasons as to why an employee quits is a function of demographics and job characteristics. They have a significant dependence on factors wherein an employee usually engages like

S.C. Satapathy and A. Joshi (eds.), *Information and Communication Technology for Intelligent Systems (ICTIS 2017) - Volume 1*, Smart Innovation, Systems and Technologies 83, DOI 10.1007/978-3-319-63673-3_35

absenteeism and late-coming [2, 3]. The current work uses data on demographics and the withdrawal behaviors like absenteeism and their interdependence to predict churn.

The Talent Attrition Analytics Model will exploit the features of predictive algorithm and data visualization tools to discover the underlying reasons for employee attrition and identify the employees at risk of leaving based on the historical employee data. The workforce analytics serves three purposes; accurately determine who is leaving to enable pre-emptively address costly departures, find out why people are leaving and determine the extent to which turnover is impacting the Business. The current scope of the work limits to the first two facets. Identification of the employee cluster at the risk of leaving allows determining the specific concerns and needs in order to retain the high valued talent. It also determines the underlying reasons for the churn and scoring the most probable factor to be the considerable reason of churn pertaining to particular employee.

2 Literature Survey

According to Abassi et al. (2000), turnover is the rotation of employees between organizations and jobs switching in states of employment and unemployment [4]. Employee attrition may be ensued in either of the turnover categories; *voluntary turnover or involuntary turnover*. Involuntary turnover occurs when an organization dissolves the term of an employee, whereas in a voluntary turnover, a worker resigns on his/her own will. Figure 1 *presents* the employee turnover trend for the year 2008–2015 collected from by the *CompData surveys* [5].

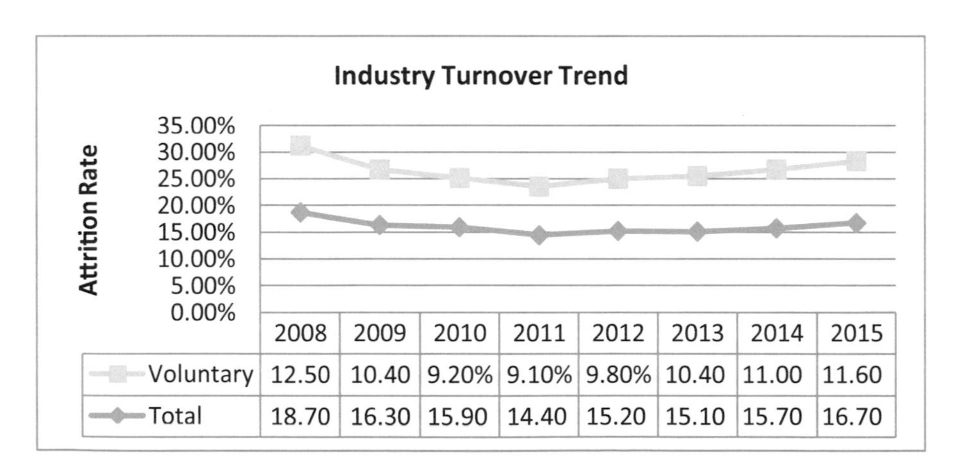

	2008	2009	2010	2011	2012	2013	2014	2015
Voluntary	12.50	10.40	9.20%	9.10%	9.80%	10.40	11.00	11.60
Total	18.70	16.30	15.90	14.40	15.20	15.10	15.70	16.70

Fig. 1. CompData's 2015 survey for industry turnover trends

Studies show that employees switch their jobs every six years on an average [6]. There has been a trend of substantial increase in voluntary turnover rate from the previous year. A research report by Society of Human Resource Management shows the average voluntary attrition rate to be 13% which was 44% increased percentage as compared to the previous year attrition rate [7].

Considering the increase in employee churn, it is very important to determine the predictors of attrition. With the current work limited to voluntary turnover, it has been found that there are various employee demographics and job characteristics that determine the reasons of attrition and may be counted to be the strongest predictors of turnover. The research findings suggest that age, gender, tenure, compensation, marital status, native place and work location, education and total work experience were some of the strongest predictors of voluntary attrition [8]. Based on the predictors of attrition, the reason of an employee quitting the job may be anticipated. Analyzing the predictors of attrition, a combination of reasons like poor compensation, weak hiring practices, lethal working environment, burnout, limited growth opportunity, etc. may appear [5, 9].

Predictive Analytics employs data mining techniques and statistical analysis to predict the results based on historical data, Data mining plays a vital role in Human Resource Management System (HRMS) for scrutinizing the hidden patterns and relationships existing in HR data [10]. A lot of study has been done on data mining and its applications in HR Systems. Pertaining to varied application in HRMS various data mining techniques may be applied to get quantified results from qualitative data [11]. Relationship and effect of the withdrawal factors like late coming, absenteeism, employee demographics, tenure with the organization on turnover has been studied with various data mining techniques(artificial neural networks, classification trees (C5.0), classification and regression trees (CART) and discriminant analysis) [12]. CART was the recommended technique for the work. To deal with the nonlinear classification and regression problems to prediction of employee attrition, logistic regression model and regression model (*Logit* and *Probit* model) and feasibility of their application have been studied [13]. Binomial logit regression technique has been worked upon to determine and analyze the factors impacting voluntary attrition [14]. Employee attrition has been analyzed employing multiple decision tree algorithms (C4.5, C5, REPTree, CART) with varied level of accuracy [4]. Comparing the decision tree algorithms, C5 decision tree gave better results and hence was recommended for the work. For predicting the employee attrition various data mining techniques can be applied to the same data set to obtain different results. In the research work with comparison of various techniques like Naïve Bayes, Support Vector Machines, Logistic Regression, Decision Trees and Random forests, Support Vector machines was recommended for the data set employed [15].

3 Research Methodology

The solution prospect of the workforce analytics for predicting the employee churn based on the historical data is confined to building the predictive model and exploiting the features of data visualization to pose the patterns of attrition in a meaningful manner employing the dashboard feature of the tool. The methodology taken herewith is divided into four stages: raw data collection, data transformation, predictive modeling and model deployment. Figure 2 shows the solution overview for the talent attrition analytics.

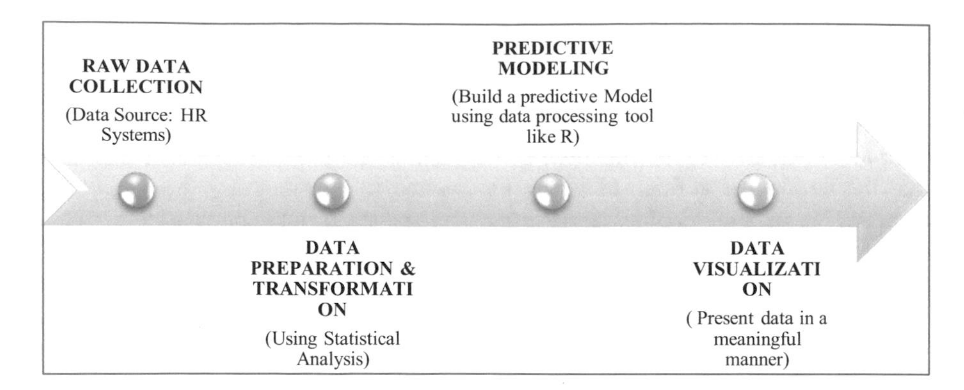

Fig. 2. Solution overview for talent attrition analytics for predicting the employee churn

3.1 Facet 1-Identifying the Employee Cluster at the Risk of Leaving

The data source for workforce analytics is generally categorized as the HR Data Source and the Business Data Source. The nature of dummy data collected is the facets comprising of employee demographics and job characteristics. Figure 3 shows the sample data having variables capable of determining the employee churn.

	A	B	C	D	E	F	G	H	I
1	Emp ID	HR.STATUS	AGE	GENDER	MARITAL.STATUS	TOTAL.EXPERIENCE	BAND	NO.OF.PRIOR.EMPLOYERS	LEAVES.TAKEN
2	U100038	A	61	M	Divorced	14.42	B2	4	64
3	U100041	I	NA	U	NA	0.04	A0	0	NA
4	U100177	I	37	M	Married	9.42	A3	3	14
5	U100223	I	27	M	Single	5.52	A2	2	52
6	U100225	I	27	F	Single	4.07	A1,A2	0	98
7	U100275	I	25	M	Single	2.75	A1,A2	0	74
8	U100334	A	27	M	Single	3.93	A3	1	16
9	U100347	A	30	M	Married	6.5	A1,A2,A3,	0	205
10	U100472	I	66	M	Unknown	8.9	C1	1	8
11	U100503	I	35	F	Married	8.3	B1	4	147

Fig. 3. Sample employee data with variables capable of determining employee churn

Based on market survey 28 variables (data points) listed in Table 1, are identified pertaining to employee characteristics and job characteristics.

To actuate the validity of the proposed approach, from the 27 identifiable variables, 19 data points have been worked upon. 19,326 rows of dummy employee data have been considered herewith. There are two categories of variables; categorical variables (13) and numeric variables (6). The employee record has been divided into training and test rows of data. For the prototype modeling 75% of data has been considered as the training data and remaining as the validation data. To build the predictive model Artificial Neural Network predictive technique has been considered. RStudio has been

Table 1. Data points/variables for predicting employee churn

S. no	Variables	Comments
1	Age	Number
2	Gender	0/1
3	Marital status	0/1
4	No. of dependents	Count
5	Nationality	Country code
6	Address (location)	Pin code
7	Department	Dept. code
8	Total prior jobs	Count
9	Position changes	Count
10	Engineering type	Flag
11	Degree level	Flag
12	Tenure	In years
13	Compensation	Annual income
14		
15	Employee engagement score	Score
16	Access card swipe count	Frequency of swipes
17	Time since last promotion	In months
18	Last performance rating	Score
19	No of leaves taken	Count
20	Frequency of leaves	Max leave/month
21	No. of projects completed	Count
22	Overtime pool	In hours
23	Reason of leaving (results from exit interviews)	Top 1–10
24	Foreign travel count	No of Trips
25	Number of trainings attended	Count
26	Experience	In months
27	Feedback count (negative/positive)	0/1
28	Work schedule flexibility	0/1

employed to build the predictive model. The R package, *neuralnet* serves the purpose. The sample code is shown in Fig. 4.

Observations: To infer the model accuracy, 10,000 of employee data have been taken into account. Based on the existing employee record, the attrition count was observed to be 3941. The validate data considered was 1000 rows. The Artificial Neural Network Technique gave 164 correct predictions which counts to 40% accuracy for the data. The accuracy chart is shown in Fig. 5. Extrapolating the model on the present workforce 20% of employees were observed to be at high "Flight Risk". 49% employees are susceptible to leave in the mid time future. 31% employees are stable and can be expected to continue long term. Figure 6 shows the business impact based on the considered data.

```
1  #Read the data
2  a=read.csv("C:/users/C52534/Desktop/ConvertedDataTest.csv", sep=",", heade
3  str(a)
4
5  #Partition the data
6  set.seed(500)
7  index <- sample(1:nrow(a), round(0.75*nrow(a)))
8  train <- a[index,]
9  test <- a[-index,]
10
11 #Apply the neural network algorithm on the train data
12 library(neuralnet)
13 nn <- neuralnet(HR.STATUS~GENDER+MARITAL.STATUS, data=train, hidden(5,3),
14 nn
15 plot(nn)
16
17 #Prediction with gthe test data
18 pr.nn <- compute(nn, test[,2:3])
19 pr.nn <- pr.nn$net.result
20 nn1 <- ifelse(pr.nn_>0.375,1,0)
21 sink("C:/users/c52534/Desktop/output.csv")
22 print(nn1)
23 sink()
24
```

Fig. 4. Sample R code employing ANN predictive technique

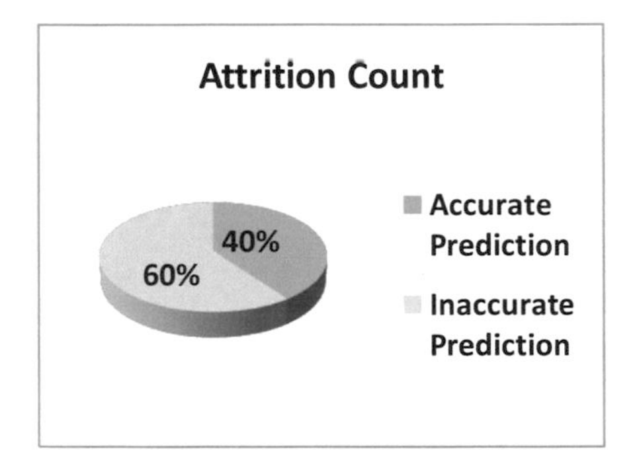

Fig. 5. Model accuracy

Fig. 6. Business impact- extrapolating the model on the current workforce

3.2 Facet 2-Discover the Underlying Reasons of Attrition

There are several factors that may be associated with the employee attrition. In order to retain the high valued employee with an organization, it is of critical importance to discover the underlying reasons or factors which may lead to employee quitting the job over a period of time. Figure 7 shows the sample association of reason an employee quits the job and the data points. Each factor may be associated with different data points. Employing the statistical analysis and exploratory factor analysis each factor may be reckoned with scores. The scores are then evaluated in ascending order. The top scores may be considered to be the most probable reason for an employee quitting the job.

A	B	C	D	E
Drivers of Attrition				
Factor	Reason for Individual Attrition(Factor Name)	Variable Code	Predictor Variable	Value
1	They are not growing professionally	V7	Department	A12
		V12	Tenure	3
		V17	Time since last promotion	12
		V24	Foreign Travel Count	0
		V25	Trainings Attended	3
2	They are not paid well	V7	Department	A12
		V8	Total Prior Jobs	2
		V9	Total Position Changes	3
		V12	Tenure	3
		V13	Compensation	3.5
		V17	Time since last promotion	12

Fig. 7. Association of drivers of attrition with the data points

4 Conclusion and Future Work

The proposed approach of *Talent Attrition Analytics* will be able to build a predictive model to predict the employee churn in an organization employing the employee demographics and job characteristics. The predictive Analytics uses the historical data to generate a trend of attrition. Based on the trend observed for the historical data, the active talent cluster at the risk of leaving may be identified. Using Statistical Analysis and exploratory factor analysis the various reason of attrition may be determined. The current approach serves to provide solution for the two facets of employee attrition thereby reducing the business risk by predicting the employee churn. The Talent Attrition Analytics framework may be able to predict the employee churn accurately within an organization utilizing the various predictive modeling techniques.

Data visualization and interpretation of results may be considered for the future work of this research. The processed data can be presented in a meaningful manner by exploiting the dashboard feature of data visualization tools and reports can be generated based on the data set. The attrition trend may be represented graphically and the results can be scrutinized to deduce the probability of an employee leaving an organization. The data points of an employee can be tracked over time and strategies can be employed by the management to minimize the rate of attrition.

References

1. The Cost of Employee turnover (2014). http://www.zenworkplace.com/2014/07/01/cost-employee-turnover/
2. Strohmeier, S., Piazza, F.: Domain driven data mining in human resource management: a review of current research. Expert Syst. Appl. **40**(7), 2410–2420 (2013)
3. Alao, D., Adeyemo, A.B.: Analyzing employee attrition using decision tree algorithms. Comput. Inf. Syst. Dev. Inf. Allied Res. J. **4** (2013)

4. Abassi, S.M., Hollman, K.W.: Turnover: the real bottom line. Public Pers. Manag. **2**(3), 333–342 (2000)
5. Bares A.: Turnover rates by industry, 08 April 2016. http://www.compensationforce.com/2016/04/2015-turnover-rates-by-industry.html
6. Kransdorff, A.: Succession planning in a fast-changing world. Manag. Decis. **34**(2), 30–34 (1996)
7. Society for Human Resource Management (SHRM). Employee job satisfaction and engagement the road economic recovery, 11 June 2014. http://www.shrm.org/Research/SurveyFindings/Documents
8. Cotton, J.L., Tuttle, J.M.: Employee turnover: a meta-analysis and review with implications for research. Acad. Manag. Rev. **11**, 55–70 (1986)
9. Allen, D.G., Griffeth, R.W.: Test of mediated performance-turnover relationship highlighting the moderating of visibility and reward contingency. J. Appl. Psychol. **86**, 1014–1021 (2001)
10. Jayanthi, R., Goyal, D.P., Ahson, S.I.: Data mining techniques for better decisions in human resource management systems. Int. J. Bus. Inf. Syst. **3**(5), 464–481 (2008)
11. Hamidah, J., AbdulRazak, H., Zulaiha, A.O.: Towards applying data mining techniques for talent managements. In: 2009 International Conference on Computer Engineering and Applications, IPCSIT, vol. 2. IACSIT Press, Singapore (2011)
12. Nagadevara, V., Srinivasan, V., Valk, R.: Establishing a link between employee turnover and withdrawal behaviours: application of data mining techniques. Res. Pract. Hum. Res. Manag. **16**(2), 81–99 (2008)
13. Wie-Chiang, H., Ruey-Ming, C.: A comparative test of two employee turnover prediction models. Int. J. Manag. **24**(2), 216–229 (2007)
14. Marjorie, L.K.: Predictive Models of Employee Voluntary Turnover in a North American Professional Sales Force Using Data Mining Analysis. A&M University College of Education, Texas (2007)
15. Saradhi, V.V., Girish, K.P.: Employee churn prediction. Expert Syst. Appl. **38**(3), 1999–2006 (2011)

File Distribution Preparation with File Retrieval and Error Recovery in Cloud Environment

Shital Mehta and Gaurang Panchal(✉)

Chandubhai S Patel Institute of Technology,
U & P U. Patel Department of Computer Engineering, Changa, India
shitalmehta2@gmail.com, gaurangpanchal.ce@charusat.ac.in

Abstract. Many studies have derived multiple ways to achieve security in the server and integrating the data in multiple servers by detecting the misbehavior in the server. The data is secured on server using encryption techniques before dividing into fragments before storing on virtual cloud. The fragments in which data is divided could be big or small. For encryption technique I had used Reed Solomon codes to encrypt fragmented data. After using encryption they are stored into different servers. Whenever user will challenge for its particular file, Cloud Service Provider will provide that file in original form. If that data chunk from particular file or file is corrupted on any server so using Reed Solomon Code we can repair our corrupted data or file. So, whenever attacker might attack on the server he might get small portion of that file in encrypted form which is not useful for him. He did not get original file from one server because that particular file fragments are dispersed on different servers. This ensures data security to cloud servers, cloud customers important information. It will also provide security from cloud service provider. And if any misbehavior of server is occurred or any file fragment is corrupted can be detected with automated tools and it can be also traced that on which server the file fragment is corrupted. Hence, in any way the data stored on server is more secure on cloud platform.

Keywords: Cloud computing · Error localization · Distributed storage · Reed-Solomon code

1 Introduction

In today's era Cloud computing is a new and Internet based technology which provides a services like infrastructure as a service, Platform as a Service and Software as a Service [1,5–8]. Cloud Computing provides a facility to end user for on demand accessing resources which is virtualized on the cloud by Anytime Anywhere [5–10]. In this technology Cloud data storage is used for storage of data which offers an on demand data outsourcing as a service. The main use of Cloud data storage is data written once by the user and rarely read by the user so it is must necessary to ensure its integrity of data. So the cloud computing

S.C. Satapathy and A. Joshi (eds.), *Information and Communication Technology for Intelligent Systems (ICTIS 2017) - Volume 1*, Smart Innovation, Systems and Technologies 83, DOI 10.1007/978-3-319-63673-3_36

is beneficial due to low cost and high efficiency [12–18]. But the Security is a major issue of Cloud data storage which contains the some sensitive data [2–8].

Cloud computing gives us services. Like application, platform and infrastructure [12]. In this model different services and products for business of individual user or consumer around the world are offered in each layer. So in today's word security of cloud computing is most important, this security refers to many policies, technologies and control used to protect sensitive information [19–33].

In this paper, the research work goal to provide higher security will use Reed Solomon code approach. Using Reed Solomon code we can fragment the particular file and then convert into codewords. So it will provide security opposite attacker. After partitioning the data will be stored in cloud servers.

2 Related Work

C. Wang *et al.* [1] proposed a concept of Ensuring Cloud Data Storage correctness and recognize misbehaving server. To guarantee the accuracy of clients' information in the cloud, they propose a successful and adaptable distributed scheme with explicit dynamic data support. It also overcomes the loopholes of file distribution techniques by decreasing the communication and storage overhead. They provide one approach for storage correctness and also data error localization by using the homomorphic token with distributed verification of erasure-coded data. So they proposed a system which include the file distribution using Reed-Solomon code, Challenge Token Pre-computation, Correctness verification for identify external or internal threats and error localization. Due to burden of user they providing the auditing system to audit the data and identify misbehaving server. Also provides the dynamic data operation like Update, Delete, Append, Insert operation.

O. Khan *et al.* [11] proposed the system where awareness of failure groupings and failure bursts are necessary for data placement strategies. They additionally contend that, in the presence of correlated failures, codes more fault tolerant than RAID-6 are expected to decrease exposure to data loss for that they consider Reed-Solomon codes. Reed-Solomon codes is utilized by Windows Azure storage for the same reasons. Using Erasure correcting code you can prevent from loss of file or data and easily correct failure of file. They suggest to get easy recovery performance, codeword symbol should be large enough which will enlarge memory consumption.

S. Khedkar *et al.* [12] proposed the system to enhance cloud data storage security. They develop the technique for data storage with token pre-computation and AES algorithm how it is stored in cloud. For achieving the data quality, availability, integrity of dependable storage services, distributed scheme is utilized. RSA does security analysis to encode the data. They had work on Data Partitioning technique and users can also store their file securely in cloud storage. They had make on method which can detect threats and misbehaving server; and this method also provides security from internal and external attacks. As per this technique file can be retrieved from the storage as per the cloud consumer request. They had used MD5 algorithm for better performance to ensure data security and retrieval of data in cloud. So in this paper they had used RSA, MD5algorithms and token precomputation technique for data storage security in cloud.

X. HAIPING *et al.* [16] proposed the system In order to make cloud storage service reliable, several research endeavors on utilizing erasure codes. They propose an reconstruction code which can fragments the data. Their approach stores data in geographically separated servers after dividing them into parities. In this paper, they augment on secure and fault-tolerant model of cloud information storage. For better secure and fault cloud storage systems, they followed RAID and XOR parity concept.

3 Problem Statement

The following three different network entities can be identified for cloud data storage (Fig. 1):

1. User: Enables the access of data which is already stored on the cloud data storage. Also user can enable to storing data on the cloud data storage
2. Cloud Service Provider (CSP): CSP provides cloud computing services and resources for storage servers.
3. Third Party Auditor (TPA): an optional TPA, is trusted to evaluate and expose the risk of cloud storage services on behalf of the users upon request and has expertise and capabilities that users might not have.

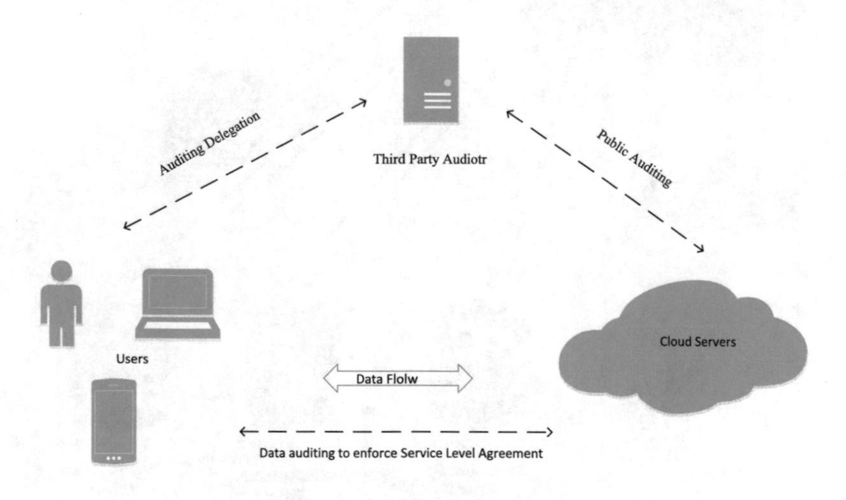

Fig. 1. Cloud data storage architecture

Two different sources can threaten the security of cloud data storage.

1. A CSP can be employee of the organisation or untrusted. He might move the data or modified the data for monetary gains.

2. Existence of economically motivated adversary, who stays undetected by CSPs for some specific period and has capability to compromise a number of cloud data storage servers.

4 Proposed Solution

This Proposed model provides the data security and faster error recovery; also provides the mechanism for identify misbehaving server against the different types of attacks and unauthorized access of data. So this model contain Reed-Solomon erasure correcting code which is used to partition the file into data fragments and then encoded. Now these fragments will be uploaded to different Cloud database server via cloud service provider. Now when user request (challenge) for his/her original file to retrieve will send one request to cloud service provider. A pictorial representation of the proposed architecture is shown in Fig. 2.

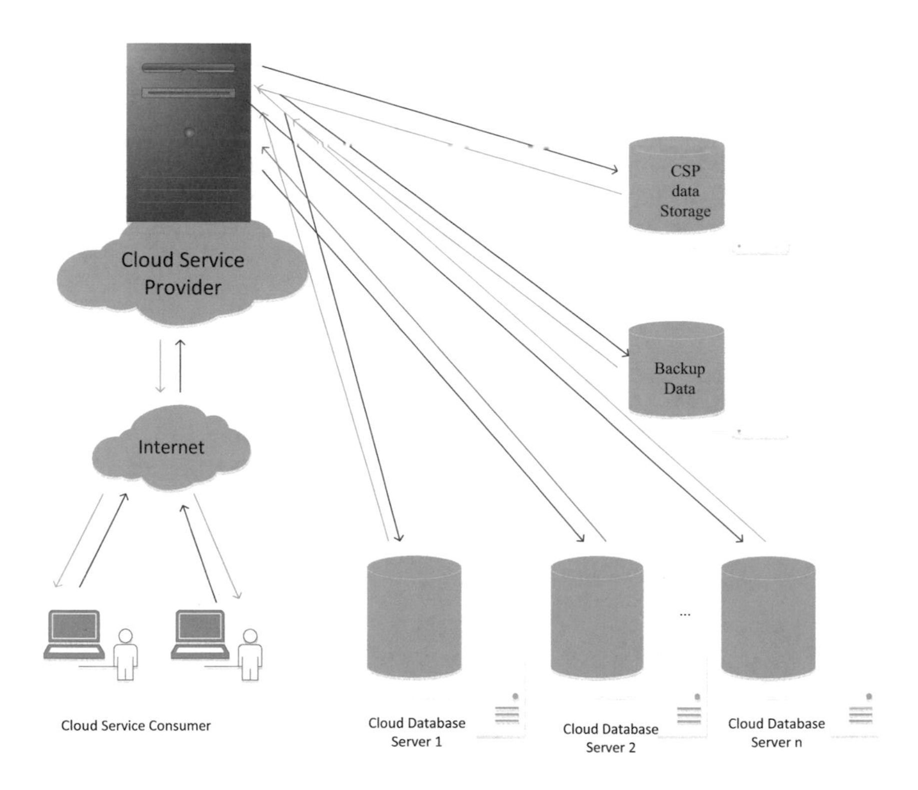

Fig. 2. Proposed architecture of cloud storage

Steps:

1. Cloud Service Consumer will create one file which will be partitioned into data fragments and then encoded by Reed Solomon Code. (File Distribution Preparation).
2. Now these fragments will be uploaded to different Cloud database server via cloud service provider.
3. Now when user request (challenge) for his/her original file to retrieve will send one request to cloud service provider.
4. If no error or modification is there the file will be generated in original form as per user request.
5. If error is occurred then we can recover lost data using Reed Solomon code.
6. So, using Reed Solomon Code user is capable to download his/her original file.

5 Conclusion

In this paper, I derived one approach which is used for faster error recovery and identify misbehaving server. For this approach contains erasure code, fragmentation which will provide higher level security. Using Reed- Solomon code we can do file distribution preparation easily and error recovery in cloud storage servers. In future work I will try to trace that file that where it is corrupted and stored in cloud storage servers. So using this concept we can easily identify misbehaving server.

References

1. Wang, Q., Wang, C., Ren, K., Lou, W., Li, J.: Enabling public auditability and data dynamics for storage security in cloud computing. IEEE Trans. Parallel Distrib. Syst. **22**(5), 847–859 (2011)
2. Ashalatha, R., Agarkhed, J., Patil, S.: Data storage security algorithms for multi cloud environment. In: 2nd International Conference on Advances in Electrical, Electronics, Information, Communication and Bio-Informatics (AEEICB 2016) (2016)
3. Azougaghe, A., Kartit, Z., Hedaboui, M., Belkasmi, M., El Marraki, M.: An efficient algorithm for data security in cloud storage. In: 15th International Conference on Intelligent Systems Design and Applications (ISDA) (2015)
4. Gupta, R.R., Katara, S., Sarkar, M.K., Mishra, G., Agarwal, A., Das, R., Kumar, S.: Data storage security in cloud computing using container clustering. In: IEEE 7th Annual Ubiquitous Computing, Electronics and Mobile Communication Conference (UEMCON) (2016)
5. Wu, S., Xu, Y., Li, Y., Yang, Z.: I/O-Efficient scaling schemes for distributed storage systems with CRS codes. IEEE Trans. Parallel Distrib. Syst. **27**(9), 2477–2491 (2016)
6. Liu, J.K., Liang, K., Susilo, W., Liu, J., Xiang, Y.: Two-factor data security protection mechanism for cloud storage system. IEEE Trans. Comput. **65**(6), 1992–2004 (2016). IEEE

7. Shen, S.-T., Lin, H.-Y., Tzeng, W.-G.: An effective integrity check scheme for secure erasure code-based storage systems. IEEE Trans. Reliab. **64**(3), 840–851 (2015)
8. Mishra, V., Pateriya, R.K., Bhopal, M.: Efficient data administration with reed-Solomon code. Int. J. Sci. Res. Manag. (IJSRM) **4**, 4929–4935 (2016)
9. Jogdand, R.M., Goudar, R.H., Sayed, G.B., Dhamanekar, P.B.: Enabling public verifiability and availability for secure data storage in cloud computing. Evolving Syst. **6**(1), 55–65 (2015)
10. https://downloads.cloudsecurityalliance.org/
11. Khan, O., Burns, R., Plank, J., Pierce, W.: Rethinking Erasure Codes for Cloud File Systems: Minimizing I/O for Recovery and Degraded Reads. http://dwww.cs.jhu.edu/~okhan/fast12.pd
12. Khedkar, S.V., Gawande, A.D.: Data partitioning technique to improve cloud data storage security. Int. J. Comput. Sci. Inf. Technol. **5**(3), 3347–3350 (2014)
13. Stallings, W.: Cryptography and Network Security Principles and Practices, 4th edn.
14. Westall, J., Martin, J.: An Introduction to Galois Fields and Reed-Solomon Coding
15. Erl, T., Mahmood, Z., Puttini, R.: Cloud Computing Concepts, Technology and Architecture
16. Haiping, X., Bhalerao, D.: Reliable and secure distributed cloud data storage using reed-Solomon codes. Int. J. Softw. Eng. Knowl. Eng. **25**, 1611 (2015)
17. Ganatra, A., Panchal, G., Kosta, Y., Gajjar, C.: Initial classification through back propagation in a neural network following optimization through GA to evaluate the fitness of an algorithm. Int. J. Comput. Sci. Inf. Technol. **3**(1), 98–116 (2011)
18. Panchal, G., Ganatra, A., Kosta, Y., Panchal, D.: Forecasting employee retention probability using back propagation neural network algorithm. In: IEEE 2010 Second International Conference on Machine Learning and Computing (ICMLC), Bangalore, India, pp. 248–251 (2010)
19. Panchal, G., Ganatra, A., Shah, P., Panchal, D.: Determination of over-learning and over-fitting problem in back propagation neural network. Int. J. Soft Comput. **2**(2), 40–51 (2011)
20. Panchal, G., Ganatra, A., Kosta, Y., Panchal, D.: Behaviour analysis of multilayer perceptrons with multiple hidden neurons and hidden layers. Int. J. Comput. Theor. Eng. **3**(2), 332–337 (2011)
21. Panchal, G., Panchal, D.: Solving NP hard problems using genetic algorithm. Int. J. Comput. Sci. Inf. Technol. **6**(2), 1824–1827 (2015)
22. Panchal, G., Panchal, D.: Efficient attribute evaluation, extraction and selection techniques for data classification. Int. J. Comput. Sci. Inf. Technol. **6**(2), 1828–1831 (2015)
23. Panchal, G., Panchal, D.: Forecasting electrical load for home appliances using genetic algorithm based back propagation neural network. Int. J. Adv. Res. Comput. Eng. Technol. (IJARCET) **4**(4), 1503–1506 (2015)
24. Panchal, G., Panchal, D.: Hybridization of genetic algorithm and neural network for optimization problem. Int. J. Adv. Res. Comput. Eng. Technol. (IJARCET) **4**(4), 1507–1511 (2015)
25. Panchal, G., Samanta, D.: Comparable features and same cryptography key generation using biometric fingerprint image. In: 2nd IEEE International Conference on Advances in Electrical, Electronics, Information, Communication and Bio-Informatics (AEEICB), pp. 1–6 (2016)

26. Panchal, G., Samanta, D.: Directional area based minutiae selection and cryptographic key generation using biometric fingerprint. In: 1st International Conference on Computational Intelligence and Informatics, pp. 1–8. Springer (2016)
27. Panchal, G., Samanta, D., Barman, S.: Biometric-based cryptography for digital content protection without any key storage. In: Multimedia Tools and Application, pp. 1–18. Springer (2017)
28. Panchal, G., Kosta, Y., Ganatra, A., Panchal, D.: Electrical load forecasting using genetic algorithm based back propagation network. In: 1st International Conference on Data Management, IMT Ghaziabad. MacMillan Publication (2009)
29. Patel, G., Panchal, G.: A chaff-point based approach for cancelable template generation of fingerprint data. In: International Conference on ICT for Intelligent Systems (ICTIS 2017), p. 6 (2017)
30. Patel, J., Panchal, G.: An IoT based portable smart meeting space with real-time room occupancy. In: International Conference on Internet of Things for Technological Development (IoT4TD-2017), pp. 1–6 (2017)
31. Soni, K., Panchal, G.: Data security in recommendation system using homomorphic encryption. In: International Conference on ICT for Intelligent Systems (ICTIS 2017), pp. 1–6 (2017)
32. Patel, N., Panchal, G.: An approach to analyze data corruption and identify misbehaving server. In: International Conference on ICT for Intelligent Systems (ICTIS 2017), pp. 1–6 (2017)
33. Kosta, Y., Panchal, D., Panchal, G., Ganatra, A.: Searching most efficient neural network architecture using Akaikes information criterion (AIC). Int. J. Comput. Appl. **1**(5), 41–44 (2010)

Data Security in Recommendation System Using Homomorphic Encryption

Kajol Soni and Gaurang Panchal(✉)

U & P U. Patel Department of Computer Engineering,
Chandubhai S Patel Institute of Technology, Changa, India
kajolsoni145@gmail.com, gaurangpanchal.ce@charusat.ac.in

Abstract. Cloud computing is like a daily routine now a day. Even though it has numbers of advantages in technical and business view, still there are some challenges there like data storage security, confidentiality and integrity. Main risk in cloud data is about to trust on cloud owner. Encrypted data is not useful for any computational process, so we cannot store as encrypted data. In recommendation system cloud plays very important role. Using homomorphic encryption, we can perform cloud data analyzation. This paper discusses about different homomorphic encryption technique and solution to recommendation system.

Keywords: Cloud data storage · Homomorphic encryption · Data security · Data confidentiality · Data integrity · Recommendation system · Collaborative filtering

1 Introduction

Cloud computing is a most popular architectural model in the field of Information Technology. It is combination of Distributed Computing, Parallel Computing and Grid Computing Architectures. It provides following kinds of services: Infrastructure-as-a-Service (IaaS), Platform-as-a-Service (PaaS) and Software-as-a-Service (SaaS). Basic Computing Resources and Storage Network Services can be categorized under IaaS. PaaS provide service to develop and run application without any worry about its complexity and maintenance. SaaS is a service in which we can provide features like subscribe and software licensing. For all these services, there is no need for users to manage or control the cloud infrastructure, including network, server, operating system (OS), storage and even the functions of applications [1, 2]. In other words, we can say cloud computing is a third-party service which can be used for delivery of the applications [3]. Some well-known service providers like Rack space, Microsoft, IBM. The buzz 'cloud computing' word way back in 2006 with the launch of Amazon EC2, gained traction in 2007 [11–29].

The research paper is divided into various sections. Section 2 introduce Recommendation system & literature review followed by Collaborative filtering and Homomorphic encryption in Sect. 3. Section 4 contains the proposed work and in last we will try to conclude the study and its future scope.

S.C. Satapathy and A. Joshi (eds.), *Information and Communication Technology for Intelligent Systems (ICTIS 2017) - Volume 1*, Smart Innovation, Systems and Technologies 83, DOI 10.1007/978-3-319-63673-3_37

2 Recommendation System

Recommendation system (RS) is a one kind of information filtering system that leads to predict some information regarding products, items or preferences [5–9]. RS became very popular in recent years and useful in many areas like movie, music, news, books, social tags, research articles, search queries and products in general. Another popular RS are restaurant, life insurance, online dating, and Twitter pages.

Figure 1 Shows overview of RS. Based on user past history and rating, system will try to match them. After that it will recommend some information. Recommendation system can be divided into two technique: Profile Based and Collaborative Filtering (CF) [4].

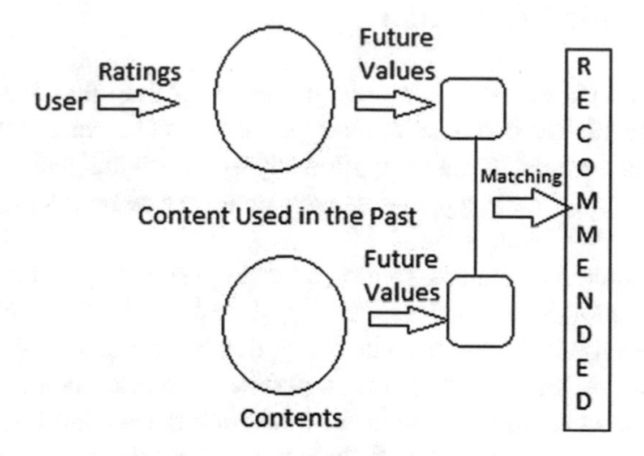

Fig. 1. Recommendation system

CF can further divide into two types: Item based CF and User Based CF. Item based CF will suggest items based on user's previous preferences. In User based CF, system will suggest items based on other user's activity who have similar kind of preferences.

Table 1. Homomorphic encryption & it's application

Algorithm	Nature of algorithm	Cloud storage	Application
Paillier [10]	Partially additive HE	No	E-voting
RSA [10]	Partially multiplicative HE	No	Internet banking
Elgamal [10]	Partially multiplicative HE	No	Hybrid system
EHC [10]	Fully HE	No	MANETS
NEHE [10]	Fully HE	No	E-commerce
AHEE [10]	Fully HE	No	Mobile cipher
BGV [10]	Fully HE	Yes	Security of integer polynomials

Here HE = Homomorphic Encryption, RSA = Rivest, Shamir and Alderman, EHC = Enhanced Homomorphic Encryption, NEHE = Non-interactive Exponential Homomorphic Encryption, AHEE = Algebra Homomorphic Encryption scheme based on updated Elgalmal, BGV = Brakerski, Gentry and Vaikuntanathan

As per research paper study and observation from theory we would like to share some views on homomorphic encryption that there are numbers of homomorphic algorithms are available. Among them we can choose any algorithm as per our requirement. For example, we require lightweight encryption to reduce complexity and power consumption specially in Internet of Things (IOT) application, we can use partially additive HE. For better understanding consider Table 1 which shows comparisons of some homomorphic encryption and its applications.

As Shown in Table 1, many algorithms are available but only BGV can be implemented on Cloud storage. As per our recommendation point of view we will choose BGV algorithm.

3 Homomorphic Encryption

Homomorphic Encryption (HE) is a one kind of encryption that allow to process on encrypted data. In HE we first need to encrypt data using secret encryption key. After processing on data we need to use decryption key to get original data. Advantage of HE is that after processing on encrypted data we can get the same result as it was applied on plain text or original data.

Advantage of homomorphic encryption over simple encryption is we do not need to worry about data integrity and privacy. In Fig. 2 we had shown simple mechanism of homomorphic encryption. Homomorphic encryption has major two operations: addition and multiplication. According to operation performed on the data it can be classified into mainly two sub categories. One is fully homomorphic encryption and another is partially homomorphic encryption. In fully homomorphic encryption both addition and multiplication is performed, while in partially homomorphic encryption any of them is used.

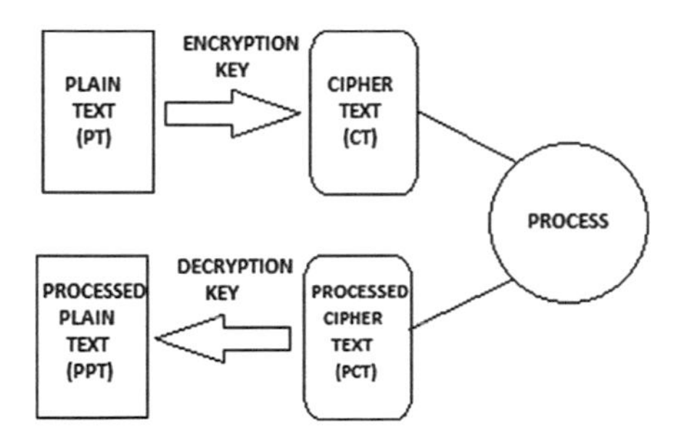

Fig. 2. Homomorphic encryption

4 Proposed Work

To Ensure data privacy we need to encrypt data. For that we proposed a better encryption scheme over simple encryption and it is HE. As we know that recommendation system will recommend items from stored data in cloud. So before putting data on cloud encrypt them using BGV algorithm. If we are supposed to share that information with third party's recommendation system, there will be no issue of data privacy and data integrity.

Figure 3 shows basic diagram of proposed system. From the figure it's clear that it will solve issue of data privacy, data confidentiality and data integrity.

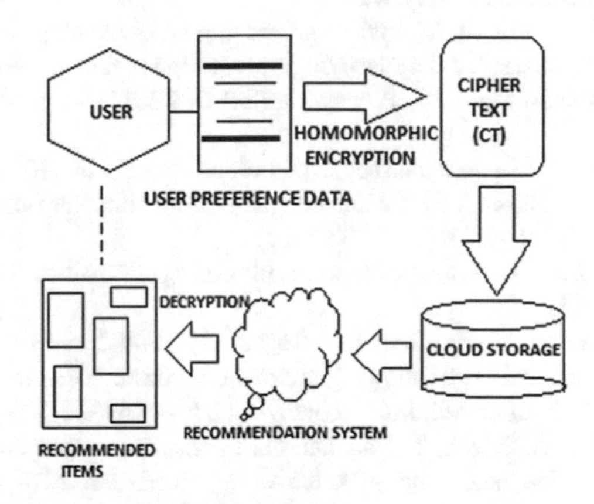

Fig. 3. Proposed homomorphic encryption for recommendation system

5 Conclusion

Providing security in cloud storage is a biggest challenge. Here in we presented our views and a way to provide data privacy and security to cloud data and it can be used for further process like recommender system. In this paper, we proposed a demo level of mechanism of homeomorphism encryption on recommender system, but in real time it require high computational power as cloud has big data storage. In future work, we will try to provide a solution with better efficiency so that it can deal with big amount of data.

References

1. Schafer, J.B., Konstan, J., Riedi, J.: Recommender systems in e-commerce. In: Proceedings of the 1st ACM conference on Electronic Commerce, pp. 158–166. ACM Press, New York (1999)
2. Kargupta, H., Datta, S., Wang, Q., Sivakumar, K.: On the privacy preserving properties of random data perturbation techniques. In: Proceedings of the 3rd IEEE International Conference on Data Mining (ICDM 2003), Melbourne, Florida, USA, pp. 99–106. IEEE, November 2003
3. Lemire, D., Maclachlan, A.: Slope one predictors for online rating-based collaborative filtering. Society for Industrial Mathematics (2005)
4. Lemire, D., Maclachlan, A.: Slope one predictors for online rating-based collaborative filtering. In Proceedings of the SIAM Data Mining (SDM 2005), Newport Beach, California, USA, April 2005

5. Aggarwal, C.C., Yu, P.S.: A General Survey of Privacy-Preserving Data Mining Models and Algorithms, Chapter 2, pp. 11–52. Springer, New York (2008)
6. Han, S., Ng, W.K., Yu, P.S.: Privacy-preserving singular value decomposition. In: Proceedings of the 25th IEEE International Conference on Data Engineering (ICDE 2009), Shanghai, China, IEEE, March–April 2009
7. Basu, A., Kikuchi, H., Vaidya, J.: Privacy-preserving weighted slope one predictor for item-based collaborative filtering. In: Proceedings of the International Workshop on Trust and Privacy in Distributed Information Processing (TP-DIS 2011), Copenhagen, Denmark, July 2011
8. Zhang, X., Hong tao D.: Ensure data securiy in cloud stoarge. In: NCIS 2011, pp. 284–287
9. Vaidya, J., Yakut, I., Basu, A.: Efficient integrity verification for outsourced collaborative filtering. In: Data Mining (ICDM), IEEE (2014)
10. Kangavalli, R., Vagdevi, S.: A mixed homomorphic encryption scheme for secure data storage in cloud, IEEE (2015)
11. Ganatra, G., Kosta, Y.P., Panchal, G., Gajjar, C.: Initial classification through back propagation in a neural network following optimization through GA to evaluate the fitness of an algorithm. Int. J. Comput. Sci. Inf. Technol. **3**(1), 98–116 (2011)
12. Panchal, G., Ganatra, A., Kosta, Y., Panchal, D.: Forecasting employee retention probability using back propagation neural network algorithm. In: IEEE 2010 Second International Conference on Machine Learning and Computing (ICMLC), pp. 248–251. Bangalore, India (2010)
13. Panchal, G., Ganatra, A., Shah, P., Panchal, D.: Determination of over-learning and over-fitting problem in back propagation neural network. Int. J. Soft Comput. **2**(2), 40–51 (2011)
14. Panchal, G., Ganatra, A., Kosta, Y., Panchal, D.: Behaviour analysis of mul-tilayer perceptrons with multiple hidden neurons and hidden layers. Int. J. Comput. Theory Eng. **3**(2), 332–337 (2011)
15. Panchal, G., Panchal, D.: Solving np hard problems using genetic algorithm. Int. J. Comput. Sci. Inf. Technol **6**(2), 1824–1827 (2015)
16. Panchal, G., Panchal, D.: Efficient attribute evaluation, extraction and selection techniques for data classification. Int. J. Comput. Sci. Inf. Technol **6**(2), 1828–1831 (2015)
17. Panchal, G., Panchal, D.: Forecasting electrical load for home appliances using genetic algorithm based back propagation neural network. Int. J. Adv. Res. Comput. Eng. Technol. (IJARCET) **4**(4), 1503–1506 (2015)
18. Panchal, G., Panchal, D.: Hybridization of genetic algorithm and neural network for optimization problem. Int. J. Adv. Res. Comput. Eng. Technol. (IJARCET) **4**(4), 1507–1511 (2015)
19. Panchal, G., Samanta, D.: Comparable features and same cryptography key generation using biometric fingerprint image. In: 2nd IEEE International Conference on Advances in Electrical, Electronics, Information, Communication and Bio-Informatics, pp. 1–6. AEEICB (2016)
20. Panchal, G., Samanta, D.: Directional area based minutiae selection and cryptographic key generation using biometric fingerprint. In: 1st International Conference on Computational Intelligence and Informatics, pp. 1–8. Springer, New York (2016)
21. Panchal, G., Samanta, D., Barman, S.: Biometric-based cryptography for digital content protection without any key storage, pp. 1–18. Springer (Multimedia Tools and Application), New York (2017)
22. Panchal, G., Kosta, Y., Ganatra, A., Panchal, D.: Electrical load forecasting using genetic algorithm based back propagation network. In: 1st International Conference on Data Management, IMT Ghaziabad. MacMillan Publication (2009)

23. Patel, G., Panchal, G.: A chaff-point based approach for cancelable template generation of fingerprint data. In: International Conference on ICT for Intelligent Systems (ICTIS 2017), p. 6 (2017)
24. Patel, J., Panchal, G.: An IOT based portable smart meeting space with real-time room occupancy. In: International Conference on ICT for Intelligent Systems (ICTIS 2017), pp. 1–6 (2017)
25. Soni, K., Panchal, G.: Data security in recommendation system using homo-morphic encryption. In: International Conference on ICT for Intelligent Systems (ICTIS 2017), pp. 1–6 (2017)
26. Patel, N., Panchal, G.: An approach to analyze data corruption and identify misbehaving server. In: International Conference on ICT for Intelligent Systems (ICTIS 2017), pp. 1–6 (2017)
27. Bhimani, P., Panchal, G.: Message delivery guarantee and status update of clients based on IOT-AMQP. In: International Conference on Internet of Things for Technological Development (IoT4TD-2017), pp. 1–6 (2017)
28. Mehta, S., Panchal, G.: File distribution preparation with file retrieval and error recovery in cloud environment. In: International Conference on ICT for Intelligent Systems (ICTIS 2017), p. 6 (2017)
29. Kosta, Y., Panchal, D., Panchal, G., Ganatra, A.: Searching most efficient neural network architecture using Akaikes information criterion (AIC). Int. J. Comput. Appl. **1**(5), 41–44 (2010)

A Novel Machine Learning Based Approach for Rainfall Prediction

Niharika Solanki and Gaurang Panchal(✉)

U & P U. Patel Department of Computer Engineering,
Chandubhai S Patel Institute of Technology, Changa, India
15pgce035@charusat.edu.in, gaurangpanchal.ce@charusat.ac.in

Abstract. The climate changes effortlessly nowadays, prediction of climate is very hard. However, the forecasting mechanism is the vital process. It is also a valuable thing as it is the important part of the human life. Accordingly to the research, the weather forecast of rainfall intensity conducted. The remarkable commitment of this proposal is in the implementation of a hybrid intelligent system data mining technique for solving novel practical problems, Hybrid Intelligent system data mining consists of the combination of Artificial Neural Network and the proper usage of Genetic Algorithm. In this research, Genetic algorithm is utilized the type of inputs, the connection structure between the inputs and the output layers and make the training of neural network more efficient. In ANN, Multi-layer Perceptron (MLP) serves as the center data mining (DM) engine in performing forecast tasks. Back Propagation algorithm used for the trained the neural network. During the training phase of the proposed approach, it gains the optimal values of the connection weights which, in fact, utilized as the part of the testing phase of the MLP. Here, the testing phase is used to bring about the rainfall prediction accuracy. It may be noted that the information/data is used to cover the information from the variables namely temperature, cloud fraction, wind, humidity, and rainfall.

Keywords: Rainfall prediction · Genetic algorithm · Artificial neural network

1 Introduction

Machine learning is an advanced mechanism for analysis is various data (i.e. data sets) that computerises explanatory which are use to build the models. The knowledge/information gain by the various calculations which are performed based on the information, the Machine Learning (ML) allows the systems to find out the hidden bits (i.e. information) of knowledge without applying alternative solutions [1–10].

There are various utilizations of machine learning. It's quite to acknowledge how much machine learning has accomplished in real world applications.

S.C. Satapathy and A. Joshi (eds.), *Information and Communication Technology for Intelligent Systems (ICTIS 2017) - Volume 1*, Smart Innovation, Systems and Technologies 83, DOI 10.1007/978-3-319-63673-3_38

Machine learning is regularly connected in the disconnected training phase. Thus machine learning is utilized to enhance the applications such as face recognition, face detection, speech recognition, genetics, image classification, weather forecast etc. [3–15]. Machine learning is connected in weather forecasting programming to enhance the nature of the estimate. Machine learning makes it moderately less demanding to create complex programming frameworks without much exertion on the human side [10–18].

Weather prediction is a confused methodology that incorporates various particular fields of mastery [6]. Weather prediction has been a standout amongst the most deductively and mechanically troublesome issues the world over in the latest century [18–28].

Since its conception Weather Prediction has made tremendous progress [6]. In any case, numerous new issues have developed. Therefore, critical topics in the community as a whole are in fact considered as a lack of timely. There are few issues to solve weather prediction which described as below;

1. The most difficult issue is enhancing accuracy.
2. High speed of the streaming data and scale it for the high dimensions.
3. Reducing the computational complexity and the memory requirement.
4. To tackle the over fitting issue, a standout amongst the most widely recognized assignment is to fit a model to a set of training data.

To solve this problem we propose a hybrid genetic algorithm based neural network rainfall prediction system which will be given in Fig. 1.

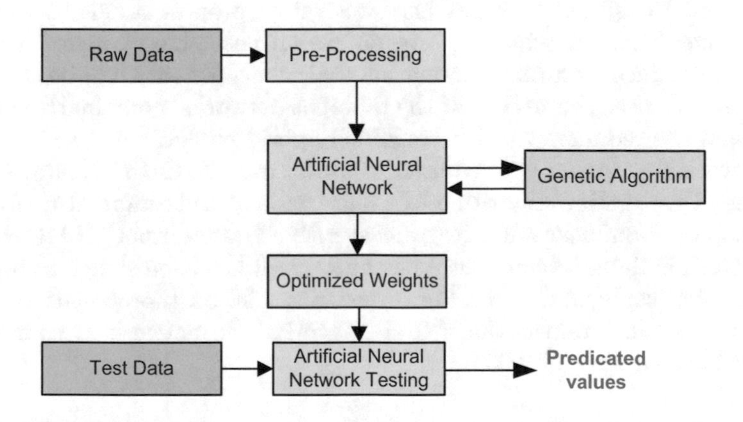

Fig. 1. Proposed approach.

2 Related Work

The work by Kumar Abhisek *et al.* [7] used Layer Recurrent Network (LRN), Artificial Neural Network in which Back-Propagation Algorithm (BPA) and Cascaded back Propagation (CBP) algorithm utilized for rainfall prediction. For the best performance in this paper, executes rainfall prediction by utilizing data set and train it. The test data sets and setting up the number of neurons in the hidden layer is done in this phase. In the present research, Ann models analyzed possibility of predicting average rainfall. In formulating, ANN constructed predictive models three layered network. BP algorithm used for the train the normalized data. Finally compare the predicted output with desired output.

The work by Ahmet Erdil *et al.* [2] utilised the concept of Neural Network along with the addition process of Back propagation Algorithm (BPA) for Prediction of meteorological factors that is radiation of secondly solar, minimum and maximum atmospheric pressures. In the Real-world scenarios, the issues and the problems are solved effectively using the artificial neural network by straightforward use. The objective of the proposed algorithm the minimization of the level of the global error. For example, the mean error and root mean square error.

The work by Kumar Abhishek *et al.* [8] used Artificial Neural Network as a part of which Feedforward with back propagation algorithm used for forecast maximum temperature. In these paper, investigates the pertinence of ANN approach by creating viable and reliable nonlinear predictive models for climate examination also additionally assess and look at the execution of the developed models utilizing different transfer functions, hidden layers and neurons to forecast maximum, temperature for 365 days of the year.

The work by Abbas Ahmadi *et al.* [1] utilized ensemble of neural network using Multi-Layer Perceptron (MLP), Radial basis function (RBF), and General Regression Neural Network (that is, GRNN) and Time Delay Neural Network (that is, TDNN) for Temperature Prediction (Dew Point, Humidity, Sea Level Pressure and Visibility.). For weather forecasting, propose a new hybrid model in this research paper, which is depends on an ensemble of neural networks. Likewise, introduce a mutual information (MI) concepts that handle the various challenges that uses the results of the different networks' combinations and try to minimise the redudancy in the combined hybrid model.

The work by Morten Gill Wollsen *et al.* [9] used Artificial Neural Network using Back-Propagation Algorithm for Solar irradiate and temperature forecasts. In this paper, (Nonlinear Auto Regressive with eXternal input) NARX network is used which is the artificial neural network which has been shown to be appropriate for the displaying of non-linear systems. The back-propagation training algorithm used for trained the neural network. The network is trained until accomplishing an error of 0.05%.

3 Proposed Approach

The proposed approach includes different process namely collection of Rrw data, pre-processing of the raw data, training GA-MLP, testing GA-MLP and the final rainfall prediction. In pre-processing first remove the missing value by the taking the centre measure value like mean of the given variable or if missing value is small in size, then remove the missing value by erasing. After the removing missing data the rainfall data will be normalised between the range of $[0 \cdots 1]$ which minimise the error.

After the pre-processing, there will be a training of GA-MLP. In this stage, genetic algorithm [5] and multi-layer perceptron [4] is used. Genetic algorithm is utilized the type of inputs, the connection structure between the inputs and the output layers. This, in fact, represents the simplification of Multilayer Perceptron structure in the Model, and makes the training process more efficient. The evaluation of the fitness function for the different combination of the variables are resolved by the MSE (means square error) when such an input combination is utilised as a part of a MLP to perform the classification/prediction task.

Also, in pre-processing missing value and normalization is carried out to improve the accuracy of the algorithm. Missing value in the data set can be given as "missing" data. To solve this missing value by, if missing value is small in size then remove the missing value by erasing or remove the missing value by the taking the centre measure value like mean of the given variable. After the removing the procedure of missing data the rainfall data will be normalised between the various ranges (i.e., $[0 \cdots 1]$) which, in fact, used to minimize the overall error and produces the better performance. The equation of normalization given as follow;

$$X_{norm} = \frac{X_t - X_{min}}{X_{max} - X_{min}} \tag{1}$$

Here, X_{norm} is normalized data, X_t is actual data, X_{max} is maximum value of x variable and X_{min} is minimum value of x variable.

After the pre-processing, there will be a training of GA-MLP. In this stage, genetic algorithm [5] and multi-layer perceptron [4] is used. The proposed architecture is trained using the multi-layer perceptron in combination of the genetic algorithm. Genetic algorithm is utilized the type of inputs, the connection structure between the inputs and the output layers.

This simplifies the MLP structure in Model, and makes the training process more efficient. In Genetic algorithm, initial population of the chromosomes is randomly generated. Then from each chromosome, the weights are extracted based on the number of genes a chromosomes having. The error and the fitness value are calculated. After the weights are extracted successfully, the initial weights are combined together and again given as an input to the network for the calculation of the actual output. These outputs when compared with the desired output, error values are obtained.

The assessment of the fitness function for every input variable combination is resolved by mean squared error (MSE) when such an input combination is utilised as a part of a MLP to perform the classification/prediction task. Here, the fitness function is examined by the calculating the RMS error. Then generate a new population, the crossover operator is applied. These steps are calculated in repetition until the stopping criteria has been satisfied. The chromosomes archives same value of fitness then the algorithm get stopped. The weights which are obtained at the end of the training phase are used to predicts the temperature, wind speed, humidity.

4 Conclusion

The use of the Genetic Algorithm to optimise the structure parameters, the connection weights between the inputs and the output layers. This, in fact, represents that the MLP structure in Model, and enables the classifier training phase more efficient and computationally less expensive. The evaluation of the problem fitness functions for the different input combination is calculated by the mean squared error (MSE) when such an input combination is exploited as a part of a multilayer perceptron to perform the classification/prediction task. The training must be repeated a few circumstances keeping in mind the end goal to get satisfactory results.

References

1. Ahmadi, A., Zargaran, Z., Mohebi, A., Taghavi, F.: Hybrid model for weather forecasting using ensemble of neural networks and mutual information. In: 2014 IEEE Geoscience and Remote Sensing Symposium, Quebec City, QC, pp. 3774–3777 (2014)
2. Erdil, A., Arcaklioglu, E.: The prediction of meteorological variables using artificial neural network, pp. 1677–1683 (2012)
3. Bishop, C.M.: Pattern Recognition and Machine Learning. Springer, New York (2006)
4. Du, K.L., Swamy, M.N.S.: Neural Networks in a Soft Computing Framework. Springer, London (2006)
5. Goldberg, D.E.: Genetic Algorithms in Search, Optimization and Machine Learning. Addison-Wesley Longman Publishing Co., Inc., Boston (1989)
6. Upadhaya, J.: Assam University, Climate Change and its impact on Rice productivity in Assam
7. Abhishek, K., Kumar, A., Ranjan, R., Kumar, S.: A rainfall prediction model using artificial neural network. In: 2012 IEEE Control and System Graduate Research Colloquium (ICSGRC), Shah Alam, Selangor, pp. 82–87 (2012)
8. Abhishek, K., Singh, M.P., Ghosh, S., Anand, A.: Weather forecasting model using Artificial Neural Network. Procedia Technol. **4**, 311–318 (2012)
9. Wollsen, M.G., Jørgensen, B.N.: Improved local weather forecasts using artificial neural networks, pp. 75–86 (2015)
10. Ganatra, A., Panchal, G., Kosta, Y., Gajjar, C.: Initial classification through back propagation in a neural network following optimization through GA to evaluate the fitness of an algorithm. Int. J. Comput. Sci. Inf. Technol. **3**(1), 98–116 (2011)
11. Panchal, G., Ganatra, A., Kosta, Y., Panchal, D.: Forecasting employee retention probability using back propagation neural network algorithm. In: IEEE 2010 Second International Conference on Machine Learning and Computing (ICMLC), Bangalore, India, pp. 248–251 (2010)
12. Panchal, G., Ganatra, A., Shah, P., Panchal, D.: Determination of over-learning and over-fitting problem in back propagation neural network. Int. J. Soft Comput. **2**(2), 40–51 (2011)
13. Panchal, G., Ganatra, A., Kosta, Y., Panchal, D.: Behaviour analysis of multilayer perceptrons with multiple hidden neurons and hidden layers. Int. J. Comput. Theory Eng. **3**(2), 332–337 (2011)

14. Panchal, G., Panchal, D.: Solving np hard problems using genetic algorithm. Int. J. Comput. Sci. Inf. Technol. **6**(2), 1824–1827 (2015)
15. Panchal, G., Panchal, D.: Efficient attribute evaluation, extraction and selection techniques for data classification. Int. J. Comput. Sci. Inf. Technol. **6**(2), 1828–1831 (2015)
16. Panchal, G., Panchal, D.: Forecasting electrical load for home appliances using genetic algorithm based back propagation neural network. Int. J. Adv. Res. Comput. Eng. Technol. (IJARCET) **4**(4), 1503–1506 (2015)
17. Panchal, G., Panchal, D.: Hybridization of genetic algorithm and neural network for optimization problem. Int. J. Adv. Res. Comput. Eng. Technol. (IJARCET) **4**(4), 1507–1511 (2015)
18. Panchal, G., Samanta, D.: Comparable features and same cryptography key generation using biometric fingerprint image. In: 2nd IEEE International Conference on Advances in Electrical, Electronics, Information, Communication and Bio-Informatics (AEEICB), pp. 1–6 (2016)
19. Panchal, G., Samanta, D.: Directional area based minutiae selection and cryptographic key generation using biometric fingerprint. In: 1st International Conference on Computational Intelligence and Informatics, pp. 1–8. Springer (2016)
20. Panchal, G., Samanta, D., Barman, S.: Biometric-based cryptography for digital content protection without any key storage. In: Multimedia Tools and Application, pp. 1–18. Springer (2017)
21. Panchal, G., Kosta, Y., Ganatra, A., Panchal, D.: Electrical load forecasting using genetic algorithm based back propagation network. In: 1st International Conference on Data Management, IMT Ghaziabad. MacMillan Publication (2009)
22. Patel, G., Panchal, G.: A chaff-point based approach for cancelable template generation of fingerprint data. In: International Conference on ICT for Intelligent Systems (ICTIS 2017), p. 6 (2017)
23. Patel, J., Panchal, G.: An IoT based portable smart meeting space with real-time room occupancy. In: International Conference on Internet of Things for Technological Development (IoT4TD 2017), pp. 1–6 (2017)
24. Soni, K., Panchal, G.: Data security in recommendation system using homomorphic encryption. In: International Conference on ICT for Intelligent Systems (ICTIS 2017), pp. 1–6 (2017)
25. Patel, N., Panchal, G.: An approach to analyze data corruption and identify misbehaving server. In: International Conference on ICT for Intelligent Systems (ICTIS 2017), pp. 1–6 (2017)
26. Bhimani, P., Panchal, G.: Message delivery guarantee and status Up-date of clients based on IOT-AMQP. In: International Conference on Internet of Things for Technological Development (IoT4TD 2017), pp. 1–6 (2017)
27. Mehta, S., Panchal, G.: File distribution preparation with file retrieval and error recovery in cloud environment. In: International Conference on ICT for Intelligent Systems (ICTIS 2017), p. 6 (2017)
28. Kosta, Y., Panchal, D., Panchal, G., Ganatra, A.: Searching most efficient neural network architecture using Akaikes information criterion (AIC). Int. J. Comput. Appl. **1**(5), 41–44 (2010)

Opinion Formation Based Optimization in Audio Steganography

Rohit Tanwar(✉) and Sona Malhotra

UIET, Kurukshetra University, Kurukshetra, India
rohit.tanwar.cse@gmail.com, sonamalhotrakuk@gmail.com

Abstract. In the present scenario where so many computational problems are being solved using nature inspired optimization algorithms. Human being the most the intelligent creature, deserves to inspire researchers by its method of opinion formation. Steganography is proving to be the ultimate tool of data security. The technique can be improved further by applying optimization. In this paper, a new technique is proposed where optimization is done on the basis of human opinion formation.

Keywords: Social impact theory · Optimization · Substitution technique · Human opinion formation

1 Introduction

Nature has always been a good source of inspiration for researchers. In context of problem solving, various researchers have find their clues from some natural behaviors. In computer science, there are many optimization problems whose solutions are inspired from nature like; Ant Colony Optimization, Bee Optimization, Genetic Algorithm, Particle Swarm Optimization, etc.

Steganography is a technique of sending messages in a secure way [4]. The technique has proven to be more secure because the existence of secret message got concealed by embedding it into an appropriate cover file. On the basis of cover file, steganography is of different types. The file chosen for work here is an audio file. The requirement of optimization in steganography comes into focus when the embedding of secret message is supposed to produce noticeable distortion in cover file.

2 Opinion Formation Based Optimization

Opinion formation model in society is a consequence of Bibb Latane's dynamic theory of social impact [7]. According to this theory, the behavior and opinions of an individual are dependent on the society to which he belongs. If the same individual moves into some other social environment he will adapt to the new society and hence its other members.

S.C. Satapathy and A. Joshi (eds.), *Information and Communication Technology for Intelligent Systems (ICTIS 2017) - Volume 1*, Smart Innovation, Systems and Technologies 83, DOI 10.1007/978-3-319-63673-3_39

The social impact on an individual influenced by his social environment can be measured using following variables [1]:

(1) **Strength**, It is the attribute which determines how much an individual can affect others. Strength can be of two types which may or may not be correlated.
Persuasiveness is the ability to convince someone with opposing opinion to change it.
Supportiveness is the ability to help those sharing the same opinion to resist influence from others.
(2) **Immediacy,** It is the term used for showing group structure. It can be stated as relations in the society of physical distance of the individuals. Generally, it expresses the possibility of communication between the two individuals.
(3) **Number of Other Individuals,** It is the number of individuals that contribute to the social impact on given individual. Usually all individuals in the population are not considered but only those whose immediacy to the given individual is above some threshold value.

3 Literature Review

Dr. Mazdak worked on use of GA in image stegnography wherein the limitations of substitution techniques were identified and possible solution using Genetic Algorithm was proposed. He tried to increase the capacity while maintaining imperceptibility using Genetic Algorithm [3].

Social impact theory optimizer method was applied by M. Macas et al. in 2007 and 2011 for feature extraction [4].

In 2011, research team at Central Scientific Instrument Organization, Chandigarh (India) use SITO optimizer in their project of impedance tongue. The motive was to classify samples of black tea. Their results showed that SITO outperformed Genetic Algorithm and Binary Particle Swarm Optimization in this case [5].

Similar research was carried out in 2012 by the same team where the aim was enhancing the performance of an electronic nose. At this time too, the results were inspiring [6].

In 2013, Rishemjit Kaur et al. in their paper discussed mathematical background for carrying optimization using human opinion dynamics. They also proposed an algorithm for the same [2].

4 Motivation

The motivation is derived from following two factors:

(1) The limitations of Steganography, substitution technique are covered to much extent when the substitution bits for embedding message are preferably MSB's. However, this action would result in more distortion because the contribution of MSB's in statistical values of audio samples is more as compare to LSB's. Thus, there is a requirement of Optimization so as to achieve robustness while maintaining imperceptibility and data hiding capacity.
(2) Social impact theory optimizer is based on theory of social impact given by Bibb Latane's. This is a population based optimizer which evolves with time until some criteria or threshold is achieved.
    Since human are considered to be the most intelligent creature of God, so the theory inspired from human would be much more effective.
    In the work done by Giri Pytela [1], the SITO has been proven better than PSO.

5 Opinion Formation Based Optimization

On the basis of opinion formation model, The Optimizer is designed with following characteristics [1]:

(1) Vector Representation of Individuals: Each individual is represented as a binary vector so as to increase dimensions of the solution space.
(2) Representation of Strength: As described already, strength is of two types; Persuasiveness and Supportiveness. Both these are represented by a single variable so as to maintain simplicity.
(3) Fitness Function: The concept of fitness function is introduced so as to evaluate individual on the basis of their strength. Strength of an individual is proportional to the fitness value. More the fitness value, more strength it has.

The Working of optimizer is divided in various modules which are explained as follows:

(1) Population: Each individual is a binary vector representation of solution of length d.

$$S_i = \{S_i^1, S_i^2, \ldots\ldots\ldots, S_i^d\}$$

(2) Fitness function: An application based fitness function is to applied on each individual so as to evaluate it and get fitness value.

$$f_i = f(S_i^1, S_i^2, \ldots\ldots\ldots, S_i^d)$$

(3) Impact of Neighbors: Those individuals (i) who are supposed to impact the opinion of other individual (j) are called neighbors of individual j and the impact of neighbors on j can be given as:

$$q_{ij} = \max(f_i - f_j, 0)$$

It is clear from the equation that a better fit individual will have positive impact on less fit neighbor.

(4) Computation of Social Impact: It is represented by a bit vector

$$I_j = \{I_j^1, I_j^2, \ldots\ldots.., I_j^d\}$$

It is derived from the social strength of j's neighbor on j.

Overall impact on bit m of an individual is calculated by the following equation:

$$I_j^m = \sum_{i \in P_i^m} a_{ij} - \sum_{i \in S_i^m} a_{ij} + \varphi_j$$

Where

P_i^m is group of persuaders,

S_i^m is group of supporters

And φ_j is the noise parameter added which can be calculated as:

$$\varphi_j = S \sum_{i=1}^{n} e^{-|f_j - f_i|}$$

S is the strength of disintegrating forces and

n is the no. of individuals

(5) Update Rule: This step is crucial in generation of new population as follows:

$$S_j^m(t+1) = \begin{cases} 1 - S_j^m(t); & \textit{if } I_j^m > 0 \\ S_j^m(t); & \textit{otherwise} \end{cases}$$

This shows that the individual will change its opinion if impact function has positive value.

(6) Mutation: Mutation is done with probability $k \ll 1$. K determines the mutation rate. This is done in order to get it out of local minima or maxima.

6 Algorithm for Optimizing Audio Steganography

The embed algorithm takes secret message and an audio file as input and gives stego file (audio file embedded with secret message) as output.

```
embed(Secret_Msg , Cover_Audio)
  {
  1.  Convert Secret_Msg into binary Vector M{}
      Initialize len=Length(Secret_Msg)
  2.  Convert Audio file in binary
      Intialize las=No_of_Audio samples
      count=0;
  3.  while count<len
          Substitute 4th and 5th LSB of audio sample with message bits;
          count=count+2;
          audiosample.next;
      end while
  4.  popsize=4
 popdone=0;
 do
     optimizer();                    //optimizer function is called here
     popdone=popdone+popsize;
     If((len-popdone)>=4)
         popsize=4;
     Else
         popsize=len-popdone;
 While(popsize>0)
}
```

Optimizer function tries to optimize the existing set of solutions. The optimization is being done using social impact theory.

```
optimizer()
{
 Initialize all Si(0)                        // initial set of population
 While (condition is true)
   For all i
      Evaluate fi(t)=f(Si(t))                // Assign fitness value to each individual
   End for
   For all i,j
      Compute strength values qij          // calculate impact of all other individuals
   End for                                   // on an individual
   For all i,d
      Compute Idi(t) and then compute Sdi(t+1) //Next generation
   End for
   For all i,d
      Apply mutation on Sdi(t+1)
   End for
t=t+1
End while

}
```

7 Conclusion

The proposed algorithm is supposed to give better results than Genetic Algorithm and Particle Swarm Optimization. Social Impact Theory Optimizer has proven to be a good optimizer in various researches in past. The Future work will be to implement the proposed algorithm and compare the results with that of genetic algorithm. The work can be extended further to increase capacity while maintaining the robustness.

References

1. Jiri, P.: Implementation and testing of social optimization algorithms. Czech Technical University, Prague (2012)
2. Kaur, R., Kumar, R., Bhondekar, A., Kapur, P.: Human opinion dynamics: an inspiration to solve complex optimization. Sci. Rep. **3**, 3008 (2013)
3. Mäs, M., Flache, A., Helbing, D.: Individualization as driving force of clustering phenomena in humans. PLoS Comput. Biol. **6**, e1000959 (2010)
4. Mazdak, Z., Azizah, A.M., Rabiah, B.A., Akram, M.Z., Zeki, M.: A Genetic Algorithm Based Approach for Audio Steganopgraphy, pp. 360–363. World Academy of Science, Engineering and Technology, Paris (2009)
5. Bhondekar, A.P., Kaur, R., Kumar, R., Vig, R., Kapur, P.: A novel approach using dynamic social impact theory for optimization of impedance-Tongue (iTongue). Chemometr. Intell. Lab. Syst. **109**(1), 65–76 (2011)
6. Bhondekar, A.P., Kaur, R., Kumar, R., Gulati, A., Ghanshyam, C., Kapur, P.: Enhancing electronic nose performance: a novel feature selection approach using dynamic social impact theory and moving window time slicing for classification of Kangra orthodox black tea (Camellia sinensis (L.) O. Kuntze). Sens. Actuators B **166–167**, 309–319 (2012)
7. Nowak, A., Szamrej, J., Latane, B.: From private attitude to public opinion: a dynamic theory of social impact. Psychol. Rev. **97**(3), 362–376 (1990)

Improving Network Lifetime by Heterogeneity in Wireless Sensor Networks

Sukhkirandeep Kaur(✉) and Roohie Naaz Mir

Department of Computer Science and Engineering, National Institute of Technology, Srinagar, India
kirangill0189@gmail.com, naaz310@nitsri.net

Abstract. Network lifetime is an important parameter to be considered to evaluate wireless sensor network performance. Lifetime is considered in different contexts i.e. some considered network lifetime when first node dies or some percentage of nodes dies or when all the nodes deplete their energy. In WSN, energy conservation becomes important to improve network lifetime. In a network with even distribution of nodes, nodes that lie near the base station deplete their energy faster and hole is created which leads to disconnected network. Incorporating Heterogeneity in the network increases energy efficiency and improves network lifetime. Heterogeneity is included in terms of energy, links and computational resources. Finding optimal number and positions of Heterogeneous nodes to improve network performance is the main concern of this study.

Keywords: Wireless Sensor Network (WSN) · Heterogeneity · Clustering · Load balancing

1 Introduction

Energy conservation is the key goal in wireless sensor network since sensor nodes are battery powered. WSN is application specific and depending upon the applications different parameters are considered i.e. network deployment, coverage and connectivity, network lifetime etc. Network deployment can be deterministic or random. Deterministic deployment is mainly considered in surveillance, security, factory automation etc. where human intervention is possible and random deployment is considered in inhostile environments such as in applications of forest monitoring, disaster management etc. Energy is consumed in sensing, processing and communication processes but the major source of energy consumption is the communication in form of radio transceiver energy.

Position of base station also plays an important role. If nodes are far away from BS they will deplete their energy faster due to increased communication distance. To avoid this, if we consider multi-hop communication then nodes near the base station will deplete their energy as a fact that they will bypass whole of the network traffic. So, Load Balancing is needed in multi-hop networks. Load balancing will improve the lifetime of WSN in different ways i.e. by incorporating clustering in the network, by distributing

S.C. Satapathy and A. Joshi (eds.), *Information and Communication Technology for Intelligent Systems (ICTIS 2017) - Volume 1*, Smart Innovation, Systems and Technologies 83, DOI 10.1007/978-3-319-63673-3_40

traffic load evenly, by varying transmission range, considering heterogeneity, by varying number of nodes near the sink, by incorporating mobility. Energy conservation is the main objective in most of the wireless sensor networks applications.

Due to non-uniform traffic pattern of WSN's, in multi-hop transmissions most of the traffic load falls on the nodes near BS. Nodes near Base-Station deplete their energy faster compared to other nodes that leads to creation of holes near BS which further leads to network disconnection. Energy conservation is an important concern in multi-hop networks due to limited power of sensor nodes. To increase network lifetime, energy conservation is needed that can be achieved in multi-hop networks by efficient MAC protocols, load balancing etc. Load Balancing (LB) technique reduces energy consumption, if LB is included in cluster based networks, it increases scalability. Different network lifetime definitions exist in literature that include when first sensor node dies, percentage of sensor node dies or when connectivity is lost. We will consider network lifetime as a total time from network deployment to the time when first node dies. Load Balancing is considered in this work in terms of uneven clustering and heterogeneity.

1.1 Applications of WSN

Wireless Sensor Network is application oriented and applications can be categorized based on their deployment i.e. Deterministic and Random network deployment. It is important to relate the algorithms and protocols in WSN to corresponding applications. Here, few applications that belongs to both type of deployment are discussed and the work is related to applications in Deterministic network because energy heterogeneity in terms of line powered nodes is possible in deterministic network.

1.2 Deterministic Network Applications

Smart Home. Smart homes involves equipping home parts with different sensors that monitor and collaborate so that everyday house activities are automated without user intervention in an easier, more convenient, more efficient, safer, and less expensive way. A variety of sensors can be used in smart homes to monitor different activities that may include light control, intrusion detection system, gas leakage detection etc. Placement of sensors in these applications is performed in deterministic manner.

Healthcare. In healthcare, WSN is used for monitoring the patients in clinical settings, Monitoring for chronic and elderly patients and Collection of long-term databases of clinical data. For monitoring patients, Different sensors used are Memscap, Millicore, Novosense, Novelda, VTT, SINTEF [1]. These sensors are used according to the disease, treatment and life cycle of patient. These applications have real time requirements and security concerns also. In case of Emergency, data should reach as soon as possible to avoid mishappening and patient's data should not be tempered. So, these applications demands QoS and security measures.

Smart Parking. One of the application of Deterministic WSN is smart parking where sensors like ambient light sensors, magnetic sensors etc. can be deployed to detect presence and movement of moving vehicles. This allows vehicle drivers to efficiently find the space and park their vehicles. Different smart parking applications in WSN is discussed in [2] that detects vehicles and send information to CSS which in turn produces billing information. In [3], Wireless based smart parking system using zigbee is developed that notify visitors of empty and non-empty parking lot.

Security and surveillance. These applications may consist of motion sensors, acoustic sensors, PIR sensors and cameras that transmit multimedia information when intrusion occurs. These applications demand real time requirements so that important data reaches without any delay. As these applications contains camera or video sensors, these sensors need to be carefully placed to avoid damage. So, deterministic deployment is performed in this case. WSN based solutions for security and surveillance are discussed in [4].

1.3 Random Network Applications

Environmental Monitoring. These are deployed in inhostile environment where sensors are deployed randomly, so to provide coverage large number of sensors are needed that increases cost and complexity. Different sensors that are used in these applications include temperature, humidity, barometric pressure, mid-range infrared etc. and applications are Great Duck Island [5], North temperature lake [6], Syracuse Project [7] etc.

Disaster management. WSN applications for disaster management includes forest fire detection, flood detection etc. These applications are deployed in random manner and sensors used are temperature, humidity, atmospheric pressure etc. Localization techniques are used to determine positions of node. These applications have strict delay requirements as in case of forest fire detection, if temperature value exceeds certain threshold, then this information must be delivered to sink in no time. Energy efficiency is also a critical requirement.

Military Applications. These are the traditional applications of WSN. WSN was earlier developed by US force for military surveillance. Military applications for WSN includes battlefield surveillance, peacekeeping, disaster relief, force protection. Sensors used may consist of light, pressure, sound, radar, acoustic sensors etc. These applications suffers from a lot of challenges that includes security, flexibility, coverage and connectivity.

2 Related Work

Heterogeneity can be described in terms of energy, link and computational resources. Load balancing approach to improve lifetime in WSN is discussed in [8] by incorporating backup nodes that balance load among clusters. Heterogeneity of nodes is considered in terms of energy and high energy nodes are elected as CH and other nodes with comparable energy known as backup nodes are put to sleep mode. Using multi-hop communication, data is transmitted and if cluster head's energy falls below threshold, then backup node is elected as new cluster head. This alleviates the problem of electing new CH among all the nodes. Simply, the node elected as backup will act as a new CH and the old CH now becomes general node. Effectiveness of approach is proved by comparing with LEACH.

A load balancing clustering is discussed [9] that considers gateways that are high energy nodes and network is clustered around these gateways. After discovering the nodes that are in range of gateways communication cost is calculated and exchanged with other gateways. Exclusive nodes i.e. nodes associated with only one gateway and other nodes that communicate with more than one gateway are evaluated based on distance and range. Network load is balanced by balancing number of nodes in a cluster and common energy required by gateway and efficiency of algorithm is proved by comparing it with shortest distance clustering approach. Node Heterogeneity is considered in The Stable Election Protocol (SEP) [10] by providing few nodes with extra energy capability known as advanced nodes and other nodes as normal nodes. Random selection of CH is done based on probabilities according to each node's energy value. Efficiency of protocol is evaluated in terms of stability, network lifetime and throughput.

Heterogeneity in terms of energy is considered in Energy efficient heterogeneous clustered scheme for WSN (EEHC) [11]. EEHC considered three types of nodes i.e. one with highest energy called super nodes, one having lower energy than super nodes called advanced nodes and with normal value of energy called normal nodes. Based on residual energy, cluster head is elected and data is transmitted in multi-hop fashion. Distributed Energy Balance Clustering (DEBC) [12] protocol selects CH based on probability of the ratio between remaining energy of node and average energy of the network. Nodes having high value of energy are selected as CH. Effectiveness of the protocol is evaluated by comparing it with SEP and LEACH and it provides better results in terms of energy consumption and network lifetime.

3 Energy Heterogeneity

Energy Heterogeneity can be achieved by line powering critical nodes in the network that consume more energy. This will increase energy efficiency of the network. In cluster based network, clusters near the sink or clusters that bypass whole traffic should be line powered if possible. This concept of line powering nodes is applicable in indoor applications and in deterministic applications. This work consider normal sensor nodes that connect to their respective CH's in one hop. Cluster head number is chosen to be 24% of total nodes. Energy Heterogeneity is provided only to Cluster heads and number of cluster heads that are provided heterogeneity are varied to find the optimal number of heterogeneous nodes that provides efficient network performance and increases network lifetime

Green color of Heterogeneous nodes relates to green communication that improve energy efficiency of the network. Green Wireless Communications finds innovative solutions to reduce energy consumption and improves system performance while providing required Quality of Service to users. Using green Communication concept in Wireless Sensor Networks can improve the power optimization while maintaining the Quality of Service (Figs. 1, 2, 3, 4 and 5).

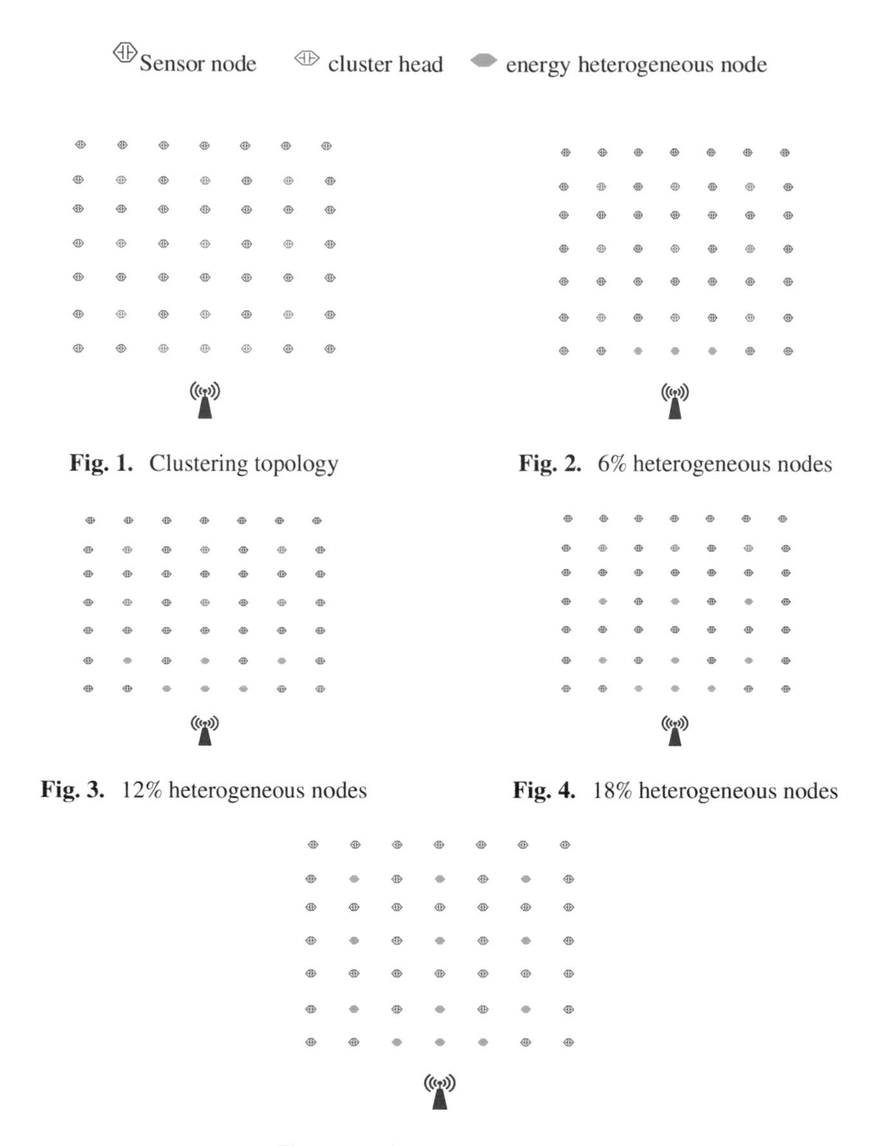

Fig. 1. Clustering topology

Fig. 2. 6% heterogeneous nodes

Fig. 3. 12% heterogeneous nodes

Fig. 4. 18% heterogeneous nodes

Fig. 5. 24% heterogeneous nodes

Simulation is performed using ns-2 Simulator. Network is considered for 50 nodes and BS is placed at downward position. Load balancing is achieved by assigning very few nodes to the CH that are near BS and nodes keep on increasing as we move upwards. 25% of total nodes are considered cluster heads and sensor nodes send data to their respective CH's in one hop and CH's further transmit data to BS in multi-hop fashion. Initial energy of nodes is set to 1 J. Energy Heterogeneity is applied to CH only. Results are evaluated for different percentage of heterogeneous nodes.

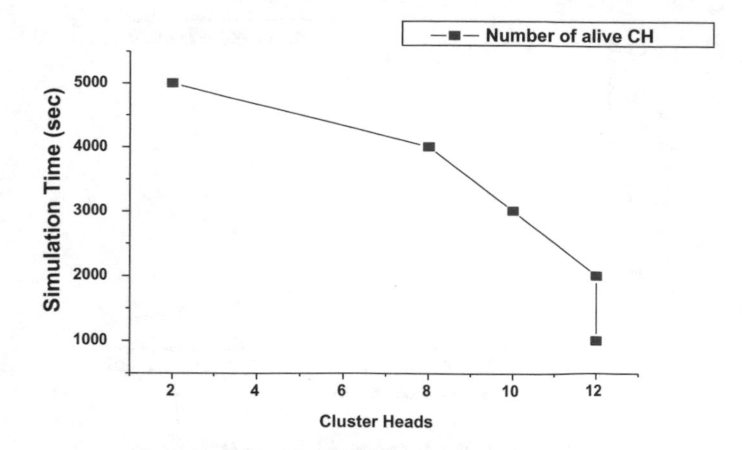

Fig. 6. Number of alive cluster heads

As cluster heads transmit data to Base Station, they drain energy faster than normal sensor nodes. Number of alive cluster heads vs. simulation time is analyzed in Fig. 6. At end of simulation, 2 CH's are alive that are located farther away from base station. So, heterogeneity is applied to prolong lifetime and results are discussed below.

For different percentage of heterogeneous nodes, energy consumption is evaluated and it is found that without heterogeneity more energy is consumed and it keeps on decreasing as we include more heterogeneous node. Sensor node energy consumption is very low compared to cluster head energy consumption, so applying energy heterogeneity to cluster heads increases energy efficiency of the network.

4 Link Heterogeneity

In WSN, Heterogeneity can be provided by varying transmission power levels of different nodes. Transmission power can be adjusted based on packet reception ratio (PRR), link quality that depends upon Received signal strength indication (RSSI) measured in traditional radios and Link Quality Indicator (LQI) found in latest radios. In cluster based networks, as CH are responsible for transmitting data to BS via multi-hop communication. Various methods of adjusting transmission power exist in literature. In [13], Packet Reception Ratio (PRR) is measured by each node at various transmission power levels. Minimum value of PRR is found and if it is greater than threshold, then that value is set as threshold otherwise it is set to maximum transmission power. Transmission power is adjusted based on number of neighbors in [14]. If number of neighbors is less than threshold, transmission power is increased.

In this work, as load balancing is considered based on load of each cluster therefore transmission power is adjusted. Sensor nodes are provided same transmission power as they reach their respective cluster head in one hop. As cluster head near sink has to transmit more data, considering load from upper layer cluster heads. So increasing their transmission power will lead to more energy drainage. CH's in top layer although have more number of sensor nodes but they do not transmit data of other CH's. So their

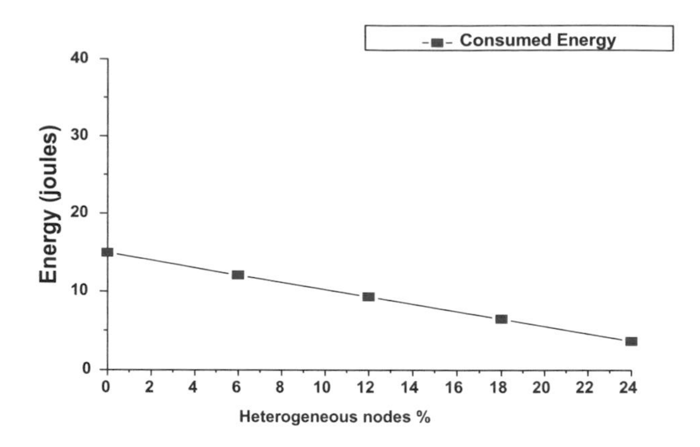

Fig. 7. Energy consumed vs. Heterogeneous nodes

transmission power can be increased. Effect of varying transmission power of different cluster heads in load balanced network is evaluated (Figs. 7 and 8).

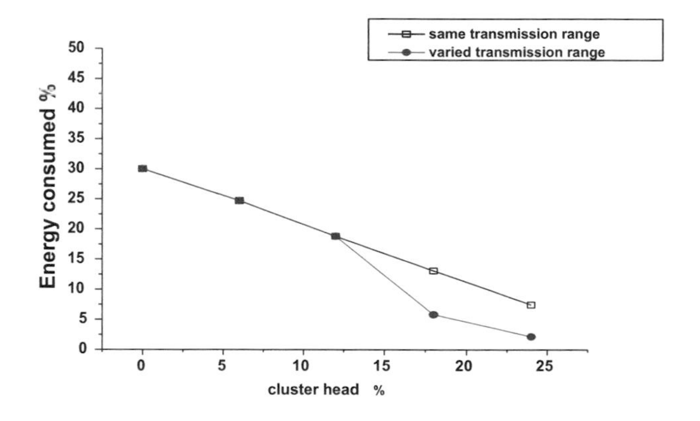

Fig. 8. Energy consumption for link heterogeneity

Above results are analyzed by varying transmission range of cluster heads near the sink. Transmission range is reduced for cluster heads near the sink and remains same for cluster heads that are farther. Decrease in energy consumption is observed for CH's near the sink and it is found from the results that significant energy savings can be achieved by varying transmission range across the network. Nearly 12.58% decrease in energy consumption is observed when transmission range of cluster heads is varied.

5 Conclusion

Heterogeneity plays an important role in improving network performance. Heterogeneity in terms of energy that is provided by line powered nodes can be applied to different deterministic applications. In this work heterogeneity in terms of energy and link is discussed and it is found that significant energy can be saved if we provide the indefinite amount of energy to the most intensive nodes i.e. cluster heads and by varying transmission range of the cluster heads near the sink.

References

1. Minaie, A., Sanati-Mehrizy, A., Sanati-Mehrizy, P., Sanati-Mehrizy, R.: Application of Wireless Sensor Networks in Health Care System. In: 120th ASSE Annual Conference and Exposition. vol. 3, pp. 21–24. Atlanta, Georgia (2013)
2. Jeffrey, J., Patil, R., Kumar K. Narahari, Didagi, Y., Bapat, J., Das, D.: Smart Parking System using Wireless Sensor. In: Sensorcomm The Sixth International Conference on Sensor Technologies and Applications, pp. 306–312. Rome, Italy (2012)
3. Sulaiman, H., Afif, M., Othman, M., Misran, M., Said, M.: Wireless based Smart Parking System using ZigBee. IJET **5**, 3282–3300 (2013)
4. Viani, F., Oliveri, G., Donelli, M., Lizzi, L., Rocca, P., Massa, A.: WSN-based solutions for security and surveillance. In: 40th European Microwave Conference, pp. 1762—1765, Paris (2010)
5. Habitat monitoring on great duck island. http://www.greatduckisland.net/
6. Remote sensor network Accenture prototype helps pickberry vineyard improve crop management. Technical Report (2005). http://www.accenture.com
7. Kates, W.: Robot Sensors May Protect Drinking Water. Associated Press, New York (2004)
8. Gupta, G., Younis M.: Load-balanced clustering of wireless sensor networks. In: ICC 2003, IEEE International Conference, vol. 3, pp. 1848–1852 (2003)
9. Wajgi, D., Thakur, N.V.: Load balancing based approach to improve lifetime of wireless sensor network. J. Wirel. Mobile Netw. (IJWMN) **4**, 155–164 (2012)
10. Smaragdakis, G., Matta, I., Bestavros, A.: SEP: A Stable Election Protocol for clustered heterogeneous wireless sensor networks. SANPA. (2004)
11. Kumar, D., Aseri, T.C., Patel, R.: EEHC: Energy efficient heterogeneous clustered scheme for wireless sensor networks. Comput. Commun. **32**, 662–667 (2009)
12. Duan, C., Fan, H.: A distributed energy balance clustering protocol for heterogeneous wireless sensor networks. In: International Conference on Wireless Communications. Networking and Mobile Computing, pp. 2469–2473, Shanghai (2007)
13. Son, B.K., Heidemann, J.: Experimental Study of the Effects of Transmission Power Control and Blacklisting in Wireless Sensor Networks. IEEE SECON. Santa Clara, California (2004)
14. Jeong, J., Cullar, D., Oh, J.: Empirical Analysis of Transmission Power Control Algorithms for Wireless Sensor Networks. Technical report, No. UCB/EECS-2005–16, University of California at Berkeley (2005)

A Chaff-Point Based Approach for Cancelable Template Generation of Fingerprint Data

Gangotri Patel and Gaurang Panchal(✉)

U & P U. Patel Department of Computer Engineering,
Chandubhai S Patel Institute of Technology, Changa, India
gangotripatel77@gmail.com, gaurangpanchal.ce@charusat.ac.in

Abstract. In the recent biometric community security of biometric information has accepted basic significance. Cancelable biometrics is the one technique to do this, which manage before biometric information putting away in the database transforming the biometric information, in a manner that relative minutiae data is not defiled in the transformed template. This paper shows an investigation of few methods of generating cancelable biometric templates. The security level of current biometric framework is upgraded by the structure of fuzzy vault as far as concealing secret key and ensuring the template, by applying a binding strategy on biometric template and cryptographic key. Apply cyclic redundant code(CRC) to distinguish a real polynomial from an arrangement of competitors in light of its straightforwardness. In CRC based fuzzy vault scheme to overcome issue of blend substitution attack two new module is proposed chaff point verifier along with generator. The systems dispense with a blend substitution assault to enhance general security and, subsequently it can distinguish any alteration in vault.

Keywords: Cancelable template · Template transformation · Biometric cryptosystem · Fingerprint

1 Introduction

As a matter of first important thing is that, what is biometrics, "Biometrics are robotized techniques for perceiving a man in light of a physiological or behavioral trademark". Biometrics are used to distinguish a person on the basis of its characteristic. The main goal of utilizing biometric system is to give non-reputable validation. Verification implies that (I) only legitimate assets secured by the biometric system or legitimate approved clients can get to the physical and (II) pretenders are kept against getting to the ensured assets. A person who gets to specific assets later can't deny to utilizing it that guarantee by non-repudiation [6].

There are began new challenges by expanding utilization of biometrics in various environments. Import thing is that, biometric information is imperative. Accordingly, biometric template database involves critical security dangers

S.C. Satapathy and A. Joshi (eds.), *Information and Communication Technology for Intelligent Systems (ICTIS 2017) - Volume 1*, Smart Innovation, Systems and Technologies 83, DOI 10.1007/978-3-319-63673-3_41

which are remarkable to the individual user. Password which can be re-issued is beneficial but biometrics can't be re-issued. On the off chance that a password or a token is lost or stolen, it can be supplanted by another variant. This thing is not actually accessible in biometric. Fingerprint can't be re-issued if somebody's fingerprint is traded off from the database [13].

Fuses security and substitution highlights into biometrics, cancelable biometrics is only way. Designing a robust one way transformation is one way to deal with managing this issue. Cancelable biometrics implies that it is refers to the intentional and systematically repeatable distortion of biometrics features with a specific end goal to protect sensitive user specific data because about the original biometric information the transformations should not uncover much data. For securely saving biometric authentication, cancelable transforms was one of the original solutions (see Fig. 1). In place of storing the original biometric, using one-way function it is transformed [12].

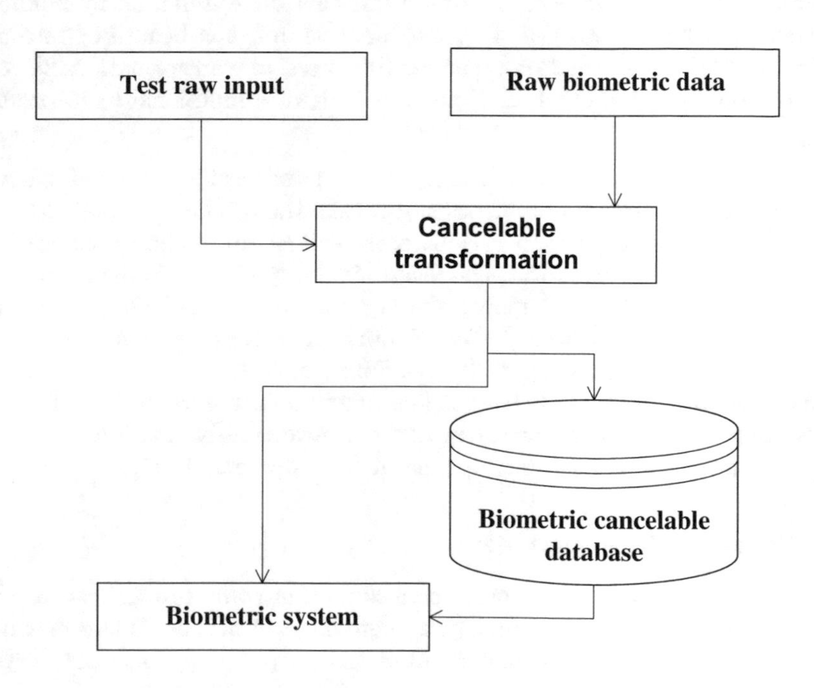

Fig. 1. Block diagram of cancelable biometric system

Because of intra class variation in biometric information a little change in the input raw biometric data, an extensive change in resulted encrypted data, leading to degradation in system prison. While keeping up the template security, to solve the intra-class variation problem proposed to combine two approaches:

1. The transform based approach
2. The biometric cryptosystem approach.

Both approaches have principle thought is that the encrypted/transformed template is stored rather storing the original template, that is more secure.

1.1 Transformation Based Approach

Using "one way" transform Template is created in transformation based approach. Transform presented in signal domain and feature domain transform. High-order polynomials are used in feature domain along with the block permutation and grid morphing is used in signal domain. From the cancelable template, the original template could not be reproduced using these transforms. So proposed another cancelable biometrics on fingerprint in light of three diverse changes to be specific Cartesian transformation, radical transformation, and functional transformation. According to places of singular points, the fingerprint minutiae is enhanced into rectangular space in Cartesian transform. Each cell is moved to another position which contains some minutiae by noninvertible transform [2]. Except the variation that the image is beautified into polar portion, radical transformation is similar to Cartesian transformation [2]. A mix of electric potential field and 2D Gaussians is used in functional transformation [2–15].

Salting: For expanding the class variety key or user specific password is applied in salting. As a source of info, a transform take the original template and user specific key. Diverse keys from various users will isolate a change comes about with the ultimate objective that the discriminability of transformed template is improved. By method for changing the key the transformed templates can be effortlessly reissued and drop [6–10]. Noninvertible transform: A many to one function f is used in non-invertible transform, it is intended to change a raw biometric picture into another shape. The function f serves as an operator which allows template reusability, diversity and non-invertibility. The function f does not need kept to be secret is the up side of this approach [6–15].

1.2 Biometric Cryptosystem

For handling intra-class variations, error-correcting coding techniques are used in this approach. Fuzzy vault and fuzzy commitment scheme is two prominent framework. Security level is high because the output is in form of encrypted template [15–20].

In this approach, from the biometric signal key is acquired and there is no requirement for user specific keys as needed in biometric salting which shows its versatility.

2 Literature Review

We have carried out the literature review [1–30] as follows.

For generating cancelable template from fingerprint Lee et al. proposed another procedure. Without requiring prearrangement of fingerprint, cancelable template can be created by this technique by giving a simple mean. Main goal of this technique is mapping of minutiae into pre-characterized 3D array which consists small cells along with discover cells which include minutiae. For that, one of the minutiae is selected as a kind of perspective minutiae and others are rotated and translated for mapping the minutiae into cells with respect to the position of reference minutiae. After mapping, if cells contain more than one minutiae then a cells are set to 1 otherwise the cells are set to 0. By sequentially visiting a cells 1D bit string is generated in 3D array. By sort of reference minutiae and user's pin the position of 1D bit string can be changed so that we can recreate new templates when we need them. By permuting the reference minutiae into another minutiae, cancelable bit-strings are generated [2].

A pair-polar coordinate based template design technique produced by Ahmad et al. which does not require enrollment. Investigate the relative relationship of minutiae in a shift along with rotation pair polar by this strategy. To assure the recovery of raw templates many-to-one mapping is applied [8].

A technique using prime numbers is proposed by Lalithamani et al. using the user pin as the key large number is created randomly, and afterward every coordinate is translated by this number. At that point using the user pin as the key an arbitrary matrix is created, prime numbers can be discovered in the area of translated coordinates can be discovered by using entries in this matrix. Unique values among sorted distance matrix are selected for generating the template. Further it can be used for generating cryptographic key [9].

Limitation of this methods: Since these strategies was executed on a much smaller database. There are some points which can be analyzed from this chart. On the off chance when the user PIN is distinctive, the execution of the strategy is satisfactory. Be that as it may, when the user PIN is traded off then the execution is disastrous. The value of FAR is too high. So this methods are insufficient for generate cancelable biometric template. So new methods are proposed using fuzzy vault and fuzzy commitment.

Fuzzy Vault: In cryptosystem, challenge is secrecy of cryptographic keys and in biometric system securing the template of user which is stored in database. The development of fuzzy vault scheme is a biometric cryptosystem which protects the biometric template along with the secret key by binding them into cryptographic framework.

Fuzzy vault is an encryption schema which leverage some of the concept of error-correcting code, to encode the information in such a way as to difficult to obtain without the key used to encode it, even if methods used for encoding are publicly known. Fuzzy vault is intended to work with biometric features which represented as an unordered set. Fuzzy vault scheme is best suitable for biometric cryptosystem due to its managing capacity of intra-class variation [3].

Fuzzy vault schema to discover genuine point chaff points are most important. So what is chaff point, it is a noise points which are not on polynomial function and generated randomly to protect genuine point.

3 Proposed System

A blend of cryptographic and biometric systems, the fuzzy vault system improve the security level of current biometric system by performing particular binding method on cryptographic key and biometric template regarding protecting template and hiding secret key. In spite of the fact that to recreate the original polynomial, it is recommended to use error-correction strategy. Rather, to distinguish a real point from the set of candidate CRC is applied. In fuzzy vault system number of points per vault is predefined parameter of encoding/decoding. So any variation on results in critical changes of this number on specific vault will be easily identified. There is Normally perform blend substitution attack on the system. The fundamental target of blend substitution attack is without being detected to bypass the authentication system. While maintaining the current authentication behavior of genuine user, without altering excessive number points the attacker apply fake injection into vault to accomplish this objective. The attacker produce fake key along with comparing CRC to attack CRC based fuzzy vault system. To bypass an authentication system, after randomly remove some vault's point by applying fuzzy vault encoding algorithm on biometric and fake key, generate the fake genuine point and injected into current vault. Propose another approach using mix of CRC and Lagrange interpolation along with chaff point verifier and generator that shown in Fig. 2.

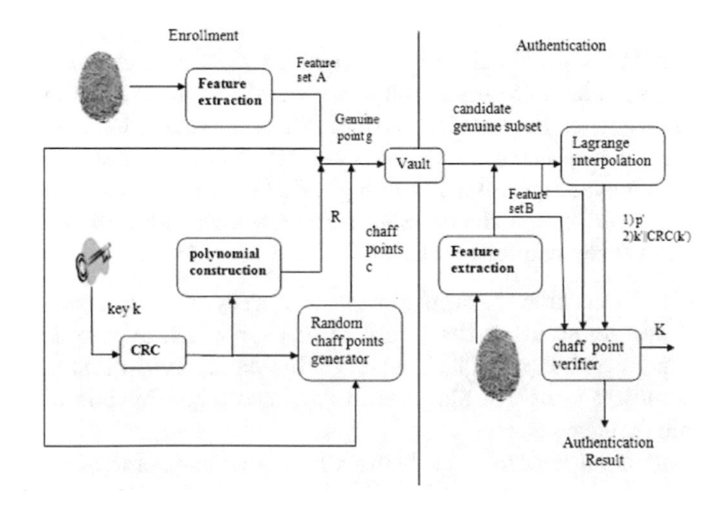

Fig. 2. CRC based fuzzy vault system

These two modules are:

1. Provide a list of capabilities extricated from biometric template associated with secret key as input, after that gives the relating set of chaff points by chaff point generator module.
2. Similar algorithm is applied with a chaff point generator to chaff point verifier module to verify chaff points after recreation of the secret key.

Inside the new system, for genuine points and secret key chaff points can be treated as virtual "CRC". At an authentication phase, by checking chaff point we are able to verify secret key and also distinguish any unapproved alterations to the vault thus we can efficiently prevent blend substitution attack.

4 Conclusion

This paper helped me to get more knowledge on various area related to our study. We came to know about the current research problems concerned with my area of work. In this work, principle commitment is to describe working of proposed CRC based fuzzy vault and the blend substitution attack on this system. As an outcome, ideational two new modules are introduced like chaff point generator and verifier, they can be differentiate any changes in original vault by applying consistent hashing along with linear projection, so through this system we can eliminate blend substitution attack on traditional CRC based fuzzy vault system.

References

1. Ala, A.-F., et al.: Internet of things: a survey on enabling technologies, protocols, and applications. IEEE Comm. Surv. Tutorials **17**(4), 2347–2376 (2015)
2. Karagiannis, V.: A survey on application layer protocols for the internet of things. Trans. IoT Cloud Comput. **3**(1), 11–17 (2015)
3. Luzuriaga, E.J.: A comparative evaluation of AMQP and MQTT protocols over unstable and mobile networks.In: 12th Annual IEEE Consumer Communications and Networking Conference. IEEE (2015)
4. Fernandes, L.J.: Performance evaluation of RESTful web services and AMQP protocol. In: Fifth International Conference on Ubiquitous and Future Networks. IEEE (2013)
5. Rostanski, M., Grochla, K., Seman, A.: Evaluation of highly available and fault-tolerant middleware clustered architectures using RabbitMQ. In: Federated Conference on Computer Science and Information Systems. IEEE (2014)
6. Xiong, X., Fu, J.: Active status certificate publish and subscribe based on AMQP. In: International Conference on Computational and Information Sciences. IEEE (2011)
7. Subramoni, H.: Design and evaluation of benchmarks for financial applications using advanced message queuing protocol (AMQP) over InfiniBand. In: Workshop on High Performance Computational Finance. WHPCF. IEEE (2008)
8. Vinoski, S.: Advanced message queuing protocol. IEEE Internet Comput. **10**(6), 87 (2006)
9. AMQP: Advanced Message Queuing, version 0.8, AMQP working group protocol specification (2006). http://www.iona.com/opensource
10. Programming WireAPI. http://www.openamq.org/
11. Pivotal Software Inc., Messaging that just works. https://www.rabbitmq.com (2014)
12. Ganatra, A., Panchal, G., Kosta, Y., Gajjar, C.: "Initial classification through back propagation in a neural network following optimization through GA to evaluate the fitness of an algorithm". Int. J. Comput. Sci. Inf. Technol. **3**(1), 98–116 (2011)

13. Panchal, G., Ganatra, A., Kosta, Y., Panchal, D.: Forecasting employee retention probability using back propagation neural network algorithm. In: IEEE 2010 Second International Conference on Machine Learning and Computing, Bangalore, India, pp. 248–251 (2010)
14. Panchal, G., Ganatra, A., Shah, P., Panchal, D.: Determination of over-learning and over-fitting problem in back propagation neural network. Int. J. Soft Comput. **2**(2), 40–51 (2011)
15. Panchal, G., Ganatra, A., Kosta, Y., Panchal, D.: Behaviour analysis of multilayer perceptrons with multiple hidden neurons and hidden layers. Int. J. Comput. Theor. Eng. **3**(2), 332–337 (2011)
16. Panchal, G., Panchal, D.: Solving np hard problems using genetic algorithm. Int. J. Comput. Sci. Inf. Technol. **6**(2), 1824–1827 (2015)
17. Panchal, G., Panchal, D.: 11 Efficient attribute evaluation, extraction and selection techniques for data classification. Int. J. Comput. Sci. Inf. Technol. **6**(2), 1828–1831 (2015)
18. Panchal, G., Panchal, D.: Forecasting electrical load for home appliances using genetic algorithm based back propagation neural network. Int. J. Adv. Res. Comput. Eng. Technol. **4**(4), 1503–1506 (2015)
19. Panchal, G., Panchal, D.: Hybridization of genetic algorithm and neural network for optimization problem. Int. J. Adv. Res. Comput. Eng. Technol. **4**(4), 1507–1511 (2015)
20. Panchal, G., Samanta, D.: Comparable features and same cryptography key generation using biometric fingerprint image. In: 2nd IEEE International Conference on Advances in Electrical, Electronics, Information, Communication and Bio-Informatics, pp. 1–6 (2016)
21. Panchal, G., Samanta, D.: Directional area based minutiae selection and cryptographic key generation using biometric fingerprint. In: 1st International Conference on Computational Intelligence and Informatics, pp. 1–8. Springer, New York (2016)
22. Panchal, G., Samanta, D., Barman, S.: Biometric-based cryptography for digital content protection without any key storage. In: Multimedia Tools and Application, pp. 1–18. Springer, New York (2017)
23. Panchal, G., Kosta, Y., Ganatra, A., Panchal, D.: Electrical load forecasting using genetic algorithm based back propagation network. In: 1st International Conference on Data Management, IMT Ghaziabad. MacMillan Publication (2009)
24. Patel, G., Panchal, G.: A chaff-point based approach for cancelable template generation of fingerprint data. In: International Conference on ICT for Intelligent Systems, p. 6 (2017)
25. Patel, J., Panchal, G.: An iot based portable smart meeting space with real-time room occupancy. In: International Conference on Internet of Things for Technological Development, pp. 1–6 (2017)
26. Soni, K., Panchal, G.: Data security in recommendation system using homomorphic encryption. In: International Conference on ICT for Intelligent Systems, pp. 1–6 (2017)
27. Patel, N., Panchal, G.: An approach to analyze data corruption and identify misbehaving server. In: International Conference on ICT for Intelligent Systems, pp. 1–6 (2017)
28. Bhimani, P., Panchal, G.: Message delivery guarantee and status up- date of clients based on IOT-AMQP. In: International Conference on Internet of Things for Technological Development, pp. 1 6 (2017)

29. Mehta, S., Panchal, G.: File distribution preparation with file retrieval and error recovery in cloud environment. In: International Conference on ICT for Intelligent Systems, p. 6 (2017)
30. Kosta, Y., Panchal, D., Panchal, G., Ganatra, A.: Searching most efficient neural network architecture using Akaikes information criterion (AIC). Int. J. Comput. Appl. **1**(5), 41–44 (2010)

Student Feedback Analysis: A Neural Network Approach

K.S. Oza[1(✉)], R.K. Kamat[1], and P.G. Naik[2]

[1] Department of Computer Science, Shivaji University, Kolhapur, Maharashtra, India
{kso_csd, rkk_eln}@unishivaji.ac.in
[2] Department of Computer Studies, CSIBER, Kolhapur, Maharashtra, India
pgnaik@siberindia.edu.in

Abstract. With a specific end goal to excel in teaching-learning the students and the educator must be responsible to each other. The trust built in such an environment will permit the educator to create a conducive teaching-learning environment. A fundamental tool which can be for the above said purpose is the feedback of students on various aspects of teaching-learning. Feedback is ultimate necessity to ensure effective learning. It helps the students to comprehend the subject being examined and gives a clear direction on the most proficient method to enhance their learning. The present paper depicts the analysis of the feedback using Artificial Neural Network (ANN). The feedback of the students in the form of text messages is converted to numerical vectors by considering the positive and negative keywords. Thereafter the ANN is trained using the above said input to predict whether the feedback is positive or negative. We could really enhance student's accomplishment and achievement all the more viably in an effective way utilizing the aforesaid approach. It paper conveys the advantages and effects of this approach to the students, educators and scholarly establishments.

Keywords: Artificial neural network · Text mining · Feedback · Classification

1 Introduction

Feedback both from students to teachers and vice versa, when compelling, is broadly thought to be fundamental to learning. Individuals learn speedier and all the more profoundly in the event that they comprehend what the qualities and shortcomings of their learning are and in particular, how to enhance future performance. Various tools and techniques are used to systematically analyze the aforementioned process of feedback. Constructive feedback can likewise assume a fundamental part in streamlining educating, learning and evaluation. At the point when unequivocally connected to evaluation undertakings, learning results and stamping plans, feedback capacities to make and keep up importance for educators and students alike through a support of the reason for appraisal and how it identifies with learning results.

The present paper explores data mining in specific the technique of text mining, the instruments and methods which once were bound to research laboratories.

S.C. Satapathy and A. Joshi (eds.), *Information and Communication Technology for Intelligent Systems (ICTIS 2017) - Volume 1*, Smart Innovation, Systems and Technologies 83, DOI 10.1007/978-3-319-63673-3_42

The text mining technique in the present work is being embraced by employing ANN for making strides in teaching-learning process. Data mining is one of the most seasoned area of research drawing in analysts on the structured and unstructured data. Here structure of database is settled and can be mined effectively utilizing existing information mining calculations. Content mining a.k.a. text mining is another domain under data mining which manages unstructured information. Here quality data is mined from the content. Here quality information is mined from the text. This text may be a document, a comment, reviews or feedback etc. Text mining has numerous applications in summarizing the documents, document clustering and classification etc. Another application area of text mining is opinion mining and sentiment mining. Here customer's reviews are analyzed to predict their opinion about the particular product or service. The present paper explores the text mining aligned with the classification approach which is one of the supervised techniques in data mining. In this approach class labels are predefined and known. The tool used for classification and also further for performance prediction in this paper is Artificial Neural Network (ANN). It consists of input layer, hidden layer and output layer. Here input is divided in two parts one for training and other part for testing. Using activation functions the ANN is trained to recognize a pattern. Once trained its accuracy is checked using test data set. As portrayed in this paper ANN tool is quite reliable and accurate for classification and prediction for student feedback. The paper presents training of the said ANN tool to classify students feedback for faculty deliberations interms of positive or negative.

The paper is structured as follows: After introducing the concept a brief review of the work done by the leading research groups in this domain is taken. Thereafter the paper put forth the details of dataset preparation followed by application of ANN and the implementation details. The concluding part presents results and comments on the effectiveness of the approach.

2 Literature Review

Feedback of the students on the teaching learning process has been regarded as the most valuable tool in strategizing the teaching learning process. There are many leading research groups [1–5] presently working on different issues related to student feedback. In the light of the research presented in this paper there are quite a few researchers found to strive their best to use technological means for meaningful analysis of the feedback. A gist of the same is presented in this section.

New classifiers for real time data has been developed using ANN for high performance and real time response. Different classifiers like decision trees, RCC classifiers, feature-map etc. have been discussed in detail by [9]. A four layer neural network has been proposed by [7] which can map any input to any number of classifications. It works both on continuous and binary input.

A recurrent convolution neural network has been introduced by [8] for text classification. It captures contextual information with recurrent structure for representation of text. Due to poor performance of student's admissions for higher education are also low. A ANN model has been developed by [9] to predict performance of students for being considered for admission into the university.

Classification of incomplete data is a challenge specially if it's a survey data. A Hopfield based neural network model has been proposed to classify incomplete data. Here incomplete patterns are transformed into fuzzy patterns [10]. Another approach for classifying incomplete data is using multiple-values construction method which is implemented using fuzzy set theory by [11].

There are many techniques for text mining where some are character based similarity techniques and some are term based similarity techniques. Text mining has wide applications in almost all the domains. Many researchers have carried out surveys to study these techniques and applications in detail [12–14]. With the basic literature survey in place we now present our approach of text mining for the student feeback analysis.

3 Dataset Preparation

Data set consist of students feedbacks collected for particular lecture series. After each lecture, tutor feedback is collected from the students which are usually one line text comment. These comments are collected in text file named as feedback.txt. A java program is written to remove stop words, stemming word and to convert all the comments to lowercase in feedback.txt. To identify positive words and negative words two text files consisting of positive words (positive.txt) and negative words (Negative. txt) is created for similarity check and substitution. Here each line of txt file is read into an array and each word in a line is read and stored as separate array item. This item is then matched with the words in positive.txt and negative.txt to find out whether the word is positive or negative. If the word is positive it is replaced with 1 and if it is negative then replaced with −1 otherwise it is replaced with 0 indicating neutral word.

All the words in the line are replaced one by one and updated words in the form of 0, 1, −1 are written back to the feedback.txt. Now feedback.txt is consisting of only 0, 1 and −1 as text.

Once feedback.txt is preprocessed a class label is appended at end of each line indicating it as positive or negative comment. Here class label is categorical with number 1 indicating positive feedback and 0 indicating negative feedback. If the feedback is neutral then also it is classified as positive feedback. This file is used as input for artificial neural network.

4 Artificial Neural Networks

It's a concept based on working of human brain. Human brain learn by experience and this idea is being simulated using ANN. They work as good predictive models consisting of functional entities called neurons which are connected by weights in the network. Figure 1 shows a schematic representation of ANN. This network has 'n' number of inputs. Weight is associated with each input and it changes according to training algorithm. Input function 'η' is product of input and weights associated with inputs and forms the net input. This net input is an input for activation function 'S' to produce output 'Y' i.e., S(η) = Y. Output depends on activation function.

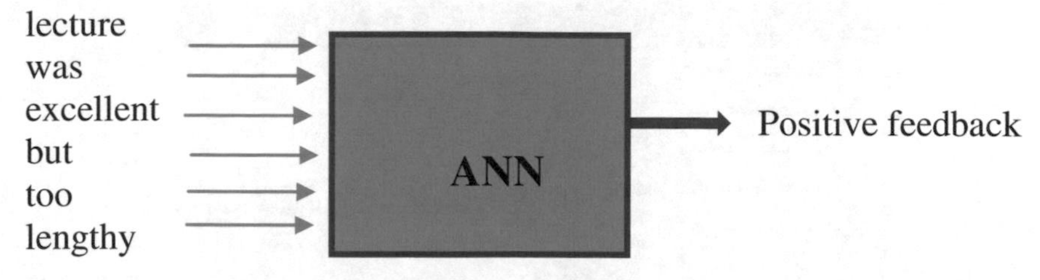

Fig. 1. ANN model for feedback classification

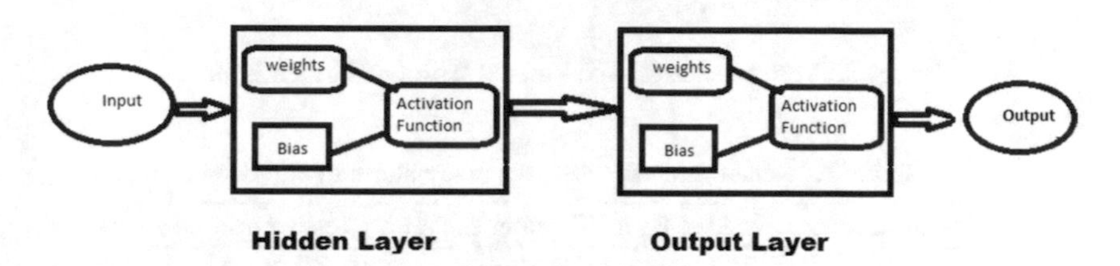

Fig. 2. Feed-forward artificial neural network with hidden layer

Non linear sigmoid activation function is used for present work and ANN is trained using back propagation algorithm. Feed-forward approach is used to train the network as shown in Fig. 2. Network consists of one input layer, one hidden layer and one output layer. Each layer consist of multiple nodes to carry out computation.

5 Implementation Details

Simulation of ANN model is carried out using R studio. This model has multiple inputs and single output as shown in Fig. 1. Inputs to the model are comments stored in the feedback.txt file which is converted to.csv file for easy processing. Feedback.csv file consist of 101 comments. There are 10 inputs to the model and one output. These inputs are positive, negative and neutral words in the comment converted to 1, −1 and 0 after preprocessing. Output is binary indicating positive or negative comment. Here ANN is used for classification which is supervised technique. Dataset which is fed as input to ANN is ready with the class label. ANN will be trained using some part of this dataset and rest of the dataset will be used for testing ANN. ANN is first trained using simple sigmoid function available in R.

Dataset is split in two parts for training (70 records) and testing (31 record). Partition of dataset is done randomly. Figure 3 shows ANN model with one node in hidden layer. ANN model accuracy is shown by confusion matrix. Confusion matrix gives accuracy of classification. Diagonal entries give correctly classified records. In training dataset 2 records which are classified as positive are actually negative feedbacks (false negative). Similarly 6 positive feedbacks are classified as negative (false positive). Accuracy for training is 0.88 and for testing it is 0.87. To get good training dataset partition is fixed and numbers of neurons in the hidden layer were increased. Following table gives the details (Table 1).

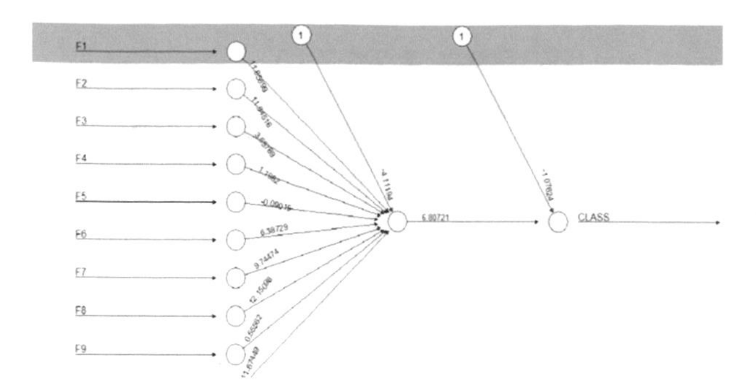

Fig. 3. Performance of ANN model with one hidden layer

Table 1. Relationship between accuracy and hidden nodes

No. of neurons in hidden layer	Training accuracy	Testing accuracy
1	0.8857	0.871
2	0.9	0.9032
3	0.9	0.9032
4	0.9	0.9032
5	0.9	0.9032

With two neurons in the hidden layer accuracy of training and testing is 0.90. This accuracy remains even when the number of neurons is increased to 5 and more. To get more accuracy increase in number of hidden layers and neuron resulted in over fitting of the model.

To get still better classification ANN model was trained with traditional back propagation algorithm available in R. In this model accuracy of testing dataset is increased but training dataset is reduced a bit. Figure 4 shows ANN model with five neurons in hidden layer and back propagation activation function applied.

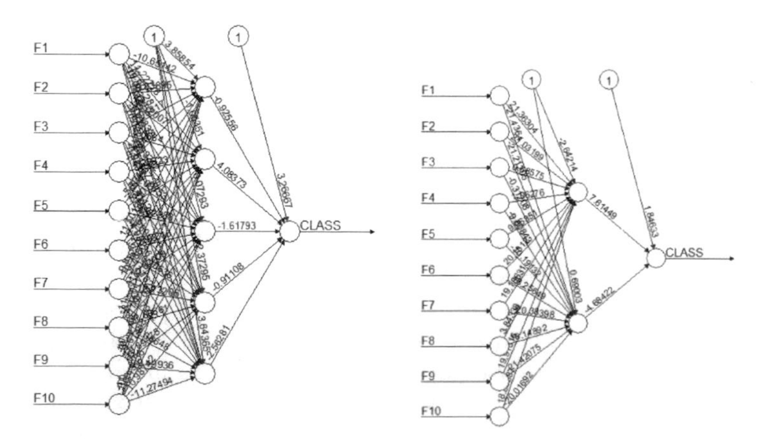

Fig. 4. ANN model with five and two neurons in hidden layer

As the data set is small accuracy of training and testing remains constant with number of neurons from 2 to 20. After 20 it starts over-fitting. Following table shows accuracy as the number of neurons increases in the hidden layer.

ANN model with back-propagation learning algorithm gives testing accuracy as compared to simple sigmoid function model. Experiments were also done with two hidden layers but results were not satisfactory. ANN model trained with back propagation algorithm and with two neurons in hidden layer was accepted for student feedback classification. Confusion matrix shown in the Table 2 validates the classifier

Table 2. Confusion Matrix for testing and training with two neurons in hidden layer

	0	1
0	14	3
1	6	47

	0	1
0	9	1
1	0	21

6 Conclusion

Student feedback are classified using ANN model which is trained using back-propagation learning algorithm and feed forward network. Present ANN model uses very less number of hidden neurons (only two) thus making the model efficient. This model has good number of applications in academic institutes and also in corporate training assessment. The results suggest that ANN model can be used by tutors to improve or update their teaching abilities and also institutes can use it to grade their teachers automatically

References

1. Cleary, M., Happell, B., Lau, S., Mackey, S.: Student feedback on teaching: Some issues for consideration for nurse educators. Int. J. Nurs. Pract. **19**, 62–66 (2013)
2. Van der Kleij, F., Adie, L., Cumming, J.: Using video technology to enable student voice in assessment feedback. Br. J. Educ. Technol. (2016)
3. Zou, D., Lambert, J.: Feedback methods for student voice in the digital age. Br. J. Educ. Technol. (2016)
4. Okumuş K., Yurdakal, I.: Peer Feedback through SNSs (social networking sites): student teachers' views about using Facebook for peer feedback on microteachings. İlköğretim Online **15**(4) (2016)
5. Mayhew, E.: Playback feedback: the impact of screen-captured video feedback on student satisfaction, learning and attainment. Eur. Polit. Sci. (2016)
6. Lippmann, R.: Pattern classification using neural etworks. IEEE Commun. Mag. **27**(11), 47–50 (1989)

7. Specht: Probabilistic neural networks for classification, mapping, or associative memory. In: IEEE International Conference on Neural Networks (1988)
8. Lai, S., Xu, L., Liu, K., Zhao, J.: Recurrent convolutional neural networks for text classification. In: Proceedings of the Twenty-Ninth AAAI Conference on Artificial Intelligence, pp. 2267–2273 (2015)
9. Oladokun, V.O., Adebanjo, A.T., Charles-Owaba, O.E.: Predicting students' academic performance using artificial neural network: a case study of an engineering course. Pac. J. Sci. Technol. **9** (2008)
10. Wang, S.: Classification with incomplete survey data: a Hopfield neural network approach. Comput. Oper. Res. **32**, 2583–2594 (2005)
11. Wang, Hai, Wang, Shouhong: Towards optimal use of incomplete classification data. Comput. Oper. Res. **36**, 1221–1230 (2009)
12. Gupta, V., Lehal, G.S.: A survey of text mining techniques and applications. J. Emerg Technol. Web Intel. **1**, 60–76 (2009)
13. Jusoh, S., Alfawareh, H.M.: Techniques, applications and challenging issue in text mining. IJCSI Int. J. Comput. Sci. Issues **9**(6), 431–436 (2012)
14. Gomaa, W.H., Fahmy, A.A.: A survey of text similarity approaches. Int. J. Comput. Appl. **68(13)** (2013)

Comparative Analysis of Mobile Phishing Detection and Prevention Approaches

Neelam Choudhary(✉) and Ankit Kumar Jain

Computer Engineering Department, National Institute of Technology, Kurukshetra 136119, Haryana, India
neelamchoudhary197@gmail.com, ankitjain@nitkkr.ac.in

Abstract. Mobile phones have taken a crucial part in today's transferable computer world. Mobile devices are more popular these days because of their small screen size, lower production cost, and portability. Because of their popularity, these devices seem to be a perfect target of harmful malicious attacks like mobile phishing. In this attack, the attackers usually send the fake link via emails, SMS message, messenger, WhatsApp, etc. and ask for some credential data. Mobile phishing is fooling the users to get the sensitive personal information. This paper presents a comprehensive analysis of mobile phishing attacks, their exploitation, some of the recent solutions for phishing detection. Our survey provides a good understanding of the mobile phishing problem and currently available solutions with the future scope to deal with mobile phishing attacks conveniently.

Keywords: Phishing attack · Mobile phishing · Android · Smart phone · Mobile malware

1 Introduction

In early days, there was a trend of personal computers, but with the change in time mobile devices replaces personal computers. These days mobile security is a major concern because attackers have diverted their mind from personal computers to cell phones because with the increase in technology people are more attracted towards mobile phones as it is the small and multi-functioning device, so attackers found more interest in hacking these devices. Mobile phones seem to be an ideal target for attackers [1]. There are various risks associated with the mobile device out of which phishing is more widespread. Phishing is an online identity theft. The aim of phishing attack is to steal sensitive personal information like username, password, and credit/debit card details by fooling the user to visit fake links, apps or webpages [2, 3].

Mobile phishing is an emerging threat in which malicious person sends an SMS message to the user, and that SMS contains links to malicious applications and WebPages [4] and which if visited may ask the user for sensitive information like username, password, credit/debit card details. In 2012, 4000 Phishing URLs were found by Researchers from Trend Micro, and these URLs were specially designed for Mobile WebPages [5]. Moreover, in 2012, a survey had reported 267, 259 malicious apps out of which 254, 158 are on the Android platform. With the increase in the use of

S.C. Satapathy and A. Joshi (eds.), *Information and Communication Technology for Intelligent Systems (ICTIS 2017) - Volume 1*, Smart Innovation, Systems and Technologies 83, DOI 10.1007/978-3-319-63673-3_43

smartphones, the number of malware also has risen by 614% [6]. Mobile malware is identified by their malicious attack behavior, remote control, and procreation behavior. The malicious behavior indicates the way by which malware attacks the mobile devices after infecting user's device. The remote control behavior indicates that how remote server is used to exploit the infected mobile device [7]. The procreation behavior indicates that how malware is transmitted from one victim to another. Through malware, it is possible to launch different types of attacks. Figure 1 shows the attacks initiated by mobile malware.

Fig. 1. Types of mobile malware

In 2014, a report found that for different mobile platforms, mobile phishing attacks have been increasing day by day [8]. It was reported that in the Android's Google Play store, malicious apps have increased by approximately 388% in between the year 2011 and 2013 [9]. Recently in 2015, Kaspersky Lab had detected 7030 mobile phishing Trojans, 884,774 malicious mobile programs and 2,961,727 malicious installation packages [10]. In this paper, we have discussed different kinds of Mobile Phishing Attack Methods and Mobile Phishing Defense mechanisms. Moreover, we discuss various research issues and challenges that need further research work.

The rest of the paper is organized as follows. Section 2 presents the classification of mobile phishing attack. Section 3 presents the taxonomy of various types mobile phishing detection and filtering techniques. Section 4 presents the open issues and challenge in phishing detection and protection. Finally, Sect. 5 concludes the paper.

2 Classification of Mobile Phishing Attacks

The attacker performs the phishing attack by utilizing the technical subterfuge and social engineering techniques. In social-engineering techniques, attackers carry out this attack by sending bogus e-mail. Attackers often convince recipients to respond using names of banks, credit card companies, e-retailers, etc. Technical subterfuge strategies install malware into user's system to steal credentials directly. Figure 2 shows different types of mobile phishing attacks. These attacks are the way by which attackers are attacking mobile devices these days [11].

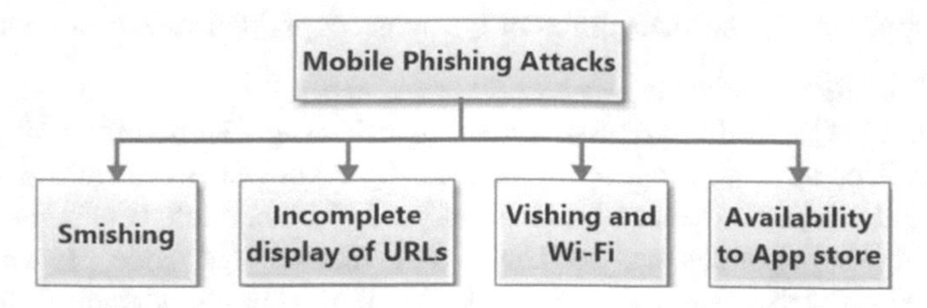

Fig. 2. Mobile phishing attack types

The different types of mobile phishing attacks are explained below in detail.

- *Smishing* - Phishing done by sending SMS messages is known as Smishing [12]. In this user receives an SMS message that contains fake links to some malicious WebPages and asks for sensitive information like username, password and banking credentials.
- *Incomplete display of URLs* - Mobile devices are smaller in size and have a smaller screen due to which it is not possible to see the complete URL when the user visits some website, and it becomes difficult to identify whether he is on the official website or some fake website.
- *Vishing and Wi-Fi* - Vishing is a voicemail phishing to get user's sensitive information [13]. Mobile phishing can also be carried out when users use public Wi-Fi hotspots to connect to the internet. Usually public Wi-Fi hotspots are not secure and by connection to these hotspots user might loss some important information and maybe a victim of phishing attacks.
- *Availability to the app store* - Phishing through installing malicious apps on mobile devices is known as application phishing. In this user installs some app from play store and he is not aware that whether this app is genuine or fake because that seems to be genuine but in reality, it starts stealing information from user's mobile device.

3 Taxonomy of Mobile Phishing Detection Approaches

In this section, we present currently available mobile phishing detection approaches. Phishing detection schemes are broadly classified in five ways. Figure 3 shows different kinds of mobile phishing detection and prevention approaches available.

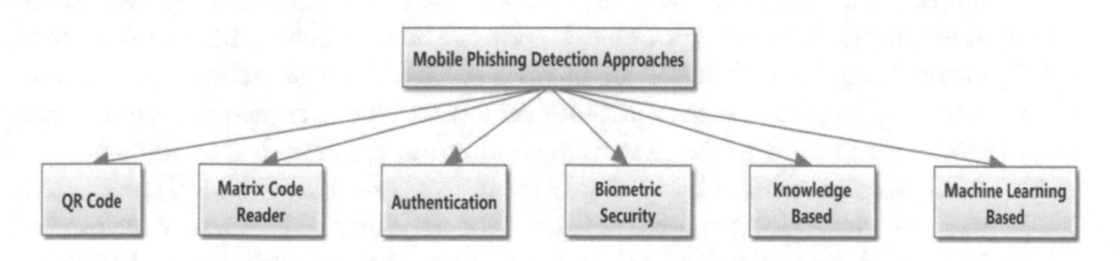

Fig. 3. Types of mobile phishing detection approaches

These approaches and their different types are explained below in detail.

(a) *QR Code Based Schemes*

(1) Simple QR Code - To protect mobile devices against harmful phishing attacks Choi et al. [14] proposed a secure single sign-on model using QR (Quick Response) code. QR code is basically a 2-D barcode that contains information in both directions i.e. horizontal and vertical. QR code prevents the user against web-phishing. In this scheme, if the user's personal information is exposed to others then also attacker can't obtain the mobile information as the user information in encrypted by the mobile device and hacker can't get access to user information only through mobile information.

(2) Mobile OTP with QR Code - Lee et al. [15] proposed a secure authentication system for online banking. This system uses both QR code and Mobile OTP together. In this authentication system a QR code is being generated by the bank as per the details provided by the user. The user reads that QR code on his/her mobile device. After that an OTP code is being generated with the help of user's entered details and hashed mobile serial number. Thus at the end to complete the whole transaction process, user enters the OTP code.

(b) *Matrix Code Reader Based Scheme*

Tanaka et al. [16] presented a user authentication method. In this method to prove the identity of the individual, there is no need to type user login credentials like username, email id, card details, and password. In this method, a matrix code is being used. The user reads the information on his/her mobile device with the help of matrix code reader. And after reading the information, it converts and transmits to the creator via a trusted network. This method is named as SUAN. SUAN is abbreviated as Secure User Authentication with Nijigen code. SUAN is supposed to be a secure user authentication as for SUAN there is no need to install any additional software.

(c) *Authentication Based Schemes*

(1) Simple Authentication Based - Jonvik et al. [17] presented a simple, secure, strong and user friendly authentication mechanism. It doesn't require any extra cost. The basic idea behind this scheme is to use both mobile phones and it's SIM to prove the identity user. The proposed method combines different kinds of authentication schemes like GBA, ISIM, and strong SIM authentication method. Here GBA stands for Generic Bootstrapping Architecture and ISIM stands for IP Multimedia service identity module.

(2) Online Banking Authentication - Online Banking plays a vital role in the online internet banking industry. Fang et al. [18] presented a secure, strong authentication method for online banking. The idea behind this mechanism is to store digital certificates of clients. It provides transaction confirmation functionality. The reason behind providing this transaction confirmation facility is to prevent the user from signing the counterfeit transactions.

(3) Anti-phishing scheme by storing LUI information - Han et al. [19] proposed a method where mobile phone pre-stores the attributes of LUI and at the time of login, a plug-in of web portal will first verify the authentication of LUI. This verification is done by already stored LUI information. Though this method

protects the authentication information but still it is not that much efficient that it protects the authentication information and LUI stored at smartphones.

(d) *Biometric Security Based Scheme*
Gordon et al. [20] presented a biometric mechanism-fingerprint. This mechanism provides a high level of security and privacy for online mobile payment. In this method user's finger print impression is captured on the mobile device. Then this impression is compared to the already stored finger print impression on a database server. This complete process of capturing and monitoring finger print impression is done at run time only.

(e) *Knowledge Based Schemes*

(1) Mobile Game - Asanka et al. [21] proposed a mobile game which is presented on Google App Inventor Emulator. The main motto behind this game design is to provide conceptual knowledge to home computer users behind the phishing attacks. This conceptual knowledge may help the users in avoiding phishing attacks. As home computer users are not aware of the phishing attacks and they have fewer security sources as compared to IT organizations.

(2) MobiFish - Wu et al. [22] proposed an anti-phishing scheme called as MobiFish which is able to protect mobile apps and web pages against the mobile phishing. This scheme is being implemented on a Nexus 4 mobile phone which runs on Android 4.2 OS. This anti-phishing scheme is being tested on 100 phishing URLs and legitimate URLs for these phishing URLs. MobiFish uses the OCR method to detect and prevent against mobile phishing attacks. In OCR technique text is being extracted from the screenshot of the login page. This screenshot is used to detect phishing in both apps and WebPages. The tool used for this method is Tesseract which considers being most efficient open source OCR engine. Tesseract extracts text from the images or the screenshot taken on the mobile phones. MobiFish consists of two major components. For Mobile web pages it consists of WebFish and for Mobile applications, it consists of AppFish. MobiFish seems to be lightweight as in this scheme there is no need of any machine learning algorithm, and it doesn't use any external search engines.
The comparative analysis of various anti-phishing approaches is shown in Table 1.

(f) *Machine Learning Based Schemes*

(1) KAYO - Amrutkar et al. [23] proposed a mechanism named KAYO which is being designed and implemented. KAYO is a static analysis technique. KAYO differentiates between the malicious and genuine mobile WebPages. It detects mobile malicious pages by measuring 44 mobile features from WebPages. Out of these 44 features, 11 are newly identified mobile specific features. KAYO's 44 feature set is divided into four classes namely HTML, mobile specific, URL and JavaScript features. Kayo detects mobile suspicious and malicious pages that are not detected by existing techniques like Google safe browsing and virus total. It provides real-time feedback to users. KAYO is considered to be the first technique to identify mobile malicious WebPages. KAYO provides 89% true positive rate and 90% accuracy.

Table 1. Comparative analysis of various anti-phishing approaches.

Authors	Approach	Advantages	Limitations
Choi et al. [14]	QR code	Scalable on the communication channel between computer and mobile devices	High cost, not efficient
Tanak et al. [16]	Matrix Code Reader	No need to install additional software on user terminals	No use of public key encryption method
Han et al. [19]	Using LUI information	Simple and efficient method to protect authentication information	Not suitable to protect information when stored at mobile device
Gordon et al. [20]	Biometric mechanism	Provides high level of security	Processed at run time only
Asanka et al. [21]	Mobile Game	Provides conceptual knowledge behind mobile phishing attacks	Not suitable to detect zero-hour phishing attack
Wu et al. [22]	MobiFish	Uses OCR to extract text from the screenshot captured	Not suitable for Windows and Ios

(2) MP Shield - Bottazi et al. [24] proposed a framework or method for detection of phishing in Android smartphones. MP Shield is an Android application which is being implemented on the top level of TCP/IP stack as a proxy service. The main aim of MP Shield is to inspect IP packets originated from the mobile application with the purpose of phishing. MP-Shield uses VPN service to inspect these IP packets. VPN service runs without root privileges and is a TCP/IP layer proxy that is provided by the Android libraries. MP Shield provides a high level of security without even disturbing users.
The comparative analysis of existing machine learning based anti-phishing schemes is shown in Table 2.

Table 2. Comparison between existing machine learning algorithm based anti-phishing approaches.

Authors	No. of feature	Algorithm	Framework	TP rate (%)
Amrutkar et al. [23]	44	Logistic regression	Scrapy	89
Bottazi et al. [24]	53	J48, SMO, BayesNet, IBk, SGD	WEKA	89.2

4 Research Issues and Challenges

Phishing is one of the major threats in today's connected world for mobile users. It is visualized that in the upcoming years malware attacks on mobile devices will be shared between the smartphones and the cloud. Major challenges in mobile phishing are spear phishing, vishing, SMS based phishing i.e. smishing and standard web based attacks. There are various tools and techniques available to prevent from mobile phishing attacks but as the behavior of malware is changing rapidly. Therefore, it becomes difficult to avoid mobile devices from the eye of attackers. So some new techniques and tools need to be developed.

5 Conclusion

These days mobile phishing seems to be one of the most profitable and popular attacks. Mobile phishing can be easily carried out by attackers because mobile devices are smaller in size and have the small screen so it becomes difficult for the user to view the complete URL and he may click on the links without complete information which results in possible phishing attacks. Apart from this users also download and install applications for which he is not sure whether it is fake or legitimate one. This paper discusses different types of mobile phishing attacks and their defense mechanisms available. A comparative table is prepared for easy glancing at the advantages and drawbacks of the available approaches. Yet no single technique is enough for adopting it for Phishing detection purposes. Detecting the phishing website in the mobile environment is still an open challenge for further research and development.

References

1. La Polla, M., Martinelli F., Sgandurra, D.: A survey on security for mobile Devices. In: IEEE Communications Surveys & Tutorials, pp. 446–471 (2013)
2. Jain, A.K., Gupta, B.B.: Phishing detection: analysis of visual similarity based approaches. Secur. Commun. Netw. (2017). Article ID 5421046. doi:10.1155/2017/5421046
3. Jain, A.K., Gupta, B.B.: A novel approach to protect against phishing attacks at client side using auto-updated white-list. EURASIP J. Inf. Secur., 1–11 (2016)
4. Mobile Phishing definition. https://capec.mitre.org/data/definitions/164.html
5. Trendlabs Security Intelligence Blog. http://blog.trendmicro.com/trendlabs-security-intelligence/when-phishing-goes-mobile/
6. Juniper report. http://venturebeat.com/2013/06/26/254158-android-apps-are-malicious-as-malware-skyrockets-614/
7. Guo, D.F., Sui, A.F, Guo, T.: A behavior analysis based mobile malware defense system. In: 6th International Conference on Signal Processing and Communication Systems (ICSPCS), pp. 1–6. IEEE (2012)
8. Ashford, W. http://www.computerweekly.com/news/2240215873/Phishing-attacks-track-mobile-adoption-research-shows

9. The CyberSecurity Source. http://www.scmagazine.com/report-malicious-apps-in-google-play-store-grow-388-percent/article/3349611
10. Kaspersky Lab Mobile malware evolution 2015. https://securelist.com/analysis/kaspersky-security-bulletin/73839/mobile-malware-evolution-2015/
11. Shahriar, H., Klintic, T., Clincy, V.: Mobile phishing attacks and mitigation techniques. J. Inf. Secur. **6**, 206–212 (2015)
12. Wilson. http://www.zcorum.com/smishing-yes-its-all-bad/
13. Foozy, C.F.M., Ahmad, R., Abdollah, M.F.: Phishing detection taxonomy for mobile device. J. Comput. Sci. **10**, 338–344 (2013)
14. Choi, K., Lee, C., Jeon, W., Lee, K., Won, D.: A mobile based anti-phishing authentication scheme using QR code. In: International Conference on Mobile IT Convergence, pp. 109–113. IEEE (2011)
15. Lee, Y.S., Kim, N.H., Lim, H., Jo, H., Lee, H.J.: Online banking authentication system using mobile-OTP with QR-code. In: 5th International Conference on Computer Sciences and Convergence Information Technology (ICCIT), pp. 644–648 (2010)
16. Tanaka, M., Teshigawara, Y.: A method and its usability for user authentication by utilizing a matrix code reader on mobile phones. In: International Workshop on Information Security Applications, pp. 225–236. Springer, Heidelberg (2006)
17. Jonvik, T., Boning, F., Jorstad, I.: Simple strong authentication for internet applications using mobile phone. In: IEEE Global Telecommunications Conference, pp. 1–5 (2008)
18. Fang, X., Zhan, J.: Online banking authentication using mobile phones. In: 5th International Conference on Future Information Technology. IEEE (2010)
19. Han, W., Wang, Y., Cao, Y., Zhou, J., Wang, L.: Anti-phishing by smart mobile device. In: IFIP International Conference on Network and Parallel Computing Workshops, pp. 295–300. IEEE (2007)
20. Gordon, M., Sankaranarayanan, S.: Biometric security mechanism in Mobile payments. In: 7th International Conference on Wireless and Optical Communications Networks (WOCN). IEEE (2010)
21. Asanka, N., Love, S., Scott, M.: Designing a mobile game to teach conceptual knowledge of avoiding “phishing attacks”. J. e-Learn. Secur. **2**, 127–132 (2012)
22. Wu, L., Du, X., Wu, J.: MobiFish: a lightweight anti-phishing scheme for mobile phones. In: 23rd International Conference on Computer Communication and Networks (ICCCN), pp. 1–8. IEEE (2014)
23. Amrutkar, C., Kim, Y.S., Traynor, P.: Detecting mobile malicious webpages in real time. IEEE Trans. Mobile Comput. (2016)
24. Bottazzi, G., Casalicchio, E., Cingolani, D., Marturana, F., Piu, M.: MP-shield: a framework for phishing detection in mobile devices. In: IEEE International Conference on Computer and Information Technology; Ubiquitous Computing and Communications; Dependable, Autonomic and Secure Computing; Pervasive Intelligence and Computing (CIT/IUCC/DASC/PICOM), pp. 1977–1983. IEEE (2015)

Implementation of Video Error Concealment Using Block Matching Algorithm

P.K. Rajani[1(✉)], Arti Khaparde[2], and Aditi D. Ghuge[1]

[1] Pimpri Chinchwad College of Engineering, Pune, India
rajani_ranjith@yahoo.co.in, ghugeadit@gmail.com
[2] Maharashtra Institute of Technology, Pune, India
artikhaparde@gmail.com

Abstract. Video Error Concealment is the error hiding technique in videos. In recent years, there is huge requirement of error concealment in video applications such as in video streaming, entertainment, advertisement, media, security, etc. The simulation on MATLAB for the error videos using Block matching algorithm (BMA) has been performed to achieve the concealed videos. From an error video, error frame is detected using Histogram and correlation. This frame is corrected using BMA. First step of BMA is to divide the current frame of a video into macroblock. Second step is to compare each of the macroblocks with a corresponding block and its adjacent neighbors in the frame or previous frame. Third step is to models the movement in a macroblock from one location to another. Last step is to calculate this movement for all the macro blocks that is comprising a frame. This error block is replaced by correct reference block. The quality of the error video and concealed video is measured using PSNR (Peak Signal to Noise Ratio) and SSIM (Structural Similarity Index Method). An improvement in quality is observed in concealed video.

Keywords: Block matching algorithm · Error concealment · SSIM · PSNR

1 Introduction

Receiving videos with least error has become one of the main challenges due to the limited bandwidth of communication channels, video data are compressed, which causes them to become near to error. For estimating the damaged Motion Vector (MV), conventional classic Boundary Matching Algorithm uses a candidate MV with maximum smoothness on the boundaries of reconstructed MB. The Outer Boundary Matching Algorithm (OBMA) with assumption of spatial and temporal prediction estimates damaged MVs. The Directional Temporal Boundary Matching Algorithm (DTBMA) adopts the existing neighboring directional difference of reconstructed MB. In recent years, many switching, and adaptive algorithms based on conventional boundary matching criterions have been proposed for TEC (Temporal Error Concealment) of video frames [3, 4]. To concentrate on the design of the TEC algorithm, after obtaining the coordinates of damaged MB, the spatial neighboring MVs in the current frame, and the temporal neighboring MV from the previous frame are extracted.

S.C. Satapathy and A. Joshi (eds.), *Information and Communication Technology for Intelligent Systems (ICTIS 2017) - Volume 1*, Smart Innovation, Systems and Technologies 83, DOI 10.1007/978-3-319-63673-3_44

1.1 Concept of Video Error Concealment

Error concealment (EC) is a very famous and useful technique at the decoder side to hide the errors. It can be achieved by exploiting the temporal or spatial information from available images or video frames. EC is used to ensure the quality of videos through the noisy environmental conditions. The main part in error concealment is to replace missing parts in video frame and or video content by decoded parts of the frame or video sequence. For eliminating all or major part of error is the one of the major challenges in video processing. The error concealment deals with the spatial and temporal correlations between the neighboring frame parts as macro blocks within the same frame or is compare with the past and current frames. Error concealment is a technique which can conceal the distorted image or video [5]. It is very important to recover distorted image or video, because they are used for various applications such as video-telephone, video-conference, TV, DVD, internet video streaming, etc. These methods works on the spatial and temporal domain of the received data to conceal the error. EC methods have no delay and no redundant data when compared with retransmission-based and resilient-based methods. It has attracted lots of research interest due to its application in various fields such as multimedia, skyping, online video gaming, video chatting etc. [1, 2] (Fig. 1).

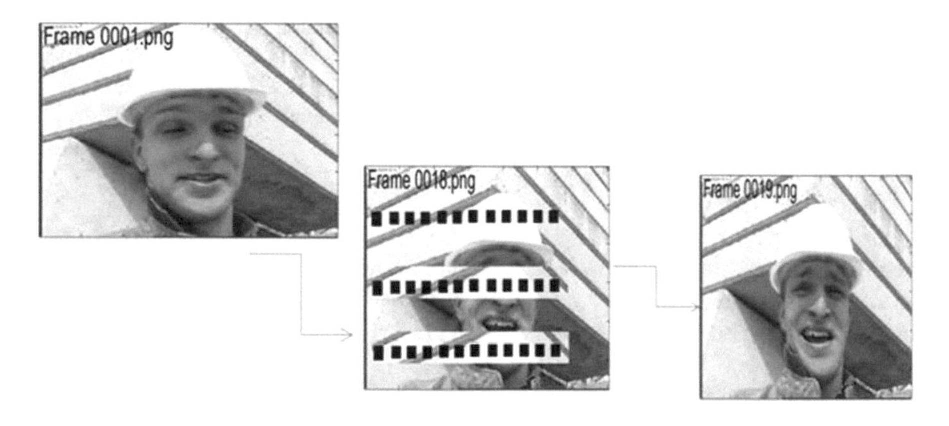

Fig. 1. Video error concealment

2 Process Model

The first step of Block matching algorithm is to divide the current frame of a video into a matrix macroblock. Second step is to compare each of the macroblocks with a corresponding block and its adjacent neighbors in the frame previous frame. Third step is to models the movement in a macroblock from one location to another. Last step is to

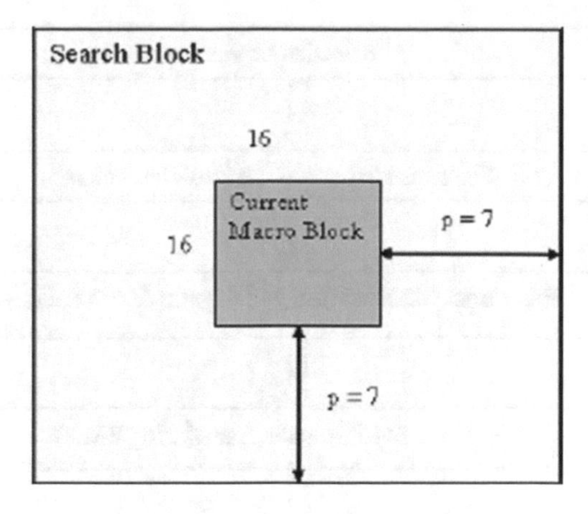

Fig. 2. BMA search parameter 7 pixel [10]

calculate this movement for all the macro blocks that is comprising a frame. The search parameter 'p' is the number of pixels on all four sides of macroblocks, which is a measure of motion. The value of p should be more for getting the larger potential motion. A full search of all potential blocks is possible which is computationally expensive task. Here the search area of p = 7 pixels and input macroblock size is 16 pixels [8]. Here 16 pixels by 16 pixels macro block is taken with a search parameter p of 7 pixel. The idea is represented in Fig. 2. There is matching of one macro block with another. The least cost one is the macro block that matches the closest to current block [6, 7] (Figs. 3 and 4).

In BMA method, error concealment is achieved in step-wise error removal method. First stage, error video of Foreman is taken as input. In this video, error is introduced at one of the frames. In the second stage, error is detected from error foreman video, by comparing the frames using Histogram matching technique. Histogram will show the difference between the error frames and error-free frames. For statistical error detection Correlation Techniques is used. This gives the correlation value difference between error frame and error-free frames. In this way error frame is recognized. Next task is error concealment of error frame using BMA algorithm and FFT

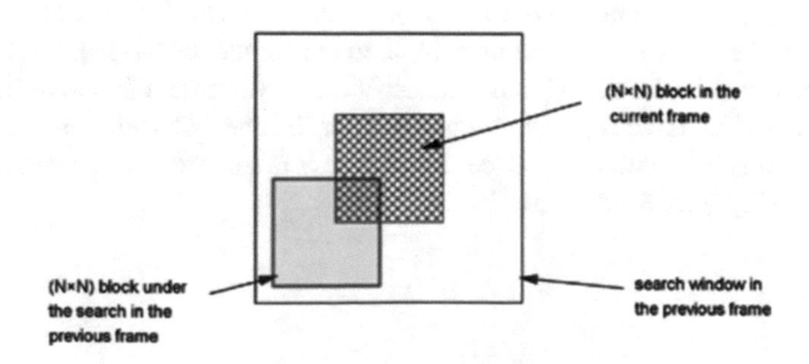

Fig. 3. Block matching illustration [10]

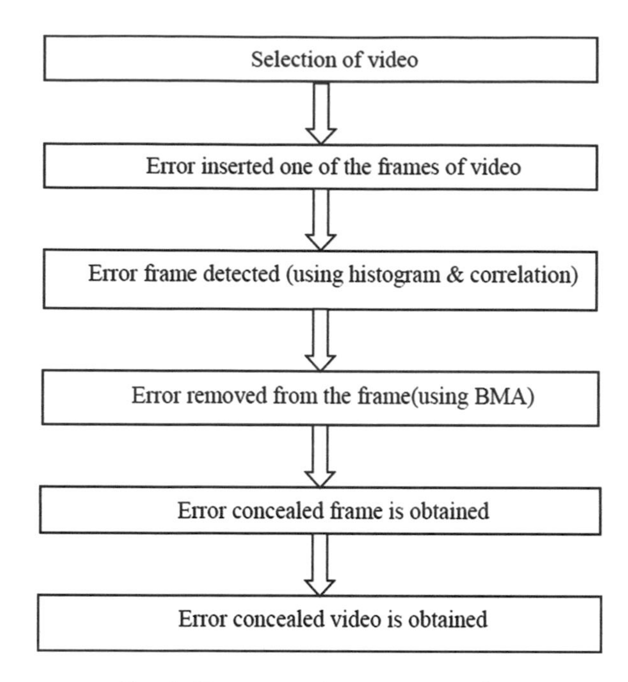

Fig. 4. Error concealment algorithm flow

(Fast Fourier Transformation). Filtering is done here and error removal is achieved. Final stage the error concealed video is displayed. This method is shown in algorithm below. Characterized motion compensated video frame is created by using motion vectors and macro blocks from the reference frame.

Foreman video is selected for the experiment which contains total 30 frames. In frame 18 of that video, error is introduced. That error should be detected and concealed, which is the objective of this work. Error is detected using Histogram and correlation. Histograms gives the idea of the density of distribution of the data and many a times for probability density estimation itself. Histogram is the distribution of numerical data in a graphical way. It is an estimation of the continuous variable probability distribution. Histogram can be created by dividing the entire range of values into a series of intervals. Then counting how many values fall into each interval. The intervals must be adjacent. They are many a times equal in size. If the intervals are of equal size, then the rectangle is erected over the bin with height proportional to the frequency [8–10]. Correlation is a statistical relationships involving dependence. It is many a time refers to as the extent to which two variables have a linear relation with each other. Correlation is the class of statistical relationships and dependence. But its use is often it refers two variables have a linear relationship with each other [3].

3 Experimental Results

3.1 Error Detection Techniques

Histogram matching is used for the detection of error. Simulation is done in MATLAB to detect error in adjacent block by using histogram matching technique and correlation function. First error frame is detected from the error video, by using histogram matching algorithm. By using graphical presentation histogram, it can be seen that there is a difference in error frame (current frame) and its previous frame.

A correlation is a function which can give the statistical correlation between random variables on the spatial or temporal domains. Correlation is done to detect exact value of changes in two frames. The difference is detected as error in frame. The results of above graph are displayed in values by using correlation and the error is detected. In the error video of foreman, there is difference in correlation value of frame 18 in comparison with nearby frames. This results in detecting frame 18 is error frame. Thus error frame is detected, as shown in the above figure (Fig. 5).

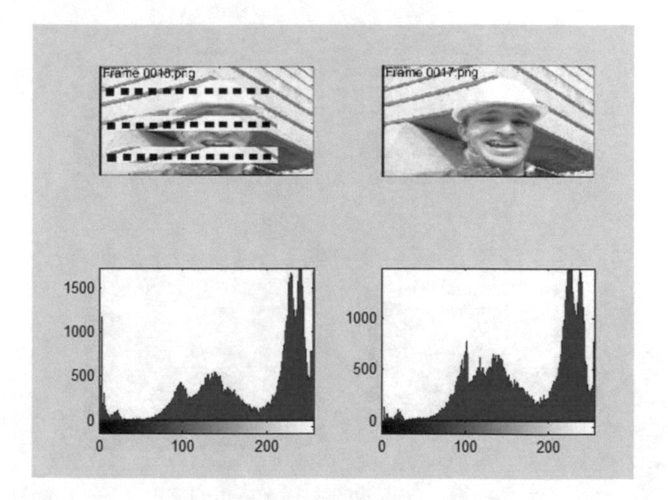

Fig. 5. Histogram result for error frame for foreman video

Once the error frame is detected, using block matching algorithm the error is removed. Here FFT is used for error removal. Figure 6 shows four subplots. First subplot shows the error frame. In second and third subplot shows block matching algorithm implementation. In last stage, error concealed image is obtained by using FFT in BMA.

In Fig. 7, the first subplot shows the error video, which is the input video. Second subplot shows, the error frame which is detected. Third is the concealed frame using BMA. Fourth subplot shows, the error concealed video as the final output.

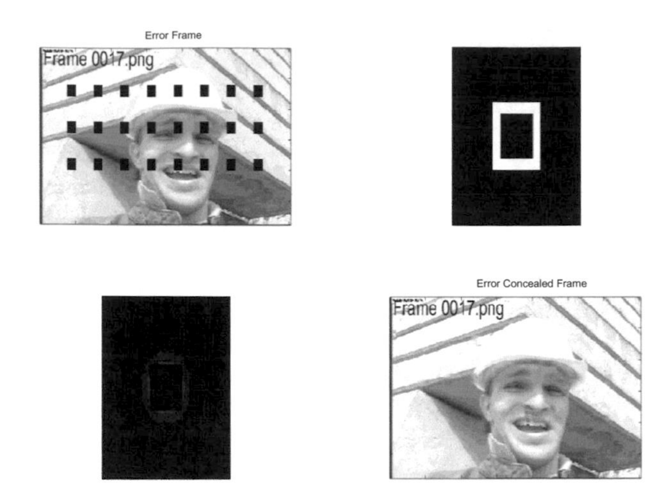

Fig. 6. Error removal from error frame for foreman video

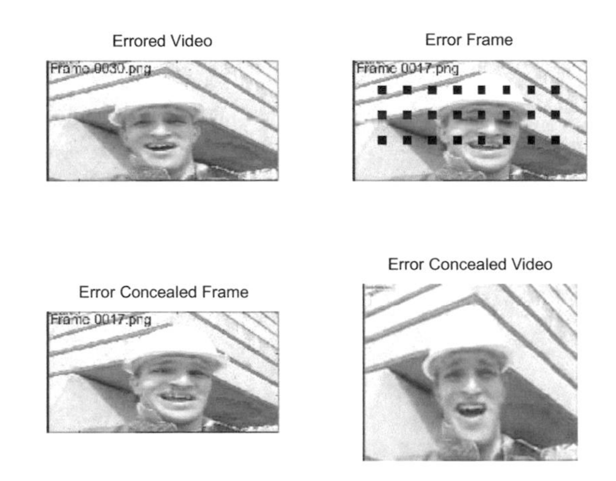

Fig. 7. Error concealed video output

3.2 Quality Analysis

Results of error and error concealed video frames are represented using SSIM and PSNR parameters. The result shows that the value of SSIM for original frame is 1. Results for error frame is 0.8256, and that of error concealed frame is 0.9042. So it shows that error is concealed up to a big extent. Result using PSNR is for error frame is 33.26 dB and for error concealed frame it is 23.79 dB. PSNR value reduced for error concealed frame. The figures below represent these results of PSNR and SSIM values. The table shows the different values for error frame and error concealed frame considering SSIM and PSNR parameters for foreman video. From that it can be concluded that, error is concealed and quality is improved from error video to concealed video (Figs. 8 and 9).

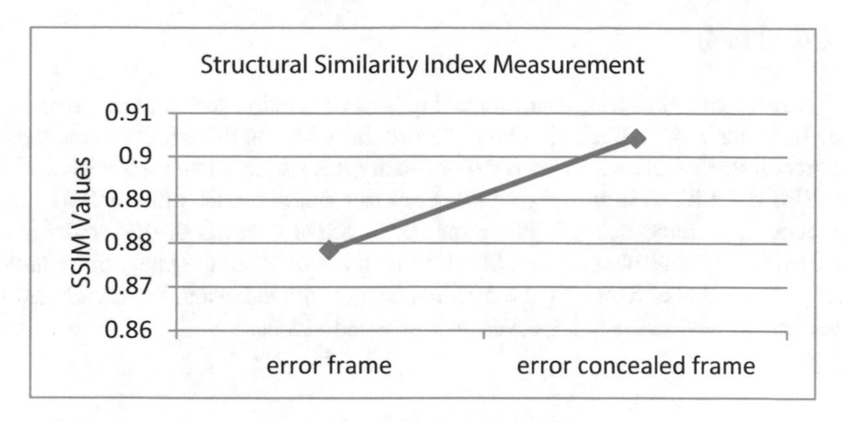

Fig. 8. Graph for SSIM values

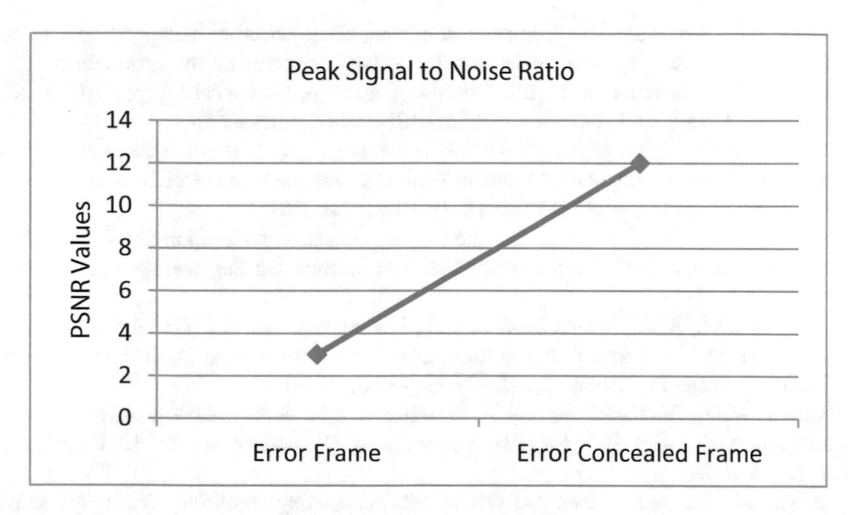

Fig. 9. Graph for PSNR values

SSIM and PSNR values for error concealed frame are approximately equal to original frame. That means, Error concealment is done by using Block Matching Algorithm is able to reconstruct the original frame without much quality reduction (Table 1).

Table 1. SSIM and PSNR vales of error frame and concealed frame

Parameters	Error frame	Error concealed frame
SSIM values	0.8784	0.9042
PSNR values	3.01	12

4 Conclusion

Error is detected using Histogram matching and correlation techniques. Error is corrected by using BMA (Block Matching Algorithm) where in the error block is replaced by correct reference block. The two major quality measuring parameters such as PSNR and SSIM are utilized in this algorithm. For error frame PSNR value is 3.01 and for error concealed frame it is 12. For error frame SSIM value is 0.8784 and for error concealed frame value obtained is 0.9042. From an error video the final error concealed video is obtained after removing the detected error. From the result, it is concluded that, quality of received video is improved in concealed videos.

References

1. Rajani, P.K., Khaparde, A.: Comparison of frequency selective extrapolation and patch matching algorithm for error concealment in spatial domain. In: Proceedings of 8th International Conference on Signal Processing Systems, New Zealand, pp. 70–74. ACM, New York, USA@2016, November 2016. ISBN: 978-1-4503-4790-7
2. Marvasti-Zadehl, S.M., Ghanei-Yakhdan, H., Kasaee, S.: A novel video temporal error concealment algorithm based on moment invariants. In: 9th Iranian Conference on Machine Vision and Image Processing. IEEE, 18–19 November (2015)
3. Shen, Z., Liu, X., Lu, L., Wang, X.: A new error concealment algorithm for H.264/AVC. In: IEEE International Conference on Computational Science and Engineering, pp. 1039–1042, December 2014
4. Marvasti-Zadeh, S.M., Ghanei-Yakhdan, H., Kasaei, S.: Dynamic temporal error concealment for video data in error-prone environments. In: The 8th Iranian Conference on Machine Vision and Image Processing, pp. 43–47, September 2013
5. Vazquez, M.G., Garcia-Ramirez, A.F., Ramirez-Acosta, A.A.: Image processing for error concealment. In: The 10th Mexican International Conference on Artificial Intelligence, pp. 133–138, December 2011
6. Karthikeyan, C.: Performance analysis of block matching algorithms for highly scalable video compression. In: 2006 International Symposium on Ad Hoc and Ubiquitous Computing, December 2006
7. Barjatya, A., Student Member, IEEE Member: Block matching algorithms for motion estimation. DIP 6620 Final Project Paper Spring (2004)
8. Suh, J.W., Ho, Y.S.: Error concealment based on directional interpolation. IEEE Trans. Consum. Electron. **43**(3), 295–302 (1997)
9. Ghanbari, M., Seferidis, V.: Error concealment based on directional interpolation. IEEE Trans. Circ. Syst. Video Technol. **3**(3), 238–247 (1993)
10. https://en.wikipedia.org/wiki/Block-matching_algorithm/orhistogram

Self-Adaptive Message Replication (SAMR) Strategy for Delay/Disruption Tolerant Network

Jitendra Patel(✉) and Harikrishna Jethva

Department of Computer Engineering, L.D. College of Engineering, Ahmedabad, Gujarat, India
jitendrabpatel@gmail.com, hbjethva@gmail.com

Abstract. Delay Tolerant Networks (DTNs) have high message delivery latency, very low message delivery probability, long delay. Delay Tolerant Networks routing schemes faces challenges of dynamically changing network topology. Due to this, routing in delay tolerant network is primary issue to consider. The main objectives of routing in Delay Tolerant Networks is to maximize message delivery ratio. A common method to maximize message delivery ratio is to replicate messages to encountered nodes. One of issues is knowing to how many replicas of a message to create? The work in this paper uses encounter history and current encounter values to dynamically finding number of message replications required. In this paper, a new message replication strategy is proposed which vary the number of message replications. Each node chooses by itself the number of message copies to create for the replication.

Keywords: Disruption tolerant network · Message replication · Routing

1 Introduction

Disruption/Delay tolerant networks (DTNs) are those class of networks which faces frequent disconnections. The networks that exhibit these characteristics are also known as 'Challenged Environments'. In 2002 Fall [1] initiated the concept of Delay Tolerant Networks which later also came to be known as Disruption Tolerant Networks (called DTN). DTN exhibits distinct characteristics like lack of continuous end-to-end connectivity, high propagation delay sometimes up to months and asymmetric data rate.

Applications of DTN include but not limited to Military network, Disaster recovery, Remote rural communication, Vehicular ad-hoc networks, Environment monitoring, e.g. lake water quality, noise and water pollution monitoring, Distribution of large scientific datasets and Distribution of data resulting from maintenance and provisioning activities [2, 3].

Routing in delay tolerant network is primary issue to consider due to dynamically changing network topologies. The main objectives of routing in DTN is to maximize message delivery ratio. In this paper, a new message replication strategy called Self Adaptive Message Replication (SAMR) is proposed which vary the number of message

S.C. Satapathy and A. Joshi (eds.), *Information and Communication Technology for Intelligent Systems (ICTIS 2017) - Volume 1*, Smart Innovation, Systems and Technologies 83, DOI 10.1007/978-3-319-63673-3_45

replications based on node's recent encounters with other nodes. The remaining of the paper is structured as: in Sect. 2, overview of related work is outlined. In Sect. 3, we brief our idea of work. Section 4 covers scenario setup for the proposed work and metrics to be evaluated for the performance. In Sect. 5, we present the simulation results comparing the performance of proposed algorithms with those of existing protocols. Conclusion and outline of the future work is given in Sect. 6.

2 Related Work

In [4], authors proposed a scheme to find forwarding probability depending on node utility and density at contact place. W. Narongkhachavana, T. Chokstaid, S. Prabhavat in [5], observed that a node should spread messages with lower probability if they are currently distributed in its local neighborhood. They used Token Bucket mechanism to implement this policy. In [6], T. Kimura, T. Matsuura M. Sasabe, T. Jonouchi, T. Takine, proposed to suppress the speed of disseminating message copies over high density areas by setting small forwarding probabilities in dense area. W. Guizhu, S. Mei, L. Run, M. Yao and W. Bingting in [7] gave the scheme to find high transfer utility of node. Utility of a node is computed using space utility and time utility. In [8], E. Kim, J. Nam, J. Choi, Y. Cho calculating data delivery probability using ProPHET to spray messages with high probability.

It is observed that DTN routing protocols lacks the support for dynamic situation. This raises a question:

(1) How many copies of messages to be replicated in network by source or relay node?

This clearly demands to contribute in routing environment. Therefore, adaptive routing protocol is required to satisfy above mentioned routing objectives.

3 Self-Adaptive Message Replication (SAMR) Strategy

Most of the schemes studied deal with improving Message Delivery ratio, Controlling Message Transmission Overhead and Minimizing Average latency.

In this work, we proposed a self-adaptive message replication routing strategy called "SAMR" that adaptively replicates varying number of messages dynamically based on node's recent encounters with other nodes.

3.1 Explanation of SAMR Strategy

In this routing strategy each node maintains "CE" (current encounter) which is incremented and updated for every recent encounter in time interval T. When the interval T expires, the encounter value L is updated using Eq. (1) and the CE is restarted.

Now each node calculate L_A no of copies of messages to be replicated in the network for each time interval T using Eq. (1) as follows:

$$L_A = L_A + CE_B. \quad (1)$$

The basic purpose of using CE is to update L_A to decide number of replicas of a message a node should create.

3.2 Flowchart of SAMR Strategy

Figure 1 shows the flow of the proposed work.

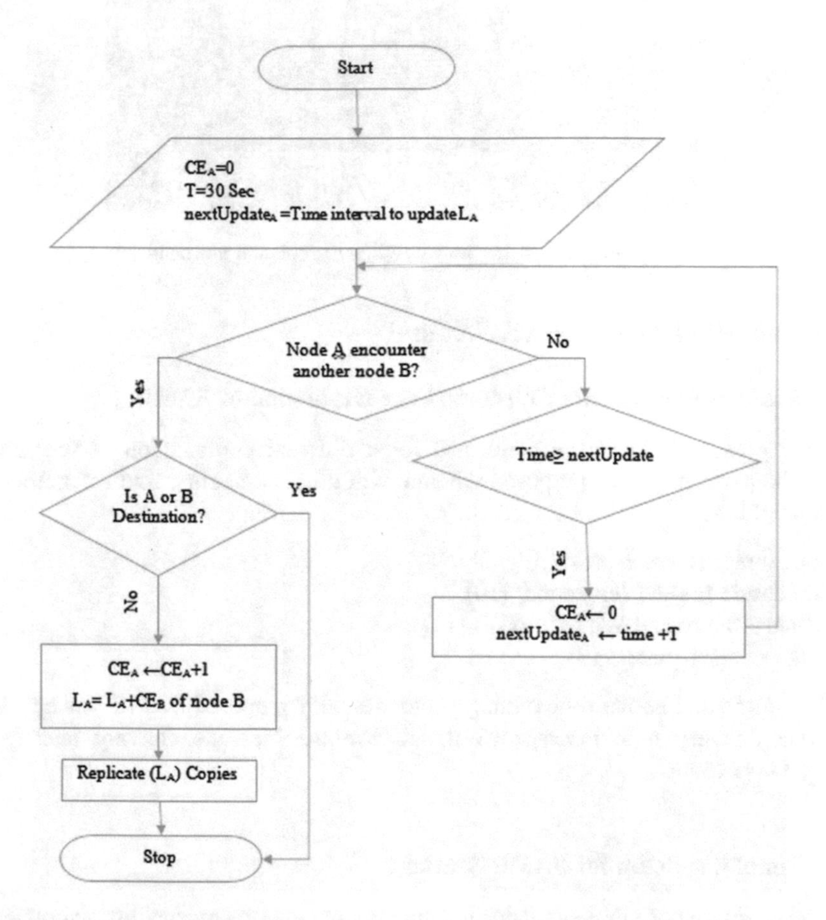

Fig. 1. Flowchart of proposed self-adaptive message replication (SAMR) strategy

4 Implementation of SAMR Strategy to Self-adapt to L_A

Proposed SAMR strategy was implemented using ONE simulator and JAVA programming language. Figure 2 (execution screenshot) shows how the proposed scheme will dynamically adapt to decide on how much copies (LA) to replicate as time elapses.

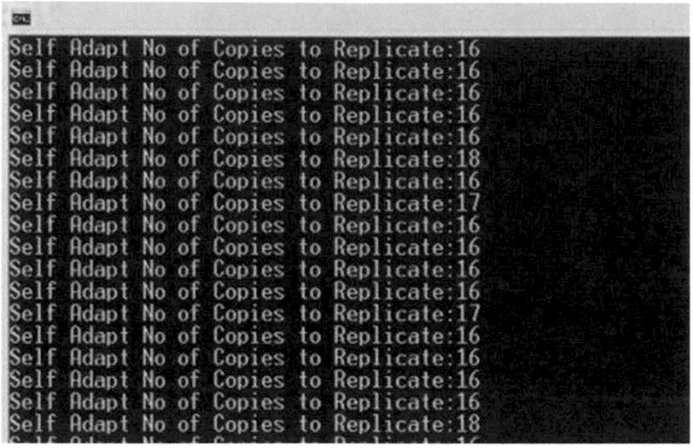

```
Self Adapt No of Copies to Replicate:16
Self Adapt No of Copies to Replicate:16
Self Adapt No of Copies to Replicate:16
Self Adapt No of Copies to Replicate:16
Self Adapt No of Copies to Replicate:16
Self Adapt No of Copies to Replicate:18
Self Adapt No of Copies to Replicate:16
Self Adapt No of Copies to Replicate:17
Self Adapt No of Copies to Replicate:16
Self Adapt No of Copies to Replicate:16
Self Adapt No of Copies to Replicate:16
Self Adapt No of Copies to Replicate:16
Self Adapt No of Copies to Replicate:17
Self Adapt No of Copies to Replicate:16
Self Adapt No of Copies to Replicate:16
Self Adapt No of Copies to Replicate:16
Self Adapt No of Copies to Replicate:16
Self Adapt No of Copies to Replicate:18
```

Fig. 2. Adapting to $L_{A,}$ no of copies to replicate automatically

5 Evaluation of Proposed Scheme

5.1 Performance Metrics Considered for Evaluation of SAMR

Number of simulations were performed for a different combination of scenarios to check the performance of proposed scheme with other competing routing protocols as mentioned here:

(1) Encounter Based Routing [9]
(2) Epidemic routing (epidemic) [10]
(3) Binary Spray and Wait (BSNW) [11]
(4) First contact router [12]

We compared above four routing protocols with proposed SAMR strategy using message delivery rate, message overhead, average message delivery latency and average hop count.

5.2 Simulator Setup for SAMR Strategy

The configuration of ONE simulator for implementation and evaluation is summarized in Table 1.

Table 1. Simulation configuration

Parameter	Configurations
Number of nodes	Default
Movement model	RandomWaypoint
Simulation time	12 h
Message TTL	300 min
Buffer size	40 MB
Wait time	0,120
No of copies	16
Alpha	0.85
$nextUpdate_A$	30
binaryMode	True

5.3 Result Analysis

After executing the proposed SAMR strategy and other related routing protocols for comparisons we obtained following results.

5.3.1 Impact of Number of Nodes on Message Delivery Ratio

From Fig. 3 it is clear that message delivery ratio is related with number of nodes deployed in the network.

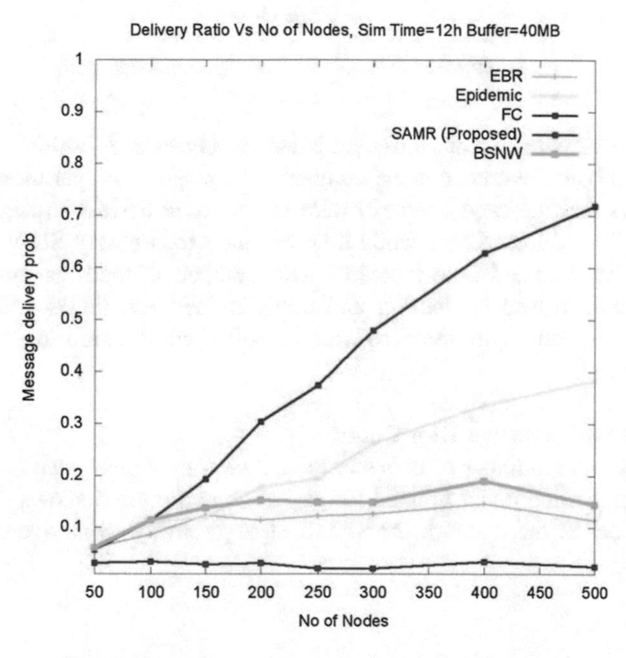

Fig. 3. Number of nodes vs. message delivery ratio

From the Fig. 3, it can also be observed that the proposed SAMR strategy out-performs in terms of message delivery ratio over all the compared schemes including epidemic routing with increase in number of nodes. Proposed SAMR scheme compared to other protocols nearly doubles message delivery when nodes are more than 300.

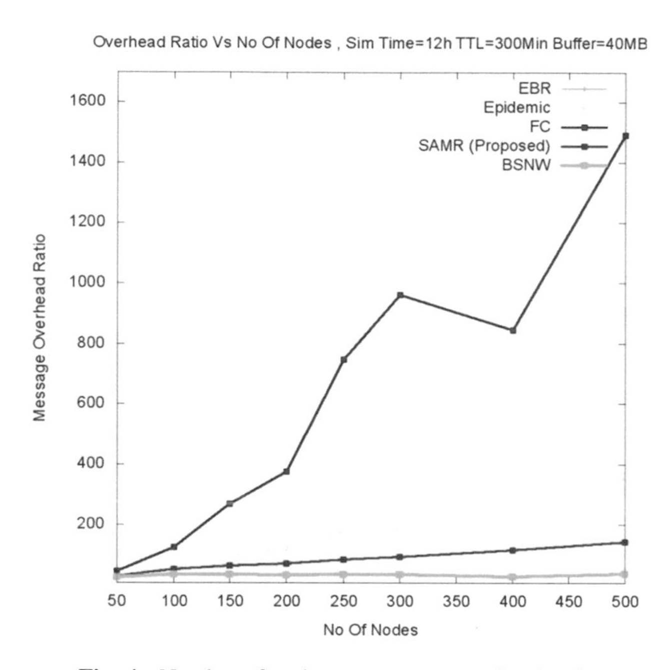

Fig. 4. Number of nodes vs. message overhead ratio

5.3.2 Impact of Number of Nodes on Message Overhead Ratio

Message transmission overhead is the number of messages used per message delivery. Figure 4 shows that message overhead ratio of proposed SAMR strategy is less than epidemic and first contact router while little bit more than binary SNW protocol.

The message overhead ratio increases with number of nodes as shown in Fig. 4. This is due to increased contention and collision between nodes with increase in density of the network. The proposed scheme still need attention to improvement in message overhead.

5.3.3 Nodes and Average Hop Count

When there are more number of nodes in the network they meet each other frequently thus increasing the hop count utilized for any scheme. Figure 5 shows that increasing Number of nodes in the network for SAMR strategy always limit average hop count below 5.0.

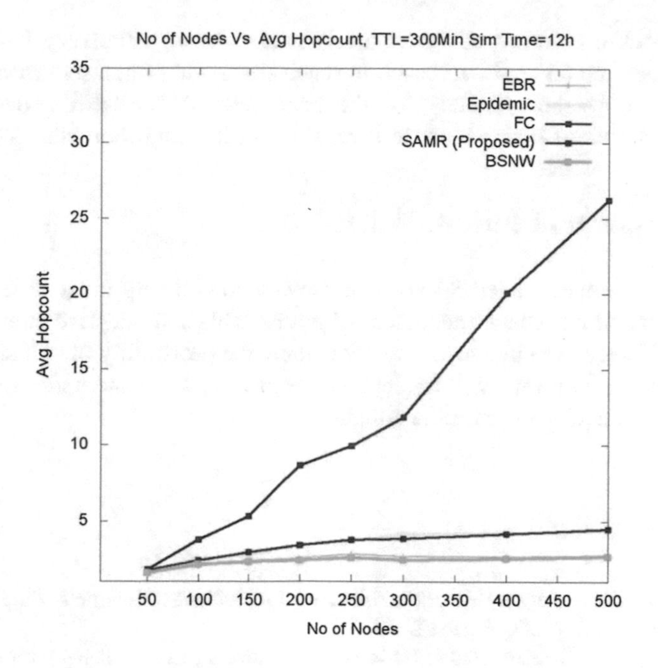

Fig. 5. Number of nodes vs. average hop count

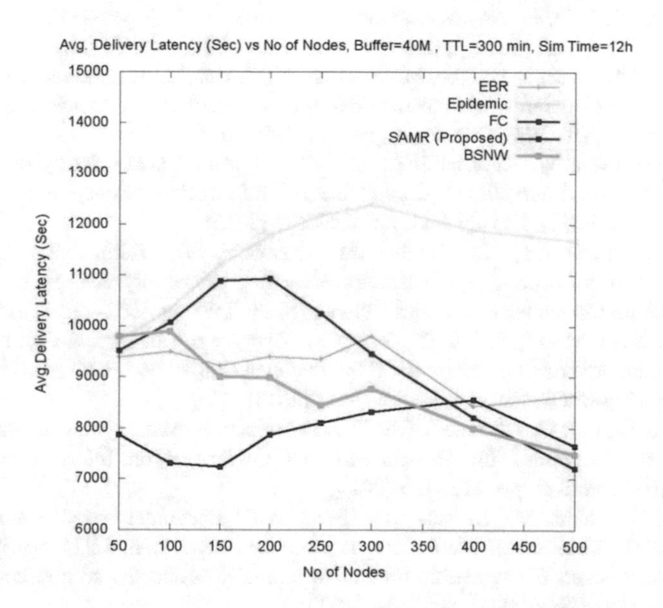

Fig. 6. Number of nodes vs. average delivery latency

5.3.4 Impact of Number of Nodes on Average Message Delivery Latency

Latency is the time taken for message to reach destination once sent from source.

Figure 6 shows how latency for the proposed SAMR scheme decreases with increase in number of nodes because it react better in congestion situation.

6 Conclusion and Future Work

We proposed a scheme called SAMR and showed how to vary the number of message replications based on recent encounters of nodes. This self-adaptive strategy helps in improving delivery ratio in dense networks where the probability of collisions between nodes is high. In future we will implement the proposed scheme under real life traces for testing its behavior in real time situations.

References

1. Fall, K.: A delay-tolerant network architecture for challenged internets. In: Proceedings of SIGCOMM 2003, p. 27, August 2003
2. Abbas, A., Shah, B., Al-Obeidat, F., Iqbal, F.: Bounded message delay with threshold time constraint in delay tolerant networks (DTNs). In: 3rd MEC International Conference on Big Data and Smart City, Muscat, pp. 1–6 (2016)
3. Abdelkader, T., Naik, K., Nayak, A., Goel, N., Srivastava, V.: A performance comparison of delay-tolerant network routing protocols. IEEE Netw. **30**(2), 46–53 (2016)
4. Kimura, T., Matsuura, T., Sasabe, M., Matsuda, T., Takine, T.: Location-aware utility-based routing for store-carry-forward message delivery. In: International Conference on Information Networking (ICOIN), Cambodia, pp. 194–199 (2015)
5. Narongkhachavana, W., Choksatid, T., Prabhavat, S.: An efficient message flooding scheme in DTN. In: 7th International Conference on Information Technology and Electrical Engineering (ICITEE), Chiang Mai, pp. 295–299 (2015)
6. Kimura, T., Matsuura, T., Sasabe, M., Jonouchi, T., Takine, T.: Density-aware store-carry-forward routing with adaptive forwarding probability control. In: International Conference on Consumer Electronics-Taiwan (ICCE-TW), pp. 182–183 (2015)
7. Wang, G., Shao, M., Li, R., Ma, Y., Wang, B.: Spray and wait routing algorithm based on transfer utility of node in DTN. In: IEEE International Conference on Progress in Informatics and Computing (PIC), Nanjing, pp. 428–432 (2015)
8. Kim, E.-H., Nam, J.-C., Choi, J.-I., Cho, Y.-Z.: Probability-based spray and wait protocol in delay tolerant networks. In: The International Conference on Information Networking (ICOIN 2014), Phuket, pp. 412–416 (2014)
9. Nelson, S.C., Bakht, M., Kravets, R., Harris, A.F.: Encounter–based routing in DTNs. Newsl. ACM SIGMOBILE Mob. Comput. Commun. Rev. Arch. **13**(1), 56–59 (2009)
10. Vahdat, A., Becker, D.: Epidemic routing for partially connected ad hoc networks. Duke Technical report CS-2000-06, pp. 1–14 (2000)
11. Spyropoulos, T., Psounis, K., Raghavendra, C.S.: Spray and wait: an efficient routing scheme for intermittently connected mobile networks. In: Proceedings of the 2005 ACM SIGCOMM Workshop on Delay-Tolerant Networking, WDTN 2005, pp. 252–259 (2005)
12. Jain, S., Fall, K., Patra, R.: Routing in a delay tolerant network. In: Proceedings of SIGCOMM 2004, pp. 145–158 (2004)

Implementation of Modified TEA to Enhance Security

Chandradeo Kumar Rajak(✉) and Arun Mishra

Computer Engineering Department,
Defence Institute of Advanced Technology (DU), Girinagar, Pune, India
{chandradeo_mce15, arunmishra}@diat.ac.in

Abstract. Tiny Encryption Algorithm (TEA) is one of the fastest Encryption Algorithms. It is a lightweight cryptographic algorithm with minimal source code. Due to its simple logic in key scheduling TEA has suffered from related key and equivalent key attacks. Therefore a modified key schedule is proposed for TEA. The new key schedule applies Boolean function based SBox to generate different round keys for TEA. The resultant Modified TEA achieves better security than original TEA. The execution time analysis of modified TEA is also presented.

Keywords: Tiny Encryption Algorithm (TEA) · Modified TEA · Enhanced TEA · SBox design · Real-time encryption · Lightweight cryptography

1 Introduction

The use of mobile devices and online applications has increased drastically during last few years. People are using these handheld devices for many critical and security sensitive applications like online banking, telemedicine, videoconferencing, Voice over IP (VOIP) and many more. All of these systems must be properly secured to protect the privacy and confidentiality of message being transmitted. Providing security for such systems demand encryption algorithms which provide high security but at the cost of very less execution time and less memory requirements. There is huge demand of encryption techniques which can encrypt user data in real time. The modern block ciphers like AES [1], DES [2] cannot be applied as it is in such applications because they use operations like substitutions and permutations which take longer time and hence may not be suitable for such applications. Therefore many lightweight encryptions algorithms have been designed and implemented such as TEA [3], XTEA [4, 5], IDEA [6], PRESENT [7], SEA [8], RC5 [9], HIGHT [10], DESL [11], DESX [11], and many variations of etc.

TEA [3] is a famous lightweight cryptographic algorithm which has very minimal code. It is very easy to implement in any programming languages such as C, C++, Java, PHP, Javascrypt, python etc. It has very small memory footprint and takes vey less time to execute. Hence it is most suitable encryption technique in real time applications, embedded devices and RFID where available time and memory is very limited. Apart from these characteristics TEA also has some weakness such as related key and equivalent Keys. These weaknesses are mainly due to very simple logic in key schedule

S.C. Satapathy and A. Joshi (eds.), *Information and Communication Technology for Intelligent Systems (ICTIS 2017) - Volume 1*, Smart Innovation, Systems and Technologies 83, DOI 10.1007/978-3-319-63673-3_46

of the algorithm. During last few years many modifications have been proposed to this algorithm [4, 5].

Present work is focused towards providing a robust key schedule to TEA to protect it from above mentioned attacks. The proposed key schedule is based on operations like rotation, XOR and substitution of key bits.

The substitution is achieved using Boolean function based SBox [12] which are computed on the fly. Generation of SBox based on cryptographically strong Boolean functions is also discussed as part of this work.

The proposed key schedule is implemented in C language and integrated with the TEA algorithm and analyzed from its security and runtime point of view. Although proposed key schedule is secure but it has additional time overhead hence it is supposed to be used in applications where higher security is required but valuable life time of the information is limited. One of such applications is to encrypt the Count Down Time (CDT) of flight vehicles in Test ranges where flight vehicles such as missile and Unmanned Aerial Vehicles are evaluated.

The paper is organized as follows: In Sect. 2 an overview of TEA algorithm is presented. Section 2.1 discusses vulnerabilities and attacks on TEA. Section 3 presents some significant works carried out related to this work. In Sect. 4 the modification to TEA is described with Proposed Key schedule of present work in Sect. 4.1 and Boolean function based SBox design in Sect. 4.2. Section 5 provides a brief description of case study of implementation. Experimentation process with Results and analysis is presented in Sect. 6 and finally the work is concluded in Sect. 7.

2 Tiny Encryption Algorithm

TEA [3] is one of the lightweight encryption algorithms. David Wheeler and Roger Needham developed TEA at Computer Laboratory of Cambridge University. It was first published at Fast Software Encryption workshop in Leuven in 1994. Tiny Encryption Algorithm (known by its convenient acronym TEA), "which is probably the most efficient—and hence fastest—software encryption algorithm ever devised" [13]. TEA is a block cipher with 64 bit block length and key of 128 bits. The key length is sufficient as recommended by modern encryption requirements. The 64 bit block is divided into two 32 bit words; let us consider as V [0] and V [1]. Similarly 128 bits key K is divided into four words (K[0], K[1], K[2], K[3]). Block diagram of TEA is shown in Fig. 2 in bold lines. It has Feistel Structure with total of 64 Feistel Rounds. Single iteration of the algorithm consists of two Feistel Rounds and hence there are total 32 iterations. The TEA can achieve complete diffusion in only six iterations but for increasing the security 32 iterations are kept. The number of iteration may be reduced where available encryption time is very small. TEA has very small footprint in terms of resource requirement. It has very minimal source code and can be implemented in almost all programming language. The algorithm uses number denote as Delta which is derived from the golden ratio [13] as shown in Eq. 1.

$$\text{Delta} = (\sqrt{5} - 1)2^{31} = 9E3779B9_{\text{h}}. \quad (1)$$

Different multiples of this number is used in different rounds so as to provide different mixing in all rounds and hence symmetry between each round is avoided.

2.1 Vulnerabilities and Attacks on TEA

Although TEA is fast and efficient it has very simple key schedule. It has 128 bit key which is divided into four words of 32 bits each. Each of the subparts denoted as K[0], K[1], K[2], K[3] is applied as show below.

```
if(Fiestel_Round is odd)
{Apply K[0], K[1]}
else
{Apply K[2], K[3]}
```

In this manner each of the 32 cycles uses the same K[0],K[1],K[2],K[3] and hence same key material is used in each cycles. Due to this weakness in key schedule TEA suffered from equivalent key attack [14] and related key attacks [14].

3 Related Works

There have been numerous significant modifications to TEA since its inception. Only some major works which have been carried out by Authors of TEA and some other researchers in this area shall be highlighted here. The two weaknesses (equivalent key and related key) of TEA were first noticed by David Wagner and soon these were corrected by the Authors of TEA by doing some small corrections. The first adjustment was there in the key schedule and second was that the key material was introduced more slowly. This improved version was named XTEA [4].Along with XTEA the authors introduced another version of TEA called block TEA [4]. This block TEA has the peculiarity that it can accept input of any block length. Later Block TEA was also found vulnerable due to lack of back propagation and it was again published as XXTEA [5]. Abdelhalim, M.B., et al. [22] have implemented MTEA (Modified TEA) in RFID system where moderate security is required. In their work they have used LFSR (Linear Feedback shift register) as a Pseudo random Number Generator (PRNG) to generate new round keys for every round of TEA and show that they achieve better avalanche effect than TEA.The works carried out on TEA by authors are remarkable. They contributed to enhancement of TEA in security, memory footprint and execution time to a large extent. The present work also is an effort to further enhance TEA security with the help of Key scheduling based on SBox.

4 The Proposed Modification in TEA

The proposed TEA (MTEA) achieves better security than TEA by *modifying the key schedule* of TEA. TEA has very simple key schedule which may face the attacks as mentioned in previous section. Therefore SBox based key schedule is proposed to enhance security of TEA.

4.1 New Key Scheduling Algorithm

The block diagram of proposed key scheduling process is shown in Fig. 1. The Key schedule has two basic components, left rotation of each 32 bit subkeys followed by substitution. The substitution is achieved using a 4 × 4 Sbox which is designed in-house using cryptographically strong Boolean Functions. This SBox accepts 4 binary bits and outputs 4 binary bits. The Proposed Key schedule accepts 128 bit cipher key from the user and generates 32 round keys each of 128 bit. These 32 round keys are used in 32 rounds of TEA. The details of key schedule are described below.

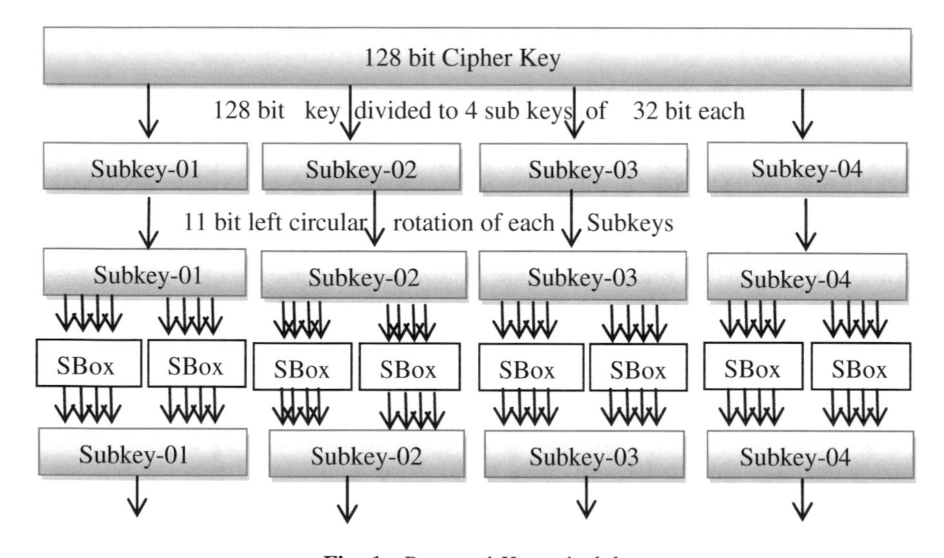

Fig. 1. Proposed Key schedule

Initially 128 bits cipher key provided by user is divided into four equal words of 32 bits each. These words are individually rotated by 11 bit positions to the left. After all four words are rotated then the 4 × 4 SBox is applied twice to each of the rotated words. The position of bits which are passed to the SBox are fixed for each 32 bit subkeys. First SBox is applied at bit positions 0,1,2,3 and second SBox is applied at bit position 28, 29, 30, 31 for each 32 bit subkeys. The position of SBox will not make any difference because wherever they are placed they will replace 4 bits with new 4 bits.

Finally one 128 bit key is ready for one round of TEA and the process is repeated in same manner where input of next round is taken from output of previous round.

This key scheduling scheme is integrated with the original TEA as shown in Fig. 2. Each round of key schedule generates 128 bit round key (K [0] to K [3]) which is applied to each round of TEA.

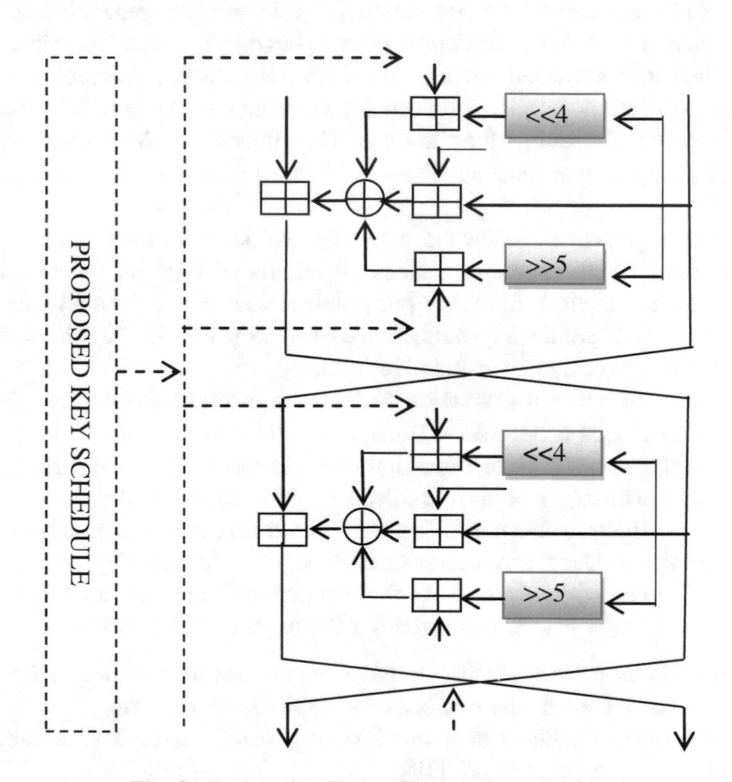

Fig. 2. Proposed MTEA with new Key schedule

4.2 SBox Design for Proposed Key Schedule

For achieving higher confusion and diffusion, nonlinearity of operations is very essential in cryptography. For creating a new cryptographic algorithm SBox design is very crucial. Hence SBox must be designed in such a way that it provides maximum nonlinearity. A cryptographically strong SBox must fulfill criteria like Strict Avalanche Criteria (SAC) [15], balanceness [15] and Bit independence criteria (BIC) [15]. SAC means that changing any bit of input must result in at least 50% of change in output bits. BIC says that change in any two bits of output of a SBox should be independent of each other; in other words if we change any bit of input then the corresponding change in any two output bits should be independent of each other. Balanceness means the output of SBox must contain equal number of zeros and ones.

The major component of nonlinearity in proposed key schedule is SBox. These SBoxes are created by using cryptographically strong Boolean functions which satisfy all criteria like Balanceness, BIC and SAC. If SBox is implemented as lookup table stored in the system then it may decrease security because it may be read by the attacker. Hence on_the_fly SBox were designed using 4 nos. of Boolean functions. For designing SBox based on Booleans functions there are two methods namely Random generation [16] and Systematic generation [17]. In random generation of SBox a Boolean function is selected randomly from a large space of all possible Boolean functions then it is evaluated against all criteria (BIC, SAC, balanceness). If these criteria are satisfied by selected Boolean function then we keep it for SBox design otherwise we discard it and go for other one. This process is tedious since the number of Boolean function of n input variables is 2^{2^n} [18] which become very large as n increases.

The other approach of systematic generation of SBox as discussed in [32] uses some algorithms which internally verifies all criteria of Boolean function and give probable Boolean functions for SBox but problem with this approach is that it starts with function which are already nonlinear. Hence to get a new nonlinear Boolaean function we must have nonlinear function in hand.

In Present work random generation of Boolean functions is adopted. The search space consisted of total 65536 (for 4 Boolean variables by formula 2^{2^n} [18]) Boolean functions out of this, half are linear function which is not suitable for creating SBox and other half has maximally nonlinear Boolean function called Bent functions and balanced nonlinear Boolean functions. The Bent functions are also not suitable because they are not 0-1 balance and hence cannot be used for creating strong SBoxes. Therefore it is required to search only those functions which are nonlinear and balanced. Following rules helped us to find out the required Boolean functions [12].

- Nonlinear Boolean functions should have at least one term of degree 2 or higher. The degree is defined as the number of variables in single term.
- For a maximally nonlinear bent function of n variables, each term should have degree less than or equal to n/2 [19]

$$f1 = x_1x_2 \oplus x_2x_3 \oplus x_3x_4 \oplus x_1x_3 \oplus x_1x_4. \quad (2)$$

$$f2 = x_1x_2 \oplus x_2x_3 \oplus x_3x_4 \oplus x_2x_4 \oplus x_1x_4. \quad (3)$$

$$f3 = x_1x_4 \oplus x_2x_3 \oplus x_3x_4 \oplus x_2x_4 \oplus x_1x_3. \quad (4)$$

$$f4 = x_1x_2 \oplus x_3x_4 \oplus x_1x_3 \oplus x_1x_4 \oplus x_2x_4. \quad (5)$$

Therefore based on above mentioned rules, all nonlinear Boolean functions are selected and out of these only those nonlinear functions were taken which are balanced. The four candidate Boolean Functions for SBox are shown in Eqs. (2), (3), (4) and (5).

The above Boolean functions are highly nonlinear since their nonlinearity is near the Bent functions. Each of the Boolean functions has maximum nonlinearity of 4 which is obtained using the formula given in expression (6) by Pieprzyk and G. Finkelstein [20].

$$\sum_{i=1/2(n-4)}^{n-4} 2^{i+2} \tag{6}$$

Where n is the number of Boolean variables which in present case is 4. This gives maximum achievable nonlinearity of a balanced Boolean function. After finalizing all four Boolean functions a SBox was designed as shown in Fig. 3.

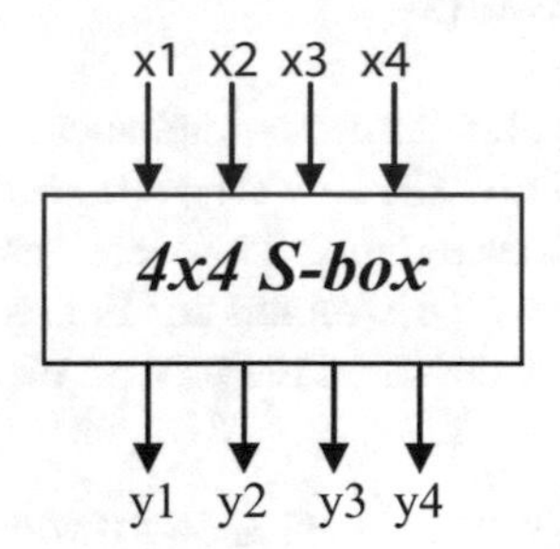

Fig. 3. Boolean function based 4 × 4 SBox

5 Implementation Case Study

As a special case study Present MTEA has been implemented for real time encryption of Count Down Time (CDT) in a Test Range. Figure 4 presents system configuration of CDT generation and dissemination in a Test Range.

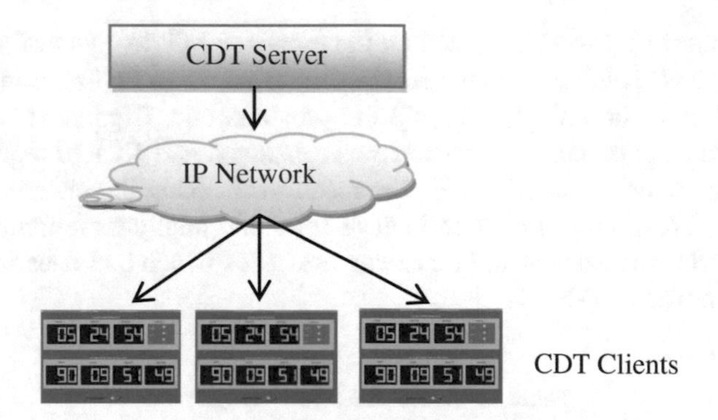

Fig. 4. Secure transmission of CDT

CDT is generated at a central location by the CDT server in UDP packet format [25] and disseminated to all tracking stations. The CDT Client deployed at tracking station receive it and display in HH:MM:SS:STATUS format.CDT is sent from the server either through IP network or by satellite links to remote clients. The new CDT packet is generated at Server every one second and displayed in all client applications every second. Every packet must reach the client within 1000 ms (1 s) for correct display of CDT at clients. This CDT data is much crucial from security point of view because if it is intercepted by adversary they may know the launch time and can intercept the flight vehicle. Therefore CDT must be encrypted before transmission. Encryption of CDT requires lightweight and fast cryptography.

6 Experiments and Results

The proposed key scheduling algorithm was implemented in C on Linux platform and then integrated with original TEA. The Key schedule successfully generated new round keys for 32 rounds of TEA. These rounds key were generated and stored in an array from where theses were used by TEA rounds for Encryption and in reverse order for decryption. The example below describes the process of encryption and decryption for a particular data.

The Plaintext: *V[0] = 1404510514, V[1] = 382078007*
CipherKey: *K[0] = 2732026831,K[1] = 3805535053,K[2] = 2865875947,K [3] = 3000351565*
Ciphertext: *C[0] = 2588239813,C[1] = 1646995231*

The decrypted plaintext is as given here.
Plaintext: *V[0] = 1404510514,V[1] = 382078007*

6.1 Completeness Test

Completeness test [21] was conducted on the proposed MTEA by creating a matrix of 64×64 elements. Total of 65 plaintexts starting from P1 to P65 were taken such that each of them differ in a single bit from the previous one. Ciphertext of each of the plaintext and its one bit changed version was calculated and XORed together. Each of the rows of the matrix was filled with the result of XORing the ciphertexts in subsequent manner. Total number of ones in each rows and finally total number of ones in the whole matrix was calculated. Percentage of entries which had ones was found and tabulated as shown in Table 1.

Table 1. Avalanche effect comparison

TEA	MTEA [22]	SBox based MTEA (Present work)
0.4892	0.5046	**0.5219**

This completeness test was done for original TEA and MTEA of Present work over the same plaintexts. The result of completeness test as carried out in LFSR based MTEA [22] was also compared with and it found that Proposed SBox based MTEA achieves highest better avalanche effect than TEA as well as LFSR based MTEA. The higher value of this percentage shows that there are more no of ones and hence ciphertexts of each plaintext and one bit changed versions are differing in more number of bit positions. This signifies that Proposed MTEA exhibits better avalanche effect which is a desirable property of quality Cryptographic algorithm.

6.2 Enhanced Security

MTEA of this work achieves better avalanche effect that original TEA and the LFSR based MTEA. This is very clear from completeness test. The completeness test reveals that making a change in plaintext causes many bits of ciphertext to be changed. Hence higher the number of ones in completeness test matrix better is the security of cryptosystem. This better security is due to nonlinear SBox and the optimized rotation operation inside key schedule. Therefore MTEA of this work provides better security than TEA or MTEA [22].

6.3 Execution Time Analysis

Although purpose of this work was to make the algorithm make more robust, a runtime comparison with TEA has also been done in order to check its suitability for encryption of CDT data. This comparison is shown in Table 2.

Table 2. Execution time comparison

Sr. No	Algorithm	Run time (microseconds)
1.	TEA	2
2.	Enhanced TEA (present work)	300

Since update rate of CDT is one second (1000 ms), the encryption delay of 300 microseconds is insignificant compared to 1000 ms (1 s is available time in encryption of CDT data). If we consider this time transmission using satellite links also then total uplink and downlink time comes to 320 ms without encryption and it will never increase beyond 500 ms. Hence CDT time can be comfortably displayed even after applying Modified encryption algorithm. The above comparison of runtime and avalanche effect has been done on HP Compaq 6200 Pro Microtower, Intel(R) Core(TM) i5-2400 CPU @ 3.10 GHz, Ubuntu 14.04.4 LTS, Linux 4.2.0-27-generic i686 with 4 GB of RAM.

7 Conclusions and Future Work

In this paper TEA has been implemented with enhanced key schedule based on SBox. Nonlinear Boolean function based SBox and optimized rotation of bits in key schedule provides new round keys for all rounds of TEA. The Present work has been compared with original TEA and LFSR based MTEA and found more secure where Present work achieves about 52% completeness. Since avalanche effect is one of the necessary and sufficient criteria of a secure cryptosystem design, MTEA of present work achieves better avalanche effect (52%) than LFSR based MTEA (50%) and original TEA (48%) hence SBox based MTEA is more secure than theses two algorithms. The run time measured shows that encryption time of proposed algorithms has increased many fold due to inclusion of key scheduling algorithm. But this provides better security than basic TEA algorithm. Above result shows that we can easily use the Enhanced TEA for encryption of CDT for flight vehicles as well as in other realtime applications.

The extra execution time overhead can be reduced by applying parallelization techniques. Independent blocks of code will be identified inside key schedule and TEA encryption logic and further these blocks can be executed in parallel.

References

1. Pub, NIST FIPS. "197: Advanced encryption standard (AES)." Federal Information Processing Standards Publication 197 (2001): 441-0311
2. FIPS, PUB.: "46-3." Data encryption standard (DES) 25 (1999)
3. Wheeler, D.J., Needham, R.M.: TEA, a tiny encryption algorithm. In: International Workshop on Fast Software Encryption. Springer, Heidelberg (1994)
4. Needham, R.M., Wheeler, D.J.: Tea extensions. Report, Cambridge University, Cambridge, UK (1997)
5. Wheeler, D.J., Needham, R.M.: Correction to xtea. Unpublished manuscript, Computer Laboratory, Cambridge University, England (1998)
6. Chang, H.S.: International data encryption algorithm CS-627-1 Fall (2004)
7. Bogdanov, A., et al.: PRESENT: an ultra-lightweight block cipher. In: International Workshop on Cryptographic Hardware and Embedded Systems. Springer, Heidelberg (2007)
8. Standaert, F.-X., et al.: SEA: a scalable encryption algorithm for small embedded applications. In: International Conference on Smart Card Research and Advanced Applications. Springer, Heidelberg (2006)
9. Rivest, R.L.: The RC5 encryption algorithm. In: International Workshop on Fast Software Encryption. Springer, Heidelberg (1994)
10. Hong, D., et al.: HIGHT: a new block cipher suitable for low-resource device. In: International Workshop on Cryptographic Hardware and Embedded Systems. Springer, Heidelberg (2006)
11. Leander, G., et al.: New lightweight DES variants. In: International Workshop on Fast Software Encryption. Springer, Heidelberg (2007)
12. Cheung, J.M.: The design of S-boxes. Dissertation, San Diego State University (2010)
13. Shepherd, S.J.: The tiny encryption algorithm. Cryptologia **31**(3), 233–245 (2007)

14. Andem, V.R.: A cryptanalysis of the tiny encryption algorithm. Dissertation, The University of Alabama TUSCALOOSA (2003)
15. Forré, R.: The strict avalanche criterion: spectral properties of Boolean functions and an extended definition. In: Proceedings on Advances in Cryptology. Springer, New York (1990)
16. Webster, A.F., Tavares, S.E.. On the design of S-boxes. In: Conference on the Theory and Application of Cryptographic Techniques. Springer, Heidelberg (1985)
17. Seberry, J., Zhang, X.-M., Zheng, Y.: Systematic generation of cryptographically robust S-boxes. In: Proceedings of the 1st ACM Conference on Computer and Communications Security. ACM (1993)
18. Burnett, L.: Heuristic Optimization of Boolean Functions and Substitution Boxes For Cryptography. Dissertation, Queensland University of Technology (2005)
19. Rothaus, O.S.: On "bent" functions. J. Comb. Theory Ser. A **20**(3), 300–305 (1976)
20. Pieprzyk, J., Finkelstein, G.: Towards effective nonlinear cryptosystem design. IEE Proc. E-Comput. Digit. Tech. **135**(6), 325–335 (1988)
21. Kam, J.B., Davida, G.I.: Structured design of substitution-permutation encryption networks. IEEE Trans. Comput. **100**(10), 747–753 (1979)
22. Abdelhalim, M.B., et al.: Implementation of a modified lightweight cryptographic TEA algorithm in RFID system. In: 2011 International Conference for IEEE Internet Technology and Secured Transactions (ICITST) (2011)
23. Mister, S., Adams, C.: Practical S-box design. In: Workshop on Selected Areas in Cryptography, SAC, vol. 96 (1996)
24. Detombe, J., Tavares, S.: Constructing large cryptographically strong S-boxes. In: International Workshop on the Theory and Application of Cryptographic Techniques. Springer, Heidelberg (1992)
25. Forouzan, A.B.: Data Communications & Networking (sie). Tata McGraw-Hill Education, New Delhi (2006)

Mean Reversion with Pair Trading in Indian Private Sector Banking Stocks

Umesh Gupta[1(✉)], Sonal Jain[2], and Mayank Bhatia[3]

[1] Department of Mathematics, JK Lakshmipat University, Jaipur, India
umeshgupta@jklu.edu.in
[2] Department of Computer Science Engineering, JK Lakshmipat University, Jaipur, India
sonaljain@jklu.edu.in
[3] Institute of Management, JK Lakshmipat University, Jaipur, India
mayank.bhatia@live.in

Abstract. While evaluating the performance of the company in the market, the stock performance of the company plays a vital role in its evaluation. In this paper, the correlation and mean reverting behaviour of various stocks of Banking (Private Banks) from Indian stock market have been examined. Five Private Sector Banks (ten combination/pairs among them) were selected for the study. Along with the correlation test, Augmented Dickey Fuller Test is conducted to test whether the time series follows the mean reverting behaviour. It has been found that three pairs from banking sector were negatively correlated and that high degree of correlation does not necessarily result in mean reversion between two time series.

Keywords: Mean reverting behaviour · Correlation · Indian stock market

1 Introduction

Overall performance of the company is assessed by measuring the return of the company stock. The performance of the company stock plays vital role in evaluating the performance of the company in the market. Where the company has not offered any dividends in past, stock split has never been exercised by the company, or information about other financial adjustments are not available, the company's common stock return is defined as the rate of change in price during a period with the price at the start of the period.

Today in the era of the globalization, the change in one economy has impact on the financial markets of other economy. At the same time, within an economy, various sectors are also affected by global and national economy. After the global turmoil in 2008, today the market for common stock investment has again become very popular. Institutional as well as individual investors have started diversifying their investments in Indian stock market. The aim of the investors is to diversify the risk associated with the investment in stocks and to enhance the chances of superior returns.

Epps, has given a phenomenon known as The Epps Effect states that, the increase in sampling frequency of data results into decrease in correlation between the return of

S.C. Satapathy and A. Joshi (eds.), *Information and Communication Technology for Intelligent Systems (ICTIS 2017) - Volume 1*, Smart Innovation, Systems and Technologies 83, DOI 10.1007/978-3-319-63673-3_47

two stocks. This is caused by non-synchronous/asynchronous trading [1] and discretization affects [2]. The stock prices do not move together in synchronization because of dissimilar environments in terms of different taxation policy, industrial growth, monetary policy, political factors and various economic factors. Certain stocks exhibits co-movement (i.e. if one stock price rises, other stock price will also rise and vice versa) that offers the investors to diversify their holding; if two stocks are correlated i.e. a statistical relationship between two stocks exists. Correlation is useful in predicting the relationship between two variables and this relationship can be exploited in future. Shapira et al. [3] proposed a model for short term stock markets' behaviour and explained several key features of the stock market.

On the other hand, if the price movement of two stocks is synchronized, they can provide opportunity of pair trading. The pair trade or pair trading is a market neutral trading strategy that provides a solution to investor to earn profit under all sorts of market conditions. If the two stocks are found to be correlated in past performance, their performance can be monitored by pair trading strategy. It states that when the correlation between the two securities temporarily weakens, investor should short the outperforming stock and create a long position in the under-performing stock. Vidyamurthy [4] discussed various pair trading quantitative methods. Ungever [5] also investigated pair trading strategy by using the cointegration method and also Gumparthi and Sarvanan [6] find that "paired share price ratio strategy" was a profitable and market neutral strategy in the banking sector scrip's of Indian financial market.

Mean reversion is considered as one of the key trading concept. Statistical Arbitrage (StatArb) refers to highly technical short-term mean-reversion strategies [7] and Dufresne and Goldstein [8] confirm that firms adjust their leverage over time to find an optimal level so that mean reversion is achieved. Herlemont [9] proposed a novel approach to find the right pair of stocks or time series for pair trading. He proposed to use Dickey-Fuller Test to find whether the log ratio of two time series is stationary or not. In 2005, Resilience Reserve also recommended for investment to consider some mean reversion in formulas for parameters such as dividend yields, real interest rates and anticipated inflation [10]. Desai and Joshi [11] applied Augmented Dickey Fuller Test along with correlation to test the mean reverting behaviour of time series of Dow Jones Industrial Average, with ten stock markets from Asia, Europe and America. The authors confirmed that high degree of correlation does not result in mean reversion between two time series.

In this paper, study has been conducted to find the correlation between different pairs of stocks from Private Sector Bank to find the relationship present among these stocks. The mean reverting behaviour of the stock pairs are also been studied and it has been found that even if the stock is having a positive relationship with another stock; it does not necessarily show the mean reverting behaviour.

2 Problem Formulation

In this study, the challenge was to identify the stocks having high correlation as well as mean reverting behaviour i.e. they tend to move together and therefore make potential pairs. The aim of this study is to identify the stocks with mean reverting relative prices.

Augmented Dickey-Fuller (ADF) test statistic of log ratio of the pair is used to identify the mean reverting behaviour of the stocks from private sector banks.

The ADF is used to testing a unit root's presence in an autoregressive time series sample. In case of mean reversion in a price series, the next price level is proportional to the current price level.

$$\Delta y_t = \alpha + \beta t + \gamma y_{t-1} + \delta_1 \Delta y_{t-1} + \cdots + \delta_{p-1} \Delta y_{t-p+1} + \varepsilon_t \quad (1)$$

where β is the coefficient of a temporal trend and α is constant,

$$\Delta y_t = y(t) - y(t-1) \quad (2)$$

Null Hypothesis

The process is not mean reverting or it is a random walk, i.e., $\gamma = 0$

3 Methodology

For Private Sector Banks, the study is conducted for the period of 2480 trading sessions starting from 1st April 2006 and ending on 31st March 2016 on five private banking stocks. This period is sufficient to examine the correlation and mean reversion as during the period many events like recession, financial meltdown have taken place. This data is tested for correlation and ADF statistic to identify mean reverting behaviour of different pair of time series. The data for ICICI Bank, HDFC Bank, Kotak Mahindra Bank, Axis Bank and IndusInd Bank was referred from www.yahoo.com.

ADF test was applied on natural log difference to pairs of time series of daily closing prices of above banking stocks. The more negative the ADF test statistic value and lesser p value enhances the chance of a time series being stationary. A 0.01 p value shows that there are 99% chances of pair of stocks being mean reverting.

By calculating the ADF test statistic for different pairs of stocks, we can find the most stationary time series. The stock pair with the most negative ADF test statistic value will be closest mean reverting time series.

The following implementation process was adopted for this analysis.

Step 1: Find logarithm of the values to normalize the data.
Step 2: Find correlation coefficient between different pairs
Step 3: Find ADF and P-value using MS Excel
Step 4: Reporting the results and analysis

4 Results and Discussion

Based on Table 1 and Fig. 1, the following observations are made.

i. The most negative ADF test statistic −3.4476 is found with Axis Bank-ICICI Bank paired time series. The p-value for this pair is found to be 0.0471. This implies that there is 4.71% chance that Axis Bank-ICICI Bank paired time series is non-stationary or there is 95.29% chance of Axis Bank-ICICI Bank paired time series being mean reverting.

ii. Highest correlation of 0.6312 is found with Axis Bank-ICICI Bank paired time series and least correlation −0.2228 was found between HDFC-IndusInd bank. Even though HDFC-IndusInd Bank pair is negatively correlated, but they are not showing mean reverting behaviour, and hence pair trading cannot be done in this pair of banking stock.

iii. There is no high correlation between any pair of stocks. Axis-ICICI and Kotak-IndusInd pairs are showing the correlation more than 0.50 and rest all are showing very weak correlation.

Table 1. Summary of results

Sr. no.	Script pair	ADF test statistic	p value	Correlation
1	Axis-HDFC Bank	−1.6726	0.7174	0.2106
2	AXIS-ICICI Bank	−3.4476	0.0471	0.6312
3	Axis Bank-Kotak	−1.9308	0.6078	0.0676
4	Axis Bank-IndusInd Bank	−2.5399	0.3495	−0.0046
5	HDFC-ICICI Bank	−1.2539	0.8950	0.1194
6	HDFC-Kotak	−2.1675	0.5074	0.0524
7	HDFC-IndusInd	−2.3369	0.4356	−0.2228
8	ICICI-Kotak	−1.7692	0.6764	0.1188
9	ICICI-IndusInd	−2.1184	0.5283	−0.2023
10	Kotak-IndusInd	−2.4716	0.3781	0.5461

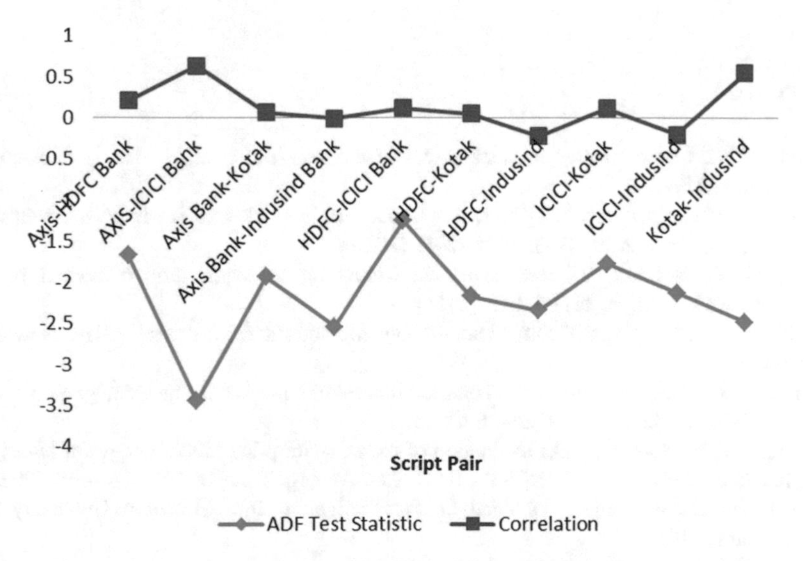

Fig. 1. Graphical comparison of ADF values and correlation coefficients

iv. Three pairs of stocks Axis-IndusInd, HDFC-IndusInd and ICICI-IndusInd are showing negative correlation, rest all the pairs are showing positive correlation. However, most of the pairs are not showing mean reverting behaviour despite having positive correlation.

5 Conclusion

There is a significantly weak correlation between IndusInd bank with other banks except Kotak bank. It signifies that investor should be cautious in pair trading by studying correlation coefficients between pairs of stocks. The good correlation and identified mean reverting behaviour on the basis of ADF test between Axis Bank and ICICI Bank provides positive hopes to the investors for pair trading in this pair.

Based on ADF test statistic, it has been concluded that Axis Bank and ICICI Bank are the most closely associated with one another and they have mean reverting present at more than 95% confidence interval. This also makes them eligible for pair trading. However, careful consideration should be given as the required level of significance for pair trading is 1%.

It is also clear that there is no high correlation between these pairs of stocks but some of them are still showing mean reverting behaviour. Therefore it is concluded that high degree of correlation does not result in mean reverting pattern. It has also been found that positive correlation between pair of scripts does not necessarily result in mean reverting pattern.

These conclusions are made based on study into the data for the period of 2480 trading sessions starting from 1st April 2006 and ending on 31st March 2016. The study suggests the investors to follow the implementation process explained in this paper for pair trading in banking stocks.

References

1. Epps, T.W.: Comovements in stock prices in the very short run. J. Am. Stat. Assoc. **74**, 291–298 (1979)
2. Münnix, M.C., Schäfer, R., Guhr, T.: Impact of the tick-size on financial returns and correlations. Phys. A **389**(21), 4828–4843 (2010)
3. Shapira, Y., Berman, Y., Ben-Jacob, E.: Modelling the short term behaviour of stock markets. New J. Phys. **16**, 053040 (2014)
4. Vidyamurthy, G.: Pairs Trading Quantitative Methods and Analysis. Wiley, New Jersey (2004)
5. Ungever, C.: Pairs trading to the commodities futures market using cointegration method. Int. J. Commer. Financ **1**(1), 25–38 (2015)
6. Gumparthi, S., Sarvanan, D.: Evaluation of paired share price ratio strategy for investments in Indian banking sector. IRACST - Int. J. Res. Manag. Technol **5**(5), 361–367 (2015)
7. Lo, A.W.: Hedge Funds: An Analytic Perspective, p. 260. Princeton University Press, Princeton (2010)

8. Herlemont, D.: Pairs Trading, Convergence Trading, Cointegration. YATS Finances and Technologies (2004)
9. Collin-Dufresne, P., Goldstein, R.S.: Do credit spreads reflect stationary leverage ratios? J. Financ. **56**, 1929–1957 (2001)
10. Resilience Reserve Taskforce.: Report, Institute of Actuaries of Australia (2005) http://www.actuaries.asn.au/Library/RESILIENCE%20TASKFORCE%20REPORT%2014%20June.pdf. Accessed 9 Dec 2016
11. Desai, J., Joshi, N.: Finding the identical twin. Int. Educ. Res. J. **1**(5), 21–23 (2015)

Detection of Diseases on Crops & Design of Recommendation Engine: A Review

Sindhu Bawage(✉) and Bashirahamad Momin

Department of Computer Science and Engineering,
Walchand College of Engineering, Sangli, India
sindhub543@gmail.com, bfmomin@yahoo.com

Abstract. It is very difficult to detect and monitor the diseases manually and also needs the expertise in the field so that process becomes time consuming. Hence, image processing can be used to detect the diseases and further giving the correct recommendation for the detected disease will be the better solution because only detection of disease will not be the helpful. Disease detection process using image processing involves: Image Acquisition from farmers, image pre-processing and enhancement, edge detection and segmentation, feature extraction, classification of extracted features. The process will not stop here a correct recommendation is very necessary to prevent losses which are faced by the farmers. So designing the recommendation system which gives the best doses for detected disease is the good solution to the farmers.

Keywords: Image processing · Machine learning · Classification · Recommendation engine

1 Introduction

Agriculture is one of the main sources of income in India. So that solving agricultural problems using different engineering techniques is the one step towards building developed India. Considering the all problems in farming, disease detection on the crops is one of the major problems which affect the quality as well as the quantity of the agricultural production which results significant reduction in global food production. Monitoring the health and diseases on the crops plays an important key in successful cultivation of crops. Now days, farmers mainly depends on the expertise in the field to do a monitoring and analysis of crop diseases which requires loads of work and processing time. So, disease detection can be made automated so that it will complete in minimum period of time and with accuracy by adding little human interaction. For crop disease management, the diseases need to be detected at their early stage to control their spread speed. Generally crop diseases have the visual symptoms which are helpful in detecting crop diseases. On crops, some general diseases are seen like leaf curl, leaf minor, mosaic, anthracnose, borers and others are viral, bacterial and fungal diseases. In this case image processing can a better option to analyze the diseases. There are many image processing techniques which are useful to process the image and get the correct result. In this paper we are going to explore different image processing techniques which are used at different steps of crop disease detection.

S.C. Satapathy and A. Joshi (eds.), *Information and Communication Technology for Intelligent Systems (ICTIS 2017) - Volume 1*, Smart Innovation, Systems and Technologies 83, DOI 10.1007/978-3-319-63673-3_48

The general steps used for disease detection are as shown in Fig. 1. It consists following steps:

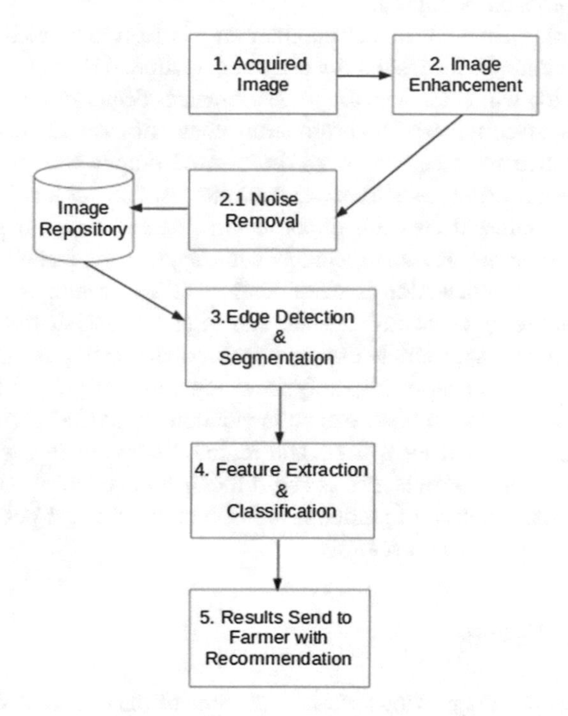

Fig. 1. Steps in disease detection

1. Image acquisition
2. Image pre-processing or enhancement
3. Image segmentation
4. Feature extraction
5. Classification

After successful detection of disease recommendations are given so that design of recommendation engine is necessary. First step is to acquire the disease image. Image can be taken as per the specified guidelines. In image pre-processing the noise from the image is removed so that it will help to detect the correct disease. There are different techniques used for noise removal. Also if there is blurring in image that can be removed in this step. The enhanced image is used for the further processing.

Next step is image segmentation; Segmentation is most important part in image processing. In Image segmentation we partition an image into different segments such as background of the image, regions on the image. The main purpose of segmentation is to simplify the image such that analysis of the image becomes easier and gives the meaningful information about disease characteristics, for that matter representation of image can be changed like RGB image is converted to binary image. Image segmentation is basically used to locate the part of the image infected by the disease,

By analyzing the different segments on the image which are having same characteristic defines a particular disease. There are Different segmentation techniques which can be used as per the problem definition.

After image segmentation, most important step is feature extraction which is also responsible to give the better results for disease detection. The main important task in this step is to decide which features are to be extracted. Features are those segments of the image which gives the useful information about disease so that one can easily analyze the extracted features and gives the desired output with the reduced representation. The visual features such as different shapes, size, color, position, contours and texture of the infected area are important in disease detection process. So considering visual symptoms, suitable techniques are applied for feature extraction.

Last step of disease detection is classification. Classification is process in which unlabeled objects are given to training set and after processing through training set objects are labeled with appropriate category whose membership is known. Depending on the extracted features classifier specify to which class image will belongs to.

The process will not stop here. Recommendation should be given to the farmers depending on the severity of the disease and region. There are variety of pesticides so the predicting which one should use is the difficult task, may be they can harm the crops as well. So there is need of guidelines for direction of use of pesticides according to diseases and quantity to be used.

2 Literature Review

Literature review contains previous work in the area of plant disease detection and the different techniques used for feature extraction and classification. Also recommendation system design current trends are discussed later in this section.

Current techniques used for the plant disease detection are mentioned in the paper [1] by Jayme Garcia Arnal Barbedo. This paper gives the comprehensive survey on the plant disease detection techniques, so that it is helpful for those who are going to start research on the issue of plant disease detection And classification. Different classification algorithms like SVM, Fuzzy Logic, Neural Network etc. are described with the advantages and the disadvantages of it, so that a clearly guide researchers to do the further work. Also Here are some papers which describe different methods for leaf disease detection and suggest various ways of implementation as illustrated and discussed here.

Revathi, Hemalatha proposed a method to detect cotton crop diseases using edge detection techniques. In [2] the proposed HPCCDD uses Sobel edge detection and Canny edge detection to perform segmentation of image. Their work consist of two phases in first phase the affected area of leaf is detected using image analysis. Image analysis on input image is done using the features like RGB pixel counting. Then the image is masked with the green pixels. Further to identify the affected area on the leaf segmentation is performed using edge detection methods like Sobel and Canny edge detection. Finally at the last stage of algorithm extracted RGB features are analyzed and classification is done with the help of computation of texture statistics. Also recommendation about use of pesticides is given to the farmers in their respective languages such as Tamil, Hindi and English along with the predicted disease. [3] This paper segmentation method is used for detection of stained part of leaf affected by anti-pathogen color products. In segmentation process, image converted to HSV color space. Further fuzzy c-means clustering is used in hue-saturation space to distinguish into different pixel classes. Further these classes are merged into two classes at the interactive stages, the searched diseased area is found one of these two classes.

In this, [10] author mentioned about their contribution, where they developed statistical classification using Neural Network. Images were segmented using K-means technique and further given to pre-trained neural network. Here leaf images are partitioned into four clusters using K-means clustering. When leaf is infected by one or more diseases then that image belongs to one or more clusters. K-means tries to cluster image into K number of classes, it finds the actual segments of the image. Further segmented image is given to feature extraction from the infected part of an image. The method used for feature extraction is the Color Co-occurrence Method (CCM method). In this method both the color and texture features of an image are considered, which represent that image. After this step, some statistical analysis of images is done to minimize the redundancy in extracted features. Finally, classification is performed using developed Neural Network algorithm. Here Feed forward back propagation Neural Network is used, the Mean Square Error (MSE) used as performance function and 10000 number of iterations were done and the maximum allowed error was 10–5. This approach of using developed Neural Network successfully detect and classify the tested diseases with a perfection around 93%.

It [9] described the method of feature extraction that applies Gabor wavelet transform technique in conjunction with Support Vector Machine (SVM) and alternate kernel function to detect the diseases that infect the tomato leaves. They have used different kernel function such as Cauchy kernel, Invmult Kernel and Laplacian Kernel to evaluate the ability of function. They acquired the images of tomato leaves in February in temperature 16°–20°, data set included 200 images of infected tomato leaves. Further, pre-processing phase include the process of enhancing image quality by removing noise, enhance smoothness for increasing image quality to increase the feature extraction efficient. In pre-processing, image were included many leaves of tomato so isolation of each leaf in image was done manually, further re-sizing of image is done for the utilization of storage space and to reduce a computational time complexity. Further the output of pre-processing is given for feature extraction phase which is completely based on Gabor wavelet transform technique. The purpose of this phase is to minimize the original data set by measuring specific features or properties of each image including texture and color. This technique is mainly used for describing textural properties of diseased tomato leaves. In order to extract feature vectors pre-processed image is divided into K non-overlapping regions of size $L \times L$. A bank of $S \times L$ Gabor filters are employed for the extracting texture features of each leaf. Final step is Classification phase which is completely based on SVM, Inputs of this phase are training data set feature vectors and the corresponding classes, whereas outputs are the diseases that infected tomato leaf. After evaluation it's found that results of SVM using Cauchy and Laplacian Kernel are 100 and 98% when comparing with Invmult kernel 78%.

This [12] paper gives the survey on different techniques used for classification and segmentation used for disease detection and described the technique with less computational efforts which gives the optimal results. Also to improve recognition rate we can use classification process like Artificial Neural Network, Fuzzy Logic, Bayes classifier and hybrid algorithms. The disease recognition process involves image acquisition, pre-processing followed by segmentation using genetic algorithm and finally segments are used to classify the disease. [13] this paper provides a method in which pixel number statistics is used to calculate leaf area to detect the disease, method which is studied here is for increasing throughput. In the proposed approach, to reduce the redundant data leaf is separated and put on a white paper to take a photo. RGB image is converted to gray scale and then to binary image. After that noise is removed from the image and the green leaf are and total area of leaf is calculated. And severity of the disease is defined.

The main approaches for recommendation system are content-based filtering, collaborative filtering, hybrid recommendation systems, and personality-based recommendation systems which is a new one. [5] This paper presents a hybrid music recommendation system in which musical pieces are ranked by considering two types of data, content-based data and collaborative data, i.e. they mentioned about the base of rating scores given by users and derived from acoustic features of audio signals. To design recommendation engine they have used three-way aspect which probabilistic generative model. Theoretically it explains mechanism for both kind of data which is observed using latent variables corresponds to genres and then observed data like users, features and pieces is decomposed into independent ones with some probability distribution over them with the help of latent variables. So the decomposition helps to adapt model efficiently for increasing number of rating scores and users. In this model users select a genre of their preference and then the genre stochastically generates a musical piece and an acoustic feature.

In [6] paper, they discussed the customized diet recommendation for every user. It observes daily nutrition essence of user and gives the general sign of user to sign indicator. Then developed model gives the recommendation daily as a service. Likely it helps to prohibit and conclude the disease and act as personalized dietitian for user. It gives recommendation based on personal information of the user, family history of disease, lifestyle he/she living and the food he/she consumes. As recommendation user receives guidelines to take care of themselves, food preference so that it help user to prevent and manage disease easily [14, 15]. There are different causes of disease on crops but fungi is most disease causing element. In this paper they focused on fungal diseases on different types of crops such as fruit crops, vegetable crops, commercial crops and cereal crops with the different techniques for each type. They divided disease detection in three parts segmentation, feature selection and classification respectively. For each type there is difference in techniques used for segmentation, feature selection and classifier used.

3 Brief Review on Techniques Used for Disease Detection

In this part we are going to explain a general methodology going to be applied while detecting the disease.

3.1 Image Enhancement

Image enhancement is necessary step in a process of disease detection. So that noise removal should be done to remove noise from an image and given for further processing.

The Gaussian Filtering
Noise can be responsible for the wrong detection of edges. So it is necessary to remove or filter noise from an image to get the accurate results. Gaussian filtering [11] is used to smooth an image, the Gaussian filter convolve with the image to reduce the noise which will help in accurate edge detection

Non-Local means De-noising
Image de-noising is based on the principle of replacing pixel color of image with the color average of pixels nearby. In probability theory, the variance law ensures that standard deviation is to be divided by three when averaged pixel count is nine. Thus, to find color of noised color nice pixels average is to be divides by nine and for average of sixteen colors one can divide it by four. De-noising is done by computing average color of favorable set of pixels. The favorable set is evaluated by comparing a whole set around every pixel, and not just the color. This new technique is called Non Local Means [11].

3.2 Edge Detection and Segmentation

Edge detection is used to distinguish image background and image body to do analysis of disease more accurately and removes redundancy. The most efficient algorithm used are Canny edge detection and thresholding [4]. We can get improved results by optimizing the non-maximum suppression stage. Also distributed canny algorithm can used for improved results [7]. Table 1 given below shows the different techniques used for segmentation and classification.

3.3 Feature Extraction and Classification Techniques

There is no superior method like the segmentation. The most efficient method for classification is depends on problem definition, objectives, image properties, size of data-set and a priori knowledge, also the experience and preference of the users. This section discusses some popular feature extraction techniques which totally depends on features, which are useful for user depending on application and classification techniques like k-Nearest Neighbor, SVM, Neural Network, Genetic Algorithm, Fuzzy Logic useful for classification of images which helps in disease detection.

As we know most of the crop diseases have the visual symptoms so that features which re taken into account are listed below:

1. Shape of objects detected after image segmentation.
2. Size, area of objects.
3. Contours detection and then find the moments of the image.
4. Texture is one of the significant symptom which helps in detection of disease etc.

Table 1. Brief review on different technologies used for crop disease detection

Paper	Methodology
Digital image processing techniques for detecting, quantifying and classifying plant diseases [1]	Present a comprehensive survey on different technologies used for disease detection using image processing
Classification of Cotton Leaf Spot Diseases Using Image Processing Edge Detection Techniques [2]	Sobel and Canny edge detection based segmentation is used and then RGB pixel counting used for image analysis. Finally proposed HPCCDD algorithm is applied for image analysis and classification
Semi-automatic method for the discrimination of diseased regions in detached leaf images using fuzzy c-means clustering [3]	Segmentation consists of image conversion to HSV and then Fuzzy c-means clustering is used to distinguish into two classes
Tomato leaves diseases detection approach based on support vector machines [9]	Feature extraction done using Gabor Wavelet Transform with the help of support vector machine which uses Cauchy kernel, Invmult Kernel and Laplacian Kernel
A Framework for Detection and Classification of Plant Leaf and Stem Diseases [10]	K-means used for image segmentation, for feature extraction Color Co-occurrence Method(CCM) is used and then neural network (feed forward back propagation NN) used for classification
Detection of unhealthy region of plant leaves using Image Processing and Genetic Algorithm [12]	Genetic algorithm is used for segmentation and segmented images are further classified. Suggested classification techniques are Artificial Neural Network, Fuzzy Logic, Bayes classifier and hybrid algorithms

So to find the different features from an image we have to perform multiple pass feature extraction techniques. Every pass in extract the different feature such as shape, size then contours in image. Texture of an image can be identified using the local binary patterns(LBP) method. Also by matching the histogram of an two images we can easily compare the texture of an two images. SIFT (Scale-invariant feature transform) can be used to analyze the local features of the image and SURF (Speeded up robust features) which is speeded up version of SIFT which overcome the slow speed factor of SIFT. These are the some well known algorithms which can be useful for feature extraction.

k-Nearest Neighbor

k-NN is the simplest classifier used for classification in machine learning, where classification is performed by selecting k neighbor which are nearest to that class by calculating the distance between them. Steps in k-NN classification method.

i. User defines sample images for every class.
ii. These samples are selected on apriori knowledge defined class definition and all samples should lie within the range of characteristics for that class.

iii. Designed approach finds different objects based on similarity with the samples and then put these samples to proper class.
iv. Classification gives the better results after few iterations.

k-NN is easy to implement and also gives better results for carefully extracted features. It works well in basic recognition problems. It more accurate because do not predict anything about data.

The main disadvantage of k-NN is that learning is slow. It may be disadvantage in some cases because it do not learn anything from data. Another disadvantage is it becomes slow when the training data is large because algorithm takes time to compute the distance with each sample in training set and sort the result.

Support Vector Machine

SVM is a supervised machine learning method used for analysis of data used for classification and regression analysis. SVM performs linear classification and non-linear classification with different kernel methods. SVM uses hyper plane to divide data into different classes. The samples which are nearest to the margin are selected to determine the hyper plane which is known as support vectors.

The main advantages of SVM is high accuracy. However, data contains error it performs good as classifier. Also the computational complexity doesn't depends n the size of the input data.

Drawbacks of SVM are it requires more training time. Also difficult to understand weight function i.e. kernel formation is difficult. It need more number of support vectors from the raining set.

Neural Network

Neural Networks are computational approach which is based on the collection of neurons (Memory Unit). Neurons are connected to the many other neurons which are contained in different layers. Multi-layer perceptron is one of the feed- forward network, back propagation neural network, probabilistic neural network are the well known neural networks.

There are many advantages of neural networks, like it requires minimum statistical training, it detects the relationships and interactions between dependent and independent variables. Also it minimizes the errors by iterating themselves using back propagation and multiple training algorithms are available. Disadvantages are it works internally so seems like black-box, greater computational complexity, training takes more time.

Fuzzy Logic

In fuzzy logic, unlike NN results can lies in between 0 and 1 which can be multi-value. It also consider the conditions like partial truth where value my range between completely true and completely false. Here we have to assign different value threshold for every class. It assign classes to image objects depending on the identified feature threshold value. Results are easy to edit and more objective than other classifier like NN. It is Useful when the classes are easily separated using one or a few features and appropriate when there is little a priori knowledge about specification of images. Disadvantages include the more computing power, it is not useful when the mathematically description of a problem is not clear and input should be crispy.

4 Conclusion

In disease detection main steps involved are image acquisition, edge detection and segmentation, feature extraction and classification. Its very important to remove noise from image to get the better results. Classification method plays an important role in disease detection. So depending on the type of training data set appropriate classifier should be chosen to get the more accurate result. Depending on the results optimal recommendation engine should be designed.

References

1. Barbedo, J.G.A.: Digital image processing techniques for detecting, quantifying and classifying plant diseases. Springer plus **2**(1), 660 (2013)
2. Revathi, P., Hemalatha, M.: Classification of cotton leaf spot diseases using image processing edge detection techniques, pp. 169–173. IEEE (2012)
3. Sekulska-Nalewajko, J., Goclawski, J.: A semi-automatic method for the discrimination of diseased regions in detached leaf images using fuzzy c-means clustering, pp. 172–175. IEEE Polyana-Svalyava (zakarpattya), Ukraine (2011)
4. Edge detection techniques. https://en.wikipedia.org/wiki/Edge_detection
5. Yoshii, K., Komatani, K.: An efficient hybrid music recommender system using an incrementally trainable probabilistic generative model. IEEE Trans. Audio Speech Lang. Process. **16**, 435–447 (2008)
6. Kim, J.H., Lee, J.H.: Design of diet recommendation system for health care service based on user information. In: IEEE Fourth International Conference on Computer Sciences and Convergence Information Technology (2009)
7. Xu, Q., Varadarajan, S., Chakrabarti, C.: A distributed canny edge detector: algorithm and FPGA implementation. IEEE Trans. Image Process. **23**(7), 2944–2960 (2014)
8. Malik, R., Kheddam, R., Belhadj-Aissa, A.: In: Toward an optimal object-oriented image classification using SVM and MLLH approaches. IEEE (2015)
9. Hefny, H., Mokhtar, U., Ali, M.A.S.: Tomato leaves diseases detection approach based on support vector machines. IEEE (2015)
10. Al Bashish, D., Braik, M., Bani-Ahmad, S.: A framework for detection and classification of plant leaf and stem diseases. IEEE (2010)
11. Buades, A., Coll, B., Morel, J.M.: A non local algorithm for image denoising. IEEE Comput. Vision Pattern Recogn. **2**, 60–65 (2005)
12. Singhi, V., Misrai, A.K.: Detection of unhealthy region of plant leaves using image processing and genetic algorithm. IEEE (2015)
13. Marathe, H., Kothe, P.: Leaf disease detection using image processing techniques. Int. J. Eng. Res. Technol. **2**(3), 2278 (2013)
14. Pujari, J.D., byadgi, A.: Image processing based detection of fungal diseases in plants. Proc. Comput. Sci. **46**, 1802–1808 (2015)
15. Cruz, A.C., Bhanu, B., Thakoor, N.S.: Vision and attention theory based sampling for continuous facial emotion recognition. IEEE Trans. Affec. Comput **5**(4), 418–431 (2015)

SCP: Skyline Computation Planner for Distributed, Update Intensive Environment

R.D. Kulkarni(✉) and B.F. Momin

Department of Computer Science and Engineering,
Walchand College of Engineering, Sangli, Maharashtra, India
rupaliwaje@rediffmail.com, bfmomin@yahoo.com

Abstract. The most promising objects of a multi dimensional dataset are identified by a skyline query. In case of a higher dimensional, distributed, large dataset undergoing the frequent updates, the response time of skyline queries becomes intolerable. It can be significantly improvised, if a proper execution plan is used for the subsequent queries. In this paper, we have proposed a skyline computation model, SCP. The model presents certain strategies which make use of results of the pre-executed queries. Using these strategies, the execution of the subsequent queries is planned in order to achieve a positive gain in response time of the overall skyline computation. The model is suitable for a distributed dataset which is update intensive.

Keywords: Skyline queries · Query profiler · Skyline computing strategies

1 Introduction

A skyline query is a multi-preference query that produces the tuples of user's interest. Specific to a dataset, often there exists a peculiar set of dimensions, which are repeatedly used in queries. This means that some of skyline queries are raised frequently and some queries are near to frequent skyline queries where query attributes get overlapped, few attributed get skipped or few new attributes get added to the query dimensions. In this era of distributed computing, the major costs incurred in skyline computations are (1) peer selection cost and (2) peer communication cost. We find that, in some of the scenarios, above two costs along with the cost of repeated computations can be avoided completely and in other certain cases the execution cost can be minimized if execution of subsequent queries is planned properly by making use of the results of the pre-executed skyline queries. Significance of this work becomes even major when the dataset experiences frequent updates. Our research effort is driven by these points. We focus on presenting a model for a distributed, update intensive dataset to optimize the overall response time of the skyline queries. This model plans the execution of the subsequent queries by making use of the preserved statistics of

S.C. Satapathy and A. Joshi (eds.), *Information and Communication Technology for Intelligent Systems (ICTIS 2017) - Volume 1*, Smart Innovation, Systems and Technologies 83, DOI 10.1007/978-3-319-63673-3_49

pre-executed queries. A data structure called 'Query Profiler' (*QP*), [1] presented earlier by us saves such statistics. For achieving the current research goal, we reuse the concept of *QP* and make new contributions which are summarized as follows:

1. For skyline computation in a distributed, update intensive environment, we have designed a novel model *SCP*, a Skyline Computation Planner.
2. Through the model, we present skyline computation strategies which are used for planning the execution of subsequent skyline queries in order a to achieve a positive shift in the overall response time.

Organization of remainder of the paper is as given next. The second section elaborates background of the concept along with related research efforts. The third section discusses the proposed *SCP* model and the last section highlights the conclusions.

2 Background and Related Work

S. Borzsonyi et. al. came up with the concept of skyline [2] for the first time. They presented *BNL*, *D&C* techniques which compute the skyline by scanning the entire dataset. To further optimize the response time, came the concept of sorting of the dataset (*SFS* [3], *LESS* [4], *SaLSa* [5]) followed by using the efficient structures like heaps, R-trees, cubes, graphs (*BBS* [6], *skyline graphs* [7], *SkyCube* [8]). When the parallel and distributed computing architectures emerged, intelligent structures like CAN and BATON were used extensively for development of skyline computation algorithms. Few other approaches present various techniques for optimizing peer selection and communication costs and they are *DSL* [9], *SSP* [10], *Skyframe* [11], *iSky* [12], *DDS* [13,14], *DiTo* [16] and *PaDSkyline* [17]. Then evolved the algorithms for more extensive computational architectures like grids, *FPGA* where focus was on using parallelism for the optimization of the response time. The examples are: *AGiDS*[15] and the efforts in [18,19]. With the evolution of new programming paradigm of *MapReduce*, various approaches for skyline computation were proposed as in [21–24].

Through the present research contribution, we aim at utilizing the concept of *QP* in the proposed *SCP* model to compute the skylines for a distributed, update intensive dataset in an optimized way. Our approach is different in couple of ways as mentioned next. At first, the proposed *SCP* model makes use of the results of pre-executed skyline queries, preserved in form of *QP*. Making use of previously available results for planning the execution of subsequent queries raised against the distribute dataset, is a totally innovative effort. Secondly, making a strategic plan fo execution of the subsequent skyline queries when updates on the distributed dataset are due, is the first effort in this research field, as per our perception.

The proposed *SCP* model has been detailed in the next section.

3 The *SCP* Model

To start the elaboration, we recall first the data structure *QP* which is used to store the statistics of all queries, executed in past. This structure comprises fields like *QId*: a unique query id, *Att*: list of query attributes, *S*: the skyline, *Sb*: list of all those query ids to which present query is a subset query, *Pr*: list of all those query ids to which present query is a partial query, *Qf*: query frequency and *Min-n* and *Max-n*: which respectively are the n-dimensional arrays which hold the minimum and maximum values respectively for the related query dimensions, from the n dimensional dataset. We also recall the query categorization like exact, subset, partial and novel query from [20] and benefits of this categorization that a skyline of an exact and a subset query is produced without a dataset access, provided the results of the previous skyline queries are available. A skyline for a subset query is served from the intersection of skylines of its parent queries and in case of a partial query, it is served fast from the initial set generated by taking union of skylines of its partial parents. The Table 1 given next provides the summary of the notations used in *SCP* model.

Table 1. Notations used by *SCP* model

SQ	Skyline query
QC	Query categorizer
C	Query category
E, *Sb*	Exact and subset query
N, *Pr*	Novel and partial query
QPM	Query profile manager
SKP	Sub skies processor
PUM	Peer and update manager
LUM	Lazy update manager
AS4	Affects skyline by strategy 4
AS5	Affects skyline by strategy 5
SN3	Skyline not affected by strategy 3
U	Set of updates

The next subsection discusses the proposed model in detail.

3.1 Overview of *SCP* Model

This model works at the centralized server, the query receiver, connected to the p peers that hold the distributed dataset (distributed either horizontally or vertically equally, whichever may be the case). Based on the observations in practice, we make the assumptions that in case of vertically distributed dataset, single attribute of the dataset is allocated to each peer and in case of the horizontally distributed dataset; equal, contiguous splits are allocated to the p peers. Also an assumption that the skyline query contains the minimum-value preferences on the attributes of the dataset. Figure 1 shows the *SCP* model.

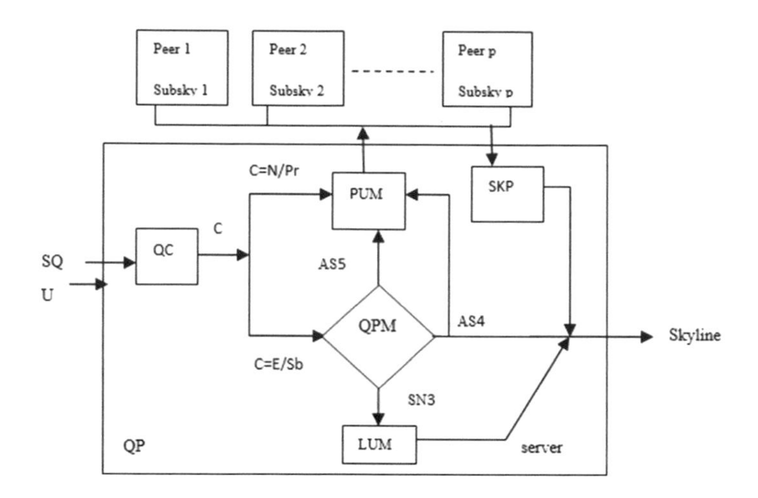

Fig. 1. The *SCP* Model.

As per this model, the centralized server which stores *QP* receives the skyline queries *SQ* from the users. The server forwards the *SQ* to the *QC* unit, which refers to *QP* and accordingly computes the category C of *SQ*. It works in two scenarios as described next.

Scenario 1: No updates on the dataset are outstanding:
In this scenario, The skyline computation strategies are as follows.
Strategy 1: For the exact and subset queries, compute the skyline referring to *QP* (without the dataset access).
Strategy 2: For the novel or partial queries, compute the skyline referring to proper peers. (by accessing distributed dataset).

The exact and subset queries are forwarded to the *QPM* unit. The *QPM* unit refers to *QP* and generates the skyline of *SQ* depending upon the category of *SQ* (as explained in Sect. 3) and immediately returns it to the user. This strategy saves the peer selection cost, peer communication cost and of course the query re computation cost as well. The *QPM* unit also updates the *QP* on relevant fields. (Qf field gets updated for an exact query, new *QP* entry is made for the subset query). All novel or partial queries are forwarded to the *PUM* unit. The *PUM* unit contains the information about the data distribution and hence it identifies the proper peers which will participate in computation of the skyline. For a novel query, the *PUM* unit finds out the proper peers and in case of a partial query, it refers to *QP* and computes the initial set. Then it sends a message (for a novel query) or the initial set (for a partial query) to all the proper peers to compute the sub sky across the data split available with each one of them. Each proper peer computes the sub sky and sends it to the *SKP* unit. In both the strategies, the *SKP* unit which receives all the sub skies merges them all to produce the final skyline. The *PUM* unit also updates the *QP* on relevant fields (new *QP* entry is made.)

The first strategy results in savings of both the peer selection and communication costs. The computational efforts are totally saved. The second strategy is followed only when dataset access is mandatory.

Now let us discuss another scenario where updates are outstanding on the dataset. We have assumed that the dataset distributed amongst p peers, undergoes the "insert" updates which take place at the pth (last) peer. In this scenario, the model works as described next.

Scenario 2: The distributed dataset undergoes the updates.

In addition to *SQ*, the centralized server receives receives U from the dataset administrator. For each single update u in U, the server hands over the current *SQ* to *QC* which refers to *QP* and accordingly generates the query category C of the current query. For the novel (N) and the partial (Pr) queries, skyline needs to be computed. The steps are quite strait: update dataset and re-compute the skyline. So, the current update u is directed to *PUM* unit for its update on the pth peer. The pth sub sky is recomputed and the *SKP* unit generates the final skyline after merging the p sub skies. However, for the cxact and subset queries, there is always a chance for response time optimization. And this major task is handled by the *QPM* unit. It is responsible for deciding whether the current update u on the dataset will affect the skyline of *SQ* or not. For making this decision, *QPM* refers to *QP* and works on following strategies.

Strategy 3: Return old skyline for *SQ* as the insertion of the tuple t does not affect the skyline of *SQ* iff all the values of the tuple t happen to be greater than their related dimensions' maximum values.

Explanation: Since the skyline will always contain the minimum values for all the query attributes, the tuple t having all values greater than the related dimensions' maximum values does not affect the skyline of *SQ*. The decision of *QPM* is *SN3*. The old skyline is returned and current update u is recorded for the lazy update on the dataset by the *LUM* unit. Here, both the peer selection and the communication costs are saved. The re computation time of *SQ* under update u is totally saved and *QP* remains unchanged.

Strategy 4: Return the skyline as t of *SQ* as the insertion of the tuple t makes t as the skyline of *SQ* iff all the values of the tuple t happen to be lesser than their related dimensions' minimum values.

Explanation: Since final skyline always contains the minimum values for all the query attributes, the tuple t having all values lesser than the related dimensions' minimum values becomes the only tuple eligible for skyline of *SQ*. The decision of *QPM* is *AS4*. And the new skyline is t. The skyline is immediately returned to the user and current update u is sent for the immediate update on the dataset to the *PUM* unit. This type of update saves the re-computation time of *SQ*. Now t gets added to the dataset split at the pth peer and the *SKP* unit updates the pth sub sky available with it accordingly. The *QP* stored at server is updated to reflect new values (as the S, Min-n field get updated).

Strategy 5: Re compute the skyline of *SQ* as the addition of the tuple t may affect the skyline of *SQ* iff one or more values of the tuple t happen to be equal or lesser than their related dimensions' minimum values.

Explanation: The tuple needs to be examined for its eligibility to contribute to the newer skyline even when a single value of the new tuple t happens to be either equal or lesser than the related dimension's minimum value. Its addition to the dataset can make other tuples eligible for contributing in the sky or can eliminate existing skyline tuples. The decision of *QPM* is *AS5*. And the new skyline is re computed. The current update u is sent for the immediate update on the dataset to the *PUM* unit and the skyline is computed as in case of a novel query. The *QP* stored at server is updated to reflect new values.

To conclude we assert that, for a *SQ* of exact or subset type, any outstanding update that fits in strategy 3, results in saving of both the peer selection and communication costs. The re computational efforts for *SQ* are totally saved. The update that fits in strategy 4, results in saving of the re computational efforts as skyline is immediately returned to the user. And that for any update that fits in strategy 5, the peer communication, selection, update and re computation costs can not be avoided.

We have performed extensive experiments for testing the performance of *SCP* model and the observations made have been discussed next.

3.2 Experimental Work

In all our experiments, a dataset of players available at www.basketball.com has been used. A server machine is used with specifications as: Intel Core i-3 2100 CPU, 3.10 GHz, 2GB RAM having Windows 7 environments. The peer machines also have the same configurations.

Our first experiment compares various skyline computation techniques when the computations are done under scenario 1 (no outstanding updates on dataset). It comprises testing the performance of four different techniques as: (1) *HNQP*: Peers have the horizontal splits, no *SCP* model used, (2) *HQP*: Peers have the horizontal splits, the *SCP* model has been used, (3) *VNQP*: Peers have the pre-sorted vertical splits, no *SCP* model used, (4) *VQP*: Peers have the pre-sorted vertical splits, the *SCP* model has been used. We observe the effect on response times when (1) cardinality of dataset and (2) total number of skyline queries are varied. The figures given next, elaborate the observations.

The experiments comprises different parameters as: t: total number of tuples in dataset,a: total number of attributes, s: total number of queries raised, p: total number of peers used.

In Fig. 2, the variations in response time is shown when the cardinality of the dataset is varied. The values set for the parameters are: $a = 5$, $s = 10$, $p = 2$. The parameter t is changed in range: 3000-18,000. From Fig. 2 it becomes clear that

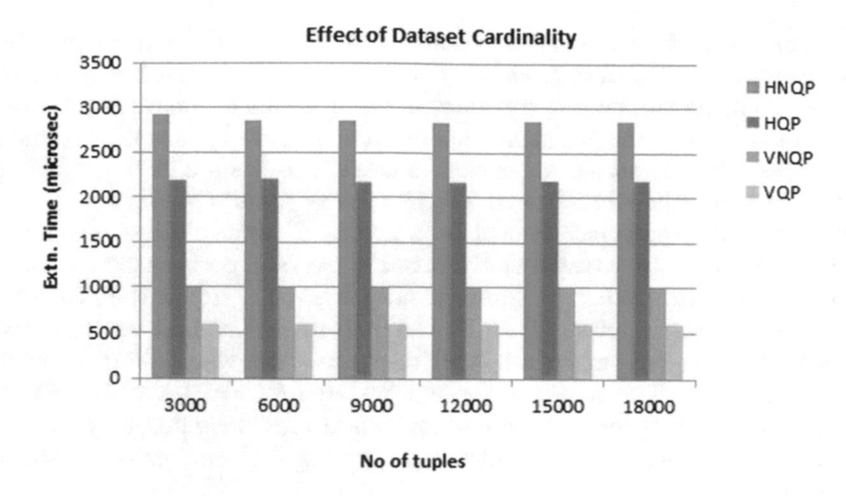

Fig. 2. Response time variations against variations in dataset cardinality.

techniques which use *SCP* model, exhibit positive gain in the response time. This gain is a result of the various strategies of the *SCP* model, that the server utilizes for either saving or minimizing the peer selection and communication costs. For the techniques *VQP* and *VNQP*, because of pre-sorted dimensions allocated to peers, returning the minimum of the dimension as demanded by the queries, becomes the constant time job. So, the best results are obtained with the *VQP* technique.

In Fig. 3, the variations in response time is shown when the total number of queries raised, are varied. The values set for the parameters are: $a = 5$, $t = 18{,}000$,

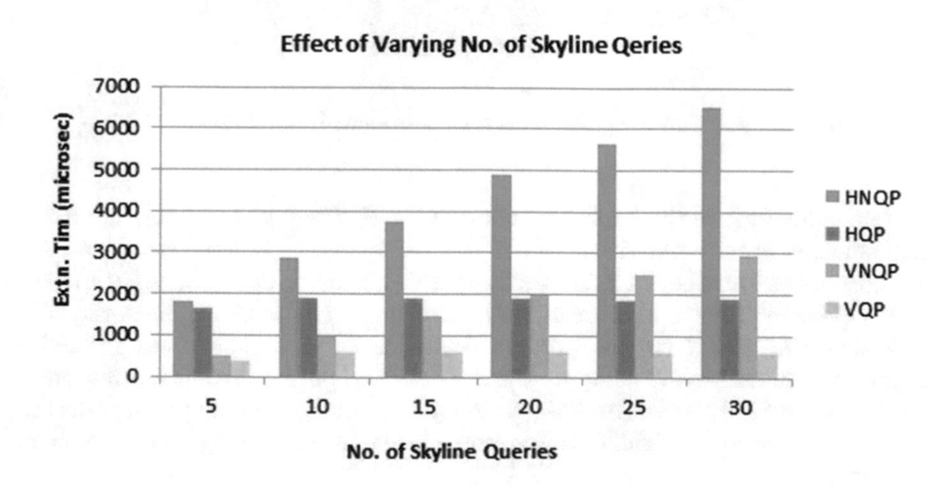

Fig. 3. Response time variations against variations of total queries.

$p = 2$. The parameter s is changed in range: 5–30. As the number of queries raised against the dataset increase, probability of occurrence of exact, subset and partial queries also increases and hence chances of optimization of the response time do also increase. Again, the optimum response time is given by the *VQP* technique as it works with pre-sorted dimensions allocated to peers and that *HQP* and *VQP* outperform their *HNQP* and *VNQP* counter parts respectively.

The second experiment focuses on scenario 2 (when dataset experiences updates). A horizontally distributed dataset is used and performance of two different techniques is evaluated. These techniques are: (1) *HQP'*: the *SCP* model has not been used at server, (2) *HQP*: the *SCP* model has been used at server. The experiment monitors variations in response times when different number of peers are used. The values of the parameters are: $a = 5$, s = 15, $t = 18000$. An additional parameter is u: number of updates for the dataset, is set to 15. The parameter p is changed in range 2–6. The Fig. 4 given next,elaborates the observations.

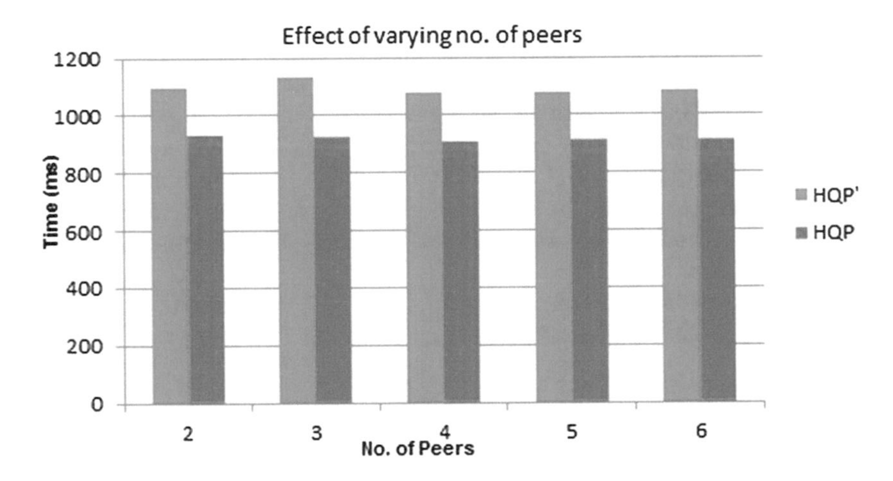

Fig. 4. Response time variations against variations of total number of peers.

Here again, due to the implementation of the *SCP* model at server, the *HQP* technique performs better than the *HQP'* technique. It is interesting to note here that as the number of peers involved are increased, decay in the response time is observed due to each peer has to test lesser tuples to compute the sub skies. However afterwards this increase causes the server to spend more time in merging the sub skies and compute the final skyline. So, a little shift in response time is observed. However, the *HQP* technique has been observed to outperform.

The next section highlights the conclusions drawn along with possible extensions.

4 Conclusions and Future Scope

Our research efforts targeted the problem of optimizing the response time of the skyline queries that are raised against distributed datasets undergoing frequent updates. To cater for this need, we have proposed *SCP* model. The model uses various strategies in order to achieve overall, positive gains in the response time by either avoiding or minimizing the peer selection and communication costs. The effectiveness of this model is asserted by the rigorous experimentations performed by us.

In future, we aim to extend this work, by further enhancing the *SCP* model to include the strategies for all possible updates on the datasets (insert, delete and update a tuple) and for the updates that may happen on any random peer. We also need to figure out what number of peers should be used so that peer communication and sub sky merging costs do not affect the positive gains in the response time.

References

1. Kulkarni, R.D., Momin, B.F.: Skyline computation for frequent queries in update intensive environment. J. Elsevier, King Saud Univ. Comput. Inf. Sci. **28**(4), 447–456 (2016)
2. Borzsonyi, S., Kossmann, D., Stocker, K.: The Skyline Operator. In: Proceedings of IEEE International Conference on Data Engineering, pp. 421–430 (2001)
3. Chomicki, J., Godfrey, P., Gryz, J., Liang, D.: Skyline with presorting. In: IEEE International Conference on Data Engineering, pp. 717–719 (2003)
4. Godfrey, P., Shipley, P., Gryz, J.: Maximal vector computation in large data sets. In: IEEE International Conference on Very Large Databases, pp. 229–240 (2005)
5. Bartolini, I., Ciaccia, P., Patella, M.: SaLSa: computing the skyline without scanning the whole sky. In: ACM International Conference on Information and Knowledge Management, pp. 405–411 (2006)
6. Papadias, D., Tao, Y., Fu, G., Seeger, B.: Progressive skyline computation in database systems. ACM Trans. Database Syst. **30**(1), 41–82 (2005)
7. Zheng, W., Zou, L., Lian, X., Hong, L., Zhao, D.: Efficient subgraph skyline search over large graphs. In: ACM International Conference on Information and Knowledge Management, pp. 1529–1538 (2014)
8. Xia, T., Zhang, D.: Refreshing the sky: the compressed skycube with efficient support for frequent updates. In: ACM SIGMOD International Conference on Management of Data, pp. 493–501 (2005)
9. Wu, P., Zhang, C., Feng, Y., Zhao, B., Agrawal, D., Abbadi, A.: Parallelizing skyline queries for scalable distribution. In: IEEE International Conference on Extending Database Technology, pp. 112–130 (2006)
10. Zhang, N., Li, C., Hassan, N., Rajasekaran, S., Das, G.: On skyline groups. IEEE Trans. Knowl. Data Eng. **26**(4), 942–956 (2014)
11. Wang, S., Vu, Q., Ooi, B., Tung, A., Xu, L.: Skyframe: a framework for skyline query processing in peer-to-peer systems. VLDB J. **18**(1), 345–362 (2009)
12. Chen, L., Cui, B., Lu, H., Xu, L., Xu, Q.: iSky: efficient and progressive skyline computing in a structured P2P network. In: IEEE International Conference on Distributed Computing Systems, pp. 160–167 (2008)

13. Hose, K., Lemke, C., Sattler, K.: Processing relaxed skylines in PDMS using distributed data summaries. In: ACM International Conference on Information and Knowledge Management, pp. 425–434 (2006)
14. Hose, K., Lemke, C., Sattler, K., Zinn, D.: A relaxed but not necessarily constrained way from the top to the sky. In: ACM International Conference on On the Move to Meaningful Internet Systems, pp. 339–407 (2007)
15. Junior, R., Vlachou, J. A., Doulkeridis, C., Nørvág, K. :AGiDS: a grid-based strategy for distributed skyline query processing. In: ACM International Conference on Data Management in Grid and Peer-to-Peer Systems, pp. 12–23 (2009)
16. Vlachou, A., Doulkeridis, C., Nørvåg, K.: Distributed top-k query processing by exploiting skyline summaries. J Distrib. Parallel Databases **30**(3–4), 239–271 (2012)
17. Chen, L., Cui, B., Lu, H.: Constrained skyline query processing against distributed data sites. IEEE Trans. Knowl. Data Eng. **23**(2), 204–217 (2011)
18. Woods, L., Alonso, G., Teubner, J.: Parallel computation of skyline queries. In: IEEE 21st Annual International Symposium on Field-Programmable Custom Computing Machines, pp. 1–8 (2008)
19. Papapetrou, O., Garofalakis, M.: Continuous fragmented skylines over distributed streams. In: IEEE International Conference on Data Engineering, pp. 124–135 (2014)
20. Bhattacharya, A., Teja, P., Dutta, S.: Caching stars in the sky: a semantic caching approach to accelerate skyline queries. In: International Conference on Database and Expert systems Applications, pp. 493–501 (2011)
21. Li, Y., Qu, W., Li, Z., Xu, Y., Ji, C., Wu, J.: Parallel dynamic skyline query using MapReduce. In: IEEE International Conference on Cloud Computing and Big data, pp. 95–100 (2014)
22. Park, Y., Min, J., Shim, K.: Parallel computation of skyline and reverse skyline queries using MapReduce. J. VLDB Endowment **6**(14), 2002–2013 (2013)
23. Zhang, J., Jiang, J., Ku, W., Qin, X.: Efficient parallel skyline evaluation using mapreduce. IEEE Trans. Parallel Distrib. Syst. **27**(7), 1996–2009 (2016)
24. Bai, M., Xin, J., Wang, G., Zimmermann, R., Wang, X.: Skyline-join query processing in distributed databases. J. Front. Comput. Sci. **10**(2), 330–352 (2016)

Application of Automatic Query Analysis Technique for a Web Learning System in DSP

Snehlata Deshpremi[1] and Suryakant Soni[2](✉)

[1] Department of Electronics and Communication,
Medi-Caps Institute of Science and Technology, Indore, Madhya Pradesh, India
snehlatadeshpremi@gmail.com

[2] Department of Computer Science,
Medi-Caps Institute of Science and Technology, Indore, Madhya Pradesh, India
Suryakant.soni@medicaps.ac.in

Abstract. Question answering learning systems based on AI have been subject of study since long. AI helps in fast access to most suitable answer in a database of question/answer pairs.

This paper presents a web learning system (WLS) which answers the frequently asked questions in DSP. The system uses a web Data Store for the answers. The system uses natural language generation (NLG), natural language processing (NLP), question logs and question classification to search out a certain answer to a submitted question. WLS also includes a self-test subsystem.

Keywords: Questing answering · WLS · Query analysis

1 Introduction

A website is a collection of web objects and data stores, which are accessed using World Wide Web and the web browser [1].

Question Answering (QA) may be an approach for mechanically responsive an issue display in terminology. Related to keyword-based research organization, it highly helps the communication between peoples and computers by naturally stating human motive in simple sentences. It conjointly excludes the conscientious searching of an enormous amount of knowledge contents came back by search engines for the right answer. After all, totally machine-controlled QA still faces defiance that are not straightforward to control's, like the vast understanding of complicated queries and therefore the refined grammar, linguistics and discourse process to get answers. It search that, in most circumstances, machine-controlled approach cannot acquire results that square measure pretty much as good as those generated by users intelligence [2].

Many applications on the online need accessing the structure outlined by the links. Example, typical tasks in computing machine management embrace checking for supporting links, finding often used ways, finding inaccessible documents. This paper presents a web application for Frequently Asked Question (FAQ) in Digital Signal Processing (DSP).

The term list dispatch to a documentation format that number of queries, as they are, or can be, inquiry by the users, and therefore the Compatible appropriate answers [3].

S.C. Satapathy and A. Joshi (eds.), *Information and Communication Technology for Intelligent Systems (ICTIS 2017) - Volume 1*, Smart Innovation, Systems and Technologies 83, DOI 10.1007/978-3-319-63673-3_50

2 Related Work

There square measure alternative reasons why queries that mix content and structure square measure vital.

An anon analysis of QA organization started in Sixties and chiefly targeted on skilled organization in conspicuous domains. Recital rooted mostly QA has obtain its analysis quality since the institution of a QA track in TREC within the late Nineteen Nineties. form of queries and required respondence, we are able to approximate compress the kinds of QA into Open-Domain QA, Restricted-Domain QA, Definitional QA and number of QA. After all, in spite of the action as delineated on top of, automatic QA still has problems in responsive complicated queries [4].

QA may be a massive and various questioning-answering forum, executive as not solely a corpus for exchange technical data however conjointly an area wherever one will obtain recommendation and ideas. After all, nearly all of the prevailing Community QA systems, like Yahoo! Answers, Wiki-Answers and Ask-Meta-filter, which cannot give intuitive and adequate info. Few analysis try are placed on transmission QA, that aims to answering queries victimization transmission information. An anon organization named Video QA was given in. This method expand the recital-based QA terminology to support factoid QA by investment the visual list of stories video still because the recital transcripts.

Chua et al. [5] planned a generalized viewpoint to increase recital-based QA to transmission QA for a variety of factoids, definition and "how-to" queries. Their system was designe to search out transmission answers from web-scale media resources like fotki, Instagram, Flicker, YouTube and daily motion etc. However, this paper concerning transmission QA continues to be comparatively thin.

Multi-media QA solely performance in particular domains and might hardly control complicated queries. Completely various from these works, our viewpoint is makes supported QA. Rather than outright collection transmission information for responsive queries, this paper methodology solely finds pictures and videos to complement the matter answers provided by users [2]. This made our approached ready to subsume many general queries and to realize higher performance.

3 Prospective Methodolgy

Figure 1 Display the diagram of planned methodology.

3.1 Query

A query is variety of a questioning-answering, in an exceedingly line of inquiry.

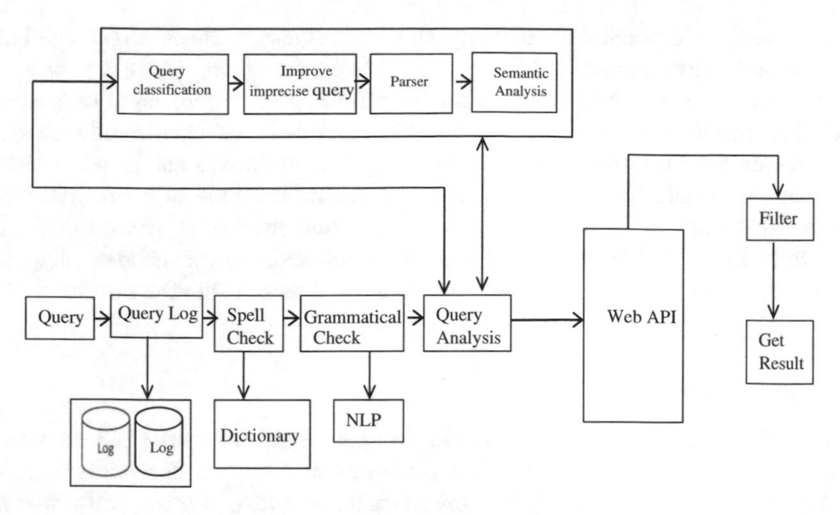

Fig. 1. Architecture of proposed query system

3.2 Query Log

Query log that serves the twin main aim of self-solving the question and after helps to group of the questions and answers. The automated complete the queries happens on the instant whereas the user is coming into the question. It produces the foremost approximately sentences in line with user's question and shows it.

3.3 Spell Analysis

This block rectifies the writing system mistake if any and the question is prepared for grammatical check. Additionally check the writing system of the words within the question and conclude if there is any mistake. This system uses a wordbook lexicon within which the errors and their corrected spellings are predefined.

3.4 Grammatical Check

This block concludes the grammatical errors in sentences if any and the question is prepared for analysis. This paper additionally checks the synchronic linguistics of the sentences within the question and rectifies them if there is any mistake. This block uses an information science within which the errors and their corrected synchronic linguistics are predefined.

This paper additionally presents synchronic linguistics check some small-small words is automatic corrected. For e.g.: “is”, “are” etc. Also, we’ve got else some common errors created by the user within the ofttimes used correct nouns or names and given their correct writing system. A synchronic linguistics checker is software system or a program feature found in an exceedingly application and is employed to seek out improper synchronic linguistics. Associate in nursing example of a software system program that features its own synchronic linguistics checker is Microsoft Word or Microsoft stand out. MS Word and MS stand out underlines synchronic linguistics errors with an inexperienced crooked underline as shown within the document [6].

3.5 Query Analysis

Query analysis may be methods utilized in databases that build use of SQL to work out a way to additional optimize queries for performance. Question analysis is a crucial.

Aspect of question process because it helps improve overall performance of question process, which is able to speed up several information functions and aspects. To do this, a question optimizer analyses a selected question statement and generates each remote and native access plans to be used on the query fragment, supported the resource price of every arrange. This block used the ideas of tongue process, tongue generation and linguistics analysis of the question and includes four parts: question classification, improve inexact question, programmer and linguistics analysis.

3.5.1 Question Classification

A query topic divided/categorization may be a downside in informatics. The action is to assign an online searching question to at least one or additional predefined classes supported its topics. The importance of question classification is under scored by several services provided by net search. An immediate application is to produce higher search result pages for users with interests of various classes.

3.5.2 Improve Inexact Query

A user question that needs a detailed however not essentially actual match is Associate in Nursing inexact question. Answers to such a question should be hierarchical in line with their nearest to the questions and answers constraints. as an example, the question Q:- CarDB is Associate in Nursing inexact question, the answers to that should have the attribute build certain by a worth kind of like Ford [7]. Our planned approach for responsive Associate in Nursing inexact choice question over an information.

3.5.3 Parser

A programme may be a software system part that takes computer file (frequently recital) and builds an information structure – typically, some reasonably analyze tree, abstract syntax tree or different data structure – giving a structural illustration of the input, checking for proper syntax within the method. The parsing could also be preceded or followed by different steps, or these could also be combined into one-step [14].

3.6 Web API

A web API is Associate in nursing application-programming interface (API) for either an online server or an online browser.

Server-side: A server-side net API may be a programmatic interface consisting of 1 or additional in public exposed endpoints to an outlined request-response message system, generally expressed in JSON or XML, that is exposed via the net—most usually by suggests that of Associate in Nursing HTTP-based web server. Mashups are net applications that mix the employment of multiple server-side net arthropod geniuses.

Net hooks are server-side net arthropod genus that take as input a homogenous Resource symbol (URI) that's designed to be used sort of a remote named pipe or a kind of decision back such the server acts as a shopper to dereference the provided URI and trigger an occasion on another server that handles this event so providing a kind of peer-to-peer IPC.

3.7 Filtering Module

Filters immaterial answers from our analysed result and provides US the precious result

3.8 Get Result

This displays the question and answer with all the subject of DSP. This paper additionally includes some keywords as an example 'signals', 'system'. Additionally show self-tests with explained answers.

4 Results

1. The projected system is shown in fig. a pair of the question is analyses through computer programmer and also the results shown within the figure. The output of the question is visualized. Question is parsed in token and these token are searched within the information and that we get presides result per the question (Fig. 2).

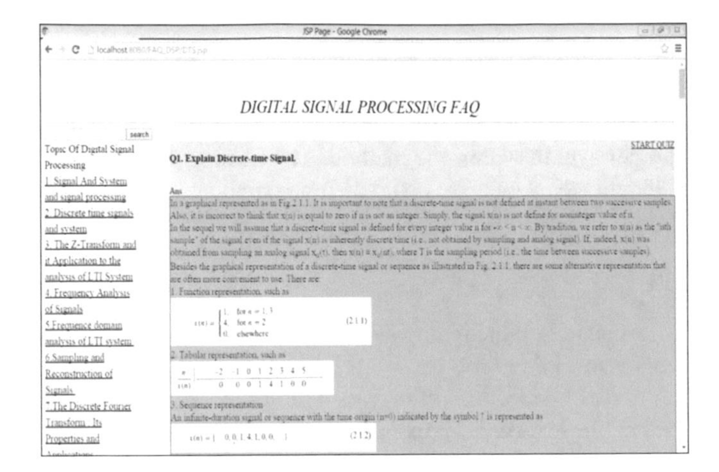

Fig. 2. DSP's topic shown

2. Quiz is additionally performed during this project per the given topics. Cut-off date is given within the quiz for finding the matter (Fig. 3).

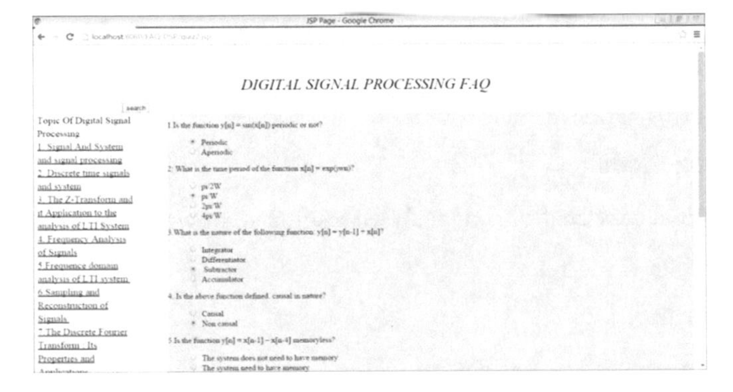

Fig. 3. Result for quiz

5 Conclusion

Our projected system includes an immense repository of question log. Question log covers numerous varieties of QAs with DSP etc. The spell-analysis checks the ofttimes typewritten mistake and it even have majority of the ofttimes-used common nouns. Synchronic linguistics check covers the ofttimes typewritten mistake. Inaccurate question are going to be improved. This method is in a position to analysis the linguistics that means of majority of the queries supported classifications and numerous conditions. It with success identifies the keyword or key phrases to be searching.

References

1. Prabhumoge, S., Rai, P., Sandhu, L.S., Sowmya Kamath, P.L.: Automated query analysis techniques for semantic based question answering system. In: IEEE, International Conference on Recent Trends in Information Technology (2014)
2. Nie, L., Wang, M., Gao, Y., Zha, Z.-J., Chua, T.-S.: Beyond recital QA: multimedia answer generation by harvesting web information. IEEE Trans. Multimedia **15**(2), 426–428 (2013)
3. Henß, S., Monperrus, M., Mezini, M.: Semi-automatically extracting FAQs to improve accessibility of software development knowledge. In: ICSE, pp. 794–797. IEEE (2012)
4. Mendelzon, A.O., Mihaila, G.A., Milo T.: Querying the World Wide Web. In: IEEE (1996)
5. Chua, T.-S., Hong, R., Li, G., Tang, J.: From recital question-answering to multimedia QA on web-scale media resources. In: Proceedings of the ACM Workshop Large-Scale Multimedia Retrieval and Mining (2009)
6. http://www.computerhope.com/jargon/g/grammarc.htm
7. Nambiar, U., Kambhampati, S.: Answering imprecise queries over autonomous web databases. In: Proceedings of the 22nd International Conference on Data Engineering (ICDE 2006). IEEE (2006)
8. Lopez, V., Pasin, M., Motta, E.: AquaLog: an ontology portable question answering system for the semantic web. In: Gomez Perez, A., Euzenat, J. (eds.) ESWC. Lecture Notes in Computer Science, vol. 3532. Springer (2005)
9. Bonino, D., Corno, F., Farinetti, L.: Dose: a distributed open semantic elaboration platform. In: ICTAI. IEEE Computer Society (2003)
10. Bonino, D., Corno, F., Farinetti, L., Bosca, A.: Ontology drive semantic search. WSEAS Trans. Inf. Sci. Appl. **1**(6), 1597–1605 (2004)
11. Celino, I., Valle, E.D., Cerzza, D., Turati, A.: Squiggle: a semantic search engine for indexing and retrieval of multimedia content. In: Proceedings of the 1st International Workshop on Semantic-Enhanced Multimedia Presentation Systems (SAMT 2006) (2006)
12. Mayeld, J., Finin, T.: Information retrieval on the semantic web: Integrating inference and retrieval. In: Proceedings of Workshop Semantic Web at the 26th International ACM SIGIR Conference Research and Development in Information Retrieval (2003)
13. Wen, J-R., Nie, J-Y., Zhang, J-R.: Clustering user queries of a search engine. Copyright is held by the author/owner. In: WWW10, 1–5 May 2001, Hong Kong. ACM 1-58113-348-0/01/0005
14. Li, Y., Wang, Y., Huang, X.: A relation-based search engine in semantic web. IEEE Trans. Knowl. Data Eng. **19**(2), 273–277 (2007)
15. https://en.wikipedia.org/wiki/Parsing#Computer-languages. Accessed Feb 2014

Statistical Study to Prove Importance of Causal Relationship Extraction in Rare Class Classification

Pratik A. Barot(✉) and H.B. Jethva

L D College of Engineering, Ahmedabad, India
hbjethva@gmail.com

Abstract. Rare class classification is important technique in real-world domains like medical diagnosis, bioinformatics, detection of oil spills in satellite images, road accident analysis etc. Imbalanced data classification resides among the top research area of current time and attracted huge interest of researcher. For rare class classification most of researchers have concentrated their study on use of methods like sampling techniques, one-class learning and ensemble based methods. But these methods suffer by some drawbacks. Here we first present an overview of imbalanced data and rare class classification, and then extract unique causes of target class using association rule mining and show importance of causal relationship in determining target class of an instance.

Keywords: Rare class · Imbalanced data · Association rule

1 Introduction

Imbalanced data does not have equal class distribution. Imbalanced data is very common in many real world domains [7, 8, 11, 13, 15], and has attract a significant attention in the fields of data mining and machine learning. However, improving the performance of imbalanced data classification has not been researched as extensively. Most existing techniques for improving classification performance on imbalanced datasets are designed to be applied directly on binary class imbalanced datasets [7].

Association rule are used to extract frequent patterns from dataset. From the extracted association rule interesting rules are determined using support and confidence value. In this paper we present a novel approach to use association rule in extraction of relevant causes of given action. We use association rule mining to extract rare class specific features value and used it to determine most relevant feature of target class.

1.1 Imbalanced Data and Rare Class Classification

Imbalanced dataset belongs to two categories: - Between-class imbalanced and Within-class imbalanced [11].

Between-class imbalanced dataset does not have equal class distribution. This means dataset has negative and positive instances and negative class instances

S.C. Satapathy and A. Joshi (eds.), *Information and Communication Technology for Intelligent Systems (ICTIS 2017) - Volume 1*, Smart Innovation, Systems and Technologies 83, DOI 10.1007/978-3-319-63673-3_51

outnumbered by positive class instances [7, 9, 15]. Imbalanced dataset which has only two classes is called binary-class imbalanced data. Class with comparatively less number of instances is called minority class or rare class [19].

As per Jerzy Stefanowski et al. [11], between-class imbalance is evaluated by global imbalance ratio (IR). There is no fix IR by which one can determines whether dataset is imbalance or not. However as per [15], if traffic accident dataset have minority examples less than 35% of dataset then it is consider as imbalance.

As per [11], in case of relative imbalance it is possible to collect more examples and increase size of dataset by maintaining global imbalance ratio. In this case the absolute cardinality of minority class is no rarer and it may be easier to learn. On the other hand, in the case of absolute rare class the minority class instances are very rare and sampling method may create large dataset in order to make rare class instances noticeable. This restricts use of sampling method for absolute rare class.

Rare class classification is the data mining task for building a model that can correctly classify both the majority and minority classes. Classifying minority or rare class is difficult because size of the rare class is too small [11, 19].

Example – From the data of 10,000 cancer probable patients only one or two person (s) might have cancer while others are lucky and cancer free. This is imbalance class distribution.

Researcher paid good attention on classification of imbalanced data. Most of classification algorithms accurately classify majority class instances but they are fails in accurate classification of minority class instances [7, 15]. Classification becomes even more complicated and less accurate if dataset is multi-class imbalanced [7]. Traditional classification algorithms with accuracy above 80% favor majority class and incorrectly classifying minority class instances.

In addition to IR, other characteristics of data like small disjunction, overlapping of classes and borderline example need to be taken care for accurate rare class classification. Overlapping and small disjunction in dataset makes cluster and outlier analysis techniques unreliable for rare class classification.

2 Related Work

In 2015, Dongmei Zhang et al. [18] proposed ensemble method for classification of imbalance sentiment data. They used under-sampling, re-sampling, and random selection of feature space. Sampling approach has disadvantage of increase computation time and lose of important information while random selection of feature space does not works for most of the real-world problems. The authors also accept the importance of feature selection and mention it as their future work. In the same year Isaac Triguero et al. [17] proposed ROSEFW-RF algorithm for extremely imbalanced big data bioinformatics problem. In their study they perform preprocessing strategies namely oversampling, differential evolution feature weighting before building a learning model. Due to oversampling the data size grows further which increases learning time [7, 9] and require more number of maps for MapReduce approach. In their study Isaac et al. [17] realizes the requirement for evolutionary feature selection approach for imbalanced data.

In Oct-2016, David Muchlinski et al. [8] proved that statistical method like logistic regression performs poorly in prediction of rare event in imbalanced data. They found that the algorithmic approach provides significantly more accurate predictions of civil war onset in out-of-sample data than any of the logistic regression models. As pre-processing step they use under-sampling. In 2013 Nayyar A. Zaidi et al. [4], founds weighted naïve Bayesian (WANBIA) as competitive alternative to the Random Forest and Logistic Regression.

Sotiris Kotsiantis et al. [9] performed a detailed review on imbalance data handling techniques. They explain sampling techniques with their disadvantages and had suggested one-class learning and SMOTE (Synthetic Minority Oversampling Technique) algorithm for imbalance classifier. Jose A Saez et al. [16], proposed SMOTE-IPF algorithm which uses iterative-partitioning filter for noise reduction from imbalanced dataset.

However as per Kun Jiang et al. (2016) [14], SMOTE uses same sampling rate for all instances of minority class which is result into poor classification. So they proposed genetic algorithm based SMOTE (GASMOTE) which uses different sampling rate for different instances.

Chomboon et al. [19], concluded that feature selection with oversampling is best method for rare class classification.

In Dec-2015, Mujalli et al. [15], proposed Bayes classifier for imbalanced accident dataset. In accident dataset instances with major injuries class are in minority compare to slight injuries class. They create separate datasets by oversampling, under-sampling and mix-sampling. Then after they used semi-naïve Bayesian classifier and Bayesian network classifier with different search method and compares results of all these classifier for differently preprocessed dataset. Authors had also list out the features which have high impact on deciding minority class.

As per Astha Agrawal et al. (2015) [7], SMOTE algorithm suffers from problem of over-fitting, and combining result from different one-class classifier trained on different sub problem may result into classification error. So they proposed new algorithm named SCUT (SMOTE and Cluster Under-sampling Technique). SCUT is a combination of under-sampling and oversampling. It handles within-class imbalance by using Expectation Maximization (EM) clustering with under-sampling. In their study they also realize future importance of cost-sensitive multi-class learning without transforming problem into binary class.

2.1 Techniques Used for Rare Class Classification

From research survey discuss above, there are two types of techniques generally used for imbalanced data classification: Data level techniques and Algorithm level techniques [9, 11, 18].

As per [9], Data level techniques are oversampling, under-sampling, feature subset selection, feature weighting. And algorithm level techniques are one class learning, cost-sensitive learning and ensemble-based method [6, 20].

2.2 Data Level Techniques

Oversampling

Random oversampling is non-heuristic method. In this method the minority class instances are replicated to balance the dataset [9]. Random oversampling may results into over-fitting and may increases computational time. SMOTE try to overcome over-fitting problem by creating synthetic sample by joining k minority class nearest neighbors instead of just replicating minority instances [9]. In [7], Astha Agrawal et al. proposed SCUT algorithm which is improvement over SMOTE and combines both oversampling and under-sampling techniques.

Under-Sampling

Random under-sampling is achieved by randomly eliminating majority class instances [9]. Under-sampling is non-heuristic method used to balance imbalanced dataset by reducing majority class example. Tomek links can be used as an under-sampling method [11]. Focused under-sampling eliminates the majority class samples lying further away [10]. The major drawback of under-sampling is loss of potentially useful data. This might results into loss of sub-concepts.
Carefully under-sampling must be done in case of within-class imbalance [7].

Feature Subset Selection

Subset of features are selected which are more relevant to target class. This is similar to a method used in improved cost-sensitive naïve Bayesian [1, 2]. Relevant features are selected for majority and minority class instances and selected features are used for classification. The selected feature set might be subset of original feature set which result into reduction of dataset.

Feature Weighting

Based on the causal relationship between features and target class appropriate weight is determined for the features. Weighted features are used for classification which is similar to the method used in weighted classifier [3–5].

Decision tree, association rule mining etc. are used for feature subset selection and feature weighting [2, 12].

2.3 Algorithm Level Techniques

Cost-Sensitive Learning

Cost-sensitive learning is used for improving imbalance data classification [13]. As per Sun et al. [13], this method assigns low misclassification cost for the majority class and high for minority class. However determination of misclassification cost is difficult and different cost might be result into different classification result.

Ensemble-Based Method

Ensemble-based method combines multiple classifiers to improve accuracy of final result [9, 14, 18]. Bagging (Bootstrap Aggregation) and Boosting based ensemble methods generally used for the class imbalance, but these methods may suffer by

over-fitting and loss of useful information as they internally used sampling methods to obtain balanced data [13].
In boosting, the classifiers within ensemble are trained serially. After each classifier operation boosting increases the weight associated with misclassified example and decreases the weights associated with correctly classified example [9]. Thus, subsequent classifier focuses more on misclassified example of previous classifier. As per [9], SMOTEBoost addresses the problems with rare class. SMOTEBoost algorithm combines SMOTE and standard boosting procedure [10].

One-Class Learning
In this method dataset is partitioned into subset and each subset is classified. Finally all result is combined. Earlier work use One Vs One (OVO) and One Vs All (OVA) approach for multi-class imbalanced data. But combining result from different one-class classifier trained on different subset may result into classification error [7].

3 Experiments: Rare Class Specific Causal Relationship Extraction

We believe that there is no action without causes. Causes are primary force to create action. Based on this philosophy we analyze three different dataset and extracted causal relationship specific to rare class. We perform detailed study to evaluate and to determine how strong causes are related to target class.

From our study we shortlist class specific most relevant features and instance value. Our result can be used to improve rare class classification.

In case of rare class dataset, the responsible causes are distributed either vertically or horizontally. Most of existing methods of feature selection works for vertically separated causes. But for horizontally separated causes instance selection is require. We design a model where we extract responsible causes no matter how it is stored in dataset.

3.1 Method

Step-1 Data preprocessing and discretization.
Step-2 Class specific association rule mining
Step-3 Rule filtering
Step-4 Identification of most relevant rule
Step-5 Priority assignment on basis of relevance to target class

Algorithm: RareCause

```
Input: D dataset, L List of rare class
Output: W Weight vector, Dr and Rf as output file
Begin
Descretize dataset D → Dr
While L is not empty
  Mine all frequent cause which has L as outcome → Rf
  While read(Rf)
    Process causes and update respective outcome-causes
    weight vector→ W
  End
End
End
```

3.2 Dataset Used

See (Table 1).

Table 1. Dataset description

Sr. No.	Dataset name	#Attribute	#Class	IR ratio
1	Glass	10	7	8.44
2	Hayes-roth	5	3	1.70
3	New-Thyroid	6	3	4.84

3.3 Result Analysis

We perform our study with three datasets. For each dataset we extract association rules specific to target class and then discover all the rules which identifies major causes of target class. We also filter out the rules and identify the tightly related feature/instance with target class. In this study we concentrate on identification of unique causes of target class.

Each dataset is first discretized to handle numeric value attributes and numeric class values are manually mapped to nominal value. Causal relationship is extracted by mining association rule using modified Apriori algorithm.

Dataset: New-Thyroid

New-thyroid is multiclass unbalanced dataset. It contains three classes {normal, hyper, hypo}. hypo is smallest class with 30 instances and normal is largest class with 150 instances.

Causal relation for minor class “hypo” and “hyper” is given in Table 2. Relativity level indicates how strongly the causes are related to target class and their influence in terms of their presence in number of instances. “Strong and major” indicates majority number of instances of given target class are due to the said causes. “Strong” indicates that although the number of instance with said causes are less but they are tightly related to target class.

Table 2. Causal relationship for rare class of new-thyroid.

Target class	Major causes	Relativity level
Hypo	thyroidstimulating = '(5.73–33.38]'	Strong and major
Hypo	thyroxin = '(–inf–2.98]'	Strong and major
Hypo	T3resin = '(96.6–104.5]' thyroxin = '(2.98–5.46]'	Strong
Hypo	thyroxin = '(2.98–5.46]' triiodothyronine = '(–inf–1.18]'	Strong
Hypo	thyroxin = '(5.46–7.94]' triiodothyronine = '(1.18–2.16]' TSH_value = '(10.7–16.4]'	Strong
Hyper	thyroxin = '(22.82–inf)' triiodothyronine = '(1.6–inf]'	Strong and major
Hyper	T3resin = '(–inf–88.7]'	Strong and major
Hyper	thyroxin = '(17.86–inf]'	Strong
Hyper	T3resin = '(88.7–96.6]' thyroxin = '(10.42–15.38]'	Strong
Hyper	T3resin = '(96.6–104.5]' thyroxin = '(12.9–17.86]'	Strong
Hyper	thyroxin = '(15.38–7.86]' triiodothyronine = '(1.18–2.16]'	Strong
Hyper	thyroxin = '(15.38–17.86]' triiodothyronine = '(3.14–4.12]'	Strong

Dataset: Glass

Table 3 shows some of causes responsible for rare classes of glass dataset. With total 214 instances and number of instance of case1 = 70, case2 = 76, case3 = 17, case5 = 13, case6 = 9 and case7 = 29, we select case6, case3, case5 and case7 as rare class and extract causes respective to these classes.

Table 3. Causal relationship for rare class of glass dataset.

Target Class	Major Causes	Relativity Level
Case6	RI = '(1.517984–1.520262]' Mg = '(2.245–2.694]'	Strong and major
Case6	RI = '(1.517984–1.520262]' Na = '(14.055–14.72]' Mg = '(1.796–2.245]'	Strong and major
Case6	RI = '(1.517984–1.520262]' Al = '(1.574–1.895]' Si = '(72.05–72.61]'	Strong
Case3	Na = '(13.39–14.055]' Al = '(0.611–0.932]' Si = '(72.61–73.17]'	Strong and major
Case3	Na = '(12.725-13.39]' Si = '(72.05–72.61]' Ca = '(8.658–9.734]' Fe = '(–inf–0.051]'	Strong and major
Case3	RI = '(1.515706–1.517984]' Na = '(13.39–14.055]' K = '(–inf–0.621]' Ca = '(8.658–9.734]' Ba = '(–inf–0.315]'	Strong and major
Case5	Na = '(12.725–13.39]' Mg = '(–inf–0.449]'	Strong and major
Case5	K = '(5.589–inf)'	Strong
Case5	Ca = '(11.886–12.962]'	Strong
Case5	Si = '(70.37–70.93]' Ca = '(6.506–7.582]'	Strong
Case7	Al = '(2.216–2.858]'	Strong and major
Case7	Ba = '(0.315–0.63]'	Strong and major
Case7	Ba = '(1.26–1.89]'	Strong and major
Case7	RI = '(1.513428–1.515706]' Mg = '(–inf–0.449]'	Strong and major
Case7	Na = '(14.055–14.72]' Al = '(2.537–2.858]'	Strong

Dataset: Hyes-roth

In Table 4 we derived all the responsible causes for target class case3. From these three causes all the instances of class case3 are covered. In hayes-roth dataset there are three classes and case3 class is in minority with thirty instances.

Table 4. Causal relationship for rare class of hayes-roth dataset.

Target class	Major causes	Relativity level
Case3	age = ‘(3.7–inf)’	Strong and major
Case3	educationalLevel = ‘(3.7–inf)’	Strong and major
Case3	maritalStatus = ‘(3.7–inf)’	Strong and major

3.4 Identification of Most Relevant Features from Extracted Causes

From the extracted causes we derived most relevant features of respective target class. Table 5 contains list of the features which are relevant to respective target class. All the features are listed in decreasing order of their influence in target class.

Table 5. Relevant features of rare classes from different dataset.

Dataset	Features	Rare Class
new-thyroid	Thyroidstimulating, Thyroxin, T3resin, Triiodothyronine	Hypo
new-thyroid	Thyroxin, Triiodothyronine, T3resin	Hyper
Glass	RI, Mg, Na, K, Fe, Ca, Si	Case6
Glass	Na, Si, A1, RI, Ca, Fe	Case3
Glass	Na, Mg, K, Ca, Si	Case5
Glass	A1, Ba, RI, Mg, Na	Case7
hayes-roth	Age, EducationalLevel, MaritalStatus	Case3

4 Conclusion

Existing methods of rare class classification is sampling techniques. But sampling techniques has drawbacks associated with it. Causal relationship extraction for target rare class opens a door for effective rare class classification without increasing size of a dataset and it also reduces learning time. Our study shows use of association rule mining to extract most relevant causes of target class. We successfully derived all the unique causes of target class and proved that each action has some specific responsible causes. Causes which have high impact on determination or rare class are classified as “Strong and major” and “Strong”. Identification of such causes helps us in deriving target class of future instances. Classification of rare class dataset using derived causes is our future work.

References

1. Kong, G., Jiang, L., Li, C.: Beyond Accuracy: Learning Selective Bayesian Classifier with Minimal Test Cost. Elsevier, Amsterdam (2016)
2. Ratanamahatana, C.A., Gunopulos, D.: Scaling up the Naïve Bayesian classifier: using decision trees for feature selection. Appl. Artif. Intell. (2003)
3. Taheri, S., Yearwood, J., Mammadov, M., Seifollahi, S.: Attribute Weighted Naïve Bayes classifier Using a Local Optimization. Springer, Berlin (2013)
4. Zaidi, N.A., Cerquides, J., Carman, M.J., Webb, G.I.: Alleviating Naive Bayes attribute independence assumption by attribute weighting. J. Mach. Learn. Res. **14**, 1947–1988 (2013)
5. Lee, C-H., Gutierrez, F., Dou, D.: Calculating feature weights in Naïve Bayesian with Kullback-Leibler measure. In: 11th IEEE International Conference on Data Mining, pp. 1146–1151 (2011)
6. Vural, M.S., Gok, M.: Criminal Prediction using Naïve Bayesian Theory, Neural Comput. and Applic. Springer, Berlin (2016)
7. Agrawal, A., Viktor, H.L., Paquet, E.: SCUT: multi-Class imbalanced data classification using SMOTE and Cluster-based undersampling, In: Proceeding of the 7th International Joint Conference on Knowledge Discovery, Knowledge Engineering and Knowledge Management (IC3K 2015. KDIR, vol. 1, pp. 226–234. SCITEPRESS (2016)
8. Muchlinski, D., Siroky, D., He, J., Kocher, M.: Comparing random forest with logistic regression for predicting class-imbalanced civil war onset data. Oxford University Press, Oxford (2015)
9. Kotsiantis, S., Kanellopoulos, D., Pintelas, P.: Handling imbalanced datasets: a review. GESTS Inter. Trans. Comp. Sci. Eng. **30**, 25–26 (2006)
10. Chawla, N.V.: Data mining for imbalanced datasets: an overview. In: Data Mining and Knowledge Discovery Handbook, chap. 40
11. Stefanowski, J.: Dealing with Data Difficulty Factors While Learning from Imbalanced Data. Springer, Cham (2016)
12. Barot, P.A., Jethva, H.B.: A study paper on importance of bitapriori over apriori in discovery of feature weight for unbalanced data classification, In: Proceeding of National Conference NCCICT (2016)
13. Sun, Z., Song, Q., Zhu, X., Sun, H., Xu, B., Zhou, Y.: A Novel Ensemble Method for Classifying Imbalanced Data, Pattern Recognition. Elsevier (2014), http://dx.doi.org/10.1016/j.patcog.2014.11.014
14. Jiang, K., Lu, J., Xia, K.: A Novel Algorithm for Imbalance Data Classification Based on Genetic Algorithm Improved SMOTE. Springer, Berlin (2016)
15. Mujalli, R.O., Lopez, G., Garach, L.: Bayes Classifiers for Imbalanced Traffic Accidents Datasets, Accident Analysis and Prevention. Elsevier, Amsterdam (2015)
16. Saez, J.A., Luengo, J., Stefanowski, J., Herrera, F.: SMOTE-IPF: Addressing the noisy and borderline examples problem in imbalanced classification by a re-sampling method with filtering. Information Science. Elsevier, Amsterdam (2014)
17. Triguero, I., del Rio, S., Lopez, V., Bacardit, J., Bentez, J.M., Herrera, F.: ROSEFW-RF: An Extremely Imbalanced Big Data Bioinformatics Problem. Knowledge-Based System. Elsevier, Amsterdam (2015)

18. Zhang, D., Ma, J., Yi, J., Niu, X., Xu, X.: An ensemble method for unbalanced sentiment classification. In: Proceedings of the 11th International Conference on Natural Computation. IEEE (2015)
19. Chomboon, K., Kerdprasop, K., Kerdprasop, N.: Rare class discovery techniques for highly imbalanced data. In: Proceeding of the International Multi Conference of Engineers and Computer Scientists, Hong Kong (2013)
20. Weiss, G., McCarthy, K., Zabar, B.: Cost-sensitive learning vs. sampling: which is best for handling unbalanced classes with unequal error costs. In: Proceedings of the International Conference on Data Mining, pp. 35–41 (2005)

Clustering and Classification of Effective Diabetes Diagnosis: Computational Intelligence Techniques Using PCA with kNN

Nimmala Mangathayaru(✉), B. Mathura Bai, and Panigrahi Srikanth

Department of Information Technology, VNR Vignana Jyothi Institute of Engineering and Technology, Hyderabad 500090, Telangana, India
{mangathayaru_n,mathurabai_b}@vnrvjiet.in, srikanth.panigrahi@gmail.com

Abstract. The fourth leading disease in the world today is Diabetes and there are number of challenges to predict and identify the disease. Data mining proposes effective approaches to identify the diabetic patients. This paper proposes clustering and classification of effective diabetes diagnosis based on computational intelligence techniques using PCA with kNN. Diabetes disease data is used to identify feature of clusters. Diabetes disease diagnosis proposes novel distribution function applied to classify each patient. This proposed procedure defines clusters and similarity measure based on classifying with each cluster using computational intelligence techniques. PCA using diabetes disease data for dimensionality reduction. Novel similarity measure is proposed in kNN for classification. Accuracy measures are computed for each patient.

Keywords: Diabetes disease · Clustering · Classification · Distribution function and PCA with kNN

1 Introduction

Today in India most of the people suffer from Diabetes. Presently 3crs persons and 2030 year will reach near 7.9crs. Worldwide Diabetes disease by 17crs peoples agonized and 2030 it will reach to nearly 36crs [13, 14].

The main contribution of this paper is to predict diabetes by classifying the patient's data. The paper discusses how to handle patient data and present days medical mining issues while developing a tool. Diabetes mellitus data has number of attributes which are categorized into two classes. The attributes are mainly either numerical nor continuous data. This data will be applied for normalization techniques to reduce noise and error rate.

Features are obtained using mining techniques based on feature of clusters are obtained. Each cluster is defined as the entropy of the maximum values. The similarity of cluster is computed using distribution function to classify them into similar classes. The performance of classified diabetic patient's diagnosis is evaluate based on accurate and efficient results.

S.C. Satapathy and A. Joshi (eds.), *Information and Communication Technology for Intelligent Systems (ICTIS 2017) - Volume 1*, Smart Innovation, Systems and Technologies 83, DOI 10.1007/978-3-319-63673-3_52

2 Related Works

The literature survey includes research problems like identification, prediction and detection of diabetic patients using data processing techniques [1, 2,28, 29]. Most of the current research of literature in the data processing with Patient data is to build knowledge base and expert system. The data base includes different types of data like numerical data, categorical data and continuous data [3, 28]. Diabetes patients data is considered to handle and maintain based on identifying features, data processing techniques using to design clustering and classification of the data. Clustering and classification algorithms are based on computed as different results of the patient, design medical tools, and applications based on predicted as patients, calculated as accurate results [4, 5, 27–34]. The computational intelligence of the Early Diabetes diagnosis review of the focus on clustering, classification, Neural networks, Neural fuzzy interface system and multilayer neural networks using reviewed of the earlier detection of diabetes [6–12]. Other related works consider as patient prediction [13–24, 27–34].

3 Dataset Description

Today's most of the people's like agonize of the Diabetes patients. Diabetes disease is the most part of similarly type-1, type-2 and gestational diabetes [10, 11, 24].

Type 1 Diabetes
Type 1 diabetes disease associate with 5% to 10% patient's disease cases. Type 1 diabetes disease of appraising of factors as Age, gender, weight, height, BMI, HB1Ac, Cholesterol, amylase, creatinine, insulin test, urine test, skin flood.

Type 2 Diabetes
Type 2 diabetes disease associate with 90% patient's disease cases. Type 2 diabetes disease of appraise of factors as Age, gender, material status, weight, height, BMI, HbA1c, cholesterol, amylase, creatinine, insulin test, HDL (High density lipoprotein), LDL (low density lipoprotein), urine test, skin flood, thyroid, fasting glucose, fasting glucose is 1 h and 2 h (100,75 and 50 g),random sugar, pp sugar, triglycerides, uric acid, BP (blood pleasure).

Gestational Diabetes
Gestational Diabetes disease associates with 3–10% patient's disease cases. Gestational diabetes appraisers of factors as Age, gender, material status, weight, height, BMI, HbA1c, cholesterol, amylase, creatinine, insulin test, urine test, skin flood, thyroid test, fasting glucose, fasting sugar, fasting glucose is 1 and 2 h (100,75 and 50 g).

4 Proposed Methodology

4.1 PCA with KNN Method Flow

See Fig. 1.

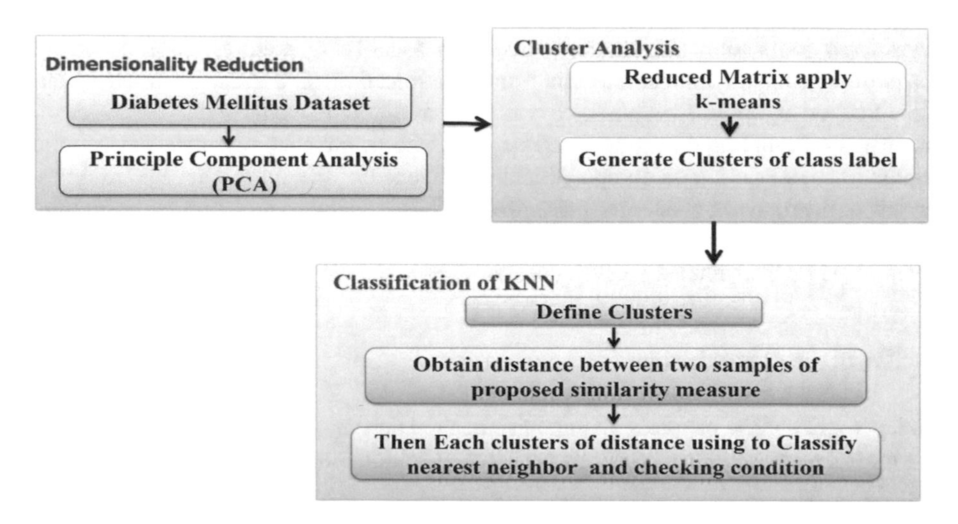

Fig. 1. Proposed method PCA with kNN method Flow

4.2 Data Generalization

Generalization is method based on numerical attributes into nominal attributes or categorical attributes used in this context. We have considered to normalizing statistical techniques for generalization [10, 11]. It is showing below

In this equation x = {$x_{1,}$ x2……. xn} dataset of Numerical attributes

With L classes of the class variable. It is showing as C = {C1…Cn}

X_i is attributing of the data set [13]

$\overline{Xi}$ is the mean

σ is a standard deviation

Let us consider as the normalization or Generalization, such that

$$\text{Generalization G } = \frac{Xi - \overline{Xi}}{\sigma} \tag{1}$$

Example

Diabetes mellitus parameters of the BMI and HbA1c 16.5,15.4 and 5.2,7.2. Those values based on normalize values are Computed generalized values 0.9538,0.9059 and 0.3006,0.4235.

4.3 Principal Component Analysis

(a) Dimensionality Reduction
Dimensionality reduction methods based on reducing the noise the dimensionality of the diabetes mellitus data. The number of methods of the reduce data of supervised and unsupervised into high to low of reduced dimensionality of the data. Whether it is using as reduce data of new method called as PCA of kNN using number samples reduced [1, 10, 11, 25, 26].

Those methods using as tasks reduction, dimensionality reduction methods and optimization problems solved to reduce data. Dimensionality reduction of reduced data then identifies the feature of the data maximum value based on identified clusters. These clusters are classified both positive and negative of the samples of patient data [1, 10, 11, 25].

(b) Calculate Features
The entropy of the distribution function with computed as datasets. It is using an efficient way to calculate the predicted data. Computed as entropy with features of the maximum value based on identified clusters [1, 10].

The entropy of a variable X. Suppose *p(x)* is the prior probabilities of X, probability of entropy donation as

$$\text{Entropy } (\mathrm{P_e}) = -\sum\nolimits_{x\varepsilon X} p(x) log_2 p(x) \tag{2}$$

Example
Diabetes mellitus parameters of the BMI, HbA1c and diabetes 16.5,15;5.2,7.2 and 1,0 each parameter and each record wise computed entropy of

matrix is = [16.5,15.4,1 and 5.2,7.2,0]
entropy value = [0.1,0.2,0.4 and 0.25,0.35]
Dimensionality matrix = [0.2333; 0.3]

4.4 Define Clusters

Define Cluster is reducing matrix applied into k-means with regularly as Euclidean measure based on generate clusters, after reducing matrix is build matrix. In this matrix applied each two samples apply the distance measure based on each cluster [1, 2, 8, 18].

Therefore, Euclidean Distance =

$$\mathrm{d}\ (\mathrm{x}, \mathrm{y}) = \sqrt{\sum_{i=1}^{n} (xi - yi)^2} \tag{3}$$

4.5 Classification Using k-Nearest Neighbor of Proposed Method

Classification of each nearest k-value of the clusters based on classified two classes. kNN using new distance measure based on classified each class is positive and negative [1, 2, 8–11, 26].

(a) Distribution Function
The distribution function is mostly two types one is Probability distribution and another one is cumulative distribution functions. Both are distributions of the random variables and multiple variables of the function [17–23].

(b) Probability distribution function
We examine dependent of the probability distribution function use idea of the realize similarity into two data sample of the diabetes mellitus data. We use similarity of distance measure using to classify the data.

$$\varphi_{\mathrm{p}} = 0.3988e^{-\left(\frac{(x-\mu x)^2}{\sigma x}\right)} \tag{4}$$

X = No of samples in the dataset, μx = average of the present data set and σx = massive volume value of data

(c) Cumulative distribution function
We examine dependents of the cumulative distribution function use idea of the realize similarity into two data sample of the diabetes mellitus dataset. We use similarity distance measure using to compute clusters of the data samples. Cumulative distribution is true.

$$\varphi_{\mathrm{C}} = 1 - e^{-\left(\frac{(x-\mu x)^2}{2\sigma x}\right)} \quad \mathrm{x} > 0 \tag{5}$$

X = No of samples in the dataset, μx = average of the present data set and σx = massive volume value of data.

(d) DM Average Function
Diabetes Mellitus (DM) average function is defined as above distribution defined. Then considered as each distribution functions based on computed as average function.

$$\text{Average function } \mathrm{F_{avg}} = \frac{\varphi \mathrm{p}}{\varphi \mathrm{C}} \tag{6}$$

(e) Distance Function
The proposed Distance function is based on distribution functions and average functions. This functions defined proposed method based on classifies each neighbor.

$$\text{Distance function (Df)} = \sqrt{\frac{(1 - \mathrm{Favg})}{2}} \tag{7}$$

4.6 Cutoff-Score

Cutoff score is defined as positive and negative, average of each positive and negative patients.

Cutoff – score (C_F)

$$C_F = \frac{NpDe + NqDf}{Np + Nq} \quad (8)$$

C_F = Cut-off Score, Np = Number of Positive Patients, Nq = Number of Negative patients, De = average score of Healthy patients (avg of positive patients) and Df = average score of Non-Healthy patients (avg of negative patients)

4.7 Proposed Algorithm of the Diabetes Diagnosis Using PCA with kNN

Input: Medical Data of the Diabetes patients with numerical and continuous attributes with the row and columns of the dataset.

Output: Identify of the patients and accurate results

Appropriate Notations of the Methodology

$\overline{Xi}$ = is the mean

σ = is a standard deviation

G = Generalization

P_e = Probability of Entropy

φ_P = probability distribution function

X = No of samples in the dataset

μx = average of the present data set

σx = massive volume value of data

φ_C = cumulative distribution function

F_{avg} = Average function

Df = distance function

Cf = Cutoff – score

Np = Number of Positive Patients

Nq = Number of Negative patients

De = average score of Healthy patients (avg of positive patients)

Df = average score of Non-Healthy patients (avg of negative patients)

Step 1: **Initialize the dataset**
Medical dataset of diabetes diagnosis of the numerical and continuous data.

Step 2: **Apply data generalization of the medical dataset, data generalization of the discussed as Sect.** 4.2

Step 3: **Apply the Entropy of the diagnosis dataset computed as features and reduced noise using PCA. Detailed discuss in Sect.** 4.3

PCA using Dimensionality reduction

1. Initialize diabetes generalized data
2. Compute probability of each row and column wise p(x)
3. Then computed as the Probability of entropy (Pe)
4. Probability of entropy (Pe) based on build matrix.

Step 4: Identify Clusters using entropy value of maximum values. Maximum values based on defined clusters are cluster1 and cluster2 as positive or negative.

1 Start
2 initialize matrix (Pe)
3 then each record wise computed feature of the reduction matrix and apply the K-mean algorithm define clusters, two clusters defined cluster1 (c1) and cluster (c2).
4 Stop

Step 5: Those clusters classified using novel distribution function based on classified clusters of the patient records. Go through this Sect. 4.5 **Build**

a. **Distribution functions of the PDF and CDF. Given that Sect.** 4.5 **(b, c)** Computed as Distribution functions of the Probability distribution function (φp) and Cumulative distribution function (φC)
b. **PDF and CDF based on defined as function average of the patient record. Section** 4.5 **(d)** Function average (Favg) is defined as PDF and CDF based on computed as average function.
c. **The distance function(Df) is defined as an average function. Section** 4.5 **(e)** **End**

Step 6: Repeat this procedure of the step (5) and step (6) for each record wise

Step 7: Compute Cf-Score based on apply the two cases of conditions (Cf value similar value). detailed discuss in Sect. 4.6

Each patient Classification cases

Cutoff score based on classify patients following approach two cases

Case 1

If Df- Score > Cf- Score, the patient is classified to the positive group (Healthy Group)

Case 2

If Dz- Score $\leq$ Cf- Score, the patient is classified to the negative group (Non-healthy group)

Step 8: Stop.

4.8 Performance Evolution Metrics

Performance evolution metrics of the computational intelligence techniques using evaluated as diabetes patients of true positive, false positive, true negative value and false negative values of the precision, recall and accuracy of the patients.

D_P = Denote as Total number of positive patients
D_n = Denote as Total number of Negative patients
$D_{P\text{-}P}$ = Denote as Diabetes patients of the positive patients correct classified
$D_{N\text{-}N}$ = Denote as Diabetes patients of the negative patients correct classified
$D_{P\text{-}N}$ = Denote as Diabetes patients of the positive patients incorrect classified
$D_{N\text{-}P}$ = Denote as Diabetes patients of the positive patients misclassified.

True Positive Rate (TPR): True positive rate is positive data of diabetes patient's data is correctly classified.

$$\text{TPR} = \frac{\text{DP} - \text{P}}{\text{Dp}} * 100$$

True Negative Rate (TPR): True negative rate is negative data of diabetes patient's data is correctly classified.

$$\text{TPR} = \frac{\text{Dn} - \text{n}}{\text{Dn}} * 100$$

False Positive Rate (FPR): False positive rate is positive data diabes pient's data is incorrectly classificd.

$$\text{FPR} = \frac{\text{Dp} - \text{n}}{\text{Dp}} * 100$$

False Negative Rate (FNR): False Negative rate is negative data of diabetes patient's data is misclassified.

$$\text{FNR} = \frac{\text{Dn} - \text{p}}{\text{Dn}} * 100$$

Precision: Precision is positive data of diabetes patients correctly classified

$$\text{Precision} = \frac{\text{DP} - \text{P}}{\text{Dp}} * 100 \quad (9)$$

Recall: Recall is negative data of diabetes patients correctly classified

$$\text{Recall} = \frac{\text{Dn} - \text{n}}{\text{Dn}} * 100 \quad (10)$$

F-Measure: Precision and Recall is using defined F-measure of the diabetes patients.

$$\mathrm{F-Measure} \ = 2 * \frac{Precision * Recall}{Precision + Recall} \tag{11}$$

Accuracy: Accuracy of the positive and negative of the complete patient data of the computed as accuracy.

$$\mathrm{Accuracy} \ = \frac{\mathrm{Dp - p + Dn - n}}{\mathrm{Dp + Dn}} \tag{12}$$

5 Results

The classification of each patient wise classify either positive or negative, based on computed as accurate results. PCA with kNN algorithm based on classified patients and computed accurate results of the sensitivity, specificity, precision, recall, f-measure, ROC and accuracy of the each record wise. Pima dataset of nearest neighbor classified each class label and k nearest (k = 1,3,5,7,9,11,13,15) value computed.

Table 1 is represents computed as in detail statistical analyzed of the pima diabetes dataset. statistical measure as count, mean, Standard deviation, median, mode, smallest, largest, standard error, variance, Margin error, confidence value, upper bound and lower bound of each attribute and record wise calculations done.

Table 2 is evaluated as each positive and negative patient k nearest neighbor of the varying k value of each positive and negative value based on computed and performance evaluation of calculated as sensitivity, specificity, f-measure, Roc curve and accuracies. Evaluated positive and negative values of the performance evaluation as k value based on computed

Table 3 represents as calculated as error rate of the each nearest values of different errors. MAE, RMSE, positive value and predictive error (PE) (PE = positive value – rmse value).

Figure 2 represents as error rate of calculated as the MAE, RMSE and Predictive error rate. In this shown Fig. 2 as two lines and two bars upper lines are positive value and lower line as shown that predictive error.

Table 4 represent as that follow us each nearest neighbor of the computed true positive, true negative, false positive and false negative values of the computed individual nearest values (k = 1,3,5,7,9,11,13 and 15) of accuracies.

Table 5 represents as 10-folds cross validation average weighted accuracies individual nearest neighbor of sensitivity, specificity, precision, recall, f-measure and ROC

Table 6 and Fig. 3 represents as 10 fold cross validation of the average accuracies of the individual nearest neighbor of sensitivity, specificity and accuracy

Figure 3 represents as individual nearest neighbor of sensitivity, specificity and accuracy

Table 1. Each attribute wise statistical analysis

	A1	A2	A3	A4	A5	A6	A7	A8	Class
Count	768	768	768	768	768	768	768	768	768
Mean	3.8450521	120.89453	69.105469	20.536458	79.799479	31.992578	0.4718763	33.240885	0.3489583
SD	3.3695781	31.972618	19.355807	15.952218	115.244	7.8841603	0.3313286	11.760232	0.4769514
Median	3	117	72	23	30.5	32	0.3725	29	0
Mode	1	100	70	0	0	32	0.254	22	0
Largest (1)	17	199	122	99	846	67.1	2.42	81	1
Smallest (1)	0	0	0	0	0	0	0.078	21	0
Standard error	0.1215892	1.1537125	0.6984425	0.5756261	4.1585097	0.2844951	0.0119558	0.4243608	0.0172105
Coefficients variance	0.8763413	0.264467	0.2800908	0.7767755	1.4441699	0.2464372	0.7021514	0.3537882	1.366786
Margin error	1.1411079	3.7811896	3.5702706	1.2873733	0.6924393	4.0578295	1.4241943	2.8265503	0.7316434
Sample variance	11.354056	1022.2483	374.64727	254.47325	13281.18	62.159984	0.1097786	138.30305	0.2274826
Kurtosis	0.1592198	0.6407798	5.1801566	−0.520071	7.2142596	3.2904429	5.5949535	0.6431589	−1.600929
Skewness	0.901674	0.1737535	−1.843608	0.1093725	2.2722509	−0.428981	1.9199111	1.1295967	0.6350166
Upper bound	4.0833669	123.15581	70.474416	21.664685	87.950158	32.550189	0.4953097	34.072633	0.3826909
Lower bound	3.6067373	118.63325	67.736521	19.408231	71.6488	31.434968	0.448443	32.409138	0.3152258
Confidence level (95.0%)	0.2386871	2.2648088	1.3710858	1.1299895	8.1634112	0.5584815	0.02347	0.8330464	0.0337853

Table 2. Evaluated each metric wise Positive (P) & Negative(N) patients

Positive (p) & Negative (N)	k = 1		k = 3		k = 5		k = 7		k = 9		k = 11		k = 13		k = 15	
	P	N	P	N	P	N	P	N	P	N	P	N	P	N	P	N
Sensitivity	0.53	0.794	0.552	0.82	0.537	0.836	0.545	0.836	0.511	0.834	0.493	0.0.836	0.5	0.862	0.515	0.862
Specificity	0.206	0.47	0.18	0.448	0.164	0.463	0.144	0.455	0.166	0.489	0.164	0.507	0.138	0.374	0.138	0.485
Precision	0.58	0.759	0.622	0.774	0.637	0.771	0.67	0.778	0.623	0.761	0.617	0.755	0.66	0.763	0.667	0.768
Recall	0.53	0.794	0.552	0.82	0.537	0.836	0.545	0.856	0.511	0.834	0.5	0.836	0.5	0.862	0.515	0.862
F-Measure	0.554	0.776	0.585	0.796	0.583	0.802	0.601	0.815	0.561	0.796	0.548	0.793	0.569	0.809	0.581	0.812
ROC	0.65	0.65	0.742	0.742	0.766	0.766	0.785	0.785	0.79	0.79	0.794	0.794	0.799	0.799	0.802	0.802

Table 3. Diabetes mellitus of kNN-error rate

	k = 1	k = 3	k = 5	k = 7	k = 9	k = 11	k = 13	k = 15	Average
MAE	0.2988	0.3092	0.3165	0.3178	0.3221	0.3254	0.3261	0.327	0.317863
RMSE	0.5453	0.4525	0.4318	0.4209	0.4187	0.4167	0.413	0.4118	0.438838
PV	0.702	0.727	0.732	0.747	0.721	0.716	0.736	0.741	0.72775
PE	0.1567	0.2745	0.3002	0.3261	0.3023	0.2993	0.323	0.3292	0.288913

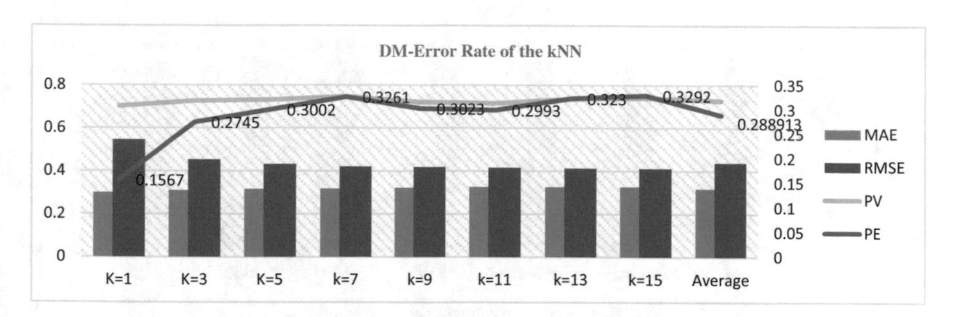

Fig. 2. kNN-diabetes mellitus error rate

Table 4. Accuracies of the nearest neighbor

Patients = 768	TP	FP	TN	FN	Accuracy
k = 1	397	103	126	142	0.6809896
k = 3	410	90	120	148	0.6901042
k = 5	418	82	124	144	0.7057292
k = 7	428	72	122	146	0.7161458
k = 9	417	83	131	137	0.7135417
k = 11	418	82	136	132	0.7213542
k = 13	431	69	134	134	0.7356771
k = 15	431	69	130	138	0.7304688

Table 5. 10 Folds cross-validation average accuracies

	k = 1	k = 3	k = 5	k = 7	k = 9	k = 11	k = 13	k = 15	Average
Sensitivity	0.702	0.727	0.732	0.747	0.721	0.716	0.736	0.741	0.72775
Specificity	0.378	0.354	0.358	0.347	0.376	0.388	0.374	0.364	0.367375
Precision	0.969	0.721	0.724	0.74	0.713	0.706	0.727	0.733	0.754125
Recall	0.702	0.727	0.732	0.747	0.721	0.716	0.736	0.741	0.72775
F-Measure	0.698	0.722	0.726	0.74	0.714	0.708	0.726	0.732	0.72075
ROC	0.65	0.742	0.766	0.785	0.79	0.794	0.799	0.802	0.766

Table 6. kNN using accuracies (sensitivity, specificity and accuracies)

	k = 1	k = 3	k = 5	k = 7	k = 9	k = 11	k = 13	k = 15	Average
Sensitivity	0.702	0.727	0.732	0.747	0.721	0.716	0.736	0.741	0.72775
Specificity	0.378	0.354	0.358	0.347	0.376	0.388	0.374	0.364	0.367375
Accuracy	0.68099	0.690104	0.7057	0.716146	0.713542	0.721354	0.735677	0.703469	0.708373

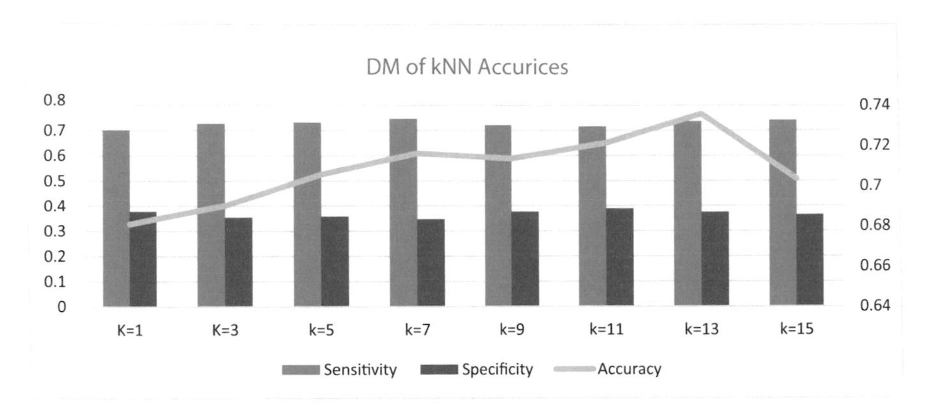

Fig. 3. kNN using accuracies (sensitivity, specificity and accuracies)

6 Conclusion and Future Work

This paper, Evaluated Diabetes diagnosed each patient wise. Clustering and classification of effective disease is diagnosed the computational intelligence techniques using PCA with kNN. We proposed novel distribution function of the classified patients with clusters. In the procedure, proposed algorithm based on predicted patients and computed accurate results of using diabetes diagnosis data.

Finally, we identified the feature based on entropy, Dimensionality reduction of the reduced noise of the using PCA and identified clusters. Those clusters of the positive and negative. Each cluster is classified to novel distribution function using classified the patients.

The future work is computed diabetes mellitus disease data collected in various diagnostic centers. Each patient record is prepared through reduced dimensionality and classified using proposed algorithm PCA with kNN, Discriminant analysis, Logit analysis and probit analysis these based on predict patients. The accurate results are calculated with performance evaluation metrics.

Acknowledgements. The proposed research work has been funded under DRDO-LSRB (Defense Research and Development Organization-Life Science Research Board)-No.CC R&D (TM)/81/48222/LSRB-284/SH&DD/2014, 8th January 2015.

References

1. Garcia, S., Luengo, J., Herrera, F.: In: Data Pre-processing in Data Mining, Intelligent Systems references library, Series vol. 72. Springer (2015)
2. Adhikari, A., Adhikari, J., Pedrycz, W.: In: Data Analysis and Pattern Recognition in Multiple Databases, series vol. 61. Springer (2014)
3. Adhikari, A., Adhikari, J.: In: Advances in Knowledge Discovery in Databases, Intelligence System References Library, vol. 79. Springer (2015)
4. Aslam, M.W., Zhu, Z., Nandi, A.K.: In: Feature Generation Using Genetic Programming with Comparative Partner Selection for Diabetes Classification, Expert System with Applications, pp. 5402–5412 (2013)
5. Patil, B.M., Joshi, R.C., Toshniwal, D.: In: Hybrid Prediction Model for Type-2 Diabetic Patients, Expert System with Applications, pp. 8102–8108 (2010)
6. Bursa, M., Khuri, S., Renda, M.E.: Information technology in bio- and medical informatics. In: 4th International Conference, ITBAM 2013, August 28, 2013, Proceedings, vol. 8060. Springer (2013)
7. Böhm, C., Khuri, S., Lhotska, L., Renda, M.E.: Information, technology in bio- and medical informatics. In: Third International Conference, ITBAM 2012, September 4–5, 2012, Proceedings, vol. 6266 Springer (2010)
8. Shankaracharya, A., Odedra, D., Samanta, S., Vidyarthi, A.S.: Computational intelligence in early diabetes diagnosis: a review. Rev. Diabetes Stud **7**, 252–262 (2010)
9. Schmitt, M., Teodorescu, H.N., Jain, A., Jain, A., Jain, S.: Computational Intelligence Processing in Medical Diagnosis, Studies in Fuzziness and Soft Computing, series vol. 96. Springer (2002)
10. Varma, V.S.R.P.K., Rao, A.A., Sita Maha Lakshmi, T., Nageswara Rao, P.V.: In: A Computational Intelligence Approach for a Better Diagnosis of Diabetic Patients, Computers and Electrical Engineering, pp. 1758–1765 (2014)
11. Varma, V.S.R.P.K., Rao, A.A., Sita Maha Lakshmi, T., Nageswara Rao, P.V.: In: A Computational Intelligence Technique for the Effective Diagnosis of Diabetic Patients Using Principal Component Analysis (PCA) and Modified Fuzzy SLIQ Decision Tree Approach, Applied Soft Computing, pp. 137–145 (2016)
12. Kramer, O.: In: Dimensionality Reduction with Unsupervised Reduction, Intelligent Systems References Library, Series vol. 51, Springer (2013)
13. Srikanth, P., Deverapalli, D.: A critical study of classification algorithms using diabetes diagnosis. In: IEEE 6th International Conference on Advanced Computing (IACC), February 2016
14. Srikanth, P., Anusha, Ch., Dharmaiah, D.: A computational intelligence technique for effective medical diagnosis using decision tree algorithm. i-Manag. J. Comput. Sci. **3**(1), 21–26 (2015)
15. Srikanth, P., Devarapalli, D.: Identification of AIDS Disease Severity Using Genetic Algorithm, pp. 99–111. Springer (2015)
16. Srikanth, P., Anusha, C., Dharmaiah, D.: Identification of Deleterious SNPs in TACR1 Gene Using Genetic Algorithm, pp. 87–97. Springer (2015)
17. Srikanth, P., Rajasekhar, N.: A novel cluster analysis for Gene-miRNA interaction documents using improved similarity measure. In: International Conference on Engineering &MIS – (ICEMIS-2016), IEEE morocco section, pp. 1–7 (2016)
18. Friedman, N., Geiger, D., Goldszmidt, M.: Bayesian network classifiers. Mach. Learn. **29**, 131–163 (1997)

19. Temurtas, H., Yumusak, N., Temurtas, F.: A comparative study on diabetes disease diagnosis using neural networks. Exp. Syst. Appli. **36**, 8610–8615 (2009)
20. Kayaer, K., Yildirin, T.: Medical Diagnosis on Pima Indian Diabetes Using General Regression Neural Networks, pp. 181–184 (2003)
21. Alonso, F.C., Perez, C.J., Arias Nicolas, J.P., Martin, J.: Computer-Aided Diagnosis System: A Bayesian Hybrid Classification Method, Computer Methods and Program in Biomedicine, pp. 104–113 (2013)
22. Beloufa, F., Chikh, M.A.: In: Design of Fuzzy Classifier for Diabetes Disease Using Modified Artificial Bee Colony Algorithm, Computer Methods and Program in biomedicine, pp. 92–103 (2013)
23. Perner, P.: machine learning and data mining in pattern recognition. In: 12th International Conference, MLDM 2016, USA, July 16–21 2016. Springer (2016)
24. Canbas, S., Cabuk, A., Bilgin Kili, S.: Prediction of commercial bank failure via multivariate statistical analysis of financial structures: the Turkish case. Eur. J. Oper. Res. **166**, 528–546 (2005)
25. Devarapalli, D., Srikanth, P., Narasinga rao, M.R.: Identification of AIDS disease severity based on computational intelligence techniques using clonal selection algorithm. Int. J. Converg. Comput.-IJCONVC (2017)
26. Kramer, O: Dimensionality reduction with supervised nearest neighbors. In: Intelligent Systems References Library, vol. 51. Springer (2013)
27. Jiang, S., Pang, G., Meiling, W., Kuang, L.: An improved K-nearest-neighbor algorithm for text categorization. Expert Syst. Appl. **39**, 1503–1509 (2012)
28. Bolon-Canedo, V., Sánchez-Marono, N., Alonso-Betanzos, A.: A review of feature selection methods on synthetic data. Knowl. Inf. Syst. **34**, 483–519 (2013). doi:10.1007/s10115-012-0487-8
29. Chandrashekar, G., Sahin, F.: A survey on feature selection methods. J. Comput. Electr. Eng. **40**, 16–28 (2014)
30. Bolon-Canedo, V., Sánchez-Marono, N., Alonso-Betanzos, A.: Feature selection and classification in multiple class datasets: an application to KDD Cup 99 dataset. Expert Syst. Appl. **38**, 5947–5957 (2011)
31. Nagwani, N.K.: A comment on "A Similarity measure for text classification and clustering. IEEE Trans. Knowl. Data Eng. **27**(9) (2015). doi:10.1109/TKDE.2015.2451616
32. Aburomman, A.A., Reaz, M.B.I.: A survey of intrusion detection systems based on ensemble and hybrid classifiers. Comput. Secur. **65**, 135–152 (2017)
33. Aburomman, A.A., Reaz, M.B.I.: A novel SVM-kNN-PSO ensemble method for intrusion detection system. Appl. Soft Comput. **38**, 360–372 (2016)
34. Arslan, A.K., Colak, C., Sarihan, M.E.: Different medical data mining approaches based prediction of ischemic stroke. Comput. Meth. Prog. Biomed. **130**, 87–92 (2016)

A Comprehensive Survey and Open Challenges of Mining Bigdata

Bharat Tidke(✉), Rupa Mehta, and Jenish Dhanani

Department of Computer Engineering, SVNIT, Surat, India
batidke@gmail.com, rgm@coed.svnit.ac.in,
jenishdhanani26@gmail.com

Abstract. Bigdata comes into big picture in early 2000, since it becomes focus of researchers and data scientist. Main purpose of research and development in the field of Bigdata is to extract and predicts meaningful information from large amount of structured as well as unstructured real world data. In this paper, systematic review of background, existing related technologies used by various big enterprises, data researchers, government officials has been discussed. In addition, presented standardized complex processes to extract useful information such as data generation, storage, modeling/analysis, visualization and interpretation. Finally discusses open issues, challenges and point out the emerging directions in which researchers can work in the age of Bigdata

Keywords: Bigdata · Data mining · Bigdata storage · Bigdata analysis

1 Introduction

Technology burst in past two decades creates imminent digital data in every field. These resulted into overabundance of data sources that generates huge amount of unstructured data, gives sudden rise to concept of Bigdata. According to McKinsey [1] Bigdata is defined as "datasets whose size is beyond the ability of typical database software tools to capture, store, manage, and analyze". Similarly on challenging issues of Bigdata Gartner IT Glossary [2] defined it as "Bigdata is high-volume, high-velocity and high-variety information assets that demand cost-effective, innovative forms of information processing for enhanced insight and decision making".

Mostly, incoming data from various sources are stored in distributed environment to get value from it. Heterogeneous and incongruent properties of such data makes integration necessary to cope up with its variety nature of data, provides meaning to data analytics. Similarly, input from various sources must be preprocessed before transfer into structured format. In addition, practitioners and researchers come with the opinion that size is not the only attribute of Bigdata. They highlighted essence of problem faced by different attributes of Bigdata as 3V's (Volume, velocity, variety), later two more attributes has been added as veracity by IBM and value by Oracle [3, 4]. Different attributes creates opportunities for various enterprise businesses to grow virtually, also gave researcher a challenges to extract valuable information from Bigdata.

S.C. Satapathy and A. Joshi (eds.), *Information and Communication Technology for Intelligent Systems (ICTIS 2017) - Volume 1*, Smart Innovation, Systems and Technologies 83, DOI 10.1007/978-3-319-63673-3_53

1.1 Motivation

Data has been generated and acquired at rapid speed involves volume with it. These phenomenons create challenge to develop methods that has to be automated and respond quickly for making decision in real time. Huge Volume of data cannot be manage using conventional databases, such data needs to move and store in distributed environment for further computation as traditional data warehouses are ill suited. Further analysis using classic OLAP cube also does not work, results in replaced by distributed storage environment such as Hadoop that normally uses master-slave architecture for storing data, also map-reduce technique for processing data in batches. Similarly NOSQL databases uses different storage techniques having columns, graphs, documents, key-value stores which can work on top of Hadoop to make it suited for real time Bigdata analytics. Some limitations of traditional system can be overcome when dealing with Bigdata using distributed data processing frameworks or tools such as

- Possible to collect and store whole data, not the sample or summaries of data
- Integration and indexing of data in real time irrespective of the format in which they came.
- Velocity with which data come can be processed using distributed algorithm in real time and stored in distributed fashion for improving existing model for further analysis.
- Analyzing existing or past data is crucial while making targeted future prediction but decision based on operational or transactional data needs real time analysis has to be processed in parallel algorithms with low latency time.

2 Bigdata Mining

Bigdata mining deals with data which scale in petabytes, beyond limit capacity of single machine. A conventional Bigdata computing framework entrust large cluster of machines having immense computing speed and performance, where a data analysis job can be performed by implementing parallel programming tools. This section explores different phases shown in figure which is essential for Bigdata mining, significantly expanded the possibilities of extracting knowledge that can be useful for different applications in different domains.

2.1 Sources of Bigdata

Concept like digital India shows transformation of physical world into virtual digitized world. Also, technologies such as Web Search Engines (Google, Bing etc.), Internet of Things (Sensors, Radio-Frequency Identification (RFID), Mobile Phones etc.), Social Networking (Facebook, Twitter), Digital libraries(Google Scholar, DBLP etc.), generates a huge amount of data flow and that at lightning speed, are main sources of Bigdata [5] (Fig. 1).

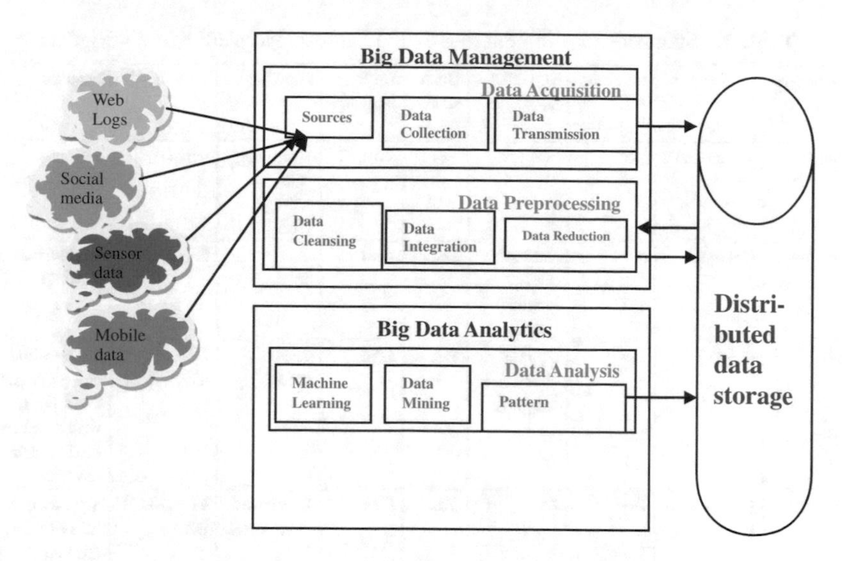

Fig. 1. Knowledge Discovery from Bigdata

2.2 Bigdata Acquisition

According to the IDC group, the quantity of world data will be 44 times bigger in the next few years [6]. With the advent of sensor technology, deployment of sensors for data acquisition has increased tremendously. These sensors are smaller, more powerful and cheaper which ultimately generates Bigdata. Similarly social media platforms provide API for data collections.

2.3 Bigdata Pre-processing

Pre-processing of Bigdata in term of data cleaning or removing duplicate data plays vital role while doing analytics of large datasets. Unremoving of human error, incorrect sensor data due to failure, irrelevant domain-specific acquisition makes data scientist work more tedious.

2.4 Bigdata Storage

Advancement in storage systems such as Hadoop distributed file system (HDFS) makes storing huge amount of unstructured data easy. HDFS is mainly based on Google file system proposed by Ghemawat et al. [7, 8] in which each file is divided into chunks and it stores those chunks in a redundant manner, as a cluster on different machine HDFS distributes large amount of data to various machines in redundant manner, which ensures in case of any node failure its running operation does not get affected. The NoSQL databases based on Bigtable [8] follows BASE (Basically Available, Soft state, eventually consistent) properties manage and handles huge amount of data

Table 1. Summary for different distributed data storage platforms for Bigdata

System/tools	Data stores	Based on ACID properties	SQL-like language	Open source	Built in language	Function	Developed at	Features
Hadoop	Master/slave	No	No	Yes	Java, C	Distributed data Storage, Processing	yahoo	Flexible cost-effective
Cassandra	Column	No	Cassandra query language (CQL)	Yes	Java		Facebook	Highly scalable no single point failure
Neo4j	Graph	Yes	Cypher query language	Yes	Java	Graph processing	Neo technology	Graph storage and processing for super-fast writes and reads flexible data modeling
MongoDB	Document store	No	No	Yes	C++	Distributed data storage	MongoDB, Inc.	Schema less index on any attribute
Hypertable	Column	No	No	Yes	C++	Distributed data storage	Hypertable Inc.	Efficient fast performance
DyanamoDB	Key/value	No	Yes	No	NA	Data storage	Amazon	seamless throughput and storage scaling

efficiently with maintaining scalability [10–12]. Some of the most known platform for data storage has been shown in Table 1.

2.5 Bigdata Analytics

Due to emergence of large storage system, complete data can be store efficiently, which shifted focus from storing sampled data to the state-of-the-arts technologies and different applications, which can do analysis of such complete data and convert it into information for making decisions. Herodotou et al. [13] proposed Starfish, a self-tuning system for Bigdata analytics based on self-tuning hadoop environment for distributing data, it chooses correct cluster, in addition also makes different configurations to handle workload and to analyze it. Interestingly user does not need to understand complex system of hadoop, since choices has been automatically made at different level of hadoop ecosystem. Nagwani [14] et al. proposed new framework for document summarization from Big textual data, based on mapreduce technique. In that they are finding semantic similarity using clustering algorithm. In addition framework consist of different stages including Latent Dirichlet Allocation (LDA) for topic modeling, ultimately improves scalability that has been evaluated using different standard measures. Patil et al. [15] presented two parallel structured ADABOOST.PL and LOGITBOOST.PL classifier algorithms, which are advancement of well known ADABOOST and LOGITBOOST boosting algorithm, also used ensemble technique for prediction. They have implemented both algorithms on mapreduce framework for distributed environment, significantly improves the speed of algorithms and also accuracy they are getting is similar or better in some cases compared to previous algorithms runs on single machine. Wu et al. [16] proposed LockStep Broadcast Tree (LSBT) model, works in pipeline manner for broadcasting data chunks from distributed data centers around the globe, so that each unit of upload link for every data node may have many connection, which can be assigned to basic unit of upload bandwidth, results in better throughput. Rathore et al. [17] introduces whole system for smart cities and urban planning in which they put different sensors at various levels to collect data, and proposed four tier architecture for data acquisition using IOT, for communication between various aspects of networks technologies, for data storage and distribution using hadoop ecosystem which include Storm, S4 etc. to store real incoming data and finally they done analysis and interpretation for making accurate decision.

2.6 Bigdata Visualization

Data visualization has been used by academicians, stakeholders from industries for analyzing and interpret their result using techniques such as plot, tables, charts and graphs According to SAS report [18], data visualization is one of key component in Bigdata analytics, they also highlighted some of problems while visualizing such large amount of data. Data visualization in tree structure has been used by many researchers for traditional datasets Lii & Fogarty [19] shows the conversion of nested tree-map to cascaded tree-map takes an advantage of 2.5 D technique to enhance the presentation of containment relationship. Cascading used limited space for representation for containment relationship. But limitation is difference between node weight and size, also the lack of stability which leads to less interactive zooming property.

3 Challenges for Bigdata Mining

3.1 Frequent Itemset Mining

Frequent itemset mining is popular technique for data mining which requires entire dataset into main memory for processing, but large datasets do not fit into main memory. Some author such as Moens et al. [20] proposed Dist-Eclat which is parallel version of pure Eclat method optimizing speed using search space distribution and BigFIM applies Apriori based method and Eclat for finding frequent itemsets and provides scalability. Riondato et al. [21] comes with PARMA algorithm which extracts approximate frequent itemsets in fewer periods using sampling method. Malek and Kadima [22] proposed a novel approach which uses clustering technique for mining of frequent itemsets using MapReduce framework which increases the performance. Zhang [23] proposes an efficient DFIMA algorithm based on distributed frequent itemset mining in which they used matrix based pruning algorithm for reducing number of candidates in an itemset and also used spark platform for distributed iterative computation to improve efficiency and scalability.

3.2 Streaming Bigdata

Stream data mining [24] is a well studied problem from many past decades and large number of publications related to stream mining can be viewed, but Bigdata stream problem is still yet to be explored fully. But, now researchers and scientists are focusing on problem facing by Bigdata community due to big real time streaming data. Vu et al. [25] proposed distributed algorithm for streaming Bigdata in which they used adaptive AM rules for regression tasks on large dataset. They used open source SAMOA platform which having distributed streaming machine learning algorithms and they compare their result on centralized implementation in MOA. Agerri et al. [26] presented new distributed and highly scalable architecture for analysis of stream textual news data using Natural language processing (NLP).

3.3 Spatial Big Data

Research on Spatial Bigdata are one of growing areas in terms of improving quality of human life [27]. The Millennium Project finds out different 15 possible challenges which has been face by human society in that 7 are related to geospatial Bigdata [28].

4 Conclusions

Volume of data operated upon by modern applications is growing at a tremendous rate, posing intriguing challenges for parallel and distributed computing platforms. These challenges ranges from building storage systems that can accommodate these large datasets to collecting data from vastly geographically distributed sources into storage systems for running a diverse set of computations on these data. Recent efforts towards addressing these challenges have resulted in scalable distributed storage systems (file systems, key-value stores, etc.) and execution engines that can handle a variety of computing paradigms. In the future, as the data sizes continue to grow and the domains of these applications diverge, these systems will need to adapt to leverage application-specific optimizations. To tackle the highly distributed nature of data sources, future systems might off load some of the computation to the sources itself to avoid the expensive data movement costs.

4.1 Unsolved Issues

To obtain accurate information based on different attributes of Bigdata, more mathematically correct frameworks need to be proposed on real context based dataset using heterogeneous hardware and software platforms, which bring critical challenges for application developers.

- A Challenges for researchers to built powerful storage models which can store large volumes of data in parallel with high dimensional data so that it can be viewed by scientist and statisticians in n*n dimensional array-based data structures.

– Distributed data mining algorithms must be develop for distributed computing and storage environments having characteristics like high accuracy, more scalability and less communication cost.
– Creating Backup storage in case of system failure and storing billions of small file in size of petabyte or more creates a challenge to build such backup infrastructure.

References

1. Baldonado, M., Chang, C.-C.K., Gravano, L., Paepcke, A.: The Stanford digital library metadata architecture. Int. J. Digit. Libr. **1**, 108–121 (1997)
2. Lohr, S.: The age of big data. New York Times **11** (2012)
3. Fan, W., Bifet, A.: Mining big data: current status, and forecast to the future. ACM SIGKDD Explor. Newsl. **14**(2), 1–5 (2013)
4. Alexandros, L., Jagadish, H.V.: Challenges and opportunities with big data. Proc. VLDB Endow. **5**(12), 2032–2033 (2012)
5. Gantz, J., Reinsel, D.: Extracting value from chaos. IDC iView, pp. 1–12 (2011)
6. Turner, V., Reinsel, D., Gantz, J.F., Minton, S.: The digital universe of opportunities: rich data and the increasing value of the internet of things. IDC Anal. Future (2014)
7. Ghemawat, S., Gobioff, H., Leung, S.-T.: The Google file system. ACM SIGOPS Oper. Syst. Rev. **37**(5) (2003). ACM
8. Jeffrey, D., Ghemawat, S.: MapReduce: simplified data processing on large clusters. Commun. ACM **51**(1), 107–113 (2008)
9. Chang, F.: Bigtable: a distributed storage system for structured data. ACM Trans. Comput. Syst. **26**(2), 4 (2008)
10. Győrödi, C., Győrödi, R., Pecherle, G., Olah, A.: A comparative study: MongoDB vs. MySQL. In: 2015 13th International Conference on Engineering of Modern Electric Systems (EMES), Oradea (2015)
11. DeCandia, G., Hastorun, D., Jampani, M., Kakulapati, G., Lakshman, A., Pilchin, A., Sivasubramanian, S., Vosshall, P., Vogels, W: Dynamo: amazon's highly available key-value store. ACM SIGOPS Oper. Syst. Rev. **41**(6), 205–220 (2007). ACM
12. Chen, M., Mao, S., Liu, Y.: Big data: a survey. Mob. Netw. Appl. **19**(2), 171–209 (2014)
13. Herodotou, H., Lim, H., Luo, G., Borisov, N., Dong, L., Cetin, F.B., Babu, S.: Starfish: a self-tuning system for big data analytic. CIDR **11**, 261–272 (2011)
14. Nagwani, N.K.: Summarizing large text collection using topic modeling and clustering based on MapReduce framework. J. Big Data **2**(1), 1–18 (2015)
15. Palit, I., Reddy, C.K.: Scalable and parallel boosting with mapreduce. IEEE Trans. Knowl. Data Eng. **24**(10), 1904–1916 (2012)
16. Wu, C.-J., Ku, C.-F., Ho, J.-M., Chen, M.-S.: A novel pipeline approach for efficient big data broadcasting. IEEE Trans. Knowl. Data Eng. **28**(1), 17–28 (2016)
17. Rathore, M.M., Paul, A., Ahmad, A., Rho, S.: Urban planning and building smart cities based on the internet of things using big data analytics. Comput. Netw. (2016)
18. SAS Institute Inc.: Five big data challenges and how to overcome them with visual analytics. Report, pp. 1–2 (2013)
19. Lü, H., Fogarty, J.: Cascaded treemaps: examining the visibility and stability of structure in treemaps. In: Proceedings of Graphics Interface, Toronto, ON, Canada, pp. 259–266 (2014)
20. Moens, S., Aksehirli, E., Goethals, B.: Frequent itemset mining for big data. In: IEEE 30th International Conference on Data Engineering, IL, Chicago, pp. 6–9 (2013)

21. Riondato, M., DeBrabant, J.A., Fonseca, R., Upfal, E.: PARMA: a parallel randomized algorithm for approximate association rules mining in MapReduce. In: Proceedings of the CIKM, pp. 85–94. ACM (2012)
22. Malek, M., Kadima, H.: Searching frequent itemsets by clustering data: towards a parallel approach using mapreduce. In: Proceedings of the WISE 2011 and 2012 Workshops, pp. 251–258. Springer, Heidelberg (2013)
23. Zhang, F., et al.: A distributed frequent itemset mining algorithm using spark for big data analytics. Clust. Comput. **18**(4), 1493–1501 (2015)
24. Joao, G.: A survey on learning from data streams: current and future trends. Prog. Artif. Intell. **1**(1), 45–55 (2012)
25. Vu, A.T., De Francisci Morales, G., Gama, J., Bifet, A.: Distributed adaptive model rules for mining big data streams. In: IEEE International Conference on Big Data (Big Data), Washington, DC, pp. 345–353 (2014)
26. Agerri, R., Artola, X., Beloki, Z., Rigau, G., Soroa, A.: Big data for natural language processing: a streaming approach. Knowl.-Based Syst. **79**, 36–42 (2015)
27. Lee, J.G., Kang, M.: Geospatial big data: challenges and opportunities. Big Data Res. **2**(2), 74–81 (2015)
28. Shekhar, S.: Spatial big data challenges. In: Keynote at ARO/NSF Workshop on Big Data at Large: Applications and Algorithms, Durham, NC (2012)

Beltrami-Regularized Denoising Filter Based on Tree Seed Optimization Algorithm: An Ultrasound Image Application

V. Muneeswaran(✉) and M. Pallikonda Rajasekaran

Kalasalingam Academy of Research and Education, Virudhunagar 626126, Tamilnadu, India
munees.klu@gmail.com, m.p.raja@klu.ac.in

Abstract. The dominant degrading factor of quality in ultrasound images is mainly due to the occurrence of speckle noise that in turn leads to false ameliorative decisions, restricts auto diagnosis and telemedicine practices. In medical image analysis speckle reduction is contemplated to be the pre-processing task that sustains decisive information and exclude speckle noise. Meta-heuristics optimization algorithm were used now a days for speckle reduction problems. Our contribution in this paper analyses the use of optimization technique in determining the best noise removing filter coefficients that removes the speckle content contributively. The proposed method comprises the use of Finite Impulse Response filter receiving the filter coefficients from Tree Seed optimization algorithm. Evaluation of noise removal with standard metrics such as Peak Signal to Noise Ratio, Correlation coefficient and Structural Similarity Index shows that the proposed method gives optimal speckle reduction score when compared with conventional filters and its superiority in despeckling medical ultrasound images. Assessment of the proposed methodology with advanced evaluation metrics ensures the ability of it in terms of preserving edges and textural features.

Keywords: Speckle noise · Tree seed optimization · Ultrasound image · Finite impulse response filter · Mean square error · Structural similarity index

1 Introduction

Ultrasonography, the preferred imaging modality and the gold standard examination procedure for soft tissue imaging of the human body, has emerged as the significant imaging tool for radiologists for the detection of cysts and cancerous tumors. It is often preferred because it is unpretentious, portable, versatile, and takes out the danger of radiation exposure [1]. Image quality is of central importance to the success of an Image analysis and understanding. One of the typical paradigm for ultrasound image processing is that the image to be analyzed must have least measure of noise. Undeniably, the presence of Speckle noise displays a coarse-grained appearance throughout the ultrasound image, which disguises the diminutive changes of gray levels and makes the discrimination between original gray level and noise a tedious process. Thus the presence of speckle makes Ultrasonography infructuous in most cases [2].

S.C. Satapathy and A. Joshi (eds.), *Information and Communication Technology for Intelligent Systems (ICTIS 2017) - Volume 1*, Smart Innovation, Systems and Technologies 83, DOI 10.1007/978-3-319-63673-3_54

Adoption of phase sensitive transducers is considered to be the main reason for the direct repercussion of speckle. In addition to the usage of phase sensitive transducers, another major determinant of speckle contamination is the distance between transducer and the maximum pressure point.

Speckle noise is the dominant factor for reduction in both the contrast and the spatial resolution values of the acquired ultrasound images, thus degrading the ability to extract information and to detect the objects that are similar in size of the speckle [3]. Reduction in speckle noise, as a direct consequence leads to reduction in false positives, makes edges clear, improves image quality and increases the possibility of detecting micro calcifications. Perceptible exploration for quality improvement through speckle suppression in ultrasound medical image has produced many different executable filtering approach each with their own merits and demerits. In 2014 Chandra and Chattopadhyay [4] optimized the 2-D low pass filter for possible reduction of Gaussian noise using differential evolution algorithm for optimizing 3×3 filter mask coefficients that was used for reducing the fitness function. The major setback in their work was that no due importance was given for the analysis of filter mask size and only a few evaluation parameters were used in proving the efficiency of their work. Alenrex in 2015 [5] described the use of soft computing techniques in filter design and analysis for speckle reduction comparatively with each of their own advantages and disadvantages. The author also tabulated the previous researches in speckle reduction in SAR and medical images. As our proposed study deals with medical images the preservation of edges becomes vital, it is essential to understand the usefulness of edges and Gupta in 2014 strategically explained the importance of preserving edges in images and experimentally verified his work using Edge Keeping Index [6]. Reddy and Sahoo attempted to reduce power delay product and increase hardware efficiency by optimizing Finite Impulse Response filter coefficients using differential algorithm [7]. In 2016 Mohammadi briefly analyzed about the effect of decreasing the population size of different optimization algorithms in designing Infinite Impulse Response filters [8]. In 2005 research group headed by Loizou made out a justifiable comparison between various despeckle filtering techniques in terms of classifying denotive and non-denotive US images of the carotid artery [2]. In the present study, an optimization based despeckling method is proposed and it is evaluated with real time ultrasound images.

2 Materials and Methods

2.1 Modeling of Speckle Noise

Speckle formation is an intrusive phenomenon, that occurs due to the interference of two or more waves that travels to the probe from the scatterer. This interference increases the amount of scattering as the fourth power of emitted ultrasound frequency that in turn produces low amplitude echo signals resulting in an appearance of a 'Speckle' pattern giving an impression of the texture that disagree to the factual tissue microstructure.

The images from the ultrasound imaging scanners seems to be log compressed [9]. Thus the speckle can be represented by

$$Z_{ij} = X_{ij} + \sqrt{x_{ij} \cdot n_{ij}} \tag{1}$$

where x_{ij} and Z_{ij} represents the gray level of the true and observed images respectively, n_{ij} represents the noise term independent of x_{ij} and X_{ij} are the values of pixel at the coordinates (i, j).

2.2 Tree Seed Optimization

Over the last few decades the growth of nature inspired optimization algorithms for solving continuous and discrete problems of various applications is increasing. Among them Tree-Seed Optimization algorithm formulated by Kiran is a population based search algorithm [10] that assumes the natural phenomena between trees and seeds. Assuming the surface covered by the seeds as search space. It is successfully applied to real time optimization problems [11]. The task of finding the location of individual tree and the seed is considered to be the optimization problem. The seed value is generated for each iteration with the help of Search Tendency (ST) and is noted as $S_{i,j}$. Search tendency is limited between the range of 0 to 1. The initial tree location $T_{i,j}$ is produced by

$$T_{i,j} = L_j + r_{i,j}(H_j - L_j) \tag{2}$$

where $r_{i,j}$ is a random number for individual location and dimensions between (0,1), L_j and H_j represents the lower and upper limit of the search region respectively. With due importance to the value of ST and $r_{i,j}$, the seed value is revised. If the value of $r_{i,j}$ seems lesser than the value of search tendency (ST), then value of seed is replaced with the value obtained from equation given below,

$$S_{i,j} = T_{i,j} + \alpha_{i,j}(B_j - T_{r,j}) \tag{3}$$

else the value of seed is updated by following equation,

$$S_{i,j} = T_{i,j} + \alpha_{i,j}(T_{i,j} - T_{r,j}) \tag{4}$$

where, $S_{i,j}$ and $T_{i,j}$ is seed value and Tree values in *j*th dimension produced by *i*th tree, B_j is the *j*th dimension of best tree location obtained so far. $T_{r,j}$ is the *j*th dimension of *r*th tree chosen from the population and $\alpha_{i,j}$ is the scaling factor in the range of $[-1, 1]$. Thus using this algorithm if the seeds produced is optimized than the former solution, the seeds replaces the tree, and the best solution B from set of trees $T_{i,j}$ is selected using Eq. (5)

$$B = min\{f(T_{i,j})\}; i = 1, 2, 3, \ldots\ldots p, j = 1, 2, \ldots\ldots n \tag{5}$$

where i and j are population and dimension of the problem respectively.

3 Design Formulation

In the proposed scheme two input image signals were used such as the noise free image $I_{org(n)}$ and $I_{noisy(n)}$ the image contaminated with multiplicative noise. In the output side the 2D FIR filter system with optimization using Tree Seed Optimization together with median filter will produce the despeckled image. The stability condition of a two dimensional filter is given in the following Eq. (6)

$$\sum\nolimits_{m=-X}^{X} \sum\nolimits_{n=-X}^{X} |h(m,n)| < X \tag{6}$$

where X is the number of element from the origin in the mask size of the filter coefficients. The proposed methodology uses Tree Seed Optimization in finding optimal filter coefficients. The steps of TSA based 2D-FIR are given below

Step 1 Select the total size of population (window size of filter), *ST* Values, and condition of termination.

Step 2 Initialize the population as per the given Eq. (2) and satisfying Eq. (6)

Step 3 Produce seed values for the existing tree values using the Eqs. (3) or (4) based on the value of $r_{i,j}$ and *ST*, and evaluate the fitness Eq. (7) and replace the tree values if the seed values produced optimized fitness than original tree values.

Step 5 Select the best value using Eq. (5) and check the termination condition.

Step 6 If the termination condition is met, assign the optimized values and store the result, else repeat step 2,3,4.

The objective of the optimization process is given by equation of Beltrami-regularized denoising [12] with data fidelity and is given as

$$fitness(I) = \left(\sum_{\Omega} \sqrt{1 + \beta^2 |\nabla \mathrm{I}|^2} \right) + \frac{\lambda}{2} (\mathrm{I} - \mathrm{I}_0)^2 \tag{7}$$

The above equation is used in image denoising problem for the reason it includes the edge information and tries to keeps vital information [13]. The image fidelity between the image being evaluated and the input image is quarantined by the term $(\mathrm{I} - \mathrm{I}_0)^2$, where I is the image evaluated and I_0 is the noisy image. The total variation $(\nabla \mathrm{I})$ is the parameter that is usually kept for regularizing whereas β and λ are balancing parameters and Ω is set of all points in the input image.

4 Experimental Results

4.1 Simulated Results

In our experiments, the simulated test image of size 128 × 128 pixels as shown in the Fig. 2a was used. The standard image is corrupted with speckle noise at different noise level. As seen in Fig. 1 the proposed scheme includes a median filter that supports the

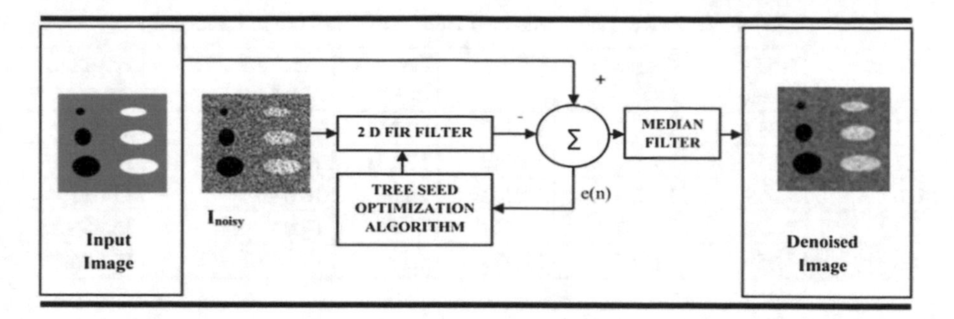

Fig. 1. Proposed denoising scheme

Table 1. 2D FIR filter coefficients obtained with (a) proposed image denoising scheme (b) ABC based image denoising scheme [14]

(a)		
a_{00} 0.0759	a_{01} 0.0967	a_{02} 0.0745
a_{10} 0.0858	a_{11} 0.2628	a_{12} 0.1083
a_{20} 0.0594	a_{21} 0.1177	a_{22} 0.0723
(b)		
a_{00} 0.1032	a_{01} 0.1047	a_{02} 0.1012
a_{10} 0.1063	a_{11} 0.1110	a_{12} 0.1059
a_{20} 0.1006	a_{21} 0.1043	a_{22} 0.1024

denoising process to preserve edges. The coefficients obtained with the Tree Seed optimization algorithm is given in Table 1a.

In addition the test image was denoised using simple and widely used filters like Gaussian [15], wiener [16], Lee [17] and ABC based 2D FIR filters [14] having coefficients given in Table 1b. As seen in Fig. 2c the output of the proposed method has edges in clear form and is visually seen even under high noise variance. The metrics Peak signal to noise ratio (PSNR) [13], structural similarity index (SSIM) [18], Normalized Cross Correlation (NK) [1] are used to compare the different algorithms. It is illustrated from Tables 2, 3 and 4 that higher PSNR, SSIM and NK values is obtained by the proposed method for a noisy image affected with noise variance 0.2 to 1.

Table 2. PSNR values obtained from noisy and filtered images

Noise variance	Noised image	Gaussian filter	Wiener filter	Lee filter	ABC-2D FIR filters [14]	Proposed method
0.2	9.317	19.712	13.369	19.211	19.406	20.59
0.4	8.088	16.727	12.649	17.131	17.257	18.151
0.6	7.599	15.325	12.251	16.081	16.641	17.037
0.8	7.222	14.687	12.04	15.56	16.021	16.456
1.0	7.059	14.143	11.979	15.2	15.854	16.296

Table 3. SSIM values obtained from noisy and filtered images

Noise variance	Noised image	Gaussian filter	Wiener filter	Lee filter	ABC-2D FIR filters [14]	Proposed method
0.2	0.063	0.291	0.307	0.371	0.412	0.48
0.4	0.041	0.191	0.224	0.265	0.304	0.337
0.6	0.032	0.156	0.199	0.227	0.27	0.311
0.8	0.024	0.144	0.174	0.206	0.243	0.268
1.0	0.024	0.127	0.173	0.2	0.233	0.259

Table 4. NK values obtained from noisy and filtered images

Noise variance	Noised image	Gaussian filter	Wiener filter	Lee filter	ABC-2D FIR filters [14]	Proposed method
0.2	0.666	0.851	0.624	0.871	0.924	0.931
0.4	0.648	0.825	0.612	0.849	0.901	0.907
0.6	0.613	0.815	0.604	0.848	0.889	0.899
0.8	0.600	0.813	0.603	0.837	0.878	0.891
1.0	0.562	0.806	0.601	0.826	0.889	0.890

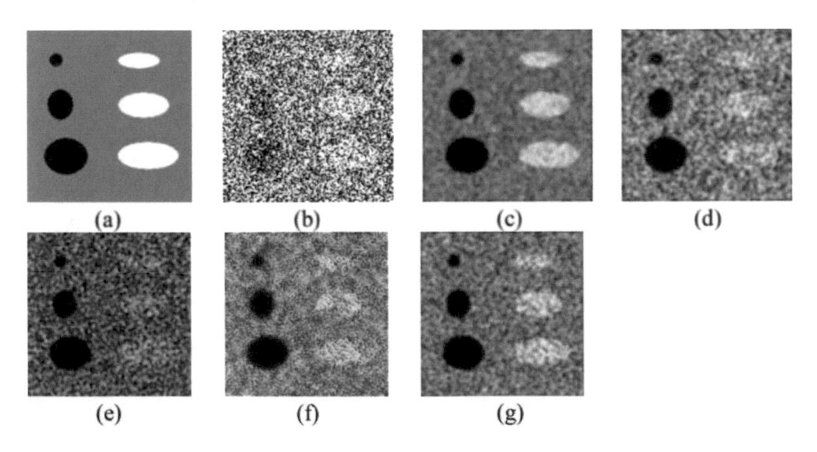

Fig. 2. Synthetic image for the proposed system and denoised images (a) noiseless image (b) noisy image with noise variance 1, (c) Denoised image with proposed method, (d) Gaussian filtered image, (e) wiener filtered image, (f) Lee filtered image (g) ABC based 2D filtered image.

4.2 Clinical Results

Figure 3a shows the noisy clinical fetal image with an gestational age of 9 weeks and 5 days, the task of speckle removal makes the head and trunk in the fetal image to be more visible as seen in Fig. 3b. Figure 3c shows the transvaginal US image acquired with the help of transvaginal probe and due to noise present the uterus and the ovary cannot be visualized, this difficulty is successfully removed after denoising process and the follicle seen in ovary can be easily visualized in Fig. 3d. Thus in all cases the usage of the proposed method reveals the important information from the image, which makes it clinically significant.

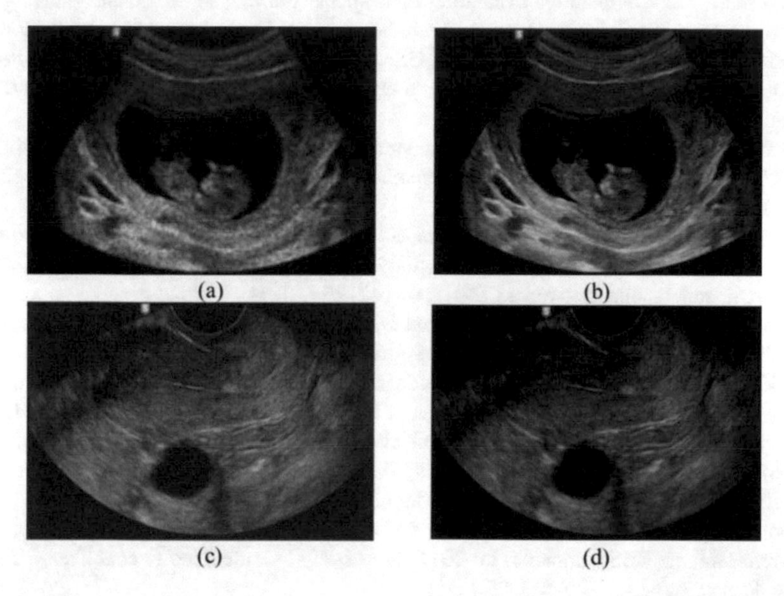

Fig. 3. (a) Noisy clinical fetal image, (b) denoised clinical fetal image, (c) noisy transvaginal US image, (d) denoised transvaginal US image.

5 Conclusion

We formulated a novel population based computational scheme for removing speckle noise using Tree Seed Optimization algorithm. Effective performance indicators such as PSNR, SSIM, NK values were used in the study. We tested the TSA based denoising scheme for synthetic image and also evaluated it for noisy ultrasound clinical Images of fetal, transvaginal and kidney. The advantage of this method is that it is easier to implement and its obtaining better results than the filters in the literature. The filter coefficients in Table 1a can be successfully used to denoise any clinical ultrasound image and it makes the proposed filter to act as convolution filter as like Gaussian and mean filters do. In our future study we focus to analyze the effect of increasing mask size with the proposed method in image quality and also to analyze the use of various optimization techniques in removing speckle noise.

Acknowledgements. The authors thank Dr. P. VIJAY BABU, M.B.B.S., D.M.R.D., Consultant Radiologist, VIJAY SCANS-Rajapalayam, Tamilnadu, for supporting the research by providing Ultrasound images and necessary patient information.

References

1. Rosa, R., Monteiro, F.C.: Performance analysis of speckle ultrasound image filtering. Comput. Methods Biomech. Biomed. Eng. Imaging Vis. **1163**, 1–9 (2014)
2. Luizou, C.P., Pattichis, C.S., Christodouluo, C.I., Istepanian, R.S.H., Pantziaris, M., Nicolaides, A.: Comparative evaluation of despecle filtering in ultrasound imaging of the carotid artery. IEEE Trans. Ultrason. Ferroelectr. Freq. Control **52**, 1653–1669 (2005)
3. Andria, G., Attivissimo, F., Cavone, G., Giaquinto, N., Lanzolla, A.M.L.: Linear filtering of 2-D wavelet coefficients for denoising ultrasound medical images. Meas. J. Int. Meas. Confed. **45**, 1792–1800 (2012)
4. Chandra, A., Chattopadhyay, S.: A new strategy of image denoising using multiplier-less FIR filter designed with the aid of differential evolution algorithm. Multimed. Tools Appl. **75**, 1079–1098 (2016)
5. Maity, A., Pattanaik, A., Sagnika, S., Pani, S.: A comparative study on approaches to speckle noise reduction in images. In: Proceedings of 1st International Conference Computer Communications and Networks CINE, pp. 148–15 (2015)
6. Gupta, D., Anand, R.S., Tyagi, B.: Ripplet domain non-linear filtering for speckle reduction in ultrasound medical images. Biomed. Signal Process. Control **10**, 79–91 (2014)
7. Reddy, K.S., Sahoo, S.K.: An approach for FIR filter coefficient optimization using differential evolution algorithm. AEU Int. J. Electron. Commun. **69**, 101–108 (2015)
8. Mohammadi, A., Zahiri, S. H.: Analysis of Swarm Intelligence and Evolutionary Computation Techniques in IIR Digital Filters Design. In: 2016 1st Conference Swarm Intelligent Evolution Computer. pp. 64–69 (2016)
9. Roomi, S.M.M., Rajee, R.B.J.: Speckle noise removal in ultrasound images using Particle Swarm Optimization technique. In: 2011 International Conference Recent Trends Information Technology, pp. 926–931 (2011)
10. Kiran, M.S.: TSA: Tree-seed algorithm for continuous optimization. Expert Syst. Appl. **42**, 6686–6698 (2015)
11. Muneeswaran, V., Pallikonda Rajasekaran, M.: Performance Evaluation of Radial Basis Function Networks Based on Tree Seed Algorithm. In: International Conference Circuit, Power Computer Technology [ICCPCT] (2016)
12. Zosso, D., Bustin, A.: A primal-dual projected gradient algorithm for efficient Beltrami regularization. UCLA CAM Report, pp. 14–52 (2014)
13. De Paiva, J.L., Toledo, C.F.M., Pedrini, H.: An approach based on hybrid genetic algorithm applied to image denoising problem. Appl. Soft Comput. (2015)
14. Latifoglu, F.: A novel approach to speckle noise filtering based on Artificial Bee Colony algorithm: an ultrasound image application. Comput. Methods Programs Biomed. **111**, 561–569 (2013)
15. Haddad, R.A., Akansu, A.N.: Class of fast Gaussian binomial filters for speech and image processing. IEEE Trans. Signal Process. **39**, 717–721 (1991)

16. Mahalanabis, A.K.: Introduction to random signal analysis and kalman filtering. Robert G. Brown. Automatica. **22**(3), 387–388 (1986)
17. Sen Lee, J.: Speckle analysis and smoothing of synthetic aperture radar images. Comput. Graph. Image Process. **17**(1), 24–32 (1981)
18. Hua, J., Kuang, W., Gao, Z., Meng, L., Xu, Z.: Image denoising using 2-D FIR filters designed with DEPSO. Multimed. Tools Appl. **69**(1), 157–169 (2014)

Thresholding Based Soil Feature Extraction from Digital Image Samples – A Vision Towards Smarter Agrology

M. Arunpandian(✉), T. Arunprasath, G. Vishnuvarthanan, and M. Pallikonda Rajasekaran

Department of Electronics and Communication Engineering, Kalasalingam Academy of Research and Education, Virudhunagar 626126, Tamilnadu, India
arunwatrap@gmail.com, arun.aklu@gmail.com, gvvarthanan@gmail.com, m.p.raja@klu.ac.in

Abstract. Soil is one of the natural material, which has the different features for the particular characteristics. In digital image processing is the principle to simplify the identification of soil features. Soil consists of both physical and chemical characteristics. These characteristics are used to find the field of soil usage. Thresholding is the conversion of colour image into binary image and that is used for shape based identification. It applicable for feature extract from curvature, valleys, and non-smoothening surfaces and it enhances the feature and get more information. Fractal dimension is one of the soil feature. A new model is proposed to assign various threshold values apply to the same sample and to determine the range and also the best image model (Red-Green-Blue, Hue-Saturation-Value, Hue-Saturation-Luminance and Hue-Saturation-Intensity) of soil samples. The device can also be modelled as most powerful tool for prediction of land usage for various fields such as agriculture and construction.

Keywords: Thresholding · Feature extraction · Fractal dimension · RGB thresholding · HSV thresholding · HIS thresholding · HSL thresholding

1 Introduction

Agriculture has more important field to the all the living organisms in the world. This field makes as much more effective to introduce Computer based digital image processing. In this method is to reduce the degradation of information and time consumption. Digital images are storing full of information about the object where we find the solution and extraction of features. In some more advancements of processing techniques is used to sharpen the features. Thresholding is the conversion is to extract the original features that relates to the fractal dimension analysis. In yearly 2010, Baveye surveyed into automatic detection of thresholding by the various methods of analysis with the best suitable analysis of object characterization likes shapes, background of image, grey level entropy, Similarity of grey levels and contrast of the image [6]. Histogram is used to identify the different shapes curvature and valleys, clustering is used for foreground and background objects, Entropy is used to entropy of the both

S.C. Satapathy and A. Joshi (eds.), *Information and Communication Technology for Intelligent Systems (ICTIS 2017) - Volume 1*, Smart Innovation, Systems and Technologies 83, DOI 10.1007/978-3-319-63673-3_55

foreground and background regions, Object attributes is used to similarity between the grey level and binary level images such as edge coincidence, fuzzy shapes etc., Spatial domain is used to correlate the pixel values and Local is used to value assign on each pixel to the local image [13].

In 2009, Camergo and Smith recommended pattern classification for plant disease causing agents. Using Support Vector Machine (SVM) enhanced and segmented features are extracted from cotton plants. Segmentation is pre-processed into RGB-HSV thresholding. It briefly announced the grey level histogram and Box counting Method (Fractal Dimension) [3]. In 2013, Simon adar et al. illustrated soil changes detection under the imaging spectroscopy sensors [9, 15]. Sensors are used to get form the alarm whenever get the high probability of change detection. But, it can work only without change in light and atmospheric [1]. In 2013, Jayme Garica and Arnal Barbedo proposed plant disease detecting, quantifying and classifying in any part. They converts grey scale image into binary image with the help of RGB and HSL thresholding. It divulges combination of RGB-HSL model is best to the object counting [2]. In 2008, Shen Weizeng et al. suggested HIS model with Sobel operator is most preferable to image segmentation. Hue component is helps to reduce illumination in vein and Disease Spotted and Segmented by Sobel operator [16]. It's only suitable for leaf disease identification. The main aim of this paper is also to identify the suitable thresholding method and with suitable image models for the using of still images. It reveals the RGB-HSI-HSL-HSV model in brief and which model is best one to calculation of fractal dimension.

2 Materials and Methodology

2.1 Input Samples

Soil samples are in the specification of square shaped 24-bit colour images. It is in the resolution of 150 × 150 pixels. Images are captured in the particular weather and light conditions to gather more information. These real time images are shown in Fig. 1a–c. The real time image samples have taken from the same cultivable land to three different places in same weather condition at Watrap, Virudhunagar District, and Tamilnadu, India. These samples are in square shape for the Automation function of Box Counting method [10].

2.2 Thresholding

Thresholding is the conversion of colour image into binary image, which enhances the colour component into 1 and the null component into 0. It is mainly applied for uniform lightening throughout the image and chosen the range by manually or automatically. There are different image models are looking into input image samples such as RGB, HSV, HIS and HSL-models [5].

In 2011, Antonia Macedo-Cruz et al. clarified sensor based oat plant assessment from frost damage. It discloses thresholding forms using (i) Otsu's method. (ii) Isodata algorithm. (iii) Fuzzy thresholding [4]. Otsu's method is to separate the various classes

(a) Real time input Sample1

(b) Real time input Sample1

(c) Real time input Sample1

Fig. 1. Real time input samples

in an image, Isodata algorithm is to find the maximum and minimum threshold value and Fuzzy thresholding fully based upon the fuzzy set theory. This combination of both to form from average of those three threshold values.

$$t = \frac{t_i + t_o + t_f}{3} \tag{1}$$

T - Average threshold value, t_i- Isodata threshold value, t_o- Otsu's threshold value, t_f- Fuzzy threshold value

In 2014, Yamini Sharma and Yogesh k. maghrajani make cleared tumour size extraction from MRI brain Images. It examines the global thresholding to form the binary image from the MRI grey scale image. In global thresholding assign the pixel value as '0' or '1' to above or below threshold values at the specific pixel value [12].

$$G = \begin{cases} 1, F(x,y) > T \\ 0, F(x,y) \leq T \end{cases} \tag{2}$$

T - Threshold value, F(x, y) – Pixel value at the position of x, y.

2.3 RGB Model

In RGB model, it enhances the pixel rates of individual Red, Green and Blue components. Input image samples are under removal of null components and to enrich the information present. It applies the value of each color intensity into the value of barrier in thresholding. In 2014, [7] Mohamad abdou barber et al. established the face recognition system using HSL based R component. In RGB model investigated by nine skin color detection. Grey scale images are under the Grey world approach with the corrected components of R and it is used to found the average reflectance of the samples. Whit batch approach is used to investigate the chromaticity of the illumination. Stretch approach is used for histogram equalization. Modified Grey world approach used to assign 128 instead of higher pixel value. Human face detection is handled in an efficient manner of using modified grey world approach.

2.4 HSL-HSI-HSV Model

In HSL model, Instead of Red-Green-Blue values are to assign Hue-Saturation-Luminance. In HSV, HIS and HSL models, Hue is the value only dominates the color components of an area. Hue value is same for all the models but saturation is different for every model and its measures the colorfulness in a specified components Intensity, Light and Value only investigates the luminance present in area. Hue value is measured in the range of 0°–360°. Saturation and value are both range from 0 to 100%. Saturation decreases the color moves towards white. But the value decreases the color moves towards black. In HSL model, luminance is more or less similar to value, but luminance 0% denotes the black and 100% denotes the white [13].

2.5 Fractal Dimension

Fractal dimension (FD) is defined as a mathematical descriptor of image feature which characterizes the physical properties of soil images. It is more insensitive in the factor of local intensity, Zooming, and local scaling of grey levels. It is more effective with insensitive to the multiplicative noise. In soil images is to detect dielectric permittivity, volumetric, gravimetric water content, bulk density and mechanical resistance [10].

$$FD = \frac{\log N(s)}{\log(\frac{1}{s})} \tag{3}$$

FD- Fractal dimension, N(s) - Number of 1's present in the box, S- Size of the box.

Equation (3) shows the identification of fractal dimension using Box Counting Method. Size of the box is be in variable like $3 \times 3, 4 \times 4, 5, 5, 6 \times 6$ and vice versa [11].

3 Results and Discussion

3.1 Threshold Conversion

Using the above real time soil samples are converting using the various models of colour images. In from the binary images having all the pixels in terms of 0's and 1's which helps to find the fractal dimension of the samples. From Fig. 2a–e are the converted colour, RGB- binary image, HSL binary image, HSV binary image and HIS binary images. In RGB model, level of threshold values of corresponding Red-Green-Blue has been calculated using Lab VIEW Vision Assistant 2014 [8]. Minimum of threshold set into 40 and the maximum of value set into 130 for all the colour components. In Fig. 2b–e, where the colour red and black shows the information and null components.

(a) Real time input sample 1

(b) RGB Binary Image

(c) HSL Binary Image

(d) HSV Binary Image

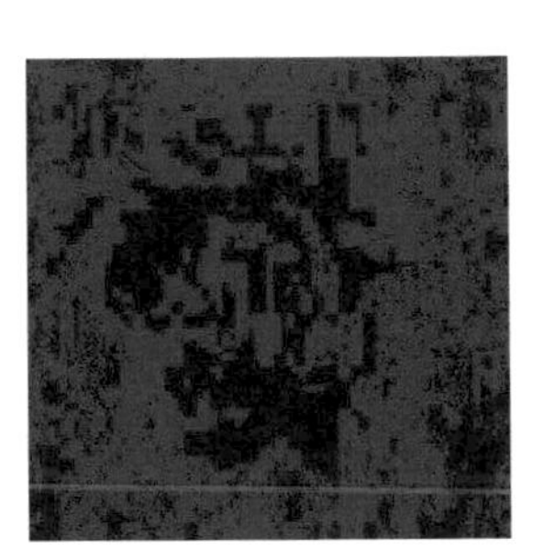

(e) HSI Binary Image

Fig. 2. Binary Images of Real time input sample 1

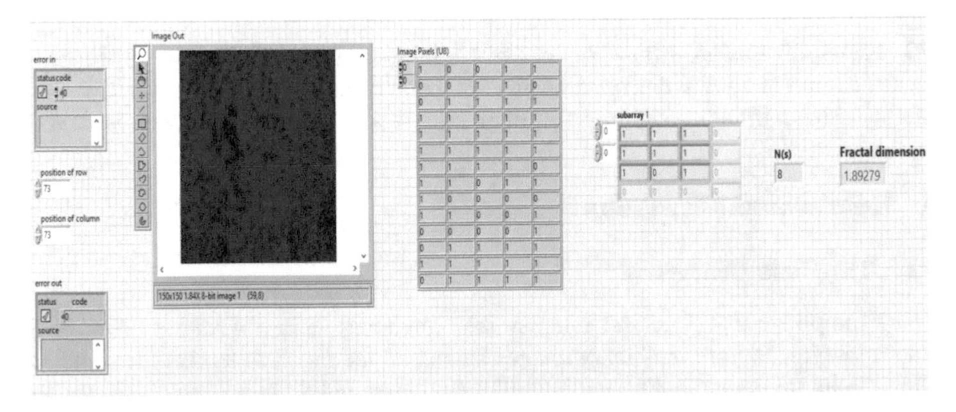

Fig. 3. Front panel of fractal dimension conversion with RGB model thresholding

In HSL model, assign the different minimum and maximum threshold value for each components. Hue value is set into min-max of 30–130, Saturation is set into min-max of 0–20 and the luminance is in the range of min-max of 40–160. This converted binary image is shown in Fig. 2c.

In HSV model, assigns the threshold values similar like to HSL model. Hue and Saturation are set into same as HSL model and the Value is set into the range as 40–160. These converted binary image is shown in Fig. 2d.

In HSI model, assigns the threshold values similar like to HSL and HSV model. Hue and Saturation are set into same as HSL model and the Intensity is set into the range as 40–160. These converted binary image is shown in Fig. 2d.

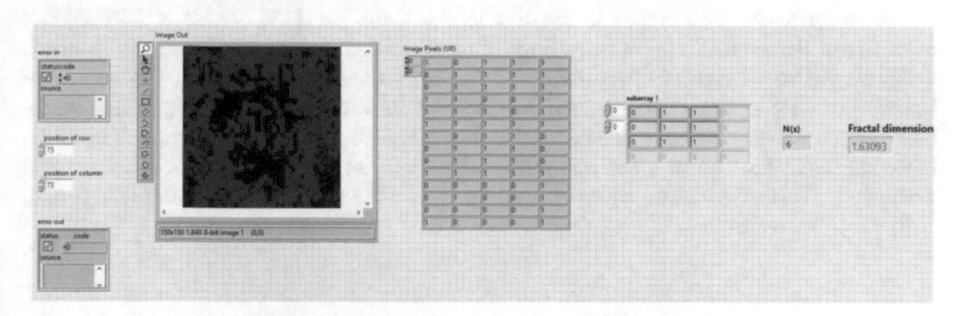

Fig. 4. Front panel of fractal dimension conversion with HSL, HIS and HSV model thresholding

Figure 3 shows the fractal dimension calculation using Box Counting Method with RGB model thresholding. This sub array displays the Centre of 3 × 3 box pixels in terms of 0's and 1's. N(s) counts the number of 1's present in the sub array. This fractal dimension icon shows the FD value of corresponding real time samples. From the Table 1 exhibits fractal dimension value of corresponding Real time samples with the suitable image models. When compared all those image models, [14] RGB image model has more impact and get higher and reliable fractal dimension as shown in Fig. 4.

Table 1. Results for fractal dimension calculation

Samples	RGB model	HSL model	HSV model	HSI model
Real time input sample-1	1.89279	1.63093	1.63093	1.6309
Real time input sample-2	1.46497	1.00363	1.00363	1.0036
Real time input sample-3	1.89279	1.46497	1.26186	1.7712

4 Conclusion

The wide range of variety of thresholding conversion have in digital image sampling. In gather more ideas from the survey of thresholding and image models, compare the image models with the fractal dimension values it concludes RGB model is better for the field of regions from gathering the input samples. In RGB model has the higher fractal dimension, when compared to all those three image models. In this RGB model considers all the colour components of (Red, brown, yellow, green, Blue and grey). But, in other models of (HSL, HIS and HSV) only considers some of the specified colour components. Soils are in all colour depending upon the nature of environment, so each colour component imports one specific characteristics. This fractal dimension makes known some kinds of physical parameters and also known to the field of usage. In future determined fractal dimension and image model help to find the physical parameters water content, grain resistant, dielectric constant, relative permittivity and field density. Using Vision builder is to make an application of mobile or computer.

Acknowledgements. The author would like to thank the Sir. C.V. RAMAN KRISHNAN International Research Centre for providing financial assistance under the University Research Fellowship. Also we thank the Department of Electronics and Communication Engineering of Kalasalingam University, (Kalasalingam Academy of Research and Education), Tamil Nadu, India for permitting to use the computational facilities available in Centre for Research in Signal Processing and VLSI Design which was setup with the support of the Department of Science and Technology (DST), New Delhi under FIST Program in 2013 (Reference No: SR/FST/ETI-336/2013 dated November 2013).

References

1. Adar, S., Shkolnisky, Y., Ben-Dor, E.: Change detection of soils under small-scale laboratory conditions using imaging spectroscopy sensors. Geoderma **216**, 19–29 (2014)
2. Garcia, J., Barbedo, A.: Digital image processing techniques for detecting, quantifying and classifying plant diseases. SpringerPlus **2**, 660 (2013)
3. Camargo, A., Smith, J.S.: Image pattern classification for the identification of disease causing agents in plants. Comput. Electron. Agric. **66**, 121–125 (2009)
4. Macedo-Cruz, A., Pajares, G., Villegas-Romero, I., Pajares, G.: Digital image sensor-based assessment of the status of oat (Avena sativa L.) crops after frost damage. Sensors (2009). doi:10.3390/s110606015
5. Moranduzzo, T., Melgani, F.: Automatic car counting method for unmanned aerial vehicle images. IEEE Trans. Geosci. Remote Sens. **52**, 1635 (2014)
6. Baveye, P.C., Laba, M., Otten, W.: Observer-dependent variability of the thresholding step in the quantitative analysis of soil images and X-ray micro tomography data. Geoderma **157**, 51–63 (2010)
7. Berbar, M.A., Laba, M., Otten, W.: Skin colour correction and faces detection techniques based on HSL and R colour components. Int. J. Sign. Imag. Syst. Eng. **7**, 104 (2014)
8. Ding, Z., Zhang, R., Kan, Z.: Quality and safety inspection of food and agricultural products by lab VIEW IMAQ vision. Food Anal. Methods (2014). doi:10.1007/s12161-014-9989-1
9. Raje, C., and Rangole, J.: Detection of leukemia in microscopic images using image processing. In: International Conference on Communication and Signal Processing (2014)
10. Oleschko, K., Korvin, G., Munoz, A.: Mapping soil fractal dimension in agricultural fields with GPR. Nonlin. Process. Geophys. **15**, 711–725 (2008)
11. Hocevar, M., Brane, S., Godes, T.: Flowering estimation in apple orchards by image analysis. Precis. Agric. (2014). doi:10.1007/s11119-013-9341-6
12. Sharma, Y., Meghrajani, Y.K.: Brain tumor extraction from mri image using mathematical morphological reconstruction. 978-1-4799-6986-9/14/$31.00/2014
13. Ghamisi, P., Couceiro, S.: Multilevel image segmentation based on fractional-order darwinian particle swarm optimization. 0196 2892/$31.00/2013

14. Minervini, M., Abde, T., Tsaftaris, S.A.: Image-based plant phenol typing with incremental learning and active contours. Ecol. Inform. **23**, 35 (2013)
15. Lloret, J., Bosch, I., Sendra, S., Serrano, A.: A Wireless Sensor Network for Vineyard Monitoring That Uses Image Processing. Sensors (2014). doi:10.3390/s110606165
16. Yachun. W., Zhanliang. C., Hongda. W.: Grading method of leaf spot disease based on image processing. In: International Conference on Computer Science and Software Engineering (2008)

MobiCloud: Performance Improvement, Application Models and Security Issues

Jayati Dave(✉), Yusra Shaikh, Tarjni Vyas, and Anuja Nair

Institute of Technology, Nirma University, Ahmedabad 382481, India
{15mcen06, 14bce136, tarjni.vyas, anuja.nair}@nirmauni.ac.in

Abstract. Recent years have seen an exponential increase in cloud computing for services such as high computational capabilities, vast storage, applications etc. that they provide. Moreover, today's world is a constant shift from desktop or laptop devices towards handheld smartphones because of the wide range of applications that are supported. This paper discusses the present state of cloud computing for mobile, termed as MCC or Mobile Cloud Computing. This paper also explains the MCC architecture, and the various existing application models, namely Energy and Performance based, constraints based, and Multiple Objectives based models and their examples. We also identify the research gaps in this area which cover the need for standardization of mobile cloud execution platforms. We further point out the vulnerability smartphone clones, and the need to address the privacy issues and security attacks against the user.

Keywords: Mobile Cloud Computing · Computational offloading · VM migration · Smartphone clones · Cloudlets

1 Introduction

Cloud Computing has gained much popularity in recent years because of its ability to provide storage, computing, applications and services over the Internet. Along with these services, reduction of capital cost, decoupling of services from concealed technology and provision of flexibility concerned with resource provisioning.

Likewise, smartphones are being more and more popular today as wide-ranging applications are supported by them like image processing, games, electronic commerce, video processing, online social media services. The demand for computing resources is increasing as the smartphone applications are getting more and more complex but battery life and advances in hardware of smartphones are slow to acknowledge the computational requirements of applications expanded throughout years. So, because of some constraints such as limited memory and battery life, low processing power and uncertain network connectivity, compatibility issues arise for many applications to get integrated with smartphones [1, 2].

Major software and hardware level changes are required to make the smartphone computationally capable and energy efficient for that, manufacturers and developers need to work together [3–5]. It is not possible to attain true limitless computational

S.C. Satapathy and A. Joshi (eds.), *Information and Communication Technology for Intelligent Systems (ICTIS 2017) - Volume 1*, Smart Innovation, Systems and Technologies 83, DOI 10.1007/978-3-319-63673-3_56

power by hardware level modifications due to size constraints. So, with partial support of hardware of smartphone, the effectiveness of software level modifications is more [6].

Migration of resource-intensive computation to the server or cloud which is full of resources takes place in computation offloading which reduces the consumption of battery power, enhances the application performance and executes the applications effectively. Storage constraints can be overcome by the cloud services as the cloud offers storage services [7]. Many applications such as social networks, file sharing, education, e-commerce, searching, healthcare, gaming etc. exist with the support of cloud for various domains.

The forum for Mobile Cloud Computing provides the definition of MCC as under [4]. 'Mobile cloud computing when simply defined is an infrastructure where computation and storage sensitive tasks are not carried out in the mobile device. Such applications shift the data computing and data storage tasks into the cloud, thus providing MC services not only to the mobile users but also to a wide range of subscribers'. Some approaches integrate cloud computing and mobile computing thus generating a new service known as Mobicloud computing [12].

'Mobile cloud' has two major aspects: (a) ad-hoc mobile cloud and (b) infrastructure based. The hardware infrastructure remains fixed and services are provided to the users in infrastructure-based mobile cloud whereas ad-hoc mobile cloud is a set of mobile devices. These mobile devices behave in a manner which replicates a cloud and access internet based or local facilities and provided them with mobile devices.

As the traditional offloading techniques need a lot of bandwidth and they are energy unaware, they cannot be used for the smartphones directly. Some applications use cloud services but its usage is limited to application-specific and storage facilities only such as Siri of apple which is a personal assistant and voice based. So an application model is required which supports computation transferring and which is used also for mobile cloud environment regarding context awareness, bandwidth, heterogeneity, application partitioning overhead, energy consumption and network data lost (Table 1).

Table 1. Cloud computing vs MobiCloud [13].

Issues	Cloud computing	Mobile cloud computing
Device energy	✗	✓
Bandwidth utilization cost	✗	✓
Network connectivity	✗	✓
Mobility	✗	✓
Context awareness	✗	✓
Location awareness	✗	✓
Bandwidth	✗	✓
Security	✓	✓

2 Mobile Cloud Computing: Architecture

The conventional architecture of mobile cloud computing is as shown in Fig. 1. The connection of mobile devices to the mobile networks is done via base stations which may be access points, satellites or base transceiver. Connections and functioning interfaces are setup and monitored by the stations between the networks and mobile devices. User requests and the information is transmitted to central processors which are connected to the cloud servers. Services such as authorization, accounting based and authentication are provided to the users by the mobile network operator then requests of the subscriber are delivered to a cloud through Internet. The requests are processed and corresponding services are provided to the cloud controllers in the cloud. The concepts such as service-oriented architecture, virtualization and utility computing are used to develop the services.

Cloud computing can be viewed as a distributed network system, implemented on a large scale, and which uses several servers in the data centers. Classification of the cloud services is easily understood by studying its layer concept. Starting from the upper layers these layers are stacked as Infrastructure as a service (IaaS), Platform as a Service (PaaS) and Software as a Service (SaaS).

Data centers layer. This layer provides the hardware and infrastructure related resources. Several servers are combined with networks with high-speed to provide such services.

IaaS. It is on top of data center layer. IaaS provides networking components, servers, and hardware. Payment is usually on a per-use basis. Thus, Cost can be saved by the clients and payments is based on the resources used by the clients. Expansion and Reduction of the infrastructure dynamically are possible. Amazon Elastic Cloud Computing and Simple Storage Service (S3) are examples of IaaS.

PaaS. It offers an environment, advanced and integrated that facilitates testing, constructing and launching custom applications. Google App engine, Amazon Map Reduce, Microsoft Azure are examples of PaaS.

SaaS. Information and application can be accessed by the users using the Internet in this layer. One of the pioneers is Salesforce that offers such a model (Fig. 2).

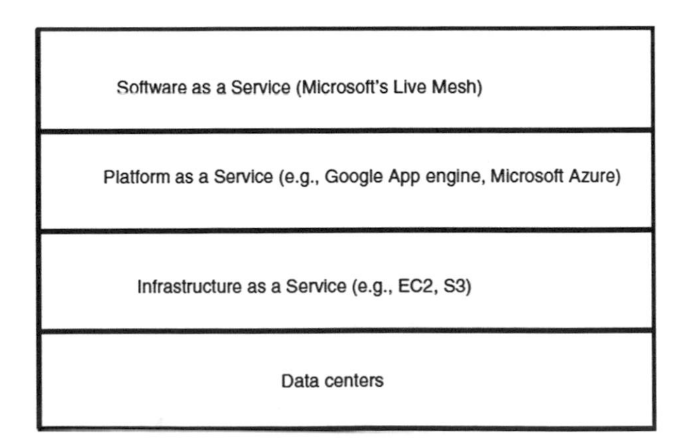

Fig. 1. Service oriented cloud computing architecture [14]

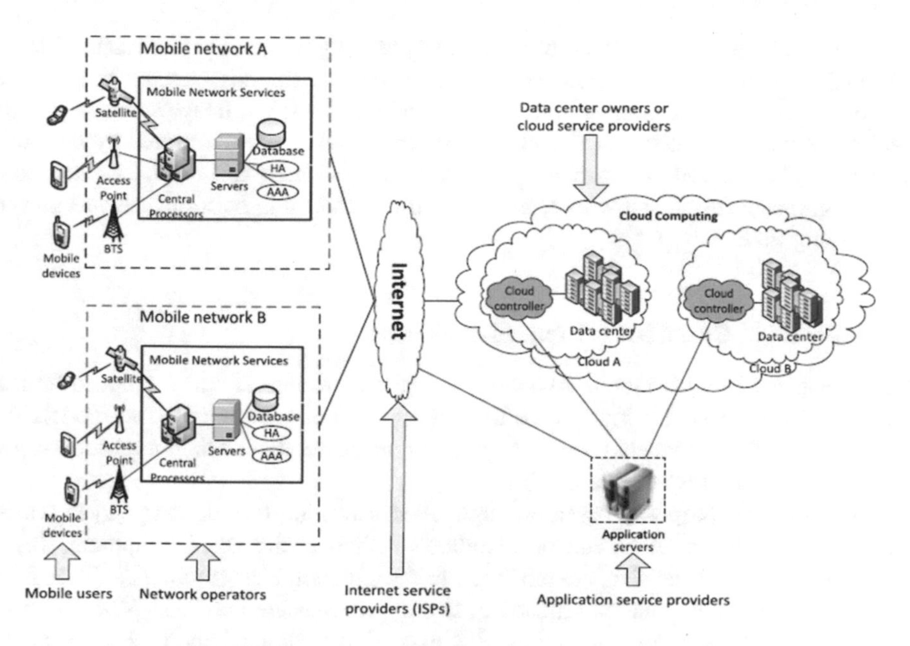

Fig. 2. Architecture of mobile cloud computing [14]

3 Mobile Cloud Computing: Application Models

To reach a specific goal like enhancing the performance of an application, achieving energy efficiency for mobile devices or running those applications deficient in proper resources needed for executing them locally, the mobile cloud computing models are designed. The mobile cloud computing models can be classified into the following categories based on the goals of goals of the application models.

3.1 Energy and Performance Based Application Model

The primary focus of this model is performance improvement of applications which are mobile, by utilizing cloud resources to the fullest. So all the computations which rely heavily on resources are shifted or offloaded to the cloud which performs computations at incredible speeds. This way the task takes no time compared to the mobile devices. As the cloud resources are utilized, the applications are executed with the enhanced performance.

One of the examples of the performance based applications is CloneCloud [9]. Programmer support for the conversion of applications is not required in the Clone-Cloud and some part of the application execution which is unmodified is offloaded from the mobile devices to the smartphone clone which resides in the cloud. Here, the importance of synchronization of the smartphone clone and smartphone is high for the consistent execution. Application process enters into a sleep state when augmentation is required and the process is transferred to the clone. Once the application execution is completed, application process comes out of the sleep state.

The database, node manager, partition analyzer, migrator and partition analyzer are the main components of the CloneCloud. The tasks of the migrator are to suspend, package, resume and merge the thread states on both the sides. Image synchronization and communications between threads which are migrated, are performed by the node manager. CloneCloud was tested by Chun et al. for three problems, i.e., behavior profiling, image search and virus scan. 21.2% improvement in performance was gained concerning time.

3.2 Based on Constraints Application Models

In a scenario where the mobile device is having local resources which are insufficient. In this scenario, where resources are limited, constraint-based application models are used. Using these models, one can perform resource demanding applications on devices which are resource constrained.

A model was proposed based on augmented execution technique by [2] in which the virtual machines which run on cloudlet which is a cluster of computers. Tasks which are resource intensive, are offloaded to the cloudlet. Here, mobile device acts as a thin client. For computation offloading, two approaches are used. I.E. VM synthesis and VM migration. In VM migration, VM execution is stopped and processor, memory, and disc states are saved. Then the virtual machine is migrated to the cloudlet and the execution is resumed. Software virtualization and process migration are more fragile compared to VM migration that is the advantage of this model. Moreover, the approach is not limited to language based virtualization where specific programming language is necessary for virtualization.

3.3 Based on Multiple Objectives Application Models

To meet the varied goals such as energy efficiency, resource constraints, and performance this model is used. As these models support many objectives unlike energy and performance based models which are meant to fulfill a singular goal, they are considered to be more effective. Cuckoo [11] offloads the applications to the nearest cloud. As it integrates the existing development tools which are familiar to developers, to makes the programming simple for the developers. Both remote and local methods are supported by it as it is mainly designed for Android platform [10]. It is the responsibility of application of smartphone to install the services from the server which are offloaded to any nearby JVM. The server address is given to the resource manager residing in a smartphone in form of resource description file and two dimension barcode when the service is installed. The address is registered by the address registrar and smartphones can use the remote resources.

Partial offloading of the application is supported by Cuckoo and well-known tools are used by used for the application development which is the main advantage of it. State transferring from remote sources and asynchronous callbacks are not supported by Cuckoo which is its one of the shortcomings. Moreover, security features aren't there to prevent users from installing codes which are malfunctioned on the server. Cuckoo's offloading decisions are contexted unaware and static (Table 2).

Table 2. Comparison of various application models [15]

Criteria	CloneCloud	Satyanarayan model	Cuckoo
Latency	Medium	High	Medium
Bandwidth utilization	High	Medium/high	Low
Execution resource	NI/CL*	NI	NI/CL
Platform	Android	VirtualBox	Android
Scalability	Low	Low	High
Privacy & security	Low	Low	Low
Complexity	High	Medium	Medium

4 Conclusion and Future Work

New ways of effective communication can be opened which can help data transmission between network operators and mobile devices by using mobile cloud applications. By using mobile cloud applications, we can reduce energy consumption and achieve energy efficiency in mobile devices with the use of compression and aggregation techniques.

Standardization of the mobile cloud execution platforms is needed to make computation offloading easy. Licensed applications and data of the users in smartphone clone are vulnerable to privacy issues and security attacks. So there is a requirement of a security mechanism to protect the mobile users from the infectious or VMs and illegal access to the private data of users. If a smartphone clone is exposed to a malicious user or a hacker, then the clone may be installed on a smartphone of the same model by the attacker and licensed applications can be used by him illegally. Hence, a better and safer mobile cloud application privacy control framework and security framework may be considered as future directions for mobile cloud computing.

References

1. Vallina-Rodriguez, N., Crowcroft, J.: Energy management techniques in modern mobile handsets. IEEE Commun. Surv. Tutorials **99**, 1–20 (2012)
2. Satyanarayanan, M., Bahl, P., Caceres, R., Davies, N.: The case for VM-based cloudlets in mobile computing. IEEE Pervasive Comput. **8**(4), 14–23 (2009)
3. Mascolo, C.: The power of mobile computing in a social era. IEEE Internet Comput. **14**(6), 76–79 (2010)
4. Barba, E., MacIntyre, B., Mynatt, E.D.: Here we are! where are we? Locating mixed reality in the age of the smartphone. Proc. IEEE **100**(4), 929–936 (2012)
5. Wright, A.: Get smart. Commun. ACM **52**(1), 15–16 (2009)
6. Kemp, R., Palmer, N., Kielmann, T., Seinstra, F., Drost, N., Maassen, J., Bal, H.: eyeDentify: multimedia cyber foraging from a smartphone. In: 11th IEEE International Symposium on Multimedia, ISM 2009, pp. 392–399. IEEE, New York (2009)
7. Amazon simple storage service. http://aws.amazon.com/s3/. Accessed 8 Dec 2011

8. Yang, X., Pan, T., Shen, J.: On 3g mobile e-commerce platform based on cloud computing. In: 2010 3rd IEEE International Conference on Ubi-Media Computing (U-Media), pp. 198–201. IEEE, New York (2010)
9. Chun, B.-G., Ihm, S., Maniatis, P., Naik, M.: Clonecloud: boostingmobile device applications through cloud clone execution (2010). arXiv preprint: arXiv:1009.3088
10. Android. http://www.android.com. Accessed 21 Nov 2011
11. Kemp, R., Palmer, N., Kielmann, T., Bal, H.: Cuckoo: a computation offloading framework for smartphones. In: Mobile Computing, Applications, and Services, pp. 59–79. Springer, Berlin (2012)
12. Huang, D.: Mobile cloud computing. IEEE COMSOC Multimed. Commun. Tech. Comm. E Lett. **6**(10), 27–31 (2011)
13. Othman, M., Madani, S.A., Khan, S.U.: A survey of mobile cloud computing application models. IEEE Commun. Surv. Tutorials **16**(1), 393–413 (2014)
14. Dinh, H.T., et al.: A survey of mobile cloud computing: architecture, applications, and approaches. Wirel. Commun. Mob. Comput. **13**(18), 1587–1611 (2013)
15. Dhivya, J., Ushananthini, M.: Application models for mobile cloud computing

EEEMRP: Extended Energy Efficient Multicast Routing Protocol for MANET

Aanjey Mani Tripathi(✉) and Sarvpal Singh

Department of Computer Science and Engineering,
Madan Mohan Malviya University of Technology, Gorakhpur 273010,
Uttar Pradesh, India
aanjeymanit09@gmail.com, spscs@mmmut.ac.in

Abstract. MANET consists of collection of mobile nodes with none infrastructure. Each node in Manet organizing yourself and with help of radio links develop a network with the objective of maximizing the mobility into wireless, mobile and autonomous domain. The area, as wherever speedy readying and vigorous reformation are needed and wired network is not obtainable are the popular applications of Manet. Most of the manet applications are depends upon multicast operations. Multicasting is often used to ameliorate the potency of the radio link whereas transmitting numerous messages to escapade integral broadcast identity of radio/wireless transmission. Remarkably the multicast performs high responsibility in MANETs. In Manet, multicast routing should address a large space of problems like low energy, low bandwidth, and mobility. MANET provides subsidiary bandwidth than their wired counterparts; consequently, gathering information about creation of a routing table overpriced and expensive from energy purpose of. Therefore, Energy is one amongst the leading problems in MANETs due to extremely vigorous and dispersed personality of nodes and its superlative use has become an essential demand as all nodes are battery power-driven and therefore the failure of 1 node might have an effect on the entire network. This paper addresses the difficulty of energy optimisation in Mobile Ad hoc networks. The reduction of energy in an exceeding node might increase the likelihood of network ripping. In this paper, we have got tried to scale back the likelihood of such ripping.

Keywords: Mobile ad-hoc network (MANET) · Energy · Energy efficiency · Wireless network · Multicast · Multicast Routing Protocol

1 Introduction

The objective of a Multicast Routing Protocol, the transmission of messages to a set or subset of nodes within the same network and also minimising the resources required for this message to delivery. It is necessary to seek out ways which will be shared by completely different destinations. In such a situation, sharing of ways is feasible because the message is same for all the destinations. Being a tree, each node incorporates a single parent, however, might have zero (itself a destination), one or additional intermediate nodes between itself and destinations. Just in case of multiple intermediate nodes before destinations, multiple divergences for destinations are

S.C. Satapathy and A. Joshi (eds.), *Information and Communication Technology for Intelligent Systems (ICTIS 2017) - Volume 1*, Smart Innovation, Systems and Technologies 83, DOI 10.1007/978-3-319-63673-3_57

attainable. The task of finding nodes wherever such divergences are created, to cut back the price of the tree is extremely tough. Too early divergences (close to the root node) result in produce shorter individual ways from the root to the destination at the expense of inflated total price of the tree. On the opposite hand, too late divergences (close to destinations) result in produce lengthy individual ways/paths, however, till suitable limit, supporting lower total price of the tree. This paper has tried to increase the prevailing energy economical protocol [EERCCP] [3] by using the unicast perimeter routing in multicast situations.

2 Related Work and Literature Review

In [1] uses localised source routing scheme, during this case once any node wishes to multicast then it is solely needed to understand the coordinates of the destinations and therefore the position of its next hops. To succeed in next hops its uses localized routing scheme that is not using the neighbors advance in contrast to traditional geographic routing schemes. In [3] author developed a protocol which reduces the congestion and creates economical just in case of energy. Within the paper [5] author divided the paper in to 3 part with the refinement in packet delivery quantitative relation and throughput, it additionally overcomes the matter of dependency on an individual receiver for congestion detection and adjustment of receiving rates. In [2, 4, 6, 8] present the literature review on multicasting techniques and wireless security challenges and their solutions of the mobile ad-hoc network. [7] Relies on location assisted routing that reduces the energy consumption of nodes by limiting the discovering area of nodes for a replacement route to smaller zone. It is additionally reducing the overhead. [9] Appraise the achievement of the prevailing multicast protocols in Manet using self-similarity traffic model is planned & rationalized. [11] Overcome the massive overhead drawback, introduces a tiny low cluster multicast scheme supported by packet encapsulation. [12] Mentioned best topology management for balanced energy consumption in wireless networks.

3 Main Differences of the Proposed Routing Protocol with Respect to the Existing Literature

The main difference between existing algorithms and extended algorithm is energy consumption. This algorithm is better because primary thing energy consumption is low because this algorithm chooses only that node as a relay node which has maximum energy in their group. And when energy is degraded of that node then again it changes relay node and choose which has maximum energy. Throughput, packet delivery ratio, end to end delay also good but overhead increases because of negative acknowledgment (NAK) send by relay node (Table 1).

Table 1. Difference of proposed algorithm with existing one.

Parameters/algorithms	[1]	[3]	[5]	[7]	[11]	EEEMRP
Energy consumption	Low	Low	Low	Low	Low	Very Low
Packet Delivery Ratio	High	High	High	High	High	High
End to End Delay	Low	Low	Low	Low	Low	Very Low
Overhead	High	Low	High	High	High	High
Throughput	High	High	High	High	High	High
Security	Medium	Medium	–	–	–	Low

4 Ease of Use

4.1 Extended Energy Efficient Tree Construction

In [EEEMRP], we structured a multicast tree and that tree has been routed from the source node respect to the receiver node. The geographical space between nodes is presumed along with the residual energy of node. Now, we sorted all the nodes from the source and rearranged in an arranged order based on their locations. A threshold value is set, such that the node having the value less than Q are unicast whereas the nodes having value more than Q are multicast. In multicasting case, the nodes possessing minimum energy per the corresponding receiver is considered as the relay node, which is responsible for the forwarding the packets to all the corresponding receivers.

4.2 Residual Energy Calculation of a Node

1. We assume a network N and divide the network in multicast groups and assume each group [NGi] has P number of nodes.
   NG1, NG2...............NGx.
   Note: - Each node occasionally calculates their remaining energy.
2. Now we calculate the consumed energy for both modes (Transmission mode and Reception mode).
   In transmission mode:

$$\text{Consumed energy} = \text{TransP} * t \tag{1}$$

   Where t = Transmission time, TransP = transmitting power
   In Reception mode:

$$\text{Consumed energy} = \text{RecP} * t \tag{2}$$

   Where t is Reception Time., RecP = reception power
3. The value of t can be examined as

$$t = DAs/DAr \tag{3}$$

DAs = Data size, DAr = Data rate

4. Therefore, By using Eqs. (1) or (2) and Eq. (3) we calculate the Residual Energy as:

$$\text{RE} = \text{Current Energy} - \text{Consumed energy} \tag{4}$$

5 Extended Energy Efficient Multicast Routing Protocol

Here, in the extended energy economical Multicast Routing Protocol, a multicast tree routed is constructed at the source towards the receivers. The geographical distance of every node is presumed and the residual energy of that node is measured. And then sorted the nodes are sorted based on their location from the source and organised in an order. A threshold worth Q is set and therefore the nodes that are but $(n < Q)$ are unicast from the source and also the nodes that have the larger than threshold $(n > Q)$ are multicast. For present work, the count of 2 hops as threshold values is taken. Just in case of multicasting the node that has the utmost energy per the corresponding receiver is set as the relay node. And after that the relay node then forwarded the packets to corresponding receiver from the source. All Nodes in every partition set, send their Energy information to their relay node sporadically. Whenever a relay node decays its energy state below a precise threshold, then it piggyback with a NACK (Fig. 1) with energy information of its connected nodes and sends it back to its previous relay or sender node to tell that this sender of NACK is not anymore able to relay packets, therefore, next relay node Nr' ought to be chosen in succeeding future rounds. The energy of such Nr' is beyond different nodes and it lies among the range of sender node (Fig. 2).

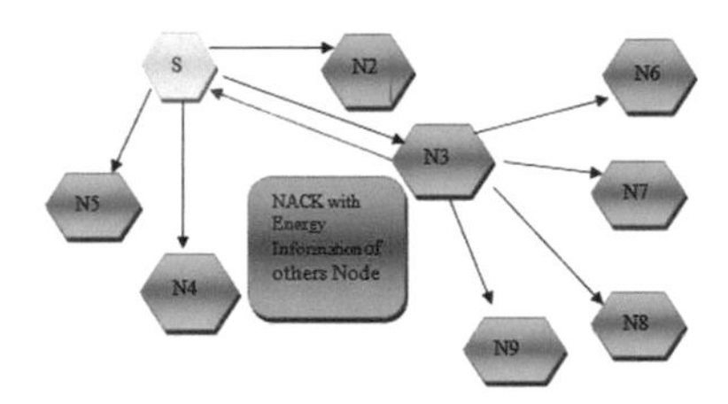

Fig. 1. Tree construction of EEEMRP with NACK.

5.1 Algorithm

1. Evaluate the separation d of every node from source (S, Ni)

$$\text{Where } i = 1, 2.\ldots\ldots\ldots n.$$

2. Make class of nodes Ni in move up order of d.

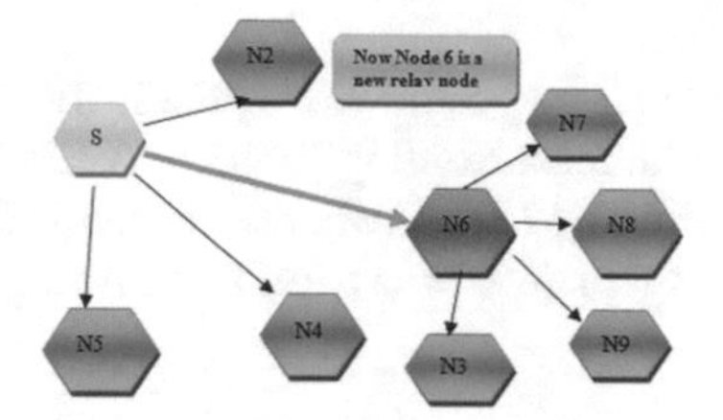

Fig. 2. Tree construction of Extended Energy Efficient Multicast Routing Protocol after the chosen the new relay node by the source for multicasting in the groups.

3. Create the partitions X1, X2....Xm of the nodes Ni such that.

$$
\begin{aligned}
X1 &= \{N1.\ldots\ldots..NQ\},\\
X2 &= \{NQ+1.\ldots..N2Q\}.\\
&\;\;.\\
&\;\;.\\
Xm &= \{NmQ+1.\ldots..Nn\}
\end{aligned}
$$

Where Q is the distance threshold

4. Source unicast the packets to X1.
5. In X2 Source finds a relay node Nr which has max (Ei), likewise in X3, X2 finds a relay node Nr' which has max (Ei) and so on.
6. Then S unicast the packets to Nr which in turn multicast the packets to the rest of the nodes in X2, and the Nr node in X2 also unicast the packet to Nr' in X3 which in turn multicast the packet to rest of node in X3.
7. The entire Nodes in each partition send its Energy information to its relay node periodically. e.g All other nodes in X2 update its energy information to Nr node.
8. Whenever a relay node decayed its Energy level below certain threshold, only then it selects an alternative relay node Nr" (future relay node) within its partition and the information of selected node Nr" is piggyback with a NACK and send it back to its previous relay node to inform that the current sender of NACK is no anymore able to relay packets so choose this Nr" in future.
9. Along with that Nr (current relay node) informs the others about Nr" too.

5.2 Simulation Results

The working of this extended algorithm [EEEMRP] has been compared against the existing energy efficient protocol [EERCCP] [3] Using NS2 simulator. The performance of both the algorithms are calculated and compared against the chosen parameters such as Energy comparison, End to End Delay, Throughput, Overhead, and Packet Delivery Ratio. All such comparisons between both algorithms have been explained in following sections:

5.2.1 Energy Comparison

In this the simulation result and graph shows that the EEEMRP survives more than the Existing Energy Efficient Multicast Routing Protocol (EERCCP) [3]. In EERCCP, the receiver loses its energy within 17.92 s. whereas in the EEEMRP it loses its energy within 21.08 s. This shows that in our version of protocol, survival rate of nodes is higher (Fig. 3).

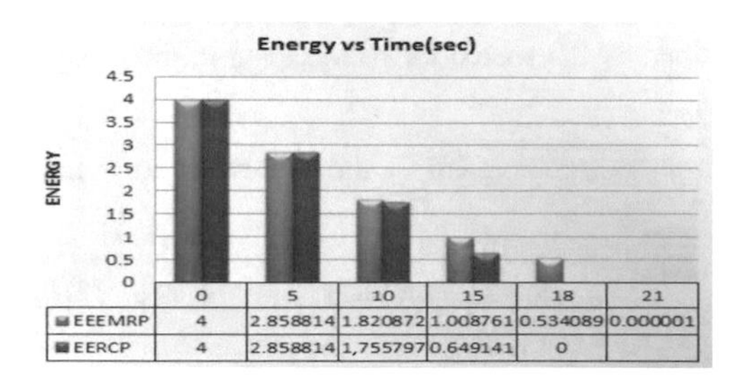

Fig. 3. Energy vs. Time

5.2.2 Energy Comparison Table

In the Table 1, the decreasing order of energy for a node 26 is shown for the both protocol Existing EERCCP and EEEMRP, and the energy of others nodes is also checked. Here table shows that the EEEMRP survives more than the EERCCP. This table shows the energy per second in which EERCCP nodes loses their energy at the time 17.92 s but in case of EEEMRP the nodes are existing till 21.08 s (Table 2).

5.2.3 End to End Delay

The end-to-end-delay is typically the average of all live data packets from the source node to destination node for both algorithms. The comparison is shown in Fig. 4. It is clear from the graph that end-to-end delay in case of the EEEMRP is less as compare to (Fig. 4).

5.2.4 Throughput

It is the number of packets received by all the nodes in the network. Our simulation result clearly shows that the throughput of EEEMRP is greater than the EERCCP protocol (Fig. 5).

5.2.5 Overhead

The down side of our EEEMRP is that the overhead associated with this is more than the existing EERCCP. This happens due the use of negative acknowledgement (NAK). However, this increased overhead contributes positively in reduction of probability of network splitting and thus reduces the overhead due to network split (Fig. 6).

Table 2. Energy Comparison

Existing Protocol (EERCCP)		Extended Protocol (EEEMRP)	
Time (S)	Energy (J)	Time (S)	Energy (J)
0	4	0	4
0.100116	3.970748	0.100116	3.970748
0.500182	3.901124	0.500182	3.901124
1.006033	3.779264	1.006033	3.779264
1.50004	3.663492	1.50004	3.663492
2.002143	3.547549	2.002143	3.547549
2.506735	3.431688	2.506735	3.431688
3.003426	3.300841	3.003426	3.300841
4.002955	3.082734	4.002955	3.082734
5.000913	2.858814	5.000913	2.858814
6.002091	2.640725	6.002091	2.640725
7.001318	2.420622	7.001318	2.420622
8.001253	2.19662	8.001253	2.19662
9.000796	1.977673	9.000105	1.986276
10.00102	1.755797	10.00064	1.820872
11.00013	1.533919	11.00211	1.657465
12.00358	1.312046	12.00255	1.494354
13.0006	1.09256	13.00002	1.332512
14.00003	0.871763	14.00432	1.171531
15.0015	0.649141	15.00079	1.008761
16.00074	0.427839	16.0002	0.846592
17.00327	0.206254	17.00124	0.68354
17.92137	0.000055	18.00088	0.52109
17.92169	0	19.00169	0.356813
		20.00014	0.182266
		21.00083	0.012812
		21.08008	0.000001
		21.0804	0

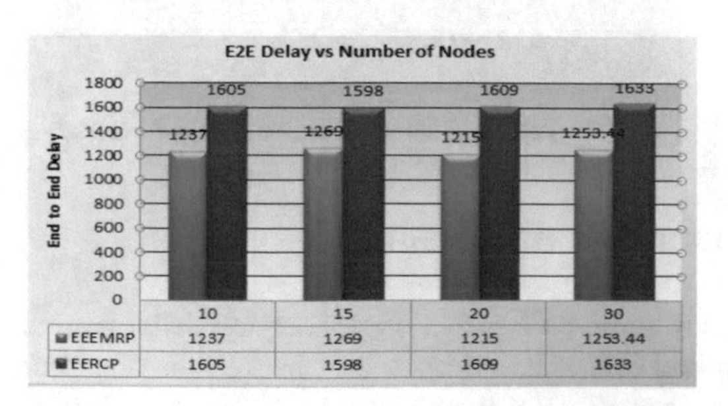

Fig. 4. E2E Delay vs. Number of nodes.

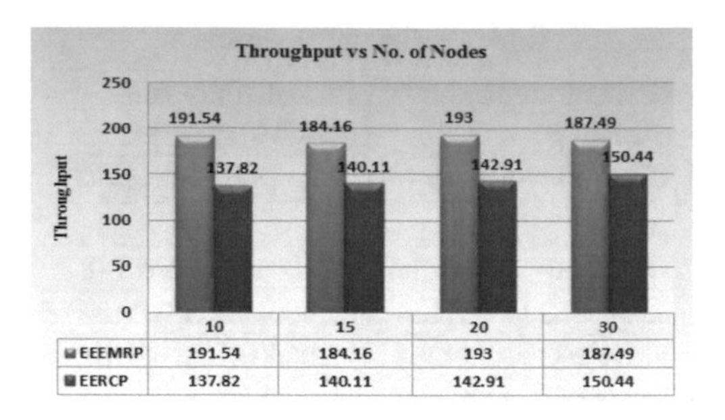

Fig. 5. Throughput vs. No. of nodes.

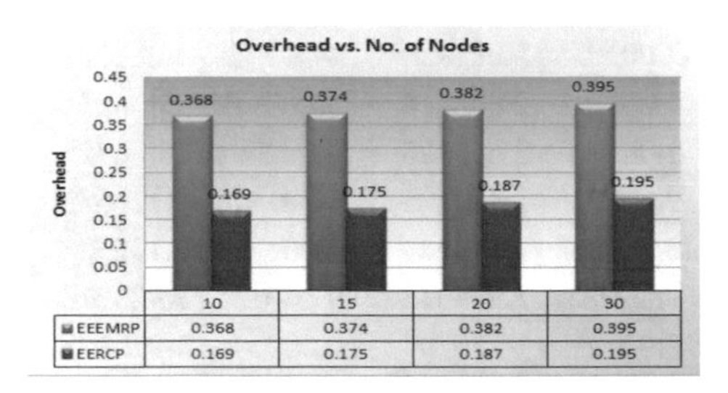

Fig. 6. Overhead vs. No. of nodes.

5.2.6 Packet Delivery Ratio

It is defined as the ratio of successfully received number of packets to the total no. of packets sent. The packet delivery ratio of EEEMRP is higher than the EERCCP (Fig. 7).

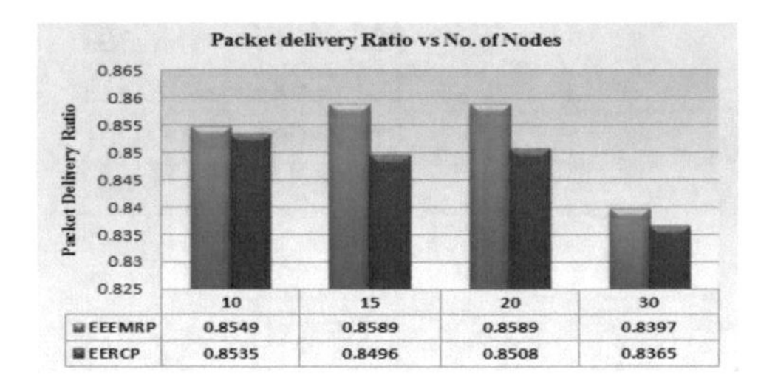

Fig. 7. Packet delivery ratio vs. No. of nodes.

6 Conclusion and Future Scope

This paper concludes that the EEEMRP is performing better than the EERCCP in case of energy, End to end delay, throughput, and packet delivery ratio. The overhead of EEEMRP is more than EERCCP but it contributes to reduction in probability of network splitting and thereby positively contributes to reduce the overhead due to this split. The promising growth in MANET generates the vast research scope in this area, especially in applications which are based on multicasting. The work presented in this paper forms a base where more such protocols may be modeled having systematic control on energy and thereby increasing the scope of networks sustainability.

References

1. S′anchez, J.A., Ruiz, P.M.: Lema: Localized energy-efficient multicast algorithm based on geographic routing. In: Proceedings 31st IEEE Conference on Local Computer Networks (LCN 2006), pp. 3–12, November 2006
2. Mishra, A., Nadkarni, K.M.: Security in wireless Ad-hoc network, in Book. The hand book of Ad Hoc Wireless Networks (Chap. 30), pp. 499–549. CRC press LLC, Boca Raton (2003)
3. Srinivasa Rao, K., Sudhistna Kumar, R., Venkatesh, P., Sivaram Naidu, R.V., Ramesh, A.: Development of energy efficient and reliable congestion control protocol for multicasting in mobile Adhoc networks compare with AODV based on receivers. Int. J. Eng. Res. Appl. **2** (2), 631–634 (2012). ISSN: 2248-9622. http://www.ijera.com
4. Chen, X., Wu, J.: Multicasting techniques in mobile Ad-hoc networks, Computer Science Department, South West Texas State University, San Marcos, The Handbook of Ad-hoc Wireless Networks, pp. 25–40 (2003)
5. Bhushana Rao, G.S., RajanBabu, M.: An energy efficient and Reliable Congestion Control Protocol for multicasting in mobile Adhoc networks. Int. J. Comput. Sci. Inf. Secur. **7**(1), 140–146 (2010)
6. Luo, W., Fang, Y.: A survey of wireless security in mobile Ad hoc networks: challenges and available solutions, Ad hoc wireless networking, pp. 319–364. Kluwer Academic Publishers, Dordrecht (2003)
7. Mohammed, M.: Energy efficient location Aided Routing Protocol for wireless MANETs. Int. J. Comput. Sci. Inf. Secur. 4(1 and 2) (2009)
8. Cheng, M., Shun, J., Min, M., Li, Y., Wu, W.: Energy-efficient broadcast and multicast routing in Multihop Ad Hoc wireless networks. Wirel. Commun. Mobil. Comput. **6**(2), 213–223 (2006)
9. Tariq, O., Greg, F., Murray, W.: On the effect of traffic model to the performance evaluation of multicast protocols in MANET. In: Proceedings of the Canadian Conference on Electrical and Computer Engineering. pp. 404–407 (2005)
10. Junhai, L., Liu, X., Danxia, Y.: Research on Multicast Routing Protocols for Mobile Ad-Hoc Networks Computer Networks, pp. 988–997. Elsevier, Amsterdam (2008)
11. Chen, K., Nahrstedt, K.: Effective location-guided tree construction algorithms for small group multicast in MANETs. In: Proceedings of the IEEE INFOCOM Conference, pp. 1180–1189 (2002)
12. Lia, Y., Chengb, X., Wuc, W.: Optimal topology control for balanced energy consumption in wireless networks. J. Parallel Distrib. Comput. **65**(2), 124–131 (2005)

Brain Subject Estimation Using PSO K-Means Clustering - An Automated Aid for the Assessment of Clinical Dementia

P. Rajesh Kumar(✉), T. Arun Prasath, M. Pallikonda Rajasekaran, and G. Vishnuvarthanan

Department of Electronics and Communication Engineering, Kalasalingam Academy of Research and Education, Virudhunagar 626126, Tamilnadu, India
rkrrajesh74@gmail.com, arun.aklu@gmail.com, m.p.raja@klu.ac.in, gvvarthanan@gmail.com

Abstract. Structural brain imaging plays an essential role in acknowledgement of variations that presence in brain relevant to Alzheimer's disease and different kind of brain disorders. Mostly MR Imaging has preferred because of its higher resolution capabilities to diagnose AD than other modalities. Magnetic resonance imaging is an efficient for visualization and diagnosing various brain disorder in brain. Pathology segmentation for differentiate the diseases affected region to make separation of necrosis and similar damaged tissues cause by disease from normal tissue using clustering principle take up in image processing. Clustering is implemented to make grouping similar characteristics pixels together as a group. In this paper k-mean clustering is performed to separate White Matter (WM), Grey Matter (GM), Cerebrospinal Fluid (CSF), Lateral ventricle, hippocampus region as different individual group with accomplice of Partial Swarm Optimization (PSO) in brain. Different k- mean cluster initialization methods were executed and an exact segmentation were done using PSO k-mean clustering. Volume of both grey matter and white matter are estimated to make comparison with the bench mark images for classifying the various stages of AD.

Keywords: Alzheimer's disease (AD) · Magnetic resonance imaging · Particle Swarm Optimization (PSO) · K-means clustering

1 Introduction

In brain disorders MRI takes place significant role in detection of disease, diagnosis and preplanning for treatment. Decisive and disorder related tissues segmentation in MRI is an efficient aid for physicians treatment arrangement of the particular disease. Throughout the world more than 35 million people are affected by various stage of AD. If the affected probability of AD has been rising double for every five years, then by 2050 it will be reached 135 million [7]. In recent the benchmark for AD analysing is based on mini-mental state examination (MMSE) and Clinical dementia Ratio (CDR). AD can be understood very rarely from the behaviour and attitude of a patient.

S.C. Satapathy and A. Joshi (eds.), *Information and Communication Technology for Intelligent Systems (ICTIS 2017) - Volume 1*, Smart Innovation, Systems and Technologies 83, DOI 10.1007/978-3-319-63673-3_58

The progressive death of brain cells starts at initially, over a course of period only, the physicians could make a judgement about the patient who has affected by AD. Additionally, still it needs an extensive focus to find ways for explicit and accurate and segmenting the brain matters accurately. Segmentation of disease affected area means to separate necrosis and other tissue damages cause by illness, from natural tissues [9]. The peculiar information about different types of tissue in brain as long as different excitation sequence could be contribute by various MRI models. This character represents them as a skilful tool. But the defection of each modal to give through information, forces adept to use different models for taking decision about the exact location, diagnosis and progress level of illness. The clustering principle is an unsupervised learning method for classification that has implemented for recognizing essentially the structure of an object by separating them into various subsets that are more informatics context [8]. Here in this work various brain subjects Grey Matter (GM), lateral ventricle, hippo campus region and white matter are segmented and grouped them as individual for estimating the volume of each. The severity of AD could be classified by comparing those volumes with standard state of art Clinical Dementia Ratio (CDR) and Mini-mental State examination (MMSE).

2 Related Works

H.T. Gorji and Haddadnia were implemented lattice computing schemes to differentiate the AD affected from healthy people and compared their status and performance with Clinical Dementia Ratio (CDR) in various AD subject and the K-NN classifier was used to classifying the stages of AD [1]. Mohamed M. Dessouky and Mohamed A.et al., were stated pattern search algorithm for extracting the feature in AD and got the nearest optimum number of features and which has given higher accuracy. Lesser metrics parameters and features only executed in this work [2]. A. ortiz, J.M. Gorriz were proposed combination Entropy gradient clustering and Self Organizing Map (SOM) for segmenting the Grey matter(GM), Cerebrospinal Fluid (CSF), White Matter(WM) of brain and the volume of about said has estimated. Segmentation and volume estimation process only performed in this paper [3]. ElahehMoradi, Antoniettapepe were implemented semi supervised learning process to differentiate Mild Cognitive Impairment (MCI) and Alzheimer's disease (AD) with the help of Random Classifier. An important drawback of this paper is Bio marker is used [4]. Saima Farhan, Muhammad Abuzar Fahiem were proposed automated processing method for identifying Alzheimer's disease and various structural features for classifying AD subjects by using combined Multilayer perceptron (MLP), SVM and J48 classifier. Grey matter (GM), Hippocampus region and White Matter (WM) volumes only have estimated [5]. ChuanchuanZheng, Yong Xia were reviewed all the standard classifiers which is being very helpful to identify efficient classifier for classifying the AD subjects with pattern based perspective. Lacking of Mild Cognitive Impairment (MCI) analysis were not discussed [6].

3 Materials and Methods

3.1 K-Mean Clustering

Measure of similarity is one of the most significant tool for clustering algorithm which is very useful to determine how much closer two patterns each other. K-Means clustering categorizes data vectors into pre-assigned number of cluster depending upon Euclidian distance in the act of similarity measure [13]. Data vectors which are having smaller Euclidian distance between one another with in the cluster and those are centroid vector linked with a centroid vector which denotes "centre point" of that cluster. The standard k-means algorithm designed as

$$d(z_p - m_j) = \sqrt{\sum_{k=1}^{N_d} (z_p k - m_j k)} \tag{1}$$

1. Random initialization of cluster centroid vector:

   How much amount of centroid vector needs (k) to form a cluster are selected from the total (N) number of samples in a random manner. And for 5 iterations k-means clustering is applied to those selected random samples.

2. Centroid vector initialization using sub sampling:

   In sub sampling principle, 5% of samples are selected from the total number of samples and then k-means clustering is implemented for those chosen random samples for 5 iterations.

3. Uniform sampling selection:

   The sample data selected in between minimum and maximum value from the N-number of samples for the centroid selection and uniformly distributed those data vector with equal distance between one another. Finally k-means clustering carried out for the taken cluster values.

4. Replication process for cluster canter selection:

   For all chosen data authorize with the nearest node of class vector.
   The distancc of centroid measured using

$$D(x_p, z_c) = \sqrt{\sum_{i=1}^{d} (x_{pi} - z_{ci})^2} \tag{2}$$

x_p is the pth feature vector and the centroid z_c, cluster c and d number of features in all individual vector.

The calculation of new centroid is measured by the below formula

$$z_c = \frac{1}{n_c} \sum_{\forall} x_p \in C_c x_p \tag{3}$$

4 Particle Swarm Optimization

Optimization can be defined as the process of selecting the best values from the available input [10] Particle Swarm Optimization the process of searching the population that happens depending upon the stochastic searching process for the total population. This principle which was designed and executed by Kennedy and Eberhart [11]. The techniques which designed after noticing the social interactions like fish schooling and bird flocking This method which preserves of a particle in a observable contribution has produced to achieve optimization problem by every particle. The term swarm represents the represents the number of convenient solutions for an optimization. Where every potential mentioned as each single particle [14]. The objective of this algorithm is to achieve best position of particle which has to carry out an exact outcome for the taken fitness function Adjustment of position towards best position identified by particle until an exact position in neighbourhood of that particular particle.

$$x_i(t+1) = x_i(t) + v_i(t+1) \tag{4}$$

W-inertia weight c_1 and c_2 are the acceleration co-efficient
A particle's personal best position i is calculated.

$$y_i(t+1) = \begin{bmatrix} y_i(t) & if\, f(x_i(t+1)) \geq f(y_i(t)) \\ x_i(t) & if\, f(x_i(t+1)) \leq f(y_i(t)) \end{bmatrix} \tag{5}$$

The equation of social component c_1 becomes

$$c_2 \ulcorner_{2,k}(t)\left(y_{j,k}(t) - x_{j,k}(t)\right) \tag{6}$$

In this above formula y_i - Best particle of ith particle in the neighbourhood. Until the given criteria gets end, PSO of (4) and (5) equations repeated.

4.1 Clustering Using PSO

The k-means algorithm which executes to coincides faster than PSO [12]. And the impact of PSO is, when in the initialization of the cluster centre selection and it reduces the accuracy of PSO [15]. Replacement of existing old cluster centre member's mean point of a cluster which leads the process of k-means as faster converge. The utilization of this merit the convergence rate of the PSO can be improved. After the updating the speed of each particle, it starts the process when it found the new position of the particle better than the optimized position which has denoted as the best position [16]. Then the new position will be transferred as the cluster centre.

5 Volume of GM, WM and CSF

The physician could take the decision about the severity of Alzheimer's disease from the changes in the volume of GM, WM and CSF regions. Number of previous works have shown the result as the reduction of GM is due to the reason of brain atrophy in Alzheimer's disease. We have estimated the volume of various tissue classes, those are GM, WM, and CSF for all the taken slices of MRI. In order to calculate the volume of tissues, the segmentation process has performed using PSO k-means clustering in pre-processed MR Images WM, GM, CSF and the help of FSL software volume estimated of each regions in terms of pixels The term "thresh" mentioned for GM intensity threshold range.

$$Volume_{WM} = \sum_{slice=1}^{n} \sum_{i=1}^{x} \sum_{j=1}^{y} f(i,j) > thres$$
$$Volume_{GM} = \sum_{slice=1}^{n} \sum_{i=1}^{x} \sum_{j=1}^{y} f(i,j) == thres$$
$$Volume_{CSF} = \sum_{slice=1}^{n} \sum_{i=1}^{x} \sum_{j=1}^{y} f(i,j) < thres$$

6 Results

Based on the comparison from Tables 1 and 2 are shown that the proposed method of Particle Swarm Optimization based K-Means Clustering is better than the other method in all cases of problem for segmentation of brain tissues and the segmentation accuracy were matched with the ground truth.

Table 1. Iterations and Closest distance for achieving fitness

S. No	Method	Maximum iterations taken to achieve fitness	Maximum closest distance
1	K-Means (Random)	27	457.649
2	K-Means (Subsample)	26	285.116
3	K-Means (Uniform)	20	281.892
4	PSO based K-Means	5	260.213

Figures 1a and 2a represent the input MRI brain images in axial view and Figs. 1b and 2b represent processed image using K-Means Algorithm (Random) and Figs. 1c and 2c represent processed image using K-Means Algorithm (Sub sampling) and Figs. 1d and 2d represent processed image using K-Means Algorithm (Uniform) and Figs. 1e and 2e represent processed image using K-Means Algorithm based on Particle Swarm Optimization. Figures 3a and 4a represent the input MRI brain images in

Table 2. PSNR, MSE, Volume of GM,WM

Images	PSNR	MSE	Volume of GM (in pixels)	Volume of WM (in pixels)
1.	39.3266	6.8952	7752	6363
2.	38.9447	7.4055	8330	6273
3.	38.1764	9.8955	8326	8310
4.	37.3575	9.8904	9477	7053
5.	37.4523	10.0985	9658	6993
6.	36.4950	14.4597	8001	8256
7.	36.4690	14.4589	8094	9383
8.	36.3946	14.1069	7874	8561
9.	36.7367	13.2495	7682	7538
10.	36.6991	13.9047	8627	9318

sagittal view and Figs. 3b and 4b represent processed image using K-Means Algorithm (Random) and Figs. 3c and 4c represent processed image using K-Means Algorithm (Sub sampling) and Figs. 3d and 4d represent processed image using K-Means Algorithm (Uniform) and Figs. 3e and 4e represent processed image using K-Means Algorithm based on Particle Swarm Optimization. Quality of the segmented image was examined with PSNR and MSE and is shown in Table 2.

The average value of the sensitivity remains 95.2381, the average value of the specificity was 97.9167, and the average value of the precision was 98.3607.

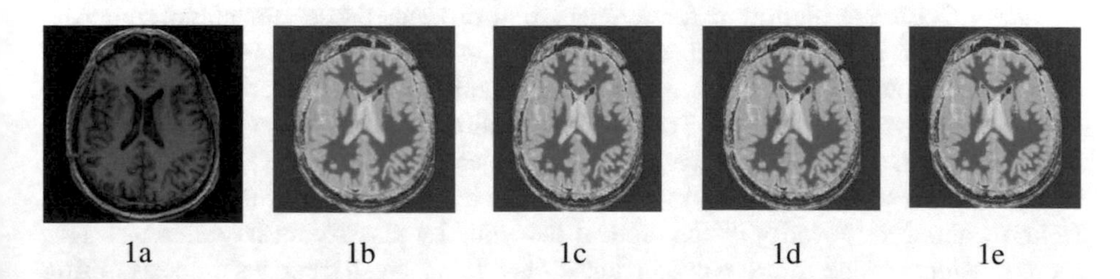

Fig. 1. Input MRI brain image 1 and segmentation results from various algorithms

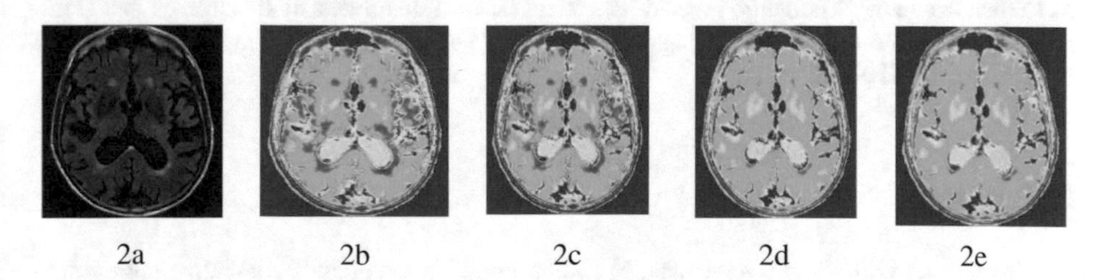

Fig. 2. Input MRI brain image 2 and segmentation results from various algorithms

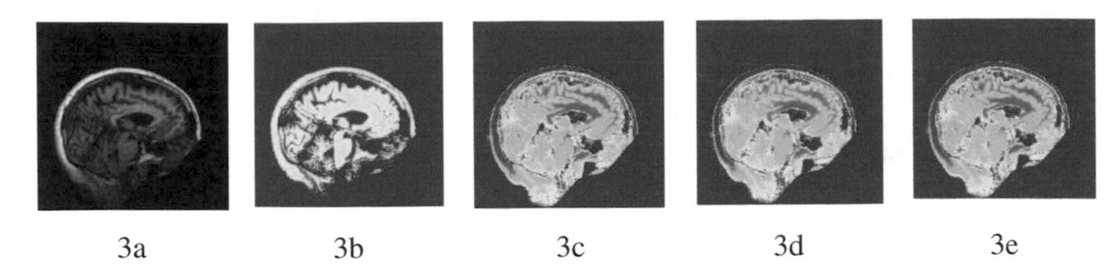

3a 3b 3c 3d 3e

Fig. 3. Input MRI brain image 3 and segmentation results from various algorithms

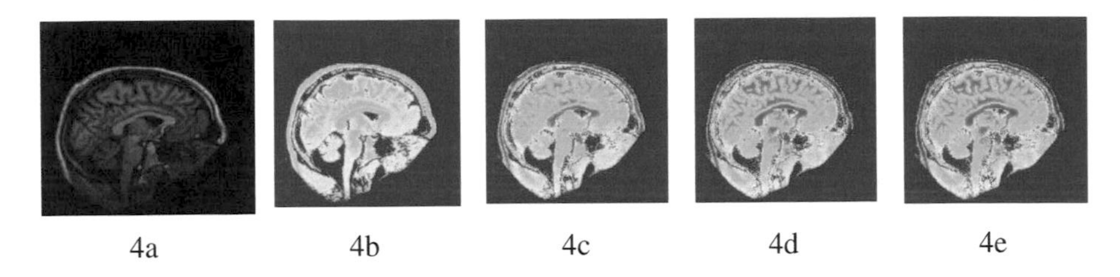

4a 4b 4c 4d 4e

Fig. 4. Input MRI brain image 4 and segmentation results from various algorithms

7 Conclusion

The present study uses Particle Swarm Optimization to improve the behaviour of K-Means Clustering algorithm for segmentation of brain tissues for classifying Alzheimer's disease. The proposed algorithm provides accurate segmentation of brain tissues when compared to K-Means Clustering algorithm based on various initialization and hence proves that Particle Swarm Optimization based K-Means Clustering provides better fitness values. The various brain subject volume were calculated and it is compared with standard ground truth images. The brain subject volume is more helpful to know about the severity of the clinical dementia by classifying its different stages more preciously. The future work includes classifying the Alzheimer's disease using various soft computing techniques.

Acknowledgement. The author would like to thank the Department of Electronics and Communication Engineering of Kalasalingam University, (Kalasalingam Academy of Research and Education), Tamil Nadu, India.

References

1. Papakostas, G.A., Savio, A., Graña, M., Kaburlasos, V.G.: A lattice computing approach to Alzheimer's disease computer assisted diagnosis based on MRI data. Neurocomputing **150**, 37–42 (2015)
2. Dessouky, M.M., Elrashidy, M.A.: Feature Extraction of the Alzheimer's Disease Images Using Different Optimization Algorithms. Alzheimers Dis. Parkinsonism **6**, 2 (2016). doi:10.4172/2161-0460.1000230

3. Ortiz, A., Gorriz, J., Ramirez, J., Salas-Gonzalez, D.: Improving MR brain image segmentation using self-organising maps and entropy-gradient clustering. Inf. Sci. **262**, 117–136 (2014)
4. Moradi, E., Pepe, A., Gaser, C., Huttunen, H., Tohka, J.: Machine learning framework for early MRI-based Alzheimer's conversion prediction in MCI subjects. Alzheimer's Disease Neuroimaging Initiative. Neuroimage **104**, 398–412 (2015)
5. Farhan, S., Fahiem, M.A., Tauseef, H.: An ensemble-of classifiers based approach for early diagnosis of Alzheimer's disease: classification using structural features of brain images. Comput. Math. Methods Med. **2014**, 11 (2014). Article ID 862307, Hindawi Publishing Corporation
6. Zheng, C., Xia, Y., Pan, Y., Chen, J.: Automated identification of dementia using medical imaging: a survey from a pattern classification perspective. Brain Inf. **3**, 17–27 (2016). doi:10.1007/s40708-015-0027-x. Springer
7. Alattas, R.: Hybrid segmentation algorithm for detecting Alzheimer's disease in MRI images. Int. J. Innov. Sci. Res. **18**, 2351–8014 (2015). 342–347
8. Sampath, R., Saradha, A.: Alzheimer 's disease Image Segmentation with Self-Organizing Map Network. J. Softw. **10**, 670–680 (2014). doi:10.17706/10.6.670-680
9. Alattas, R., Buket, D., Barkana, A.: Comparative study of brain volume changes in Alzheimer's disease using MRI scans. In: Systems, Applications and Technology Conference (2015). IEEE. doi:10.1109/2015.7160197
10. Muneeswaran, V., Rajasekaran, M. P.: Performance evaluation of radial basis function networks based on tree seed algorithm. In: 2016 International Conference on Circuit, Power and Computing Technologies (ICCPCT), Nagercoil, pp. 1-4 (2016). doi:10.1109/ICCPCT.2016.7530267
11. Arunprasath, T., Rajasekaran, M.P., Kannan, S., George, S.M.: Performance evaluation of PET image reconstruction using radial basis function networks. In: Artificial Intelligence and Evolutionary Algorithms in Engineering Systems, pp. 481–489. Springer, India (2015)
12. Chiu, C.-Y., Chen, Y.-F., Kuo, I.-T., Ku, H.C.: An Intelligent Market Segmentation System using K-Means and Particle Swarm Optimization. Expert Syst. Appl. **36**, 4558–4565 (2009)
13. Rana, S., Jasola, S., Kumar, R.: A boundary restricted adaptive particle swarm optimization for data clustering. Int. J. Mach. Learn. Cybern. **4**, 391–400 (2013). Springer
14. Cheng, M.-Y., Huang, K.-Y., Chen, H.-M.: K-means particle swarm optimization with embedded chaotic search for solving multidimensional problems. Appl. Math. Comput. **219**, 3091–3099 (2012)
15. Karami, A., Guerrero-Zapata, M.: A fuzzy anomaly detection system based on hybrid PSO-Kmeans algorithm in content-centric networks. Neurocomputing, **149**, 1253–1269. doi:10.1016/j.neucom.2014.08.070
16. Benaichouche, A.N., Oulhadj, H., Siarry, P.: Improved spatial fuzzy c-means clustering for image segmentation using PSO initialization. In: Mahalanobis Distance and Post-Segmentation Correction, Digital Signal Processing, vol. 23, pp. 1390–1400 (2013)

Estimating Software Test Effort Based on Revised UCP Model Using Fuzzy Technique

Monika Grover[1(✉)], Pradeep Kumar Bhatia[1], and Harish Mittal[2]

[1] Guru Jambheshwar University of Science and Technology, Hisar, India
grover_monika123@yahoo.com, pkbhatia.gju@gmail.com
[2] B.M. Institute of Engineering and Technology, Sonepat, India
mittalberi@gmail.com

Abstract. In software industry, testing effort estimation has always been a challenging task. To keep up with customer expectations and demands, good quality software is a must. For this purpose, a great deal of time and cost is spent on software testing during software development process. To develop an software in an increasingly competitive and complex environment, a lot of techniques or models had been introduced or experimented. Recently, Use Cases for software effort estimation has gained wide popularity. And also there are various modifications like e UCP, Re-UCP etc. that have been used for effort estimation depending upon the type of project. Like e UCP, Re-UCP etc. This paper proposes a new model for fuzzy technique by integrating fuzzy technique and Re-UCP method for reliable effort estimation.

Keywords: Effort estimation · UCP · Testing effort · Software quality

1 Introduction

Time and effort spent during software testing in life cycle of software development determines the quality of the respective software.

According to Rick Kraig and Stephan Jaskiel, "Testing refers to a concurrency life cycle process of engineering design, utilization and maintenance of testing software with an aim to measure and improve the quality of testing software" [1]. Gelperin and Hetzel [2] have characterized the growth of software testing with time. Myers discussed the psychology of testing in his work [3] The Art of Software Testing. Juran and Gyrna [4] defines quality as fitness for use. It means whether the product is actually usable by the user or not.

Software cost estimation is one of the fundamental issues making it the most important activity in software engineering. There are number of tools and techniques which can be put into practice to estimate the cost of the software.

1.1 Re-use Case Points (Re-UCP)

Parameters of eUCP and UCP are both used in the case of Re UCP so that it can adapt according to different types of projects with varying degree level of complexity and futuristic order of scalability.

S.C. Satapathy and A. Joshi (eds.), *Information and Communication Technology for Intelligent Systems (ICTIS 2017) - Volume 1*, Smart Innovation, Systems and Technologies 83, DOI 10.1007/978-3-319-63673-3_59

In Revised Use Case Point, the actors are categorized in to simple, average, complex and critical and the weights are assigned as 1, 2, 3 and 4 respectively (Table 1).

Table 1. Actor weight

Weight	Factor (wt age)	Count of actors	Weight × count
Simple	1	a	1a
Medium	2	b	2b
Complex	3	c	3c
Critical	4	d	4d
UAW			1a + 2b + 3c + 4d

Similarly, on the basis of the number of transactions, the types of the use-cases are simple, average, complex and critical and weights as 5, 10, 15 and 20 respectively (Table 2).

Table 2. Use case weight

Wt type	No. of transactions	Weight	Use case count	Weight × count
Simple	≤4	5	a	5a
Medium	5–8	10	b	10b
Complex	9–15	15	c	15c
Critical	>15	20	d	20d
UUCW				5c + 10b + 15c + 20d

Total actor and use case weight is calculated.

$$UUCP = UAW + UUCW. \tag{1}$$

The revised use case points are calculated as follows

$$\text{Re-UCP} = UUCP * TCF * EF. \tag{2}$$

whereas, TCF is Technical Complexity Factor and ECF is Environmental Complexity Factor etc.

In Re-UCP, Technical Complexity Factors has increased by 1 (in UCP, the TCFs are 13 only). The 14th parameter is scalability and is labeled as T14. Scalability of a system is defined as the ability of the system to handle increased workloads without adding resources to the existing system by repeatedly applying a cost-effective strategy for extended system capability.

All the factors of TCF are assigned values between 0–5. The value ‘0’ entails that the parameter has no impact on the technical complexity and number increases with the increase in significance. The value ‘5’ implies that parameter is treated as essential. The technical factors from T1 to T14 are multiplied by their respective weights and the

Table 3. Technical complexity factors

Factor	Description	Weight
T1	Distributed system	2
T2	Response or throughput performance objectives	2
T3	End-user efficiency	1
T4	Complex internal processing	1
T5	Reusable code	1
T6	Easy to install	0.5
T7	Easy to use	0.5
T8	Portable	2
T9	Easy to change	1
T10	Concurrency	1
T11	Includes security features	1
T12	Provides access for third parties	1
T13	Special user training facilities are required	1
T14	Scalability	2

summation of the above multiplication is called TFactor. Then, the TCF is calculated using the following formula (Table 3):

$$\begin{aligned} \text{TCF} &= 0.6 + (0.01 * \text{TFactor}), \\ \text{Whereas TFactor} &= \sum\nolimits_{\text{i}=1}^{14} TF_{\text{i}} \end{aligned} \quad (3)$$

In Re-UCP, Environmental Factors has increased by 1 (in UCP, the EFs are 8 only). The 9th parameter is Project Methodology. Here, Project Methodology describes the experience and familiarity of the developer in the selected project methodology and is labeled as F9.

All the factors of ECF are assigned values between 0–5 where value ‘0’ implies that the developer does not have any experience with that parameter and if developer does have better experience, then, the assigned value is also increased. The value ‘5’ implies that developer is considered as an expert.

The environmental factors (E1–E9), are multiplied by the respective weights and the sum of the product is called as EFactor. The complete environmental factors is calculate as EF i.e. given below (Table 4):

$$\begin{aligned} \text{EF} &= 1.4 + (-0.03 * \text{EFactor}), \\ \text{Whereas TFactor} &= \sum\nolimits_{\text{i}=1}^{14} TF_{\text{i}} \end{aligned} \quad (4)$$

The revised use case points are calculated as follows:

$$\text{Re-UCP} = \text{UUCP} * \text{TCF} * \text{EF} \quad (5)$$

Table 4. Environment factors

Factor	Description	Weight
E1	Familiar with rational unified process	1.5
E2	Application experience	0.5
E3	Object oriented experience	1
E4	Lead analyst capability	0.5
E5	Motivation	1
E6	Stable requirements	2
E7	Part-time workers	−1
E8	Difficult programming language	−1
E9	Project methodology	1

Finally, complete effort is estimated by converting UUCP to Man-Hours.

$$\text{Effort} = \text{UCP} * \text{PHperUCP} \tag{6}$$

whereas, PHper UCP is person-hours per UCP.

1.2 Fuzzy Technique

Fuzzy logic is successful methodology which helps in solving the universal realistic problems. In 1965, Prof. Lofti Zadeh provided fuzzy set theory to handle the problems having imprecise and incomplete data [4–10]. It is used to represent linguistic values like low, old and complex.

The fuzzy set shows as non-empty set A, where x is an element in X and $\mu A(x)$ is a membership function of set A.

In this paper we have made an effort to compute approximate value of STE using intermediate Re UCP and fuzzy technique. We applied fuzzy logic on the actor's weight and use case's weight (Tables 5 and 6).

Table 5. Unadjusted actor weight

Type of actors	Actor weight	TFN	Count of actors	Weight * count
Simple	1	(0, 1, 2)	n1	A1 = a1 * n1
Medium	2	(1, 2, 3)	n2	A2 = a2 * n2
Complex	3	(2, 3, 4)	n3	A3 = a3 * n3
Critical	4	(3, 4, 5)	n4	A4 = a4 * n4

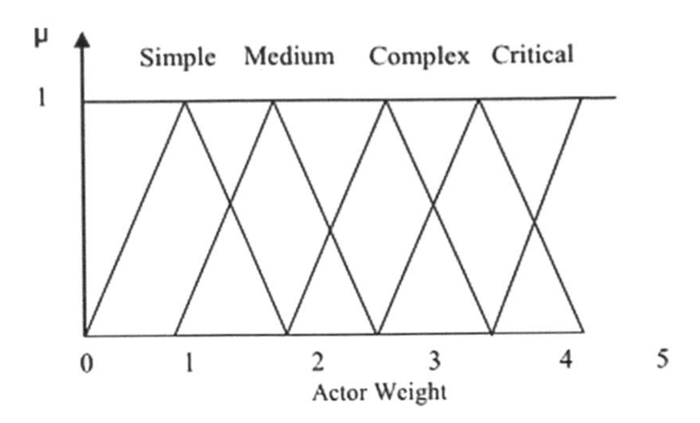

Fig. 1. Proposed framework of fuzzy actor weight

Table 6. Unadjusted use case weight

Type of use case	No. of transactions	Weight	TFN	Use cases count	Weight * count
Simple	≤4	5	(0, 5, 10)	m1	U1 = u1 * m1
Medium	5–8	10	(5, 10, 15)	m2	U2 = u2 * m2
Complex	9–15	15	(10, 15, 20)	m3	U3 = u3 * m3
Critical	>15	20	(15, 20, 25)	m4	U4 = u4 * m4

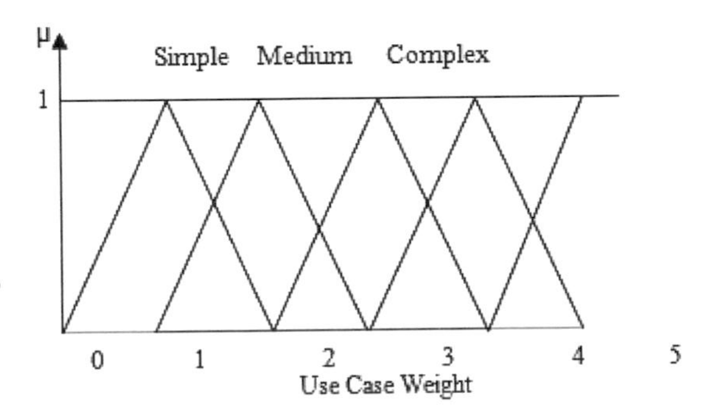

Fig. 2. Proposed framework of fuzzy use case weight

Table 7. Fuzzy rule application on use case and actor weight

	Actor weight			
	S	M	C	Cr
S	SS (L)	SM (L)	SC (M)	SCr (M)
M	SM (L)	MM (M)	MC (M)	MCr (M)
C	CS (M)	CM (H)	CC (H)	CCr (H)
Cr	CrS (H)	CrM (H)	CrC (VH)	CrCr (VH)

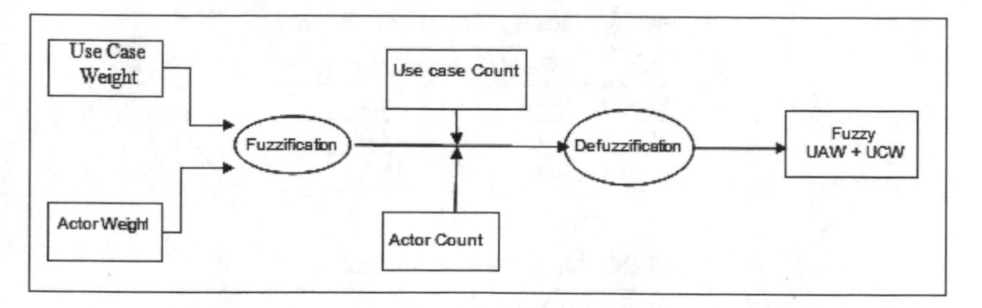

Fig. 3. Proposed framework of fuzzy use case and actor weight calculation

Table 8. Fuzzy rule base for determining the combined value of UAW And UCW

		Actor weight			
		A_1	A_2	A_3	A_4
Use Case Weight	U_1	$A_1 + U_1$	$A_2 + U_1$	$A_3 + U_1$	$A_4 + U_1$
	U_2	$A_1 + U_2$	$A_2 + U_2$	$A_3 + U_2$	$A_4 + U_2$
	U_3	$A_1 + U_3$	$A_2 + U_3$	$A_3 + U_3$	$A_4 + U_3$
	U_4	$A_1 + U_4$	$A_2 + U_4$	$A_3 + U_4$	$A_4 + U_4$

The Fuzzy base rules according to the Table 7:

1. The type of actor is simple/medium and the use case is simple, then, this case is considered as low.
2. The actor weight with the UC i.e. (complex, simple), (critical, simple), (medium, medium), (complex, medium), (critical, medium), (simple, complex), considered as medium.
3. The actor weight with the UC i.e. (medium, complex), (complex, complex), (critical, complex), (simple, critical) and (medium, critical) considered as high.
4. The actor weight with the UC i.e. (complex, critical) and (critical, critical), considered as very high.

As per the fuzzy rules and the respective values in Tables 7 and 8,

1. The value of three cells i.e. [1, 1], [1, 2], [2, 1] are considered as low.
2. The value of six cells i.e. [2, 2], [2, 3], [2, 4], [1, 3], [1, 4], [3, 1] are considered as medium.
3. The value of five cells i.e. [3, 2], [3, 3], [3, 4], [4, 1], [4, 2] are considered as high.
4. The value of two cells i.e. [4, 3], [4, 4] are considered as very high.

Table 9. Technical Complexity Factors

TCF	Weight	Count	Average
Simple	<1.0	2	0.50
Medium	1.0	8	1.0
Complex	>1.0	4	2.0

Table 10. Environmental Factors

EF factors	Weight	Count	Average
Simple	<1.0	4	0.25
Medium	1.0	3	1.0
Complex	>1.0	2	1.75

Table 11. Fuzzy rule base for determine the combined value of TCF & EF

	TCF FACTORS		
	S	M	C
S	SS (L)	SM (L)	SC (M)
M	SM (L)	MM (M)	MC (M)
C	CS (M)	CM (H)	CC (H)

On the basis of the actor's weight we classified technical complexity factors and the environmental factors as simple, medium and complex (Tables 9 and 10).

Fuzzy base rules according to the Table 11:

1. The type of technical complexity factor is simple/medium and the environmental factor is simple, then, it is considered as low.
2. The cases of technical complexity factor with the environmental factor i.e. (simple, complex), (medium, medium), (complex, medium) and (simple, simple) are considered as medium.
3. The cases of technical complexity factor with the environmental factor i.e. (complex, medium) and (complex, complex) are considered as high.

As per the fuzzy rules and the respective values in Table(s) 11 and 12:

1. The low value of three cells i.e. [1, 1], [1, 2], [2, 1] is calculated as 0.125, 0.25 and 0.5 respectively.
2. The medium value of four cells i.e. [2, 2], [2, 3], [1, 3], [3, 1] is calculated as 0.5, 0.875, 1.0 and 2.0 respectively.
3. The high value of two cells i.e. [3, 2], [3, 3] is calculated as 1.75 and 3.50 respectively.

Table 12. Fuzzy rule application to determine the combined value of TCF & EF

		TCF		
		S (0.5)	M (1.0)	C (2.0)
	S (0.25)	0.125	0.25	0.50
	M (1.0)	0.5	1.0	2.0
	C (1.75)	0.875	1.75	3.50

Table 13. Fuzzy rule base for determine the combined value of UUCP & TEF

		TEF FACTORS		
		S	M	C
UUCP	S	SS (S)	SM (S)	SC(M)
	M	MS(S)	MM(M)	MC(M)
	H	HS(M)	HM(C)	HC(C)
	VH	VHS(C)	VHM(VC)	VHC(VC)

Fuzzy base rules according to the Table 13:

1. The cases of TEF with UUCP i.e. (simple, simple), (medium, simple) and (simple, medium) are considered as low.
2. The cases of TEF with UUCP i.e. (complex, simple), (medium, medium), (complex, medium) and (simple, high) are considered as medium.
3. The cases of TEF with UUCP i.e. (medium, high), (complex, high) and (simple, very high) are considered as complex.
4. The cases of TEF with UUCP i.e. (medium, very high), and (complex, very high) are considered as very complex.

Table 14. Fuzzy rule application to determine the combined value of UUCP & TEF

	TEF FACTORS		
	S (0.30)	M (0.844)	C (2.625)
S	$0.30 * (A_S + U_S)$	$0.844 * (A_S + U_S)$	$2.625 * (A_S + U_S)$
M	$0.30 *(A_M + U_M)$	$0.844 *(A_M + U_M)$	$2.625 *(A_M + U_M)$
C	$0.30 * (A_C + U_C)$	$0.844 * (A_C + U_C)$	$2.625 * (A_C + U_C)$
VC	$0.30 *(A_{VC} + U_{VC})$	$0.844*(A_{VC} + U_{VC})$	$2.625*(A_{VC} + U_{VC})$

As per fuzzy base rules and its applications on TEF and UUCP as per Tables 13 and 14:

1. The three cells are considered as low i.e. [1, 1], [1, 2], [2, 1].
2. The four cells are considered as medium i.e. [2, 2], [2, 3], [1, 3], [3, 1].
3. The two cells are considered as complex i.e. [3, 2], [3, 3] and [4, 1].
4. The two cells are considered as very complex i.e. [4, 2], and [4, 3].

The Productive factor (PF) depends on the organization. So the value of productivity an also be distributed as low, medium and high. The value of Re-UCP evaluation is given below:

$$\text{Re-UCP} = \text{UUCW} * \text{TEF} * \text{EF} \tag{7}$$

2 Conclusion

This paper propose a fuzzy based model for estimation the software testing efforts. Actor Weight and Linguistic weights are taken as linguistic variables. As fuzzy help to cope up with uncertainty and imprecise values, this model can also produce good results. A number of extensions and application is also possible. Some soft computing techniques like neuro-fuzzy, evolutionary algorithms can also apply on this model to get more accurate results (Figs. 1, 2 and 3).

References

1. Craig, R.D., Jaskiel, S.P.: Systematic Software Testing. Artech House (2002)
2. Gelperin, D., Hetzel, B.: The growth of software testing. Commun. ACM **31**(6), 687–699 (1988). ACM, New York
3. Myers, G.J., Sandler, C., Badgett, T.: The Art of Software Testing. Wiley, New York (2011)
4. Jones, S.: Estimating Software Costs. McGraw-Hill, New York (1998)
5. Zadeh, L.A.: Fuzzy sets. Inf. Control **8**, 338–353 (1965)
6. Nassif, A.B., Luiz, F.C., Danny, H.: Enhancing use case point method using soft computing technique. J. Glob. Res. Comput. Sci. **1**(4), 12–21 (2010)
7. Mittal, A., Parkash, K., Mittal, H.: Software cost estimation using fuzzy logic. ACM SIGSOFT Softw. Eng. Notes **35**, 1–7 (2010)
8. Huang, X., Ho, D., Ren, J., Capretz, L.F.: Improving the COCOMO model using a neuro-fuzzy approach. Appl. Soft Comput. **7**(1), 29–40 (2007)
9. Kashyap, D., Shukla, D., Mishra, A.K.: Refining the use case classification for use case point method for software effort estimation. In: ACEEE, Proceedings of International Conference on Recent Trends in Information, Telecommunications and Computing (2014)
10. Kirmani, M.M., Wahid, A.: Revised use case point (Re-UCP) model for software effort estimation. Int. J. Adv. Comput. Sci. Appl. **6**(3), 65–71 (2015)
11. Gupta, D., Goyal V.K., Mittal, H.: Analysis of clustering techniques for software quality prediction. In: Second International Conference on Advanced Computing & Communication Technologies, pp. 6–9 (2012)
12. Mittal, H., Bhatia, P.K., Goswami, P.: Software quality assessment based on fuzzy logic technique. Int. J. Softw. Comput. Appl. **34**(3), 105–112 (2008)

Fault Diagnosis with Statistical Properties and Implementation Using MKSVM for Flash ADC

P. Nagaraja[1] and G. Sadashivappa[2(✉)]

[1] ECE Department, CIT Gubbi, Gubbi, India
fastnag@gmail.com
[2] Department of TC, R V C E, Bengaluru, India
sadashivappag@rvce.edu.in

Abstract. This paper focuses on the Fault Diagnosing methodologies crucial for attaining reliability and maintainability of all electronic circuits is implemented for analog to digital converter (ADC) with a wide range of faults. Fault Diagnosis (FD) is considered as the pattern recognition problem and solved by machine learning theory. Functional test is is needed instead of structural test for testing complex circuits. Fault diagnosis using Fault Dictionary, Neural Networks and Fuzzy logic are enigmatic or inconclusive diagnosis results which have more debug duration and even inaccurate repair actions that exponentially rises service overhead. The effectiveness of these methods are considered, which cover ability in detecting, identifying and localization of faults, the ability of analysing linear and nonlinear circuits, etc. Recent machine learning techniques like support vector machines (SVM) with kernel functions improve the preciseness of functional FD which reduces the product cost through correct repair process. The proposed Multikernel SVM (MKSVM) methodology gives better results than earlier methods as it works with the fundamentals of machine learning and generalization for FD.

Keywords: Structural and functional faults · ADC · Machine learning · Kernel functions and MKSVM

1 Introduction

Diagnosis of faults in circuits is the most competitive in research & development as circuits are part of every equipment. The lack of competent and efficient methodologies in FD resulted in steep rise in the cost. The efforts of research thus, are now primarily on establishing efficient methodologies and developing algorithms in FD. Component tolerances, noise and nonlinearities present will introduce difficulties in the diagnosis of given circuit. We are considering ADC which is one of the most commonly used mixed signal unit in this paper. Continuous ADC is widely used in video, metrology, and frame-stores for video signals, transient recorders, radar signal processors and image processing applications. The Staircase type ADC find its use for telecommunication applications. Successive- Approximation ADC is used in data acquisition, embedded and control applications. Sigma-delta ADCs find its use in low frequency, low level

S.C. Satapathy and A. Joshi (eds.), *Information and Communication Technology for Intelligent Systems (ICTIS 2017) - Volume 1*, Smart Innovation, Systems and Technologies 83, DOI 10.1007/978-3-319-63673-3_60

signal measurements of process controllers, thermocouple temperature measurement and programmable loop controllers. There is more of freedom for evolving new methods in ADC FD and thus, this is an emerging field of research. Advanced technical approaches to diagnose faults give a better intuition for more research and application areas as the present diagnostic methods might not be suffice neither to give competent testing methods nor did to assure the desired accuracy and reliability. This influenced us to have focus on FD of ADC.

1.1 Motivation

It is often necessary to find the cause of failures in mixed-signal systems for design debug, repair, or to cope up with the manufacturing process for maximum production with low production cost. The typical process in creating an electronic circuit is shown in Fig. 1. An electronic circuit undergoes testing during the design-prototype phase and during manufacturing before being shipped, where it is tested to verify whether the circuit satisfies its performance specifications. Typical performance specifications for an analog or mixed-signal circuit are gain and bandwidth of an amplifier, total harmonic distortion for an ADC, etc. During the prototype phase, if the circuit fails to meet its performance specifications, it is required to find the main reason for failure in order to change design or the prototype, to ensure correct operation of the circuit. During manufacturing test, if the causes of failure of a product can be easily found, the information can be used for repair, or to tune the design or the manufacturing process in order to improve yield. Once the testing is complete, the information collected during test is saved and based on this data FD and ADC fault localization are performed. In complex mixed-signal circuits, a large number of failures occur due to the interaction between analog and digital circuits. Failures in analog circuits caused by digital switching noise are an example of such a failure. More investigation into the classification of such failure modes and efficient algorithms for fault simulation are required for a comprehensive solution.

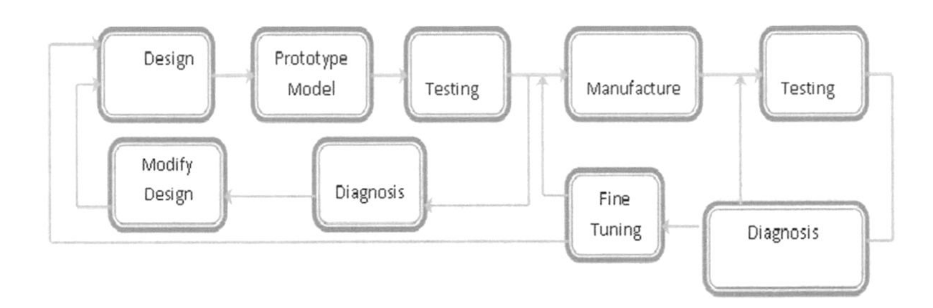

Fig. 1. Process of creating electronic circuit.

The FD has begun with detecting failure of components on printed circuit boards, to identifying defective components in circuits. During the era of discrete analog circuits, the FD mainly trusted on the exact knowledge of the testing engineers about the circuit's behaviour, operational characteristics and their response to different input conditions. FD covers all steps taken for identifying the fault, locating the fault and identification of root causes [1]. With the development of artificial intelligence [AI] technology, different methodologies to analyse circuits fault by using on fuzzy theory [6] and neural network (NN) [7] were presented till 1990s. To overcome few decisive issues of NN, like choice of network architecture, over fitting, under fitting and local minima, SVM is implemented in place of traditional NN [9].

SVM, invented by Vapnik, is a recent machine learning algorithm works with the principle of less sample statistical learning concept and follows the foundation of structural risk minimization designed to reduce the generalization error [2]. Recently, success in functional FD has been achieved using SVM, which compose of a more progressive class of machine-learning (ML) methodologies [8].

With all the experiments done till now it is observed that SVM-based functional FD exploits various kernel functions and involving incremental learning is expected to provide improved results. In Self learning, learning takes place when new example(s) arrive and automatically adjusts to that what has been learnt according to the arriving samples. The main difference between self and traditional ML is that it is not expecting large training set for the learning, but the training examples arrive over time. This new technique is expected to give maximum accuracy of diagnosis in any circuit.

2 Flash ADC

Flash ADC or Continuous ADC is very fast when compared with other types of ADCs and uses comparators and priority encoder for converting the analog input into its digital form. Strobe is a fixed voltage used as reference obtained a IC series voltage regulator as part of the converter circuit. The analog input is fed to all comparators to compare it with the reference voltage obtained by the potential divider network. When the ADC voltage is higher than the reference voltage at each comparator, the corresponding comparator output will go high. Output of all comparators are connected to priority encoder which produces binary output depending on the highest-order active

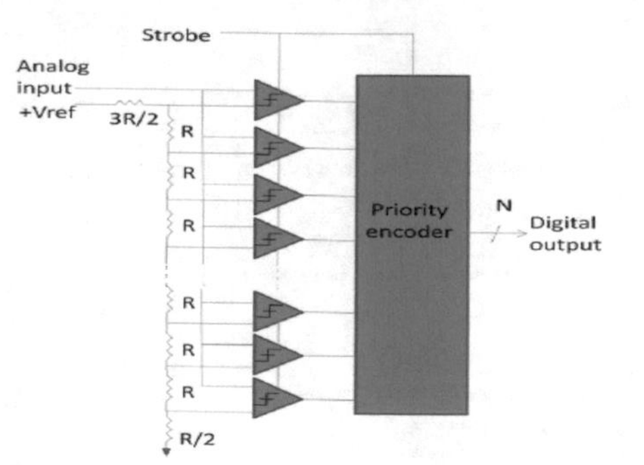

Fig. 2. Flash ADC

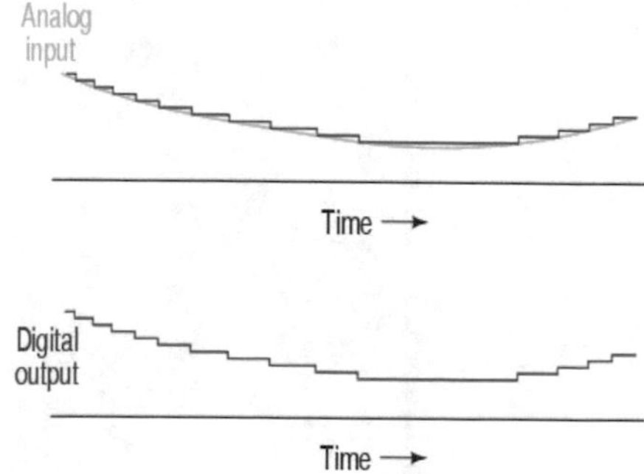

Fig. 3. Analog input & digital output of flash ADC.

input. Figures 2 and 3 shows analog input and digital output of ADC. Faults in an ADC can be in input signals, analog parts or digital parts. Some important defects are Stuck-at 0/1 at the output of comparator, Stuck-at 0/1 defect at the output of ADC, Low/high gain of comparator etc., the function of each component of ADC is indicated in the system function. The digital equivalent of the analog input is compared with calculated theoretical values to give the error and decision is taken to obtain important parameters of ADC.

3 Diagnosis Using Support-Vector Machines

SVM theory has a intellectual history to result in improved generalization capability than traditional methods, which uses Empirical Risk Minimization. Main advantages of SVM are its regularisation parameter, which averts over-fitting, using the kernel trick, simple margin interpretation and unique result. SVMs contribute a dynamic scheme to learn from repair logs and determines the exact reason for functional failures. SVM is used in many applications of bioinformatics, text mining, face recognition and FD etc.

The main purpose of SVM is to separate 2 classes by defining an Optimal Separating Hyper plane (OSH) and the margin should be maximized to be 1/||W||. One class (Class1) on left side of OSH is labeled by $y\psi = -1$ and other class (Class2) on right side is labeled by $y\psi = -1$ as shown in Fig. 4. Once the hyper plane is determined, new vectors can be classified easily using SVM. Algorithms of SVM include linear and nonlinear classification algorithm.
Hence, given the sample data:

$$\{(x_1, y_2), (x_2, y_2), \ldots \ldots (x_i, y_i)\} \in (X \times Y)^l, x_i \in X = R^n$$

x_i is the input vector, $y_i \in Y = R$ is output and i is the total number of sample data, the SVM classification function is:

$$f(x) = w \cdot \phi(x) + b = O \tag{1}$$

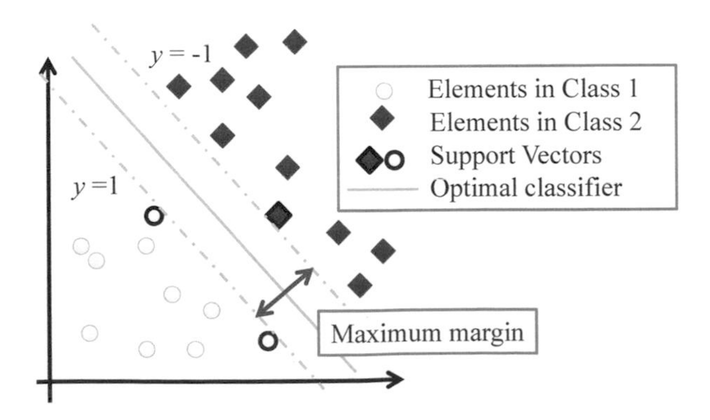

Fig. 4. The optimal separating hyperplane.

Where $\phi(x)$ is the non-linear mapping function, The following constraint should be satisfied by OSH:

$$y_i[(w \cdot \phi(x_i)) + b] \geq 1 \; i = 1, 2, \ldots. l \tag{2}$$

When the slack vector ξ_i is introduced, OSH problem is changed into optimization problem as:

$$\begin{cases} \min \phi(w, \xi) = \frac{1}{2}(w, w) + C\sum_{i=1}^{l} \xi_i \\ s.t\, y_i[(w \cdot \phi(x_i)) + b] \geq 1 - \xi_i \; \xi_i \geq 0, i = 1, 2, , \ldots. l \end{cases}. \tag{3}$$

$(\frac{1}{2})(w, w)$ is the regularization parameter, that regulates the Classification function and generalization capability. $C\sum_{i=1}^{l} \xi_i$ is a penalty factor, which balances regularization and empirical risk item. As expression (3) is solved by the Lagrange method the optimization problem (3) is transformed into another optimization problem given by (4) into its dual problem:

$$\begin{cases} \max L(a) = \sum_{i=1}^{l} a_i - \frac{1}{2}\sum_{i=1}^{l}\sum_{j=1}^{l} y_i y_j a_i a_j K(x_i, x_j) \\ s.t. 0 \leq a_i \leq c, \sum_{i=1}^{l} y_i a_i = 0 \; i = 1, 2, \ldots. l \end{cases} \tag{4}$$

Equation (4) is translated into the following constrained form when the kernel function $K(x_i, x_j) = \phi(x_i) \cdot \phi(x_j)$ is introduced.

$$\begin{cases} \max L(a) = \sum_{i=1}^{l} a_i - \frac{1}{2}\sum_{i=1}^{l}\sum_{j=1}^{l} y_i y_j a_i a_j (\phi(x_i) \cdot \phi(x_j)) \\ s.t. 0 \leq a_i \leq c, \sum_{i=1}^{l} y_i a_i = 0 \quad i = 1, 2, \ldots. l \end{cases} \tag{5}$$

Where α_i, α_j are the Lagrangian multipliers. The objective function is as follows:

$$\begin{aligned} f(x) &= sgn[(w \cdot \phi(x)) + b] \\ &= sgn\left[\sum_{i,j=1}^{l} y_i a_i K(x_i, x_j) + b\right] \end{aligned} \tag{6}$$

Where $K(x_i, x_j)$ is called the kernel function,
$K(x_i, x_j) = \varphi(x_i) \cdot \varphi(x_j)$. Radial basis kernel function is given by

$$K(x_i, x) = exp\left\{-\gamma|x_i - x|^2\right\}$$

That is the main objective of the classification function is to obtain penalty factor (C) and kernel function parameter (γ).

Slack variables are needed to allow misclassifications, which have $i\psi > 1$. The penalty factor is C is changeable. More penalty to misclassifications is assigned when C is high. The selection of $C\psi$ indicates the balance between over fitting and under fitting.

3.1 Kernelized SVMs and Multikernel SVM

3.1.1 Kernel Functions

Representation of data is selected by a kernel $K(xi^C\psi\mathbf{x}j\psi)$, where $x_i\psi$ and $x_j\psi$ are input vectors.

The optimization problem (5) is given by

$$\text{Minimize} \quad W2 = \frac{1}{2}\sum_{i,j=1}^{S} \alpha i\alpha jyiyjK(xi, xj) - \sum_{i=1}^{S} \alpha i + \mathrm{b}\sum_{i=1}^{S} yi\alpha i \tag{10}$$

with identical constraints used in non-kernel SVMs

$$W2 = \frac{1}{NS}\sum_{i,j=1}^{S}\left(yi - \sum_{i=1}^{S} \alpha jyj.K(\boldsymbol{x}i, \boldsymbol{x}j)\right). \tag{11}$$

The decision function (1) is now expressed as

$$f(\boldsymbol{x}) = sgn(\sum_{i=1}^{S} \alpha iyiK(\boldsymbol{x}, \boldsymbol{x}j) + \mathrm{b} \tag{12}$$

The success of an SVMs model depends on proper selection of kernel function. Some of the commonly used kernel functions [25] are.

1. Homogeneous Polynomial Kernel: $K(x_i^c\psi\boldsymbol{x}_j) = (x_i\psi \cdot \boldsymbol{x}_j)^d\,\psi$, where $d\,\psi \geq 1$.
2. Polynomial kernel: $K(x_i^c\,\psi\boldsymbol{x}_j) = (\boldsymbol{x}i\,\psi \cdot \boldsymbol{x}j + 1)^d\,\psi$, where $d\,\psi \geq \leftarrow 1$.
3. Gaussian kernel: $K(\boldsymbol{x}i, \boldsymbol{x}j) = \mathrm{e}^{-\gamma\|\boldsymbol{x}i - \boldsymbol{x}j\|^2}$, where $\psi = 1/2^2$ and $\psi\psi is\psi$ SD of GD.
4. Exponential kernel: $K(\boldsymbol{x}i, \boldsymbol{x}j) = \mathrm{e}^{-\gamma\|\boldsymbol{x}i - \boldsymbol{x}j\|}$

The selection of proper kernel is most important for the achievement of desired output from SVMs. In general, problems of classification are not bound to two classes. Typically hundreds of fault candidates are considered in FD (Fig. 5).

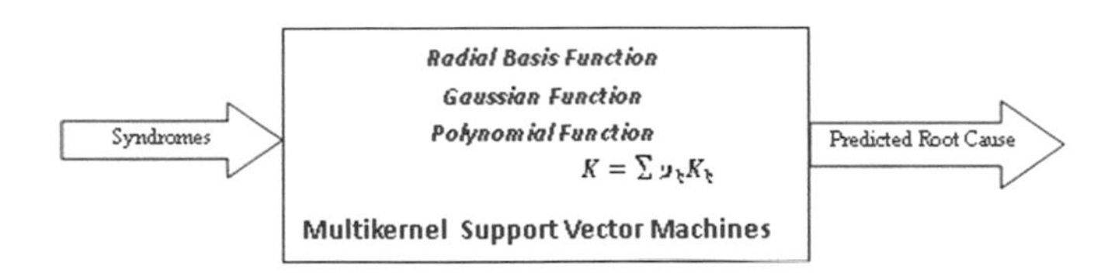

Fig. 5. Kernelized support-vector machine model.

3.1.2 Multikernel SVM

Some new bioinformatics applications have proved that use of multiple kernels instead of single kernel give better results the kernel $K(xi^C\psi\mathbf{x}j)$ is represented by linear combination of $M\psi$ *number* ψ *of* ψ basis kernels

$$\psi K\psi(\boldsymbol{x}i^{cc}\,\boldsymbol{x}j) = \sum\nolimits_{k=1}^{M} \mu_k K_k(x_i, x_j) \tag{13}$$

where $\cdot_{k\psi} \geq \leftarrow 0$ and $\sum\nolimits_{k=1}^{M} \mathrm{\mu_k} = 1$.

Each kernel $Kk\psi$ can be polynomial, Gaussian and exponential kernels. Thus the optimization problem is given by

Minimize

$$W3 = \frac{1}{2}\sum_{i,j=1}^{S} \alpha_i\alpha_j y_i y_j K \sum_{k=1}^{M} \mu_k K_k(x_i x_i) - \sum_{i=1}^{S} \alpha_i + \mathrm{b}\sum_{i=1}^{S} y_i\alpha_i \tag{14}$$

Reduced gradient method [9] is used to solve optimization problem of (14).The proposed methodology of a multikernel SVM-based FD system is shown in Fig. 6. To handle circuits with more components diagnosis system needs an adaptive kernel to attain optimum forecasting efficiency. The weights of kernels are carefully configured to fit the training set and provide improved forecasting for FD system.

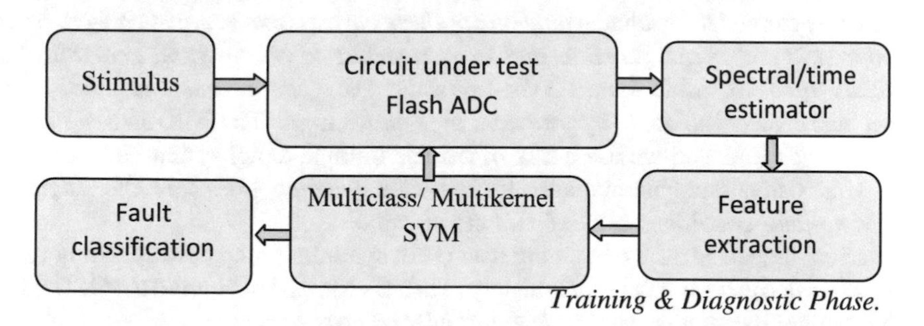

Fig. 6. SVM-based diagnosis flow.

3.2 Self Learning in Support-Vector Machines

The problem of SVM training for huge size data sets can be solved by Self learning. It can aid online learning as new data for failing units arrive to provide repair outcomes. The use of MKSVM increases complexity of computation and FD take more time to concur. This issue is handled by self learning methodology. Complexity in computation is decreased by sharing workload among different epochs during training. As support vectors decide SVM models, and the number of support vectors is very small compared to the number of training samples, SVMs can take advantage of self learning by using compression of data. Optimization problem of self learning is an expansion to (2):

$$\text{Minimize} \quad W = \frac{1}{2}\|\omega\|^2 + \mathrm{C}(\mathrm{L}\sum\nolimits_{i\in S*} \xi i + \sum\nolimits_{i\in S*} \xi_i') \tag{15}$$

Where $\psi S\psi$ is the training set ψ *and* $\psi S^* \leftarrow$ indicates support vectors derived from earlier SVM models, and Self learning is made more efficient by multiple kernels to yield optimum diagnostic accuracy of the given CUT with reduced training time.

4 SVM Diagnosis Flow

Diagnosis flow using multikernel SVM of consists of six steps that are indicated in Fig. 6.

Stimulus is applied at the input to sample the output response (time domain, frequency domain, transfer function, etc.) of the CUT which contain information on what is happening in the circuit. CUT is the electronic circuit for which fault diagnosis is to be performed. Spectral/Time Estimator estimates behavioral characteristics of the circuits that are accomplished by a specific frequency response. Frequency features are used for exploiting generating fault signatures. Raw signals are not used during analysis since they need large memory space for storage results inaccuracy of analysis. Extracting features from original signal is a method of drawing the useful information to eliminate artifacts and lower the dimensionality by preserving characteristics which denote specific the fault patterns.

SVM is very suitable for FD, as electronic systems always show nonlinear, complexity and diversity features. In multiclass classification, each system element has a particular class assigned to it, which distinguishes it from other elements. Each of the classifiers is trained to recognize only one of these classes (one-vs.-all strategy). Select the appropriate c, kernel function and its parameters to obtain good generalization capability of SVM. Multi-kernel SVM enhances the generalization capability and exploit more discriminative information in sample data. For MK SVMs training change in each time and we use a mix of kernels in multikernel system, including the linear, Gaussian and polynomial kernel. *Fault Classification gives clue about different fault or no fault conditions of CUT and classification.*

In actual diagnosis a set of training data (fault syndromes and corresponding repair solutions) is derived from the repair history. Then SVMs will determine the OSH based on the training data and system is ready to analyze new cases.

In the training stage three operations are performed as shown in Figs. 7 *and* 8. *Depending on the* faults in given *circuit* data is prepared and location of each fault are recognized. *Linear, polynomial Gaussian and exponential kernel* functions with different parameters are selected for training ANN, Decision tree or SVM. When SVM is used its efficiency depends on the values of parameters selected for each kernel. Matlab is used for determining the values of $w\psi$ and b. The output of the SVM for a new input vector can be determined from the decision function. In the diagnosis stage, rank the output of all the SVMs, and component with the maximum output is the root cause. Let us consider 3 fault candidates A, B, and C with corresponding SVM decision functions given by $f_A(x)^C \psi f_B(x)^C \psi f_C(x)$. For new input vector $x0, f_A(x0) = 0(1^C f_B(x0) = 0(8^C \psi f_C(x0) = -0(2$, the main root cause of the defect is candidate B.

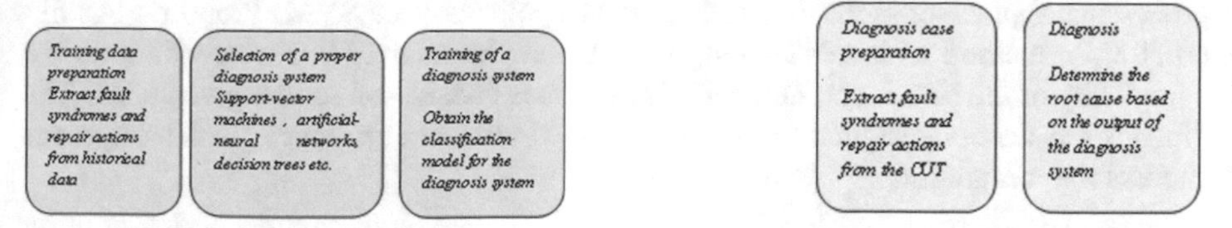

Fig. 7. Training stage

Fig. 8. Diagnosis stage

4.1 Test Setup and Test Procedure

Partial Response Maximum-Likelihood read channels and high speed Ethernet need flash ADCs. An n-bit flash ADC need 2^n-1 number of comparators and a priority encoder. Experiment is performed on the 3 bit flash ADC. Relevant information about the given circuit is obtained by applying stimulus to the CUT with given specifications. In the prototype model we have considered 16 fault syndromes from fault history. The response observed in the simulator with the details of the parameters such as voltage, current and power at different points of the circuit is used as data for diagnosis. The MATLAB toolbox is used for design and implementation of SVM algorithms. Results indicate that time needed for training is reduced after the implementation of self learning (Figs. 9 and 10).

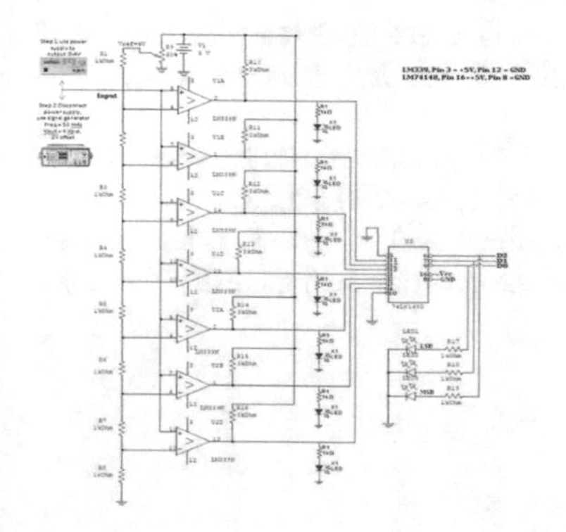

Fig. 9. Schematic diagram of test circuit: 3 bit flash ADC

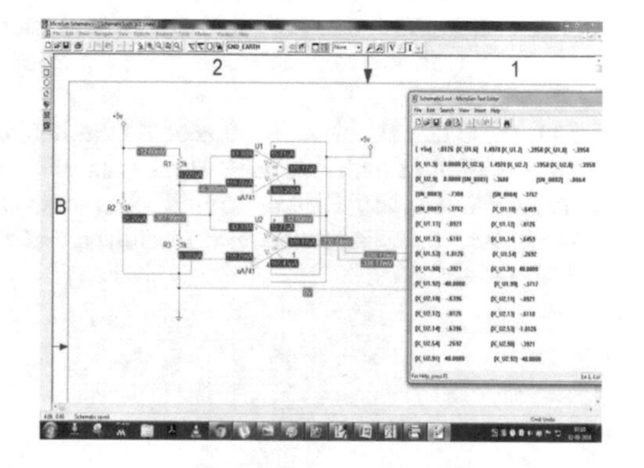

Fig. 10. Simulator output

5 Conclusion

This proposed work on FD technique has a significant improvement over the conventional diagnosis methods to result in optimum accuracy in finding faults of ADC. It is intended to diagnose both the known faults and the unknown faults effectively. Since multimodalities are available for FD, achieving higher success rate and optimum diagnosing accuracy with small training time are prime important. In most of the applications FD takes more time and ambiguous since technicians check the CUT manually or semiautomatically. SVM is used in industrial, medical and image processing applications. FD is one important application of SVM. Proposed Multikernel SVM method with self learning gave better results over 90% depending on the complexity of the circuit and number of syndromes considered for the given CUT. The diagnosis system is designed for adapting to new error circumstances which improves success rate on-the-fly.

References

1. Catelani, M., Giraldi, S.: A measurement system for fault detection and fault isolation of analog circuits. Measurements **25**, 115–122 (1999)
2. Vapnik, V.: The Nature of Statistical Learning Theory. Springer, New York (1995)
3. Abbas, M.: Fault detection and diagnoses methodology adaptive digitally-calibrated pipelined ADCs. IEEE Trans. **26**, 30–35 (2011)
4. Czaja, Z.: Using a square-wave signal for fault diagnosis of analog parts of mixed-signal embedded systems controlled by microcontrollers. In: Instrumentation and Measurement, Technology Conference (2007)
5. Singh, M., and Koren, I.: Incorporating fault tolerance in analog to digital converters. In: Proceedings of the International Symposium on Quality Electronic Design IEEE, pp. 1–6 (2002)
6. Bilski, P., Wojciechowsk, M.: Automated diagnostics of analog systems using fuzzy logic approach. IEEE Trans. Inst. Measur. **56**(6), 2322–2329 (2007)
7. Aminian, F., Modular, A.: Fault-diagnostic system for analog electronic circuit using neural networks with wavelet transform as a preprocessor. IEEE Trans. Inst. Measur. **56**(5), 1546–1554 (2007)
8. Liu, D., Quan, H., Dai, G., Zhang, Z.: An iterative SVM approach to feature selection and classification in high dimensional datasets. Elsevier J. **46**, 2531–2537 (2013)
9. Ye, F., Zhang, Z., Chakrabarty, K.: Board-level functional fault diagnosis using artificial neural networks, support-vector machines, and weighted-majority voting. IEEE Trans. **32**, 723–736 (2013)

Study of Digital Watermarking Techniques for Against Security Attacks

Snehlata Maloo[1,2(✉)], N. Lakshmi[1,2], and N.K. Pareek[1,2]

[1] Department of Computer Science, Mohanlal Sukhadia University, Udaipur, India
sneha.maloo21@gmail.com, nlakshmi@mlsu.ac.in, npareek@yahoo.com
[2] Department of Physics, Mohanlal Sukhadia University, Udaipur, India

Abstract. Digital watermarking is fragile to different authorization and copyright protection in spatial domain, So in the vast majority of watermarking strategy's transform domain is utilized. In that paper we have an review of such computerized watermarking procedures and techniques like discrete wavelet transform (DWT), singular value decomposition (SVD), discrete cosine transformation (DCT) with various methodologies alongside their applications, advantages and limitations.

Keywords: DCT · DWT · SVD · LSB · Digital watermarking

1 Introduction

Recent growth of modern technologies and Internet has made digital library an essential component of all institutional libraries. This has helped to reduce the gap between the substance offered by a customary library and the data need of people [1]. The growth of digital media and contents has also increased security problems including the need for reliable and secure copyright protection techniques [2]. In the absence of efficient protection of copyrights of digital products in a digital library, content authors may not come forward for contributing to the intellectual properties to the digital library. The digital libraries in most academic institutions lack digital contents from local authors who are reluctant to contribute to the teaching and training materials fearing about the security aspects of their content and protection from copying, re-proposal of the digital contents by others. Further there is a need of quick and easy identification of digital contents. The metadata initially attached to the content when it is created can often get lost as the files are copied and can be manipulated and transformed from one format to another. Without an effective content identification authors are hesitate to digitize and distribute their digital contents due to the fact that content can be easily copied, altered re-proposed and even shared or sold without the permission or knowledge of its owners.

Security assumes an essential part in the usage of digital libraries. As of late digital watermarking has been widely utilized for protection of intellectual properties on digital media and on Internet. Digital watermarking is one of the answers for the security issues of digital substance. It is possible to identify the digital contents by identifying the source and hence the author, creator and owners.

S.C. Satapathy and A. Joshi (eds.), *Information and Communication Technology for Intelligent Systems (ICTIS 2017) - Volume 1*, Smart Innovation, Systems and Technologies 83, DOI 10.1007/978-3-319-63673-3_61

Digital watermarking is the way toward implanting an example of bits which distinguishes the copyright data into pictures, recordings, sounds and other sight and sound information by a specific algorithm. These implanted advanced examples known as watermark are normally imperceptible and can be later removed from, or distinguished in the mixed media for security purposes (Fig. 1).

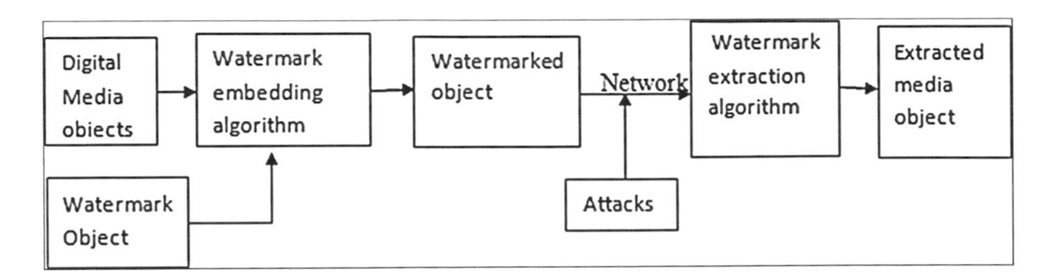

Fig. 1. Block diagram for digital watermark embedding and extraction

1.1 Spatial Domain

These are the systems, which utilizes bit level control on the given picture like least significant bit, insert the message by adjusting the coefficients of the LSB in a picture.

As per Koushik Pal, Mahua Bhattacharya, Goutam Ghosh, a biomedical picture watermarking system utilizing modified bit replacement algorithm in spatial space. In this calculation, various duplicates of a similar data are covered up in the biomedical pictures [8]. Discovery is performed by applying bit majority calculation which can reproduce the concealed information from different recovered sets subjected to some ordinary attacks. Keys for watermark recovery are additionally implanted in the cover picture. The prevalence of the modified bit replacement scheme over the LSB method is shown through assessment of picture quality measurements that shows much improvement in the visual and factual intangibility of shrouded information.

As per Shaveta Chutani, Himani Goyal steganography is the craftsmanship and procedure of concealing data in pictures, sound, video or content and other cover media has advanced and created with time. Various enhancements have been accomplished concerning security, power and limit [9].

They additionally show a survey of various LSB implanting strategies in spatial domain for picture steganography. An assortment of methods which have developed after some time and have enhanced the fundamental LSB strategy have been talked about. These strategies additionally been looked at on the premise of various parameters.

1.2 Transform Domain

These are the procedures, which implants the message by adjusting the coefficients in the frequency domain, for example, in the discrete fourier transform (DFT), discrete cosine transform (DCT), and discrete wavelet transform (DWT) cases.

Different works have been as of now completed every now and then to assess Robustness of Transform Domain Watermarking Techniques

2 Discrete Cosine Transform (DCT)

According to Alessandro Piva et al. they works in the frequency space, inserts a pseudo-irregular grouping of real numbers in a chose set of DCT coefficients. By misusing the statistical properties of the embedded sequence, the check can be dependably removed without turning to the first uncorrupted picture.

As per Juan R. Hernández, Martín Amado and Fernando Pérez-González, a spread-spectrum like DCT watermarking method for copyright security of still computerized pictures is investigated. The DCT image is connected in squares of 8 × 8 pixels as it generally in JPEG calculation. The watermark can encode data to trail unlawful mishandling. For adaptability reason, the original picture is a bit much amid the ownership confirmation handle, so it must be demonstrated by noise [10].

Two tests are incorporated into the ownership confirmation handle: watermark deciphering, with which the letter conveyed by watermark is extricated, and watermark discovery that chooses whether given picture includes a watermark created by a specific key.

They relate summed up Gaussian dispersions for measurably demonstration the DCT coefficients of main picture and show that how subsequent locator formations prompt to extensive upgrades in execution concerning the relationship recipient, which has been generally considered in the writing and makes utilization of the Gaussian noise presumption. Subsequently of their work, systematic expressions for execution measures, for example, the likelihood of mistake in the watermark translating and chances of false alert and identification in watermark discovery are inferred and diverged from test comes about (Fig. 2).

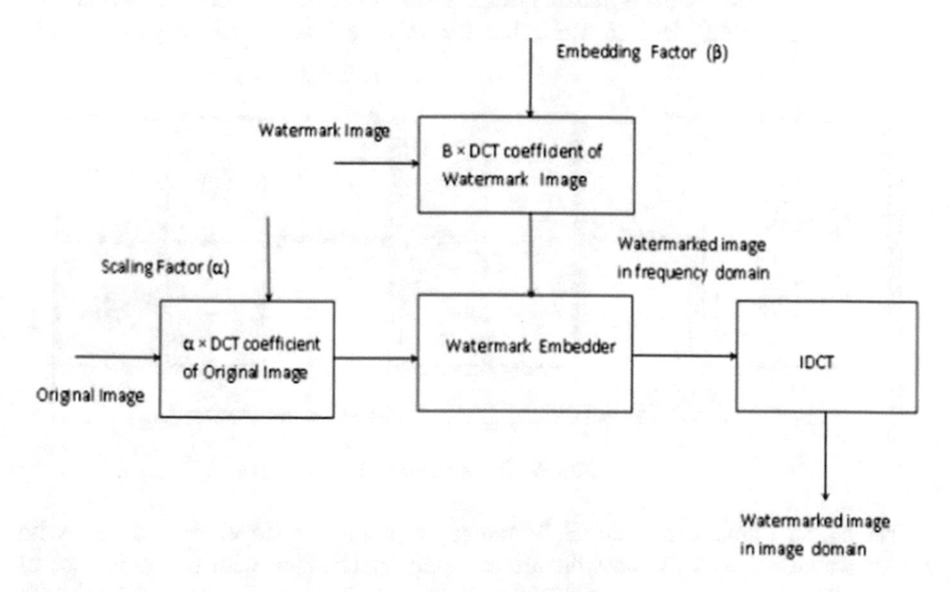

Fig. 2. DCT technique

Deepa Kundur and Dimitrios Hatzinakos demonstrated a fresh vigorous watermarking technique known as FuseMark in view of the standards of image fusion method for copy safety or robust cataloging functions. They consider the issue of logo watermarking in still pictures and utilize multiresolution information combination standards for watermark inserting and extraction. A human visual framework based on contrast sensitivity, is consolidated to shroud a higher vitality concealed logo in notable picture segments. Watermark extraction incorporates both portrayal of attacks and logo estimation using a rake-like recipient. Statistical examination shows how their extraction approach can be utilized for watermark recognition applications to diminish the issue of false negative location without expanding the false positive discovery rate. Simulation comes about check hypothetical perceptions and shows the down to earth execution of FuseMark [11].

3 Discrete Wavelet Transform (DWT)

Santi P. Maity, Malay K. Kundu examines the extent of wavelets for execution change in spread spectrum picture watermarking. Execution of an advanced picture watermarking calculation, in general, is dictated by the visual intangibility of the hidden data (subtlety), unwavering quality in the recognition of the concealed data after different normal and deliberate signal processing process (vigor) connected on the watermarked signals and the measure of information to be unseen (payload limit) with no influencing the imperceptibility and reliable property.

They propose few spread spectrum (SS) picture watermarking plans utilizing DWT (discrete wavelet change), M-band wavelets and biorthogonal DWT combined with different balance, multiplexing and flagging systems. The execution of the watermarking strategies additionally detailed alongside the relative benefits and negative marks (Fig. 3).

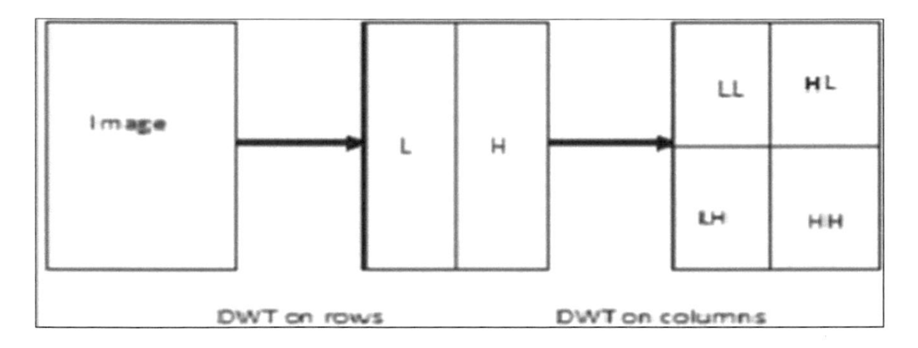

Fig. 3. DWT technique

Nallagarla Ramamurthy, Dr. S. Varadarajan proposes a novel way to deal with embed watermark into the host picture utilizing quantization with the assistance of Dynamic Fuzzy Inference System (DFIS) is proposed. The cover picture is disintegrated up to 3-levels utilizing quantization and discrete wavelet transform (DWT). A bitmap of size 64×64 pixels is inserted into the host picture utilizing DFIS lead base. The DFIS is used to create the watermark weighting capacity to install the intangible watermark [12].

Ms. Snehal Pritesh Shah proposed a algorithm based video watermarking plan in the discrete wavelet change (DWT) space. Scene change investigation initially led to disintegrate video into various scenes. Every edge of the video is changed to wavelet space by DWT. The watermark picture decayed into 8-bit planes, mixed and installed into the mid-recurrence DWT coefficients. The nature of the watermarked video is upgraded by GA.

4 Singular Value Decomposition (SVD)

As indicated by Indranil Sengupta, Vivekananda Bhat K, and Abhijit Das, quantization catalog modulation is one of the best strategies for performing blind watermarking, because of its effortlessness and great distortion-reliable tradeoffs. Authors introduce another sound watermarking algorithm in view of SVD and dither-balance quantization is displayed. The watermarking is installed utilizing dither-adjustment quantization of the particular estimations of the squares of the host audio signal. The watermark could be indiscriminately removed with no learning of the main audio signal. individual and target tests affirm heighty intangibility accomplished by the projected scheme. Besides, the plan is very vigorous against assaults including additive white Gaussian noise, low-pass filtering, MP3 pressure, resampling, requantization, editing, resound option and denoising. The watermark information payload of the code is 196 bits per second.

Meenal A. Kamlakar, Chhaya Gosavi and Abhijit J. Patankar presents, a computerized watermarking plan for color media utilizing block based SVD approach. The proposed calculation for watermarking handles the issue of false positive recognizable proof and can be utilized for shading recordings. Watermarking utilizing SVD should be possible all the more productively if watermark is implanted in the blue segment of the host casing of the video, rather than installing it in the casing without isolating the color part. Investigation and results demonstrates that the projected computation is strong safe and additional productive contrasted with our before plan where implanting is managed without isolating the color segments of the picture to be watermarked.

As indicated by Abdel Hamid Benhocine, Lamri Laouamer, Laurent Nana and Anca Christine Pascu, Watermarking innovation intends to solve multimedia copyright protection issues specifically on PC systems, for example, Internet. The innovation must show confirmation of ownership with a high level of confidence and exhibit a solid resistance against the programmer's bending like the noising, filtering, compression. The approach proposed in their paper comprises of another picture watermarks installing/extraction calculation in view of the solitary esteem decay SVD.

5 Similarity Study

Original picture of size 512 × 512 pixel and watermark picture of size 50 × 90 pixel (Tables 1, 2 and 3)

Table 1. Results of different assaults on DCT watermarked pictures utilizing PSNR

S. no.	Attack	PSNR (dB)
1	Unsharp	11.78
2	Gaussian noise	11.89
3	Salt and pepper	16.3
4	Rotation	9.8

Table 2. Results of different attacks on DWT watermarked pictures utilizing PSNR

S. no.	Attack	PSNR (dB)
1	Unsharp	12.38
2	Gaussian noise	11.98
3	Salt and pepper	24.3
4	Rotation	9.7

Table 3. Results of different attacks on SVD watermarked pictures utilizing PSNR

S. no.	Attack	PSNR (dB)
1	Unsharp	33.38
2	Gaussian noise	35.58
3	Salt and pepper	35.38
4	Rotation	34.72

6 Conclusion

Consequently, this manuscript spotlight on digital picture watermarking in the frequency domain and advanced watermarking strategies like DWT, SVD, DCT, DWT-SVD, DCT-SVD, with inclinations, shortcomings and functions. Relative examination of a variety of methods is also given in this manuscript. Inserting and extraction of watermark is being done utilizing the strategies said in this manuscript. For verifying the energy of these systems distinctive attack on watermarked picture are performed like rotation, salt and pepper noise, unsharping and Gaussian Noise. DCT-SVD exhibits improved outcomes amongst these methods took a gander at to the extent PSNR after attack on watermarked picture.

References

1. Sturges, P.: Remember the human: the first rule of netiquette, librarians and the Internet. Online Inf. Rev. **26**(3), 209–216 (2002)
2. Fifarek, A.: Technology and privacy in the academic library. Online Inf. Rev. **26**(6), 366–374 (2002)
3. Raj, Y.A., Alli, P.: An analysis and overview of modern digital watermarking. Am. J. Appl. Sci. **9**(1), 66 (2012)

4. Cox, I., Miller, M., Bloom, J., Fridrich, J., Kalker, T.: Digital watermarking and steganography. Morgan Kaufmann, New Jersey (2007)
5. Agarwal, R., Santhanam, M.S., Venugopalan, K.: Multichannel digital watermarking of color images using SVD. In: 2011 International Conference on Image Information Processing (ICIIP), pp. 1–6. IEEE (2011)
6. Zhao, J., Koch, E.: Embedding robust labels into images for copyright protection. In: KnowRight, pp. 242–251 (1995)
7. Wolfgang, R.B., Podilchuk, C.I., Delp, E.J.: Perceptual watermarks for digital images and video. Proc. IEEE **87**(7), 1108–1126 (1999)
8. Pal, K., Ghosh, G., Bhattacharya, M.: Biomedical image watermarking in wavelet domain for data integrity using bit majority algorithm. Am. J. Biomed. Eng. **2**(2), 29–37 (2012)
9. Goyal, H., Chutani, S.: LSB embedding in spatial domain. Int. J. Comput. Technol. **3**(1), 153–157 (2012)
10. Hernández, J.R.: DCT-domain watermarking techniques for still images: detector performance analysis and a new structure. IEEE Trans. Image Process. **9**(1), 55–68 (2000)
11. Andrews, H.C., Patterson III, C.: Singular value decomposition (SVD) image coding. IEEE Trans. Commun. **24**(4), 425–432 (1976)
12. EL-Emam, N.N.: Hiding a large amount of data with high security using steganography algorithm. J. Comput. Sci. **3**(4), 223–232 (2007)
13. Mohan, B.C., Kumar, S.S.: A robust image watermarking scheme using singular value decomposition. J. Multimed. **3**(1), 7–15 (2008)

Automatic and Intelligent Integrated System for Leakage Detection in Pipes for Water Distribution Network Using Internet of Things

Shikha Pranesh Gupta(✉) and Umesh Kumar Pandey

MATS University, Raipur, India
09.shikha@gmail.com, umesh6326@gmail.com

Abstract. The problem of leaking distribution system is very important issues across the world to operate and via moving steps in this direction better performance of services from water supply organization can be achieved. Even though the methods and technology used in a leakage localization are based on only one kind of sensor, therefore the leakage is not identified until the water has risen above the surface. Due to physical constraint and unique feature of water distribution network designing effectively identification of leakage is very difficult. This paper incorporates idea to propose a new effective practical approach to collect the information from the sensor and after the analysis and computation of that data; information is communicated using any technology like Bluetooth, wireless network, wired network etc. which will be helpful to fire some important decision and based on this valuable decision, leakage control parameter can be controlled using Internet. This paper aims to propose the use of technology of this era, Internet of things (IoT) integrated with recent advances in electronics embedded technology to secure the most valuable resource water for this era as well as for future generation. This paper aims for proposing use of multi-sensor fusion data and Internet of things for leakage detection in pipes for water distribution network.

Keywords: Water distribution network · Internet of things · Leakage · Sensor

1 Introduction

Water is a valuable resource for every country. Water leakages are being one of the major issues which every county is facing [1]. In underground pipeline leakage detection is a difficult task. In so many countries there is no automatic and intelligent detection method present. Leaks are detected by the public if it is raised above the surface. Government has to spend money in finding invisible leakages and fixing them.

Due to leakages in water pipelines which are underground so many countries suffer in terms of financial loss, fitness, energy and surroundings. In India 70% land is for agriculture purpose, so water is a precious commodity and we cannot ignore the fact that India is under high stress level [1] as shown in Fig. 1. Leakage percentage to various countries has been provided, so stress level for other country is also very high.

S.C. Satapathy and A. Joshi (eds.), *Information and Communication Technology for Intelligent Systems (ICTIS 2017) - Volume 1*, Smart Innovation, Systems and Technologies 83, DOI 10.1007/978-3-319-63673-3_62

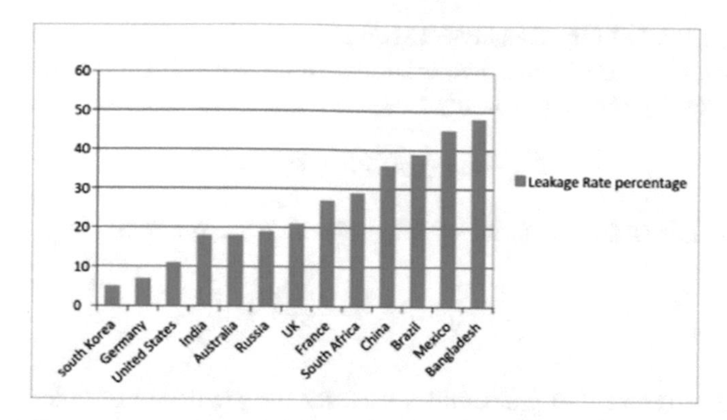

Fig. 1. Leakage percentage in water distribution system for different countries [1, Fig. 2]

Water supply network is one of the prominent networks for any county. Our purpose is to keep record of hydraulic parameter in pipes and converting this into digital information and intelligently it will locate the leakage of pipe in water distribution network (WDN). We are proposing a smart application which involves the use of sensors, actuators, microcontroller and cloud services in place of locally implementing this [2].

The whole paper is basically covering the concept of leakage detection and the idea behind the automatic and intelligent integrated system (AIIS) for leakage detection in WDN. Section 2 describes the mathematical probabilistic fusion method; Sect. 3 introduces the modelling technique for the leakage detection in AIIS, while in Sect. 4 we see the architecture used for AIIS and in Sect. 5 working strategy of AIIS is proposed Sect. 6 gives the conclusions, research and implementation directions.

2 Mathematical Methods in Probabilistic and Estimation-Theoretic for Leakage Detection

Various multi-sensor data fusion method has been listed below.

2.1 Bayes Rule

Bayesian inference is a method in which all forms of uncertainty are expressed in terms of probability. It is formulated over some factors which are not known for the model and some beliefs assumed before getting familiar with data. Then after the observation of sample data, the Bayes' Rule applies to find a posterior distribution for these unknowns, which considered the prior and the data both [3]. Based on this posterior distribution calculation of predictive distributions for future observations can be done.

2.1.1 Joint Probability Density Function

Bayesian inference considering two arbitrary events e and h [4] that is used for joint probability density function. Density function using chain rule can be represent as

$$\mathrm{P(e,h)} = \mathrm{P(e|h)\ P(h)} = \mathrm{P(h|e)P(e)} \tag{1}$$

Conditional densities in terms of Bayes theorem are obtained:

$$\mathrm{P(e|h)} = \frac{P\langle h|e\rangle P(e)}{P(h)} \tag{2}$$

P(h|e), P(e|h) and P(e) probabilities used for the prediction of outcomes. Assume that we require getting various likelihood of unknown event e. Prediction of prior beliefs for the occurrence of event e might be there. P(e) is given in terms of relative likelihood in the prior probability. Hypothesis h can be created by getting more information about the occurrence of event e. This can be observed and based on that conditional probability P(h|e) can be mathematically modelled. Product of original prior information of event occurring and hypothesis made, produced the new likelihood for the occurrence of event e. New likelihoods used in posterior probability P(e|h).

The main drawback of the Bayesian inference is that the probabilities (e) must be known. To estimate the conditional probabilities, Pan et al. [5] proposed the use of NNs, whereas Cou´e et al. [6] proposed the Bayesian programming. Hall and Llinas [7] described the following problems associated with Bayesian inference.

(i) Difficulty in establishing the value of a priori probabilities and explaining the uncertainty of the decisions.
(ii) Complexity when there are multiple potential hypothesis and a substantial number of events that depend on the conditions.
(iii) The hypothesis should be mutually exclusive.

2.2 MultiSensor Bayesian Infererence

Bayes theorem can be used for the fusion of hypothesis derived from the prior beliefs. Let's take the following set of hypothesis [4].

$$\mathrm{H^n} \triangleq \{\mathrm{h_1} \in \mathrm{H_1}, \ldots, \mathrm{h_n} \in \mathrm{H_n}\} \tag{3}$$

Equation 3 is used to calculate a posterior distribution P $(\mathrm{e}|\mathrm{H^n})$ where $\mathrm{e} \in \mathrm{E}$. Distribution function can be directly calculated through Bayes theorem as given below

$$\begin{aligned} \mathrm{P(e|H^n)} &= (\mathrm{P\ (H^n\,|\,e)P\ (e)})/\mathrm{P\ (H^n)} \\ &= (\mathrm{P\ (h_1, \ldots,\ h_n|\ e)\ P\ (e)})/\mathrm{P\ (h_1, \ldots,\ h_n)} \end{aligned} \tag{4}$$

Every possible combinations of hypothetical condition for a particular event is considered on joint distribution. This requires the prior knowledge of the joint distribution $P(h_1, \ldots, h_n | e)$. If joint distribution is not known then it's tough to perform this. Suppose that true occurrences for the event $e \in E$, and information obtained from i_{th} source is not rely on the information from other sources

$$P(h_i | e, \ h_1, \ldots, h_{i-1}, h_{i+1}, \ldots, h_n) = \ P(h_i | e) \tag{5}$$

$$P(h_1 \ldots, \ldots, h_n | e) = P(h_1 | e) \ldots P(h_n | e) = \prod_{i=1}^{n} P(h_i | e) \tag{6}$$

Substituting this into Eq. 4 gives

$$P(e | H^n) \ = \ [P(H^n)]^{-1} P(e) \prod_{i=1}^{n} P(h_i | e) \tag{7}$$

Equation 7 is specifying the relationship between Posterior probability on event e and the outcomes coming from i = 1 to n information sources. In above equation P (H^n) is a normalising constant. Equation 7 can be calculated for various hypotheses. Priori can be stored in the conditional probability P $(h_i|e)$ in terms of h_i and e. Assume that the hypothetical sequence $H^n = \{h_1, h_2, \ldots, h_n\}$ is present, and hypothetical values get instantiated, so that likelihood functions Λi (e) are created and it is only for the unknown event e.

$$P(h_1, \ldots, \ h_n) \neq P(h_1) \ldots P(h_n) \tag{8}$$

Here, each hypothesis from h_1 to h_n relies on event $e \in E$. if hypothesis is not depend on event as well as information sources, small value using it will increase the occurrences of event.

2.3 Recursive Bayes Updating

In recursive Bayes updating past information should be remembered and new information coming in the form of P $(h_k | e)$. Whole information collected till current time is used to calculate the total likelihood. In Eq. 7 recursive addition of new information is used for getting event with $H^k \triangleq \{h_k, \ H^{k-1}\}$ for posterior distribution where hypothetical sequence for conditional probability is assumed.

$$\begin{aligned} P(e, H^k) &= \ P(e | H^k) \ P(H^k) \\ &= \ P(h_k, H^{k-1} | e) P(e) \\ &= P(h_k | e) P(H^{k-1} | e) P(e) \end{aligned} \tag{9}$$

Comparing both sides of equation gives

$$\begin{aligned} P(e|H^k)\ P(H^k) &= P(h_k|e)\ P(H^{k-1}|e)P(e) \\ &= P(h_k|e)P(e|H^{k-1})P(H^{k-1}) \end{aligned} \quad (10)$$

Noting that P $(H^k)/P\ (H^{k-1}) = P\ (H_k \mid H^{k-1})$ and rearranging gives

$$P(e\,|H^k) = (P(h_k|e)\ P(e|H^{k-1}))/(P(h_k|H^{k-1}) \quad (11)$$

P $(e|H^{k-1})$ is adding all previous information. So Eq. 11 is providing the benefit to store only posterior likelihood.

3 Modelling Technique

The probabilistic methods have various limitations. In information fusion problem, limitations of these methods identified and it is listed below [4].

1. Precision of Models: In probabilities specification the necessity should be precise for quantities.
2. Inconsistency and Complexity: consistent deductions about state of interest via consistent set of beliefs in the form of probability are a tedious task and the cumbersome situation is to deal with the large volume of probability with correctly applied probabilistic reasoning methods.
3. Uncertainty about uncertainty: obscurity in when it reflects the situation of uncertainty and probability is assigned for it.

Limitation of probabilistic approach can be overcome by interval calculus, fuzzy logic, and the theory of evidence techniques

There are three mainly techniques for solving these issues; interval calculus, fuzzy logic, and the theory of evidence [4].

Prior knowledge of probability distribution is helpful for probabilistic theory. In the same way prior membership function for various fuzzy set is giving the quick understanding of fuzzy sets theory. In the next section fuzzy logic method has been given, as for fusion of data this algorithm is well suited. Furthermore, it has been often integrated with probabilistic [8, 9] and D–S evidential [10, 11] fusion algorithms in a complementary manner.

3.1 Fuzzy Logic Method

Comparing all the algorithms described above the fuzzy logic method is well suited for the leak detection component because it applies the fuzzy rules to boundary leak size instead of a complex mathematical model. Xu's fuzzy rules is consider as a reference [12]. The output of fuzzy logic will be "large", "medium" and "small". Cross correlation was used to localize the leak. In this model, two inputs were applied, pressure difference (Pressure Diff) and flow difference (Flow Diff). One output was leak size (Leak Size).

The pressure difference and flow difference were identified as negative small (NS), zero (Z), positive small (PS), positive medium (PM), and positive large (PL). The leak size was identified as zero (Z), small (S), medium (M) and large (L). For getting the exact position of leak location in pipe of WDN various acoustic method, Inverse Transient Method, Transient Damping Method, genetic algorithm can be chosen.

4 Architecture Used for AIIS

IoT is a most popular research field and research in this field is under certain consideration. Still there is no specific architecture of IOT. One of the popular three layer architecture [13] contains perception layer, Network layer and application layer shown in Fig. 2.

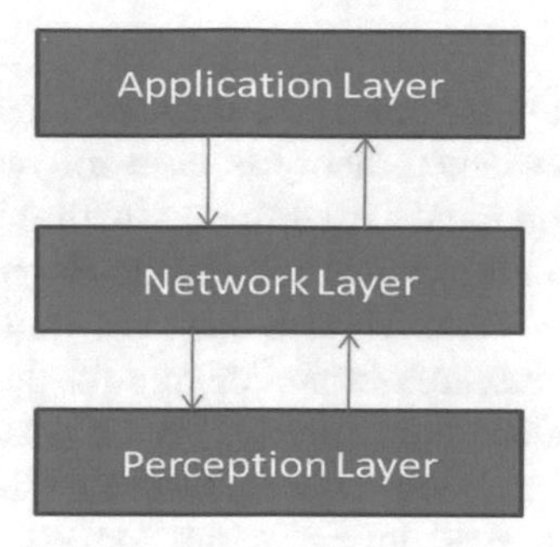

Fig. 2. Architecture of IoT

5 Working Concept of AIIS

No doubt that the modern growth in the field of technology has getting new dimension with the development of IoT. Its an era where low cost, small size, power saving sensors are utilized for the purpose of various application. Now a day, various tremendous IoT applications practically implemented in day to day life. IoT researches have given their valuable contribution and industries accepted it. Applications of IoT can be seen in various field like industries, including intelligent environmental monitoring, intelligent building, homes, and cities, smart transportation, and smart health care [14, 15]. With the grace of recent technology we are proposing the AIIS for WDN. Recent advances cloud services; big data analytics can resolve the leakage detection problem in a smart way. It comprises the following steps:

1. In WDN various sensors are assembled to measure the important hydraulic characteristic of water like pressure and flow. This measured information is converted into the digital signal which can be easily transmitted through digital communication network. Actuator is used to accept program command and then task can be performed.

2. Next step is a transmission of data. The various technology used to implement this include cellular technology 2G/3G/LTE (Long tern evolution), Wi-Fi, Bluetooth or Zigbee among other. In this process the digital information received is transmitted using any of technology to a database server or processing center.
3. It requires Gateway which gives connectivity with public network. A cloud platform to store data gathered from the real world and properly formatted to be used by end-user facilitating the so called smart applications.
4. An application on a smart phone or tablet to get access to the cloud and retrieve information gathered by the wireless network and internet with the actual sensors and actuators. Last step is to facilitate the automation, location based services, authentication and safety etc.

6 Conclusion

This paper elaborates initially, mathematical method using fusion concept for leakage detection. Probability and evidence theory is used for modeling the uncertainty and fuzzy sets theory is most commonly used to deal with the fuzzy membership. Fuzzy logic is conceptually more suited for this framework. Based on Indian framework there is an urgent need of AIIS for WDN. Though there is a various technique present for the leakage detection, then also researchers are looking for the smart solution. No method has been developed for Indian infrastructure so far that can detect leaks and take automatic smart decision using cloud services. Moreover there is so many different case scenario can be exist as physical infrastructure for WDN can be vary in terms of material used for pipelines, soil type, population etc. There is a wide scope of research can be done in this field.

References

1. Gupta, A., Mishra, S., Bokde, N., Kulat, K.: Need of smart water systems in India. Int. J. Appl. Eng. Res. **11**(4), 2216–2223 (2016)
2. Marsico, A., Broglio, A., Vecchio, M., Facca, F.M.: Learn by examples how to link the Internet of things and the cloud computing paradigms: a fully working proof concept. In: IEEE 3rd International Conference on Future Internet of Things and Cloud (FiCloud), Rome, pp. 806–810 (2015)
3. Philosophy of Bayesian inference. http://www.cs.toronto.edu/~radford/res-bayes-ex.html
4. Khaleghi, B., Khamis, A., Karray, F.O., Razavi, S.N.: Multisensor data fusion: a review of the state-of-the-art. Inf. Fusion **14**(1), 28–44 (2013)
5. Pan, H., Liang, Z.-P., Anastasio, T.J., Huang, T.S.: Hybrid NN-Bayesian architecture for information fusion. In: Proceedings of the International Conference on Image Processing (ICIP 1998), pp. 368–371, October 1998
6. Cou´e, C., Fraichard, T., Bessi`ere, P., Mazer, E.: Multi-sensordata fusion using Bayesian programming: an automotive application. In: Proceedings of the IEEE/RSJ International Conference on Intelligent Robots and Systems, pp. 141–146, October 2002

7. Hall, D.L., Llinas, J.: Handbook of Multisensor Data Fusion, pp. 487–488. CRC Press, Boca Raton (2001)
8. Escamilla-Ambrosio, P.J., Mort, N.: Hybrid Kalman filter-fuzzy logic adaptive multisensor data fusion architectures. In: Proceedings of the IEEE Conference on Decision and Control, pp. 5215–5220 (2003)
9. Sasiadek, Z., Hartana, P.: Sensor data fusion using Kalman filter. In: Proceedings of the International Conference on Information Fusion, pp. WeD5/19–WeD5/25 (2000)
10. Yen, J.: Generalizing the Dempster-Shafer theory to fuzzy sets. IEEE Trans. SMC **20**(3), 559–570 (1990)
11. Zhu, H., Basir, O.: A novel fuzzy evidential reasoning paradigm for data fusion with applications in image processing. Soft Comput. J. Fusion Found. Methodol. Appl. **10**, 1169–1180 (2006)
12. Xu, D.L., et al.: Inference and learning methodology of belief-rule-based expert system for pipeline leak detection. Expert Syst. Appl. **32**, 103–113 (2007)
13. Wu, M., Lu, T.J., Ling, F.Y., Sun, J., Du, H.Y.: Research on the architecture of Internet of Things. In: Advanced Computer Theory and Engineering (ICACTE), V5–484, August 2010
14. Vermesan, O., Friess, P., Guillemin, P., Gusmeroli, S., Sundmaeker, H., Bassi, A., Jubert, I. S., Mazura, M., Harrison, M., Eisenhauer, M., Doody, P.: Internet of Things strategic research agenda. In: Vermensan, O., Friess, P. (eds.) Internet of Things: Global Technological and Societal Trends, pp. 36–41. River Publishers, Netherlands (2011)
15. Miorandi, D., Sicari, S., De Pellegrini, F., Chlamtac, I.: Internet of Things: vision, applications and research challenges. Ad Hoc Netw. **10**(7), 1497–1516 (2012)

Proposed System on Gesture Controlled Holographic Projection Using Leap Motion

Varad Pathak[1](✉), Farhat Jahan[2], and Pranav Fruitwala[2]

[1] Ahmedabad Institute of Technology, Ahmedabad 380060, India
varad.iosmail@gmail.com
[2] Institute of Technology, Nirma University, Ahmedabad 382481, India
{14mcen08,14mcen06}@nirmauni.ac.in

Abstract. Holography is the science and practice of making holograms. Holograms currently have a very wide scope for development and is going through various changes at a very fast pace. Due to its multi-faced and multi-dimensional nature many tech pioneer companies are trying to adopt and develop it for various aspects. Since holograms are 3-Dimensional images or stream of images it creates an illusion of depth for the content it is displaying. With holograms one can create rich and immersive content which takes the user experience to a whole new level. This proposed system can be used by common people to view images and videos as mid-air holograms at home. The new implementation includes interaction with the holograms using gestures via leap motion technology. The users could use hand gestures to change, rotate, zoom in or zoom out the images or play, pause, go forward or go backward in a video.

Keywords: Mid-air display · 3D display · Holography · Interactive display · Leap motion control

1 Introduction

I.E. Sutherland quoted in his article, *The ultimate display would, of course, be a room within which the computer can control the existence of matter. A chair displayed in such a room would be good enough to sit in. Handcus displayed in such a room would be conning, and a bullet displayed in such a room would be fatal. With appropriate programming such a display could literally be the Wonderland into which Alice walked* [1].

Holograms are 3D images projected in such a way that a viewer gets a feeling of seeing a live object. This can been widely used at presentations, seminars and places where 3D objects are presented to an audience [6]. Many methods have been developed for projecting holograms [8, 9]. Gestures turn user intention into action. We used a leap motion device to interact with our 3D holograms [11], action gets detected when the user's hand is being traced over the sensor of the device which is the most natural way to interact with a hologram. It detects your hand, finger movements and gestures by controlling the pseudo-hologram. In order to control the pseudo-hologram using leap motion one needs to implement alpha SDK in his mobile application (currently only available for android).

S.C. Satapathy and A. Joshi (eds.), *Information and Communication Technology for Intelligent Systems (ICTIS 2017) - Volume 1*, Smart Innovation, Systems and Technologies 83, DOI 10.1007/978-3-319-63673-3_63

2 Related Work

2.1 Holographic Projection Screen

The projection screen is basically a screen for flight simulation where the observer will occupy a fix and defined position within the simulation cockpit and the cockpit is surrounded by the screen. Appropriate images are projected using the suitable projection system onto the screen in accordance with simulation. The figure shows an example where the screen is constructed in accordance without the cockpit. The observer is seated on the chair and is located at a fixed position with respect to projection system. Image is formed using the image projector [4] which is a part of the projection system and the image is to be viewed on the screen. Spherical shaped screen is used and has it center at position between projector and observer. As from the figure, the head of observer should be in particular to observer's eyes should remain in observation pupil of projection screen. The system which is shown in figure is designed in such a manner that the observer remains seated. The observation pupil can be considered as fixed both to the position and size and we can define a circle of 12″ diameter in vertical plane in region of observer's head (Figs. 1 and 2).

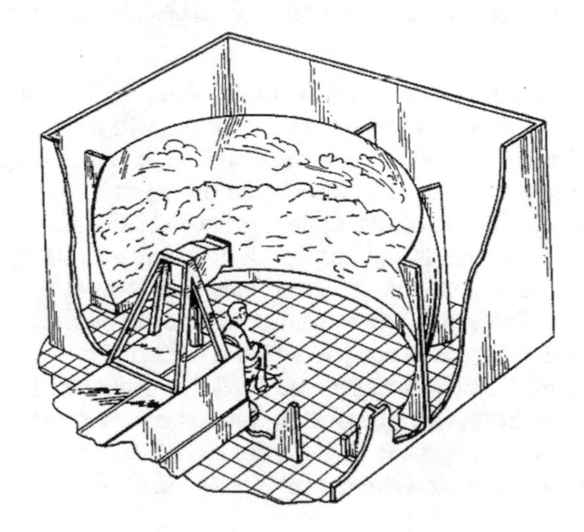

Fig. 1. Holographic projection screen [3]

2.2 Hologram Incorporating a Plane with a Projected Image

A composite display which includes static three dimensional holographic image and also includes static or/and dynamic two dimensional images. A three dimensional silhouette image of the object image is included in a silhouette hologram. A holographic diffusion screen surrounding the silhouette image can also be included in a three dimensional silhouette image. Object hologram is illuminated firstly by a light source and the silhouette hologram is illuminated by the second light source. The background for viewing the object image is provided by silhouette image. Projection of

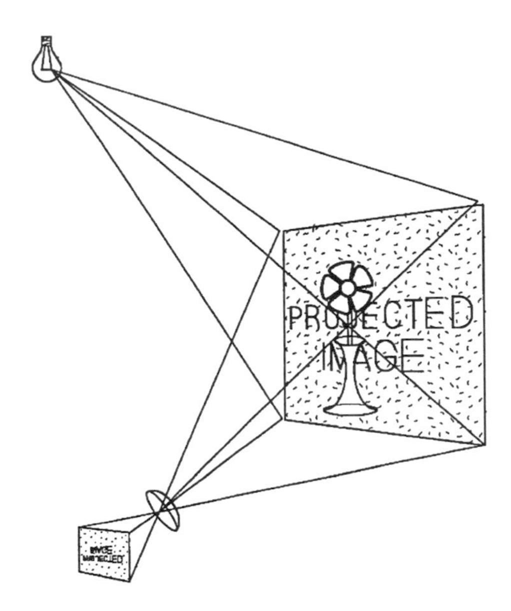

Fig. 2. 3D Silhouette hologram, projected image and the object hologram [4]

static or dynamic images on the diffusion screen whenever it is included in the silhouette hologram and due to this the composite image will include static or dynamic two dimensional imaginary which is combined with static object image. An occlusion depth cue for the object hologram is maintained by the three dimensional silhouette hologram. One feature for the current system includes black, 3D silhouette hologram which absorbs the light from the second source and acts as a background for object hologram. The above system has got many advantages over the other alternative technologies. Alternative aerial images cant provide occlusion depth without having physical objects and video screen resident which is there in the current system. Images that are generated by the current system are highly scalable which is because of the tilling technique. Another advantage is that the image can be reconstructed with little or no distortion over a wide viewing angle. Also, it is possible to combine multiple dynamic or static images to generate the composite image (Figs. 3 and 4).

Fig. 3. Three different viewpoints of example of three-dimensional silhouette hologram [4]

Fig. 4. Three different viewpoints of an example of an object hologram on a transparent substrate [4]

3 Proposed System

3.1 Building Blocks

The various building blocks [2] used for this proposed system are as follows:

- **Pyramid:** This component represents the medium where the actual hologram will be formed in 3D. The pyramid can either be made of glass sheets of high clarity or with some highly reflective plastic sheets.
- **Phone/Tablet:** The phone/tablet is a primary requirement for displaying the rendered images or stream of images that will be projected into the pyramid.
- **Display Screen:** The display screen is an optional component as we can directly position the pyramid on the phone or the tablet itself. But, it is possible to display the images on a display screen using some sort of screen casting/mirroring technology.
- **Leap Motion:** The leap motion [7] is the module which will be used to control the images or stream of images by the user. He can use pre-defined hand gestures to interact with the hologram.
- **Mobile Application:** The mobile application will help render the images/stream of images as required for the holographic projection. This application needs to be installed on the above described mobile/tablet device.
- **WiFi:** The WiFi is required when one uses the display screen to stream the images wirelessly for holographic projection.

3.2 Proposed Holographic Projection Using a Tablet or Mobile Equipped with 2D to 3D Conversion Application and Controlling It Using Leap Motion

The proposed system can be explained in the Fig. 5. The mobile phone or tablet will use the application for rendering the images or stream of images. Once the application is live and working, the pyramid will be placed on the top of the screen of the mobile device exactly at the center of it. The application will convert the images from 2D to 3D before projecting them into the pyramid. The leap motion will be connected with mobile phone or tablet so the user can interact with the hologram using hand gestures.

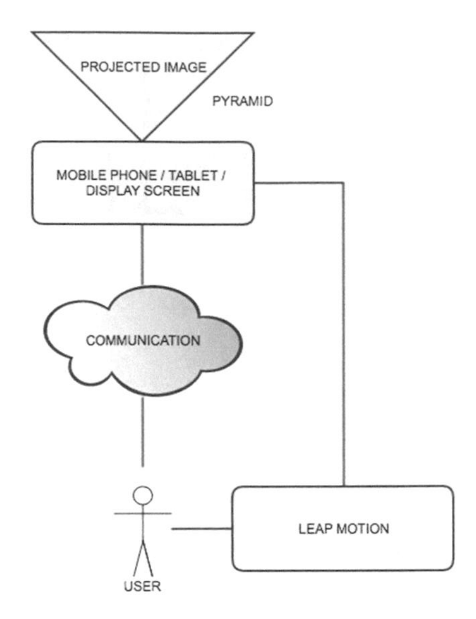

Fig. 5. User interacting with the hologram using leap motion which will be connected to a mobile application

3.3 Proposed Holographic Projection Using a Tablet or Mobile Equipped with 2D to 3D Conversion Application and Controlling It Using Leap Motion and Wirelessly Connected to a Display Screen

If the user opts for the optional display screen then the setup would be a little different than the previous one. The pyramid will have to be place on the top of the display screen exactly in the center of it. After that the Mobile phone or the table will have to be connected to the WiFi in order to mirror the screen wirelessly to the display screen to project the rendered images. The leap motion will be connected to the mobile or tablet device. This option gives the user the option of a little more mobility than the previous one as the user can now freely move around the room or space without worrying about the incorrect displacement of the pyramid while moving as it will be place on a separate display screen and wirelessly connected to the mobile device (Fig. 6).

3.4 Feasibility of the Proposed System

The feasibility study of each component in the proposed system is proposed here. The pyramid can be made of both glass sheets and highly reflective plastic sheets, but its recommended to use glass sheet. The glass sheets cost between INR 650 to INR 1,500. Also, the duration required to make the glass pyramid is not so much. The display screen is an optional component since the proposed system does not depend on this component completely. But, if the user wants to use the display then the minimum cost of display is between INR 25,000 to INR 30,000. Also, the display screen should support screen casting/mirroring feature over Wireless LAN. The mobile phone/tablet

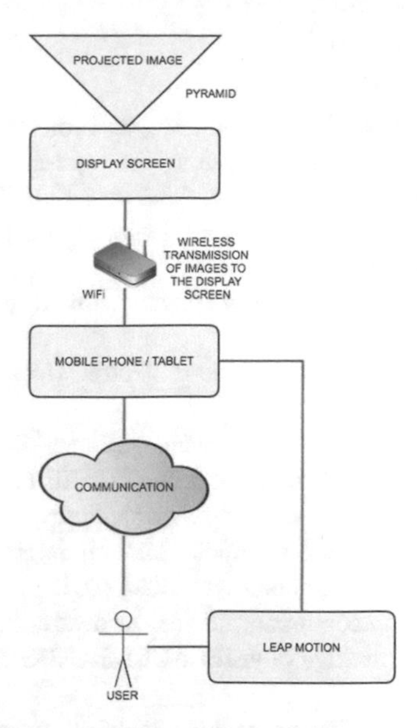

Fig. 6. User interacting with the hologram using Leap motion which will be connected to a mobile Application, which further will be wirelessly connected to a larger display screen

should be capable enough to run the application made for converting the 2D image or videos to 3D. A mobile phone with decent processing power and RAM can cost from INR 6,000 to INR 10,000. Whereas, a tablet for same specification can cost from INR 10,000 to INR 14,000. The use of WiFi depends on whether the user is using a display screen or not, since it is used to cast or mirror the phone/tablet screen using wireless network. For a WiFi, one needs a WiFi router of minimum 2.4 GHz of Frequency with IEEE 802.11 a/b/g/n capabilities. The minimum cost of a decent WiFi router is INR 800 and can range upto INR 10,000. A mobile application needs to be developed for the images and videos to be converted from 2D to 3D and projected to the pyramid. A basic application will take around 2 to 4 weeks. The Leap Motion component is an available device which costs approximately INR 5,000. In order to implement leap motion one needs to implement a SDK named Alpha SDK [8] in the application provided by leap motion itself. To get the access of the alpha SDK, a request needs to be made to the leap motion developers.

4 Conclusion

Holography is a very promising technology and the speed by which it is advancing it will sure make a place for itself in the tech market. But, the cost of implementation can vary and may be high. Our proposed system is developed by keeping the common man in focus so that it is cost effective and provides a smart way to interact with the holograms itself. The users can choose the components needed by the system according to their budget so even a person with limited budget can set up a 3D Hologram system at his house.

References

1. Sutherland, I.E.: The Ultimate Display, Multimedia: From Wagner to virtual reality (1965)
2. Fruitwala, P., Sharma, P.: Proposed system for mid-air holography projection using conversion of 2D to 3D visualization. In. J. Adv. Res. Eng. Technol. **7**, 159–167 (2016)
3. Burns, R.H., Hall, W.M., Hildebrand, B.P.: Holographic projection screen. U.S. Patent No. 4,500,163, 19 February 1985
4. Klug, M.A.: Hologram incorporating a plane with a projected image. U.S. Patent No. 6,323,971, 27 November 2001
5. Montagnino, J.G.: Scale with a holographic projector. U.S. Patent No. 6,541,714, 1 April 2003
6. Rakkolainen, I.: How feasible are star wars mid-air displays. In: 11th International Conference on Information Visualization, IV 2007. IEEE (2007)
7. Potter, L.E., Araullo, J., Carter, L.: The leap motion controller: a view on sign language. In: Proceedings of the 25th Australian Computer-Human Interaction Conference: Augmentation, Application, Innovation, Collaboration. ACM (2013)
8. https://community.leapmotion.com/t/android-sdk-hardware-requirements/2845
9. Hoshi, T., et al.: Touchable holography. In: ACM SIGGRAPH 2009 Emerging Technologies. ACM (2009)
10. Lee, J., Post, R., Ishii, H.: ZeroN: mid-air tangible interaction enabled by computer controlled magnetic levitation. In: Proceedings of the 24th Annual ACM Symposium on User Interface Software and Technology. ACM (2011)
11. www.integraf.com/a-Types_of_Holograms.htm

Mobile Cloud Forensic: Legal Implications and Counter Measures

Puneet Sharma[1(✉)], Deepak Arora[1], and T. Sakthivel[2]

[1] Department of Computer Science and Engineering, Amity University, Uttar Pradesh, India
puneetgrandmaster@gmail.com, deepakarorainbox@gmail.com

[2] Firstsoft Technologies Private Ltd., Chennai, India
sakthi@firstsoftech.com

Abstract. In recent years, the smartphone has become a powerful miniature computer owing to its computing power, portability, and flexibility. The integration of smartphone and cloud computing technology offers significant benefits and expand the capabilities of smartphones to support applications that demand high computational resources. However, the proliferation of popular technologies increases potential opportunity for misuse and illegal activities. Thus, the advancement of forensic research is inevitable to cope with the recent developments in mobile cloud platforms. The adoption of forensic methodologies to the mobile cloud investigation is still in its infancy stage. Mobile Cloud Computing (MCC) offers unique challenges compared to traditional digital forensics due to its ubiquitous nature. In this research work authors have emphasized on various solutions available for mobile cloud forensics. Authors have analyzed and discussed the scope and possibilities regarding legal implications towards forensic analysis in MCC, potential challenges, and its countermeasures.

Keywords: Mobile Cloud Computing · Mobile cloud forensic · Legal implications

1 Introduction

1.1 Mobile Cloud Computing

In the last decade, smartphones are radically emerging as a powerful handheld device, and consequently, the usage of mobile applications has significantly increased in commercial and business world. The dominant trend of mobile computing has created a significant impact on the development of several refined applications which require significant processing and storage competences. The mobile devices have the limited resource capabilities in nature due to its portability. To deal with the limitations of the mobile computing, the mobile devices rely on the cloud to extend the performance of the resource-constrained mobile devices and enable the resource-intensive mobile applications to run on the remote server over the Internet. This process is known as Mobile Cloud Computing (MCC) [1], which is growing rapidly and consequently,

S.C. Satapathy and A. Joshi (eds.), *Information and Communication Technology for Intelligent Systems (ICTIS 2017) - Volume 1*, Smart Innovation, Systems and Technologies 83, DOI 10.1007/978-3-319-63673-3_64

the cloud computing based mobile applications create lots of attention among the people. Cloud computing provides the virtualization technology to the mobile users to easily access the resource-rich cloud services in a pay-per-use manner. The remote server consists of the virtualized resource pool with the unlimited resource capabilities. The battery is one of the major constraints in mobile devices. To provide the extended battery life while executing the resource-intensive mobile applications, computation offloading is a promising method that enables resource-abundant cloud computing environment for execution [2]. MCC offers meaningful solutions for the mobile users and provides potential applications under different areas such as natural language processing, image processing, games, social networking, utilities, sensor data applications, productivity tools, and crowd computing [3].

Digital forensic methods are becoming obsolete due to the technology advancement of recent applications [4]. The currently available forensic tools are incapable of doing the forensic analysis in a forensically sound manner and unlikely to cope up with the new technologies and applications. The massive use of cloud leads to several dangerous threats to the mobile cloud environment. The ease of access to cloud computing facilitates cyber criminals to launch illegal activities in the cloud environment [5]. Even though it offers benefits to the mobile users, the malicious individuals utilize the rapidly evolving MCC technology to share the illegal information, broadcast the terrorist ideology, and facilitate the communication to perform the attack on digital information of other users. Hence, the mobile cloud forensics [6] is crucial in performing the forensic investigation in smartphone applications such as mobile credit billing, location aware commerce, money transaction, and payments. Moreover, there is a possibility to apply legal implications on forensic investigator while applying the forensic methodologies on both the mobile device and the cloud computing environment. The legal challenges related to the multi-jurisdiction and multi-tenancy are significant issues in a cloud platform. Hence, the key challenge is to follow well-established procedures and methods to ensure the chain of custody. Since most of the forensic challenges rely on the legal implications, this research work explores further legal issues, while performing the forensic investigation in a mobile cloud environment.

1.2 Mobile Cloud Forensics

MCC [7] offers ample storage and processing capabilities, and connectivity options to the mobile users, which leads the possibility to abuse the digital crimes. The rapid evolution of the mobile devices not only increases and diversifies the data during analysis but also complicates the investigation process [8]. The investigation process becomes more arduous task after inclusion of cloud, due to issues related to storage, accessing, and managing the information. Traditional digital forensics faces limitations in MCC environment due to its terminals proliferation, lack of access to physical infrastructure, lack of investigation tools for large virtualized and distributed systems, and loss of control over data. Mobile cloud forensics needs to focus on the mobile device, and the cloud server since these two environments execute the partitioned cloud-based mobile application. Consequently, to attain the effective forensics results, the two levels of investigation, including mobile level and cloud level are essential in the MCC environment [6], which includes Mobile Client, Network and Cloud Server as major components. The Mobile Client will handle mobile device forensics, which is the process of recovering the digital evidence from a mobile device by applying the forensically sound rules. The mobile device forensics investigation system extracts the evidence from the internal memory, external memory card, and Subscriber Identity Module (SIM) of the mobile device. The gathered information includes internet browsing history, call or SMS history, multimedia files, and emails [9]. The network service provider is likely to provide the evidence related to historical call records, user localization, and message information over the time. The Cloud Server will be responsible for cloud forensics, which can be defined as an application of digital forensics in the cloud environment, which is a hybrid forensic approach to generate the digital evidence. Cloud data acquisition implies two major issues such as multi-jurisdiction and multi-tenancy.

1.2.1 Scope and Motivation of Mobile Cloud Forensics

In 2010, the Federal Bureau of Investigation (FBI) arrested 202 criminals, including the World's top five cyber criminals for cybercrime intrusions. Most of the cyber threats involving the private sectors, financial systems, and intellectual property. For instance, Africa suffers 10% of global cybercrime activities in which Nigeria, Kenya, and South Africa are the hubs of malicious activity. Cyber-attacks approximately cost the country by U.S. $481 million every year in Africa [10]. Many investigators employ the diversity of devices and operating systems. It is difficult to share the evidence easily or to maintain the unified data across the investigators. For this case, the mobile cloud forensics enables the collaborative investigation anywhere at any time. Mobile cloud forensics supports dynamically changing operating systems and file formats by averting the cost of developing and maintaining the traditional forensic tools for the dynamic environment.

In MCC environment, the forensic investigation is crucial in handheld mobile devices as well as the cloud computing environment. Malicious individuals may illegally access the private information such as financial records, credit history, and other records of the mobile cloud users to commit further crime activities. The advantage of cloud flexibility, the terrorists and criminals may utilize the massive storage capacity of the cloud environment to perform the illegal activities. Hence, there is an essential requirement to extract the stored or accessed data in the cloud with the assistance of Cloud Service Providers (CSPs) [5]. With the knowledge of legal consideration, the forensic investigator needs to collect the evidence after suspecting the crime incident and ensure a chain of custody. Then, identifying the corresponding cloud provider is necessary to determine the applicable jurisdiction and law to perform the seamless forensic investigation on the cloud data. If the evidence is scattered over multiple servers, identifying and utilizing the corresponding jurisdiction is a challenging task when investigating the evidence in the cloud storage [11]. Thus, law enforcement department requires an effective forensic investigation system to deal with the latest technologies of mobile cloud computing.

2 Forensic Investigation Procedure

The digital forensic investigation involves the process of the identifying, preserving, analyzing, and presenting the admissible evidence in a court of law. It traces the human and computer-generated activities to gather the possible evidence of the crime scene [12]. Forensic investigation is categorized into four steps, including identification and preservation, acquisition, examination, and analysis, and reporting [13]. Digital evidence is extremely fragile, which is easily damaged or altered when the occurrence of improper handling and examination. In the case of failure occurs in the forensic investigation process, the evidence is inadmissible in court [14]. Hence, proper handling of proof is crucial during the forensic investigation. Table 1 shows the forensic procedures in MCC environment.

Table 1. Forensic procedures in MCC environment

Forensic phases	Mobile Cloud Computing	
	Mobile device	Cloud server
Identification and preservation	The forensic investigator needs to identify the source of evidence and focus on the state of the Smartphone	Identifies the cloud provider related to the storage of crime event
Acquisition	The acquisition phase requires the connection between the mobile device and the forensic workstation, a mobile device in an active state	The investigator needs to acquire the data from the processed VMs with the help of the CSPs, which may lead the multi-tenancy and multi-jurisdiction issues
Examination and analysis	The investigator examines the evidential data acquired from both the mobile device the cloud server	
Reporting	It is the process of maintaining and submitting the evidential records of all the activities and observations related to the crime scenes and conclusions	

3 Major Challenges in Mobile Cloud Forensics

Mobile cloud forensic investigator meets many challenges when dealing with the mobile cloud environment, which is associated with the distributed cloud infrastructure and the mobile networking specificity. Since, the mobile device stores and transmits the personal and corporate information, and performs the online transactions using the cloud platform. Cloud computing enables the mobile users to operate the mobile applications in the cloud server, which leads the malicious activities and meets multi-jurisdiction and multi-tenancy issues.

3.1 Forensic Challenges Related to Mobile Device

Mobile devices often deal with the resource scarcity issues while executing the resource-intensive applications. Moreover, it meets the challenges when accessing the mobile cloud applications involving social network applications [15–17], and storage applications [18–20]. As a client, the mobile devices comprise the part of the information about the personalized applications, which is likely to lead the privacy issues during the investigation [21]. The forensic investigator faces the inconvenience during mobile device investigation due to the lack of internal memory analysis of the Smartphones and mobile device specific standard method for data extraction [22].

According to the forensic procedure, collecting the evidence from the mobile device is a complicated task due to the hibernation behavior of the mobile device, a variety of operating systems, residing data in either volatile memory or hard disk drives. In addition, the available traditional forensic tools lack in providing the solutions to deal with the physical damage of the mobile devices. Hence, physical acquisition merely recovers the deleted data on the mobile device. The data recovery of the mobile device relies on the absence of SIM card or the use of another SIM card. Mobile Service Provider (MSP) based Mobile Number Portability (MNP) consideration is another way of identifying the mobile user when a mobile subscriber retains the same mobile number while changing the mobile operator. In mobile device investigation, the most complex challenge is the identification of the user when there is the possibility of changing the International Mobile Equipment Identity (IMEI) number by exploiting the flashing tools.

3.2 Forensic Challenges Related to the Cloud

In the cloud server, CSP grants access to collect the data from the server if the investigator has the contract or Service Level Agreement (SLA) to access the server. Otherwise, the investigator attempts to compel the provider via subpoenas, legal wrangling, and other specific tools to provide the desired information. Moreover, substantial research investment is essential to develop the sophisticated forensic tools and technologies for the mobile cloud environment. The cloud infrastructure contains the abundant data storage in a secure manner with strong SLAs, which lacks in disclosing the cloud users' data to any third party. Hence, exploiting the cloud data for the forensic investigation is the critical task [23].

The forensic challenges in the cloud mostly rely on the virtualization and distributed architecture, especially in forensic data assortment, log analysis and maintenance of chain of custody mainly due to the deficiency of cloud-based forensic tools. The utmost imperative step in the forensic examination practice is a data acquisition if any error takes place throughout data collection; it creates the data veracity and realism issue. It involves stages such as data identification, access, collection, and preservation. The physical inaccessibility of the cloud data makes the data collection is an arduous task since the vibrant flora of the cloud computing systems hardens the process of identifying the precise locality of the data at a precise time. The traditional data acquisition tools employ the physical access methods to perform evidence acquisition. Moreover, the forensic investigators have less control over the cloud data in which the data control entirely depends on the cloud service model. This distributed loose control creates the difficulties in reconstructing the crime scene and establishing the timeline [24] based on the sequence of crime events [25]. In a virtualized environment, the possibility of easily destroying data is high when the user turns off the VM after completing the required process. Hence, the data collection depends on the data volatility in which the investigator provides the first preference to collect the high volatile data such as temporary files and registry entries. Forensic investigator relies on the CSP to collect the data from the cloud, which is a crucial issue during the investigation. Accordingly, the data collection process needs to maintain a standard methodology which preserves the data integrity without compromising the confidentiality of other tenants while sharing the same VM resource.

4 Legal Implications of Forensics in MCC

Traditional digital forensic tools have been upgraded to provide the basic functions of digital media examination to the investigators in which examination involves disk imaging, data search, and analysis, and reporting [26]. The major legal issues in the mobile cloud environment engage in the lack of SLAs, privacy, ethics, and multi-jurisdiction [27]. Several characteristics are essential to perform the legally assured forensic investigation in the mobile cloud environment. The characteristics are reliability, authenticity, admissible, and believable related to the investigated crime scene. The forensic investigator often meets the legal challenges involve in obtaining search authority from the MSP and CSP, maintaining a chain of custody, and imaging the data stored in the cloud server along with the multi-jurisdiction and multi-tenancy issues. In the mobile forensics investigation ascertaining of alleged mobile customer is a convoluted task, which requires the corresponding MSP who only consists the proper identification of the subscriber.

- To handle the different operating systems, new physical devices, and different file types, the up gradation of tools is crucial in the forensic investigation.
- Acquiring the cloud data may elapse the time gap when the data is stored in another legal jurisdiction.
- Lack of a legal consideration enables the ambiguous interpretations in the area of legislative regulation. To access the data from the cloud, the investigators are required to register with the Government Gateway with the support of CSPs.
- SLA is the essential requirement for the cloud users and the providers to enable the forensic investigator. In SLA, mobile cloud users ensure that they allow the forensic investigator to access their data when the occurrence of a security incident, data breach, and intrusion.
- Loss of control over the cloud data widely hinders the crime scene reconstruction due to the lack of the knowledge of data storage location in the cloud.
- Jurisdictional issues affect the investigation procedure by introducing the delay in accessing cloud data
- The unavailability of the decryption keys prevents the forensic investigator from analyzing the acquired data from the cloud server.
- Several other shortcomings affect the forensic investigation in collecting the registry entries, temporary memory, and files, and retrieving the loss of relevant metadata such as file creation, accessing, and modification times when the investigator downloads the data from the cloud.

- Although several cloud service providers such as Amazon S3 provide the data with the authentication to the investigators, however, many of them exploit their self-authentication.
- Law practitioners like lawyers and judges and Law Enforcement Agencies (LEAs) need to have the knowledge of preserving the original data according to the new trending technology due to the advantage of utilizing the massive cloud service by the government and corporate levels.
- Lack of physical access during the investigation causes the enormous legal disruptive challenges as the nature of data processing in the distributed cloud environment.
- Ensuring that the acquired evidence belongs to the real suspect when collecting the evidence from the shared storage.
- The validity of the warrant also makes the legal implications when establishing the location of the corresponding suspect in the search warrant to collect the evidence from the cloud.
- An information security system requires proper forensic procedures with the support of legal action in terms of submitting the acquired evidence to the court with the chain of custody.

5 Feasible Solutions Towards Legal Implications on MCC Forensics

Through massive literature survey it is found that the today's forensic tools require additional features to meet the advances and the fundamental changes while investigating the mobile cloud environment.

5.1 Counter-Measures in Mobile Perspective

Mobile device forensic tool can be developed with a focus on retrieving the evidence from the mobile devices based on the hashing techniques. The mobile forensic applications are likely to perform the identification and validation process of the hash values on the specific data. In recent years the development can be seen on open source platforms as majority of the forensic research focued on the automatic data collection method in the Android forensics also [28]. Through available literature work different authors and researchers have proposed various approaches for mobile forensic which can be categorized as follows:

Metadata Analysis: Mobile forensic approach [29] explores the metadata stored on the graphical images which are captured by camera mobiles. The graphical images contain the metadata of Exif information such as timestamp information, camera model, and so on. If the metadata is associated with the consistent graphical data of the camera mobiles, the Exif information is beneficial to the investigation along with the hashing.

SD-Card Data Analysis: Later, the mobile forensic research quickly extracts the data from a Smartphone by exploiting the Android Application Programming Interface (API) on a Secure Digital Card (SD Card) which is located a specific application. However, it requires the root privileges. An open source Android forensic method employs a specialized SD Card which is expandable on the retrieved datasets. A cloud computing based Android forensic system collects the data from the SD Cards, which is the substitution of cloud computing. However, it requires the physical access to collect the data from a device and lacks in continuously acquiring the data. A native Android application method performs real-time monitoring on a Smartphone by streaming a screen of the mobile device, regardless of the root privileges in which monitoring is similar to the Virtual Network Computing (VNC). Proactive Object Fingerprinting and Storage (PROOFS) continuously perform forensic acquisition over a network in a proactive manner using a fingerprinting approach, which utilizes forensic acquisition and monitoring system [30].

Mobile Forensic Tools: To capture the information from the Android Smartphones, most of the Android forensic analysis [31] introduces a variety of tools including handheld hardware devices and software products. However, all the forensic tools require the physical access through a Universal Serial Bus (USB) connection. A forensic acquisition methodology [32] presents a tool set to acquire and analyze the volatile physical memory of the Android mobile devices. It addresses the difficulties while developing the device-independent forensic acquisition tools.

Imaging and Evidence Recovery Method: Android Forensics [33] simplifies the mobile device forensics by obtaining an image of the device memory and requires the root access. An acquisition methodology [34] employs the specialized forensic acquisition software based on overwriting the recovery partition on the SD card of the Android device to collect the data from the Smartphones. An approach [35] discusses the forensic analysis of three widely used social networking applications in the Smartphones in which the applications are Facebook, Twitter, and MySpace. It acquires the forensically sound logical image of each mobile device and analyzes the acquired data by conducting the activities of social network applications on the mobile devices.

5.2 Counter-Measures in Cloud Perspective

Cloud forensic technologies and approaches are still in its infancy stage, which contributes the solutions in different progress. The cloud forensic investigation methods enable the system to recreate the cloud service requests, renovate the unauthorized access, and search the potential evidence from the cloud server [36, 37]. It also monitors the data migration process to gather the available evidence while migrating, the manual data to the cloud storage. Presenting the evidence to the court is challenging task when the data is collected from the cloud environment [38]. Moreover, forensic examiner employs a pattern matching method and more intelligent statistical analysis tools to trace the actions and ensure the data integrity. In a mobile cloud environment, mostly a web browser is the main source of evidence for the suspected mobile device. Hence, the forensic investigator collects the information from a web browser.

Various methods and approaches from different researchers towards cloud forensics can be summarized as under.

Log Forensic Analysis: Log forensic analysis [39] tackles the difficulties in collecting and verifying the digital evidence stored in the browsing history caches in the cloud environment. Cloud application logging for forensics [40] collects the cloud log data based on the cloud application. Jurisdictional issues lead the data loss and modification of data by the malicious individuals. Thus, there is an essential requirement of identification and preservation of digital evidence promptly.

Cloud Forensic Tools: Acquiring the data from the cloud is difficult when migrating the data from one location to another and quickly disappearing the data after migration. Hence, there is an immediate requirement of developing the standard forensic tools and technologies to gather the volatile data also from the cloud server [41, 42]. As similar to the computer forensics, the mobile cloud forensic investigator deals with conventional tools such as Helix, Encase, FROST [43], and FTK tools to perform the investigation process from the data acquisition to the submission of the final report related to the particular crime scene. Securing forensic-rich data examination depends on the Encase Servlet or FTK agent to collect the data from Amazon EC2 IaaS model [44].

Due to the legal implications, the cloud service provider may not provide the response in time to the forensic investigator [45]. However, the current forensic tools and technologies are not forensically sound mechanisms for thc mobile cloud environment. Also, the existing law and governance models do not always support the emerging developments. Hence, the advancement of mobile cloud technology requires the new development of forensic tools and technologies to acquire the more forensically sound evidence from the mobile device and cloud environment without violating the legal considerations.

6 Conclusion

The seamless integration of mobile and cloud computing technology offers more potential applications to the mobile users. The increasing popularity of the mobile cloud technology also motivates the illegal activities. This research work presented a survey of existing forensic technologies, and procedures to resolve the investigation issues in mobile cloud computing. The comprehensive review and detailed analysis performed in this work significantly addresses the impact of forensic investigation in a mobile cloud platform. This work also presents the legal implications of forensic investigation, its existing solutions in the mobile device and the cloud environment while applying the forensic techniques in the mobile cloud environment.

References

1. Fernando, N., Loke, S.W., Rahayu, W.: Mobile cloud computing: a survey. Future Gener. Comput. Syst. **29**(1), 84–106 (2013)
2. Khan, A.U.R., Othman, M., Madani, S.A., Khan, S.U.: A survey of mobile cloud computing application models. IEEE Commun. Surv. Tutor. **16**(1), 393–413 (2013)
3. Wang, Y., Chen, I.-R., Wang, D.-C.: A survey of mobile cloud computing applications: perspectives and challenges. Wireless Pers. Commun. **80**(4), 1607–1623 (2015)
4. Garfinkel, S.L.: Digital forensics research: the next 10 years. Digital Investig. **7**, S64–S73 (2010)
5. Sibiya, G., Venter, H. S., Thomas F.: Digital forensic framework for a cloud environment. In: IST-Africa Conference Proceedings on International Information Management Corporation (IIMC), pp. 1–8 (2012)
6. Lee, J., Hong, D.: Pervasive forensic analysis based on mobile cloud computing. In: 3rd International Conference on Multimedia Information Networking and Security. IEEE computer society, pp. 572–576 (2011)
7. Dinh, H.T., Lee, C., Niyato, D., Wang, P.: A survey of mobile cloud computing: architecture, applications, and approaches. Wirel. Commun. Mobile Comput. **13**(18), 1587–1611 (2013)
8. Samet, N., Letafa A.B., Hamdi, M., Tabbane, S.: Forensic investigation in mobile cloud environment. In: IEEE International Symposium on Networks, Computers and Communications, pp. 1–5 (2014)
9. Barmpatsalou, K., Damopoulos, D., Kambourakis, G., Katos, V.: A critical review of 7 years of mobile device forensics. Elsevier Digital Investig. **10**(4), 323–349 (2013)
10. Harkness, T., Amole, A., Holland, E.: The growing threat of cyber attacks in Africa. World Data Protect. Rep. **15**(5) (2015)
11. Quick, D., Choo, K.: Forensic collection of cloud storage data: does the act of collection result in changes to the data or its metadata? Elsevier Digital Investig. **10**(3), 266–277 (2013)
12. Wolthusen, S.D.: Overcast: forensic discovery in Cloud environments. In: IEEE Fifth International Conference on IT Security Incident Management and IT Forensics, pp. 3–9 (2009)
13. Hegarty, R., Merabti, M., Shi, Q., Askwith, B.: Forensic analysis of distributed service oriented computing platforms. In: 12th Annual Post-Graduate Symposium on the Convergence of Telecommunications, Networking and Broadcasting, pp. 27–37 (2011)
14. Biggs, S., Vidalis, S.: Cloud computing: the impact on digital forensic investigations. In: IEEE Proceeding of International Conference for International Technology and Secured Transaction, pp. 1–6 (2009)
15. Lin, F.Y., Huang, C.C., Chang, P.Y.: A cloud-based forensics tracking scheme for online social network clients. Elsevier Forens. Sci. Int. **255**, 64–71 (2015)
16. Al Mutawa, N., Baggili, I., Marrington, A.: Forensic analysis of social networking applications on mobile devices. Elsevier Digital Investig. **9**, S24–S33 (2012)
17. Anglano, C.: Forensic analysis of WhatsApp messenger on android smartphones. Elsevier Digital Investig. **11**(3), 201–213 (2014)
18. Hale, J.S.: Amazon cloud drive forensic analysis. Elsevier Digital Investig. **10**(3), 259–265 (2013)
19. Quick, D., Choo, K.K.: Digital droplets: Microsoft SkyDrive forensic data remnants. Elsevier Future Gener. Comput. Syst. **29**(6), 1378–1394 (2013)
20. Quick, D., Choo, K.K.: Google drive: forensic analysis of data remnants. Elsevier J. Netw. Comput. Appl. **40**, 179–193 (2014)

21. Ntantogian, C., Apostolopoulos, D., Marinakis, G., Xenakis, C.: Evaluating the privacy of Android mobile applications under forensic analysis. Elsevier C Secur. **42**, 66–76 (2014)
22. Meng, Z.: Mobile cloud computing: implications to smartphone forensic procedures and methodologies, Master Thesis (2011)
23. Roussev, V., McCulley, S.: Forensic analysis of cloud-native artifacts. Elsevier Digital Investig. **16**, S104–S113 (2016)
24. Battistoni, R., Di Pietro, R., Lombardi, F.: CURE—towards enforcing a reliable timeline for cloud forensics: model, architecture, and experiments, Elsevier Comput. Commun. **91**, 29–43 (2016)
25. Reilly, D., Wren, C., Berry, T.: Cloud computing: forensic challenges for law enforcement. In: IEEE International Conference for Internet Technology and Secured Transactions (ICITST), pp. 1–7 (2010)
26. Martini, B., Choo, K.R.: An integrated conceptual digital forensic framework for cloud computing. Elsevier Digital Investig. **9**(2), 71–80 (2012)
27. Orton, I., Alva, A., Endicott-Popovsky, B.: Legal process and requirements for cloud forensic investigations. Cybercrime and Cloud Forensics: Applications for Investigation Processes (2012)
28. Grover, J.: Android forensics: automated data collection and reporting from a mobile device. Elsevier Digital Investig. **10**, S12–S20 (2013)
29. Sobieraj, S., Mislan, R.: Mobile phones: digital photo metadata (2007)
30. Shields, C., Frieder, O., Maloof, M.: A system for the proactive, continuous, and efficient collection of digital forensic evidence. Elsevier Digital Forens. Res. Workshop **8**, S3–S11 (2011)
31. Saleem, S., Popov, O., Baggili, I.: A method and a case study for the selection of the best available tool for mobile device forensics using decision analysis. Elsevier Digital Investig. **16**, S55–S64 (2016)
32. Sylve, J., Case, A., Marziale, L., Richard, G.G.: Acquisition and analysis of volatile memory from android devices. Elsevier Digital Investig. **8**(3-4), 175–184 (2012)
33. Lessard, J., Kessler, G.C.: Android forensics: simplifying cell phone examinations. Small Scale Digital Device Forens. J. **4**(1), 1 (2010)
34. Vidas, T., Zhang,C., Maloof, M.: Toward a general collection methodology for Android devices. Elsevier, Proceedings of the Eleventh Annual DFRWS Conference on Digital Investigation, 8:S14-S24, 2011
35. Al Mutawa, N., Baggili, I., Marrington, A.: Forensic analysis of social networking applications on mobile devices. In: 12th Annual Digital Forensics Research Conference on Digital Investigation, vol. 9, pp. S24–S33. Elsevier (2012)
36. Taylor, M., Haggerty, J., Gresty, D., Haggerty, R.: Digital evidence in cloud computing systems. Elsevier Digital Investig Comput. Law Secur. Rev. **26**(3), 304–308 (2010)
37. Taylor, M.H., Haggerty, J., Gresty, D., Lamb, D.: Forensic Investigation of cloud computing systems. Elsevier Netw. Secur. **3**, 4–10 (2011)
38. Birk, D.: Technical challenges of forensic investigations in cloud computing environments. In: IEEE 6th International workshop on Systematic Approaches to Digital Forensic Engineering (SADFE), pp. 1–10 (2011)
39. Thorpe, S., Ray, I.: Cloud log forensics metadata analysis. In: IEEE 36th Annual, Computer Software and Applications Conference Workshops (COMPSACW), pp. 194–199 (2012)
40. Marty, R.: Cloud application logging for forensics. In: ACM Symposium on Applied Computing, pp. 178–184 (2011)
41. Ruan, K., Carthy, J., Kechadi, T., Baggili, I.: Cloud forensics definitions and critical criteria for cloud forensic capability: an overview of survey results. Elsevier Digital Investig. **10**(1), 34–43 (2013)

42. Dezfouli, F., Dehghantanha, A., Mahmoud, R., Sani, N., Shamsuddin, S.: Volatile memory acquisition using backup for forensic investigation. In: IEEE International Conference on Cyber Security, Cyber Warfare and Digital Forensic (CyberSec), pp. 186–189 (2012)
43. Dykstra, J., Sherman, A.T.: Design and implementation of FROST: digital forensic tools for the OpenStack cloud computing platform. Elsevier Digital Investig. **10**, S87–S95 (2013)
44. Dykstra, J., Sherman, A.T.: Acquiring forensic evidence from infrastructure-as-a-service cloud computing: exploring and evaluating tools, trust, and techniques. Elsevier Digital Investig. **9**, 90–98 (2012)
45. Wittow, M.H., Buller, D.J.: Cloud computing: Emerging legal issues for access to data, anywhere, anytime. J. Internet law **14**(1), 1–10 (2010)

A Survey on Medical Information Retrieval

Shah Himani and Dattani Vaidehi(✉)

Institute of Technology, Nirma University, Ahmedabad 382481, Gujarat, India
himani.shah711@gmail.com, vaidehi.dattani@gmail.com

Abstract. Medical Science has grown widely with the advancement in technology and research. The retrieval of accurate information regarding medical terminologies has become a major requirement for progress. Everyone will need to access this information starting from layman to the expert doctors. Each one will have their set of terms of accessing same information, therefore certain techniques and methods are required to retrieve precise information. World Wide Web reduces the task of generalizing all information to a single platform. Hence the task remaining is to map the query of the user to the appropriate concept in medical science. This paper surveys on existing methods and tools that help to retrieve accurate information as per the query of the user belonging to any knowledge group.

Keywords: UMLS · Semantic network · Medical concept mapper · Medical information retrieval

1 Introduction

The medical information collected from various sources will be highly unstructured. There are various medical databases, thesaurus, standards and software that contain information of almost all medical terminologies and concepts. The main problem is organizing this huge set of data and finding relevant mapping between these concepts. Even with a structured dataset of huge size either the hardware required to fetch required information should powerful or the user should comply with speed for information retrieval. Therefore, there exists a requirement to classify and organize the information. The major problem associated with medical information structuring is vocabulary, different medical terms can refer to same concept (synonymy) and different users use different terms to seek identical information (polysemy) [3]. Polysemy can be considered as different vocabulary used by layman and a medical expert to refer to a same medical concept. [14] has provided a solution to this problem by generating a huge terminology space by extracting medical concepts from the local records and mapping them to accurate terminology.

In order to gather all the concepts to a common platform the National Library of Medicine has developed UMLS i.e., Unified Medical Language System (UMLS) [1]. It has vocabularies, standards and semantics of medical terms and concepts. It is divided into 4 parts (1) Metathesaurus (2) SPECIALIST Lexicon (3) Semantic Net and (4) Online knowledge sources server. Apart from UMLS, there are various other information sources like EMTREE, WordNet, etc. Data from these sources are extracted

S.C. Satapathy and A. Joshi (eds.), *Information and Communication Technology for Intelligent Systems (ICTIS 2017) - Volume 1*, Smart Innovation, Systems and Technologies 83, DOI 10.1007/978-3-319-63673-3_65

by various algorithms and methods. Therefore, the queries of users get appropriate results via information retrieval systems that adopt these algorithms to get results. Figure 1 is the proposed model that suggests the idea to handle huge amount of data-storage for medical information retrieval on cloud. The data is semantically aggregated using various tools/algorithms from the existing websites/sources and sent over to cloud for further processing. Once the data is on cloud, it utilizes natural

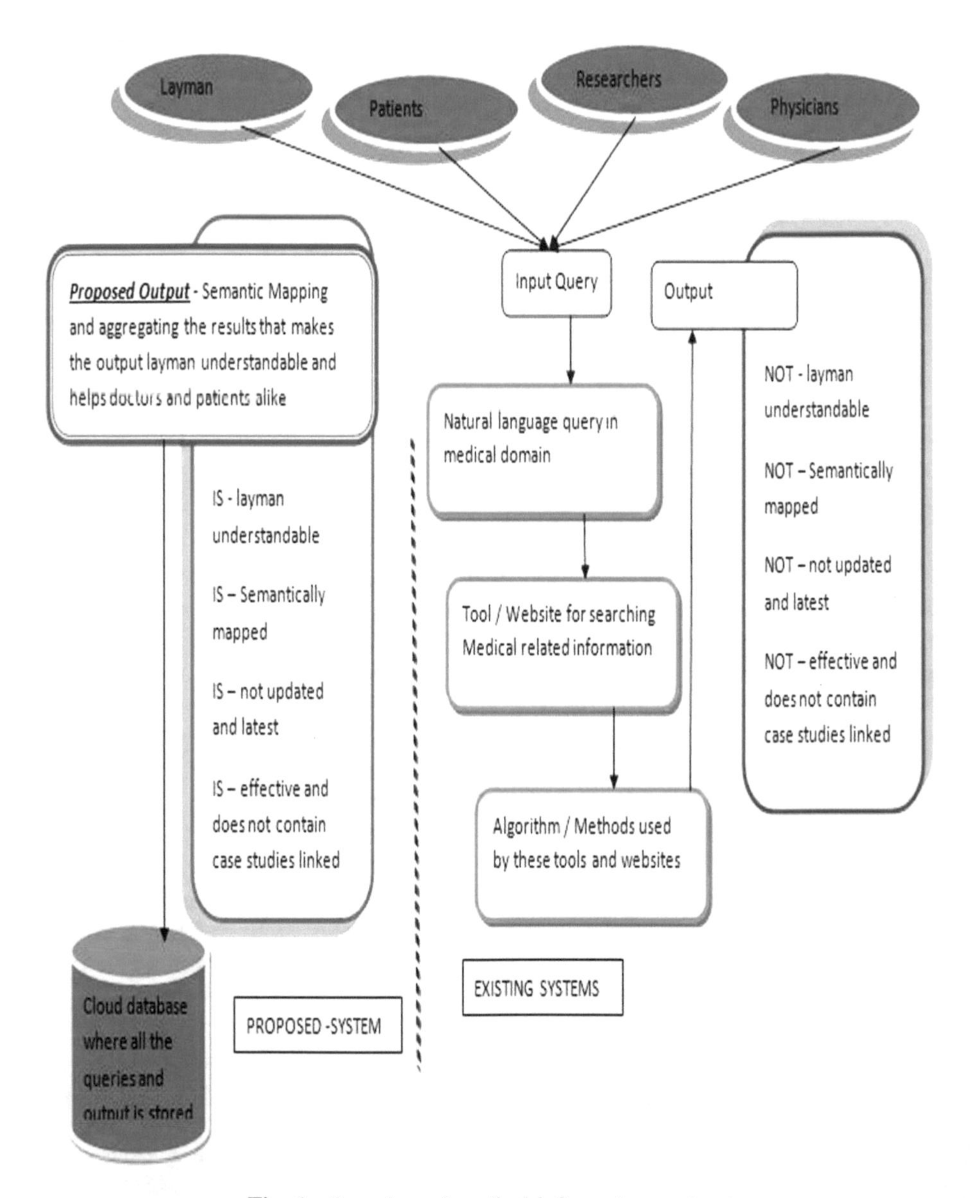

Fig. 1. Overview of medical information retrieval

processing language on the red query to retrieve proper as well as readable output. These results can also be stored for usage in future.

Section 2 describes various methods that have been proposed for structuring of medical data and its retrieval. Section 3 describes the tools that have been developed till now for retrieval of medical information as well as usage of cloud for information retrieval process. We conclude in Sect. 4 with all development done till now. In Sect. 5, we discuss the future enhancements for this field.

2 Methods

Information retrieval from a huge set of documents and database accurately requires an efficient method which can extract the data which is relevant enough for the user's query. Figure 2 represents the table containing comparison of few methods described below it.

Information Retrieval Method	Function	Drawback
Conceptual Model using UMLS [4]	Generating a conceptual graph using UMLS Metathesaurus and semantic network for extracting medical information effectively	As medical is huge domain, hence the thesaurus is very complex, therefore monitoring inconsistency is very difficult
Information Retrieval strategy in CANCERLIT [3]	Generation of Concept space by extracting terms from biomedical thesauri and their automatic indexing	Currently the system generates concept space syntactically
Automated Terminology mapping [7]	Mapping of terminologies amongst various thesauri based on name, structure and linguistic resources	Sometimes mapping found for terminology are ambiguous
Terminology server-driven search engine [9]	It uses SNOMED® CT data structure containing various medical concepts. Input query is parsed and mapped using lexical expansion as well as conceptual expansion	The concepts are arranged in hierarchical structure through a table of IS-A relationship therefore navigating dynamically in tree becomes difficult
Boolean Strategy of Information retrieval [8]	Keyword processing like(removing stop words, stemming) and indexing which retrieves the documents	It cannot retrieve keywords which do not match.
Information retrieval of Word based natural language system [8]	Keyword extraction from natural language and indexing which retrieves the documents	It cannot retrieve keywords which do not match.
Information retrieval using concept based natural language system [8]	Semantic concept mapping on the keywords extracted from natural language	It cannot retrieve keywords, which are not found in the semantic representation.

Fig. 2. Survey table of existing methods for medical information retrieval

A Conceptual model has been described in [4], which uses the Metathesaurus and Semantic network of UMLS to create a set of concepts which are later represented as conceptual graphs after obtaining relationship between various classes. An implementation for Conceptual model has been proposed by [2], which using hierarchically structured generic C++ classes provided by National Institutes of Health (NIH).

The main drawback of this model is inability to express queries in case the semantics are not available in information database. [3] has described CANCERLIT, which is actually a collection of bibliographic records containing information on all types of cancer therapy. It also describes that textual Information can be retrieved via four approaches, they are: keyword indexing and retrieval, statistically based methods, relevance feedback and semantic networks. Information retrieval strategy from CANCERLIT is based on keyword and indexing.

[7] had proposed a concept of automated approach of mapping external terminologies to the UMLS. This method tries to maps the terminologies amongst various thesauri to obtain interoperability by finding the correspondence with respect to name, structure and linguistic resources, finally evaluating these terminology mappings manually. EMTREE has been developed to index EMBASE, a biomedical and pharmacological online database. Here we map the concepts of EMTREE with respect to UMLS, initially by finding concepts in UMLS that contain a synonym with same normalized form as EMTREE term. The results of these mappings can be classified as zero mapping, one-to-one mapping and (one to many) ambiguous mapping.

[9] has proposed a terminology server to nd specialist physicians with a particular clinical expertise. Initially this system undergoes parsing of input from the user such that it determines the area of interest, next it performs the lexical expansion or conceptual expansion on the basis of the area of interest, finally linking it to the appropriate physician.

[8] has proposed a comparison of three systems utilizing distinct methods for information retrieval, they are: (1) Boolean System (2) Word-based natural language system (3) Concept-based natural language system. In this paper, the rest system for comparison was Grateful Med, it gets phrases of input via GUI which distinguishes AND and OR operations such that as per the operations and keywords the documents are retrieved. The second system is SWORD, which has features of word based automated indexing and relevance ranking, the queries obtained via natural language rest of all removes all stop words and undergoes stemming of keywords, as per the weight assigned to keywords of documents the retrieval of documents occur. The third system is SAPHIRE, it features concept based automated indexing in addition to natural language retrieval and relevance ranking, therefore this system obtains the set of keywords same as SWORD but later on performs the semantic mapping of concepts.

3 Tools

The previous section talks about different methods and algorithms that can be used for making information retrieval more precise. In this section we will discuss about the existing tools, technologies, portals and websites that help in retrieval of information related to medical domain. The advancement of internet has helped everyone to connect. Thus physicians, researchers, pharmacists, and even patients need more and more medico information. Various tools and web-sites are available to serve this purpose. Each of this existing tools and websites serve different purposes. For example, Medical Concept Mapper [1] is a terminology server, HelpfulMed [5] is a web-portal giving options to search different databases, web pages and medical related term in a simple

Name of the Resource	Platform	Source	Content	Remarks
MEDLINEPlus [11]	Website	U.S National Library of Medicine	Gives Information on health and about drugs Gives retrieval on basic queries	Would not give results for drugs that are used outside US Hence not a global library
Medscape [12]	Website	Owned by WebMD	It consists of database , drug information, news, disease information etc.	It is used for physicians and other health professionals Not very useful for normal person
ClinicalKey [13]	Website(search engine)	Published by Elsevier	It's a search engine and it has faster search because it is designed in a way to think like physicians Deep access for trusted content	You need to be registered 30 days trial available
CliniWeb [5]	Database	Supported by OHSU	Consists Database of clinically related documents Manages medical information	Just a database No specifications for query processing
PubMed [10]	Website(Search Engine)	Governed by United States National Library of Medicine	Similar to MEDLINE Except that it also includes old references to journals, recent updates and collection of books	It is not completely automated
HelpfulMed [5]	Proposed Medical Portal	Research made by Hsinchun Chen, Ann M. Lally, Bin Zhu and Michael Chau	Proposes an intelligent automated portal for medical domain which is classified as 1. User Interface 2. Search Web pages 3. Search Databases 4. Related Medical terms 5. Visual Site Browser	No room for complex queries at present Additional technology can be included
Medical Concept Mapper[1]	Terminology server	Research made by Gondy Leroy and Hsinchun	Proposes service that assists in finding access to already existing knowledge source by giving suggestions of medical terminologies to user	Precision of medical related term is increased by using concept space along-with Semantic Net Accuracy is only fair
Mayo Clinic Health Oasis [1]	Medical Web-site	Mayo Foundation for Medical Education and Research	It's a website consisting of various information on medical background like drug, symptoms, diseases etc.	Has information about general medical topic

Fig. 3. Survey table of existing tools for medical information retrieval

and understandable GUI. The following Fig. 3 contains the table of comparison of tools that are already in used for medical information retrieval.

HelpfulMed [5] is a web portal that is designed to serve purpose of a simplified and understandable GUI. This tool proposes an architecture that gives options for browsing medical related databases, medical related websites and medical related terms. Its a web portal for browsing information for physicians, researchers and other advanced medical users. It can be further enhanced for summarizing the retrieved documents.

Medical Concept Mapper is an online tool proposed in [1], that utilizes Arizona Noun Phraser, UMLS, Wordnet, computer generated concept space and Deep Semantic Parsing (DSP) algorithm [1]. In rest stage it extracts the medical phrases from natural language queries using Arizona Noun Phraser and SPECIALIST Lexicon. In second stage it retrieves the synonyms from Metathesaurus and WordNet. In third stage the related terms are retrieved based on concept space and semantic network of UMLS using DSP algorithm. The results of this tool implemented showed that appropriate medical term input resulted into higher recall whereas input from terms resulted into higher precision. Recall is number relevant documents retrieved from set of relevant documents already present in database. Precision is number of relevant documents retrieved from all the documents retrieved.

Apart from the resources given in the table and HelpfulMed [5] and Medical Concept Mapper [1] there is one more medical search engine MedicoPort [6] which is designed for medical information retrieval. The speciality of this medical search engine is that it claims of being able to seek information for people who doesn't know medical jargon. Also it uses two-level index structure [5] which helps to retrieve health information even if the keyword entered by the user is not in UMLS.

[15] has described a project that has successfully implemented technique that can retrieve published papers corresponding to a particular topic in an online clinical discussion. The strategy initially gathers messages belonging to a particular thread and processes using Metamap to get corresponding keywords from MeSH Lexicon. These keywords then retrieve the papers related to discussion from Pubmed.

[17] is an example about the existing cloud storage model for medical information. Using cloud services like amazon or Microsoft, the data can be maintained securely and used for various domains, like not only improving medical retrievals, but for analysis, diagnostic and study too.

[16] describes the evolution of cloud technology so far in medical domain. The cloud model helps medical domain a lot, since it can meet a lot of requirements like secure data-storage, easy access, high availability, multitenancy and pay-per-use model etc. also it is easy to migrate from older systems to cloud. There are many existing models and solutions on cloud these days that are much beneficial for medical domain.

4 Conclusion

As medical is a wide domain therefore for each query of the user cannot be mapped terms lexically, there exists a requirement to map them semantically as well. This concludes that medical retrieval systems still lack the efficient semantic mapping methodologies. A fully matured and semantically sound mapping for medical domain will allow NLP to achieving better query results. Also, this huge amount of data can be stored at a unified source using cloud technologies. A good retrieval system can help patients to view their medical information online and analyze it with other cases. A good retrieval system can help doctors all around the world to get updated with all the current and latest information on certain diseases, their diagnosis, causes, symptoms, cases of patients suffering from same, etc. There by resulting in fast and effective treatment. Also, it helps to keep biomedical researchers updated with latest development in medical domain.

5 Future Enhancements

Medical information retrieval is a domain where this natural language processing can be applied. The methods and tools defined here are capable enough to provide partially relevant results to searching however we can enhance these tools in future for vernacular languages with a good natural language processing technique. Ontologies and semantic mapping is applied to improve the results but apart from this neural networks can also be applied to develop algorithms for effective concept space generation and their retrieval with higher accuracy.

References

1. Leroy, G., Chen, H.: Meeting medical terminology needs-the ontology-enhanced medical concept Mapper. IEEE Trans. Inf. Technol. Biomed. **5**, 261–270 (2001)
2. Robert, J.-J., Joubert, M., Nal, L., Fieschi, M.: A computational model of information retrieval with UMLS. In: Proceedings of Annual Symposium on Computer Applications in Medical Care, pp. 167–171 (1994)
3. Houston, A.L., Chen, H., Schatz, B.R., Hubbard, S.M., Sewell, R.R., Ng, T.D.: Exploring the use of concept spaces to improve medical information retrieval. Decis. Support Syst. **30**, 171–186 (2000)
4. Joubert, M., Fieschi, M., Robert, J.J.: A conceptual model for information retrieval with UMLS. In: Proceedings of Annual Symposium on Computer Applications in Medical Care, pp. 715–719 (1993)
5. Chen, H., Lally, A.M., Zhu, B., Chau, M.: HelpfulMed: intelligent searching for medical information over the internet. J. Am. Soc. Inf. Sci. Technol. **54**, 683–694 (2003). (Special Issue: Web Retrieval and Mining)
6. Can, A.B., Baykal, N.: MedicoPort: a medical search engine for all. Comput. Meth. Programs Biomed. **86**(1), 73–86 (2007)
7. Taboada, M., Lalin, R., Martnez, D.: An automated approach to mapping external terminologies to the UMLS. IEEE Trans. Biomed. Eng. **56**(1598), 1605 (2009)
8. Hersh, W.R., Hickam, D.H.: An evaluation of interactive boolean and natural language searching with an online medical textbook. J. Am. Soc. Inf. Sci. **46**(7), 478–489 (1995)
9. Cole, C.L., Kanter, A.S., Cummens, M., Vostinar, S., Naeymi-Rad, F.: Using a terminology server and consumer search phrases to help patients nd physicians with particular expertise. Stud. Health Technol. Inf. **107**, 492–496 (2004)
10. Home PubMed, http://www.ncbi.nlm.nih.gov/pubmed

11. Health information from the national library of medicine, https://www.nlm.nih.gov/medlineplus/
12. LLC: Latest medical news, clinical trials, guidelines today on Medscape, http://www.medscape.com/
13. ClinicalKey, https://www.clinicalkey.com/
14. Nie, L., Zhao, Y.L., Akbari, M., Shen, J., Chua, T.S.: Bridging the vocabulary gap between health seekers and healthcare knowledge. IEEE Trans. Knowl. Data Eng. **27**, 396–409 (2015)
15. Stewart, S.A., Abidi, S.S.R.: An Infobutton for web 2.0 clinical discussions: The knowledge linkage framework. IEEE Trans. Inf. Technol. Biomed. **16**, 129–135 (2012)
16. Calabrese, B., Cannataro, M.: Cloud computing in healthcare and biomedicine. Scalable Comput. **16**(1), 1–18 (2015)
17. Doukas, C., Pliakas, T., Maglogiannis, I.: Mobile healthcare information management utilizing cloud computing and Android OS. In: 2010 Annual International Conference of the IEEE Engineering in Medicine and Biology Society (EMBC), pp. 1037–1040 (2010)

An Agricultural Intelligence Decision Support System: Reclamation of Wastelands Using Weighted Fuzzy Spatial Association Rule Mining

Mainaz Faridi[1](✉), Seema Verma[2], and Saurabh Mukherjee[1]

[1] Department of Computer Science, Banasthali University, Banasthali, Rajasthan, India
mainaz.faridi@gmail.com, mujherjee.sourabh@rediffmail.com
[2] Department of Electronics, Banasthali University, Banasthali, Rajasthan, India
seemaverma3@yahoo.com

Abstract. The increase in GDP of the country has given a flight to industrialization and urbanization, causing more and more utilization of agricultural lands for non-agricultural purposes. Since the availability of agricultural lands is limited, requisite measures must be taken to restore wastelands for cultivation. Therefore to filter out the suitable wastelands for reclamation and predict their level of utilization, this paper proposes the agricultural intelligence decision support system. The proposed system has two phases. The first phase consists of the mining technique in which required attributes are selected, intersection is applied as spatial predicate and weights are assigned to linguistic terms for obtaining weighted fuzzy rules. In the second phase the fuzzy inference system is constructed in accord of the weighted fuzzy spatial rules mined in the previous phase. This will assist agriculture-related organizations and persons to take well informed decisions for effective utilization of wastelands.

Keywords: Agricultural Intelligence · Weighted Fuzzy Spatial Association Rule Mining · Utilization of wastelands · Mamdani Fuzzy Inference System

1 Introduction

The country has witnessed an overall economic growth in all the sectors but agriculture sector although being the largest is facing a gradual drop in its share in the country's GDP. Moreover, country's burgeoning population has caused drastic changes in the ratio of land utilized for agricultural activities, urbanization, and other industrial development. To add to this, the intensive agricultural practices (requiring heavy use of chemical fertilizers and pesticides), cultivation of crops heavily dependent on water, improper use of irrigation practices cause land degradation and deterioration causing the biggest threat in providing food security to the country. Land degradation casts a direct impact on the productive capacity of the soil, its vulnerability to sustain flood and drought influence, growing of fodder, timber and fuel wood. Thus, to enhance the

S.C. Satapathy and A. Joshi (eds.), *Information and Communication Technology for Intelligent Systems (ICTIS 2017) - Volume 1*, Smart Innovation, Systems and Technologies 83, DOI 10.1007/978-3-319-63673-3_66

agricultural productivity there is an imperative need to extend cultivable lands, by reclamation and restoration of degraded and wastelands using sustainable agricultural practices, without causing adverse effects on the environment [1, 2]. Even if the wastelands can't be used for farming they can be used for afforestation, growing fodder, timber or fuel. These all factors promote the use of recent advancement in technology into the agricultural domain.

1.1 Agricultural Intelligence

Agricultural Intelligence (AI) is a method where collection of information, events and issues from various agricultural activities is done, and analyzed to be used in making well-informed decisions [3]. It is an architecture that encompasses various technologies, decision–support applications or databases for providing easy access of agricultural data to agricultural community. Any organization related to agriculture (like those involved in agriculture production, animal husbandry, plant and animal health, government agencies, insurance companies, agricultural investment and consultancy agencies, etc.) can provide its input to AI. The information and data provided by the contributing organizations are analyzed and turned into intelligence which is then utilized to assist them in taking efficient and timely decisions. The process of converting raw data and information to intelligence includes collecting, filtering, evaluating and analyzing information into value added product. Thus, the ultimate aim of AI is to provide valuable information in the most accurate and timely manner. AI can be used especially where accurate and timely forecasts of the future tendencies or agricultural conditions can be obtained. AI allows for better trend and market analysis, prediction of crop production, forecasting cost of agricultural products thereby helping in making proactive business decisions [4–6].

Agricultural activities have a close association with natural resources like soil, ground water, land, rainfall etc., that have apparent spatial properties which, Geographic Information Systems (GIS) deal primarily in. The advancement in remote sensing, satellite imaginary, GPS, sensors, data storage technologies and computing capabilities have paved the path for easy, timely and fault free acquisition, storage and processing of agriculture related data. Thus, agriculture and its allied sectors produce a huge amount of spatial data where spatial data mining can find its applicability [7]. Spatial data mining is a subset of data mining which deals specifically with spatial data and provides methods for extraction of implicit knowledge, spatial relations or other patterns not explicitly stored in spatial databases [8].

1.2 Weighted Fuzzy Spatial Association Rule Mining

Spatial association rule mining is one of the most researched and widely used methods of spatial data mining. A spatial association rule (SAR) is an association rule containing at least one spatial predicate [9]. A spatial predicate constituting a spatial association rule may furnish different information like distance information (close_to, far_away), topological relations (like overlap, intersect, and disjoint), and spatial orientation (like north_of, right_of, etc.) [8, 10]. Just as with conventional association rules the strength and reliability of an association rule is measured by two factors: support and confidence. A spatial association rule can be stated as:

$$\text{is\_a}(x, \text{city})^{\wedge}\text{close\_to}(x, \text{poles}) \rightarrow \text{is\_cold}(x)\ (98\%)$$

Spatial data mining alone is not able to deal with the uncertainty in spatial data. Spatial association rules work on crisp data and suffer with sharp boundary problem. Hence, fuzzy logic could be used to solve the above issues. To deal with uncertainties and impreciseness in spatial data, address the issue of sharp boundaries in categorical data and provide linguistic treatment to data, the fuzzy set approach can be fused with spatial data mining. The fuzzy set concept is used with spatial association to generate the so called fuzzy spatial association rules [11].

In traditional association rule mining, each item has same significance and given equal importance. The items appearing fewer than the minimum threshold are discarded. In contrast, the real life data contains items with varying importance. For example, a frequently occurring item may not incur large profits which a very less occurring item may produce. Therefore, to address the issue of varying importance attached to items in a transaction, weighted fuzzy association rules have been evolved [12]. Therefore, keeping in mind the above scenario, weighted fuzzy spatial association rule are proposed for identification of those wastelands which possess good amount of ground water and the type of soil that could be used for farming and/or growing fuel and timber.

2 Study Area and Datasets

For this research, a sub-arid region of Rajasthan state such as Jodhpur district is chosen as the study area. It is the second largest city of Rajasthan having a geographical area of 2,256,405 ha. For finding the suitable wasteland for reclamation, data about wasteland distribution, groundwater quality and type of soils were required in GIS format. Therefore, the required datasets were collected from ISRO center of Jodhpur.

3 An Agricultural Intelligence Decision Support System

Agricultural Intelligence is a method that assists agricultural experts to make well-informed decisions and applied where precise forecasts of the future trends, tendencies or agricultural conditions can be obtained [4]. The Agricultural Intelligence system starts with the data collection step. Agricultural related data collected from various sources (like those involved in agricultural activities like farmers, government organizations, agricultural firms and markets, agricultural investment and consultancy agencies, etc.) is in variety of formats and may contain some missing, incomplete values or extraneous values. Therefore, it is preprocessed to extract the needed and required data. In the next phase, the extracted data is transformed into a format that could be used for loading the data into the data marts or data warehouses. This whole process is known as ETL (Extract, Transform and Load). The data from warehouses and marts are used for generating useful and interesting knowledge from hidden

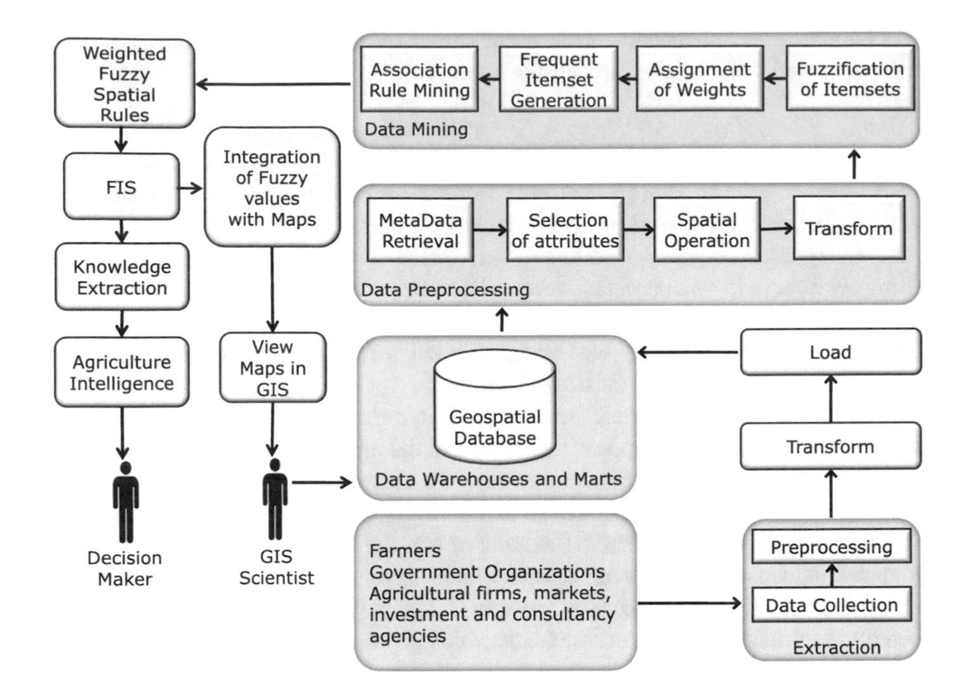

Fig. 1. Framework for agricultural intelligence decision support system

information. This knowledge can be later used to create some useful tools for decision makers to be used for enhancing agriculture productivity [13]. The overall framework for the proposed system is shown in the Fig. 1. The steps involved in the proposed decision system are explained below.

3.1 Data Preprocessing and Selection of Required Attributes

The original datasets collected from ISRO were preprocessed to extract only the required and useful information and converted into a format that could be used for mining association rules. Only those attributes of the datasets were selected that could help in wasteland reclamation as shown in the Table 1. Figure 2(a)–(c) show their corresponding maps containing thematic layers. A spatial predicate "intersection" is chosen to earmark those wastelands which have substantial groundwater and possess soil orders that can support agriculture (Fig. 2(d)).

3.2 Generating Weighted Fuzzy Spatial Rules

This is the most important step in the proposed decision system. Fuzzy set concept is used to define linguistic terms for defining the attributes instead of splitting up quantitative attributes which makes a rule more understandable for the human user. Weights are then assigned to each of these linguistic terms to reflect their significance. The

Table 1. Selected attributes of the datasets after preprocessing

S. no	Dataset	Attributes before preprocessing	Selected attributes after preprocessing
1	Wasteland	Sandy-desertic, salt affected, mining/industrial waste, land without scrub, land with scrub, gullied/ravenous land, barren rocky/stony waste land.	Salt-affected, land without scrub
2	Ground-water	Good, good but saline, good to moderate, moderate, moderate to poor, poor, poor to nil, saline, settlement, very good to good	Very good to good, good, good but saline, good but moderate, moderate
3	Soil suborder	Calcids, cambids, fluvents, gypsids, orthents, psamments, salids	Cambids, orthents, fluvents

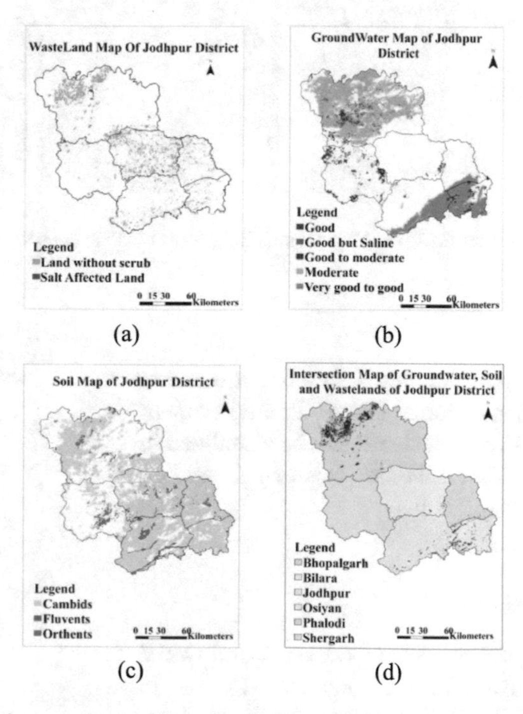

Fig. 2. Jodhpur district map for (a) Waste-land, (b) ground-water, (c) soil and (d) intersection of all three datasets

mining process can be carried in two steps. In the first step, frequent itemsets are generated by using support and then rules are mined using confidence in the next step. Selection of the algorithm for mining the weighted fuzzy association rules can depend on the user preference, mining criteria, and efficiency of the algorithm being selected. Support and confidence thresholds could be decided by the domain expert. Here, WFSARM algorithm is selected for mining rules [14].

3.3 Creation of Fuzzy Inference System

The association rules provide only statistical analysis. Therefore, for deriving further inferences, fuzzy logic is incorporated with the association rules. Mamdani Fuzzy Inference System is created to predict the utilization of waste-lands by using weighted rules (Fig. 3). The proposed decision support system as shown in Fig. 3 contains 3 inputs (viz., waste-land, ground water and soil), and one output of waste-land). Here, triangular membership function is used for the fuzzification for each input fuzzy set and centroid is used as defuzzification method (Fig. 4(a)–(d)).

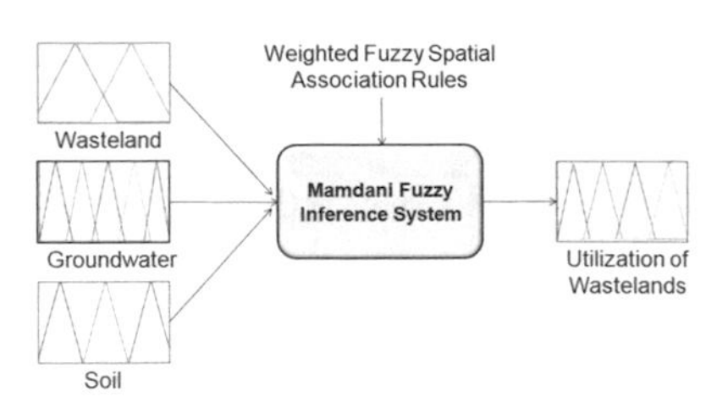

Fig. 3. Mamdani FIS

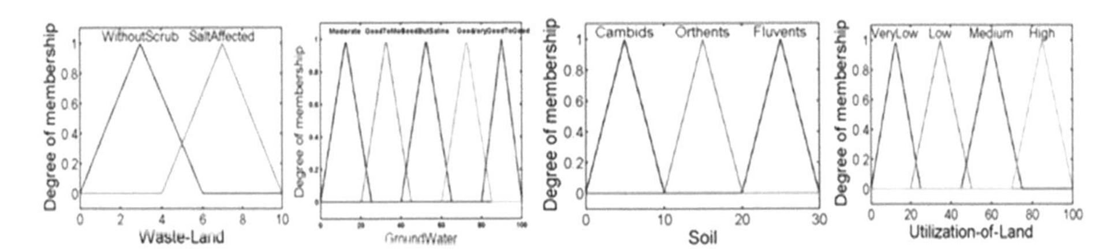

Fig. 4. Membership functions for (a) wasteland, (b) ground water, (c) soil, and (d) Utilization of wastelands

The fuzzy values of the output variable, utilization of waste-land are then exported into a GIS environment so that the utilization of wastelands can be visualized in form of a map (Fig. 5). At the last, this map is studied for the statistical interpretation.

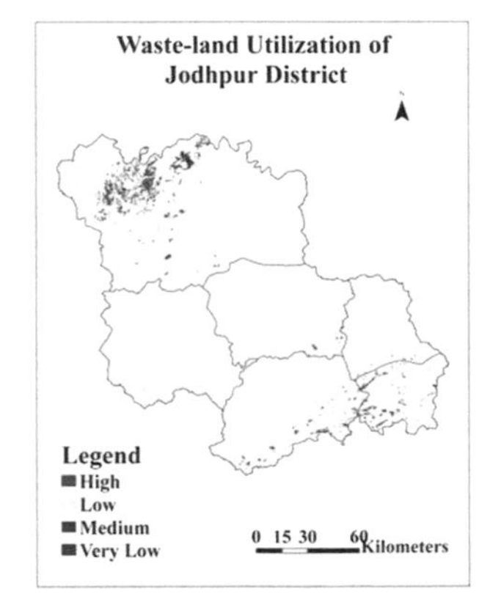

Fig. 5. Map showing utilization of wasteland

3.4 Implementation

The proposed AIDSS is developed on Windows 7 operating system using Java Swings in JDK 1.8 in Eclipse Luna IDE. ArcGIS Desktop 10.1 is used for pre-processing of spatial data and carrying intersection operation. Mamdani Fuzzy Inference System is created in Matlab R2013a. Snap shot of the developed DSS is shown in Fig. 6.

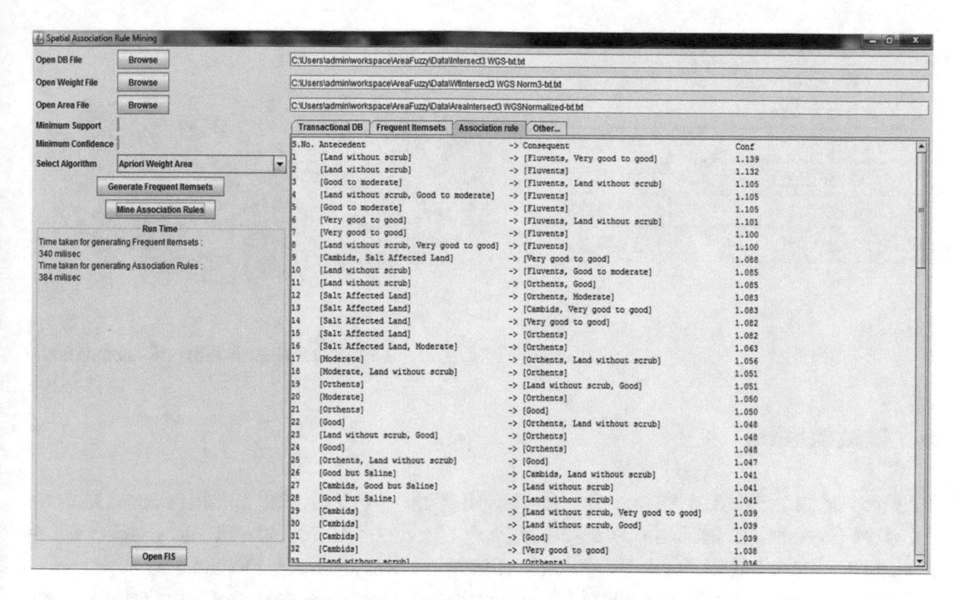

Fig. 6. Implementation of the proposed DSS

4 Results and Discussion

The mined pattern produced from the output map has a total of 36,062.71 ha of land. The analysis reveals that Phalodi taluk contains the largest wasteland area while Shergardh taluk contains the smallest area of wasteland. The distribution of wastelands with respect to taluks of Jodhpur is showed in the Fig. 7. Utilization of waste-lands among all the taluks of Jodhpur district is shown in the Table 2 and a graph depicting the same is shown in Fig. 8.

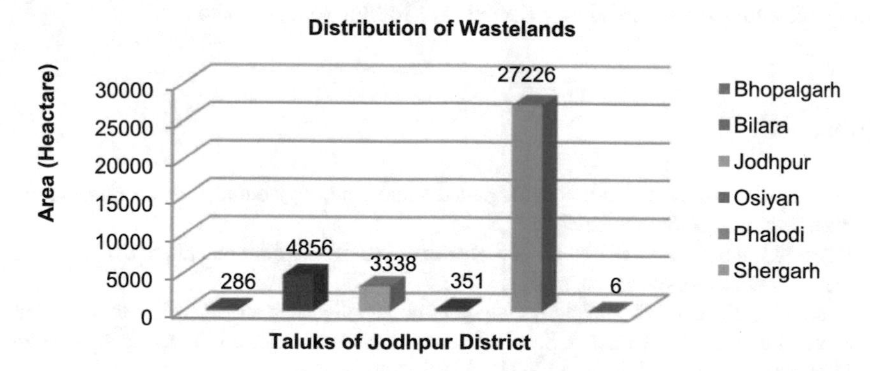

Fig. 7. Talukwise distribution of wastelands

Table 2. Utilization of wastelands.

S. no	Utilization of waste-land	Area (ha)
1	High	6032.35
2	Medium	22,976.44
3	Low	3929.35
4	Very low	3124.57

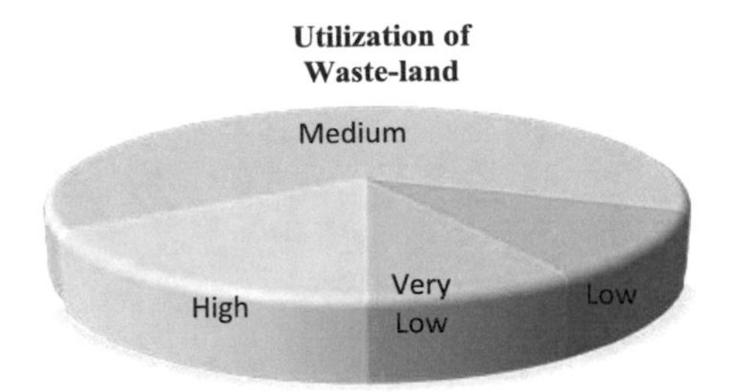

Fig. 8. Talukwise distribution of wastelands

5 Conclusion

The paper presented a frame for developing an Agricultural Intelligence Decision Support System. It uses the weighted fuzzy spatial rule to extract the hidden information and used to predict wasteland utilization for Jodhpur District. The proposed system would help environmentalist, government agencies, farmers, agriculture advisory firms, agricultural researchers and analysts etc. to strain the waste-lands having substantial ground water underneath and analyze their utilization for reclamation. Furthermore, the analysis of the mined area shows that majority of wastelands has medium utilization. The majority of waste-lands is the lands without scrubs and has considerable groundwater underneath; these lands can be used for cultivation of crops. The waste-lands which have moderate ground water can be used to produce firewood and fodder for animals. Plant species like *Acacia jacquemontii, Prosopis cineraria, Salvadora oleoides, Tecomella undulata, Leucaena leucocephala, Tephrosia purpurea and Crotalaria medicaginea* can be grown. Thus, the research helps in determining the utilization of wastelands thereby help in increasing the crop production, along with providing an additional source of income to the rural people, helping in maintaining ecological balance and providing timber and fodder for local use.

References

1. Development of Wastelands and Degraded Lands: http://planningcommission.nic.in/plans/planrel/fiveyr/10th/volume2/v2_ch5_3.pdf
2. Degraded and Wastelands of India-Status and Spatial distribution: http://www.icar.org.in/files/Degraded-and-Wastelands.pdf
3. Ghadiyali, T., Lad, K., Patel, B.: Agriculture intelligence: an emerging tool for farmer community. In: Proceedings of Second International Conference on Emerging Application of Information Technology, pp. 313–316. IEEE, Kolkata (2011)
4. Cojocariu, A., Stanciu, C.O.: Data warehouse and agricultural intelligence. Agric. Manage./Lucrari Stiintifice Seria I, Manage. Agricol. **11** (2009)
5. Munroe, F.A.: Integrated agricultural intelligence—a proposed framework. Vet. Ital **43**(2), 215–223 (2006)

6. Faridi, M., Verma, S., Mujherjee, S.: Integration of GIS, spatial data mining and fuzzy logic for agricultural intelligence. In: Proceedings of International Conference on Soft Computing: Theories and Applications. Springer, New York, Accepted on October 2016
7. Mai, K., Murali, C., Krishna, I.V., Reddy, A.V.: Data mining of geo spatial database for agriculture related application. In: Proceedings of Map India. New Delhi (2006)
8. Koperski, K., Han, J.: Discovery of spatial association rules in geographic information databases. In: International Symposium on Spatial Databases, pp. 47–66. Springer, Heidelberg (1995)
9. Shekhar, S., Huang, Y.: Discovering spatial co-location patterns- a summary of results. In: International Symposium on Advances in Spatial and Temporal Databases, pp. 236–256. Springer, Berlin (2001)
10. Han, J., Kamber, M.: Data mining—concepts and techniques. Morgan Kaufmann Publishers, Burlington (2001)
11. Ladner, R., Petry, F.E., Cobb, M.A.: Fuzzy set approaches to spatial data mining of association rules. Trans GIS **7**, 123–138 (2003)
12. Gyenesei, A.: Mining weighted association rules for fuzzy quantitative items. In: European Conference on Principles of Data Mining and Knowledge Discovery, pp. 416–423. Springer, Heidelberg (2000)
13. Ghadiyali, T., Lad K., Patel B.: ETL techniques and challenges in agriculture intelligence. In: Proceedings of AIPA, India, pp. 85–91 (2012)
14. Faridi, M., Verma, S., Mukherjee, S.: A novel algorithm of weighted fuzzy spatial association rule mining (WFSARM) for Reclamation of Wastelands. In: Accepted in First International Conference on Smart Technologies in Computer and Communication Smart Tech17, Springer, New York (2017)

Geographical Information Assisted e-Commerce (GIAE) for Leveraging Sustainability for Farmers'

Abhishek Chawda(✉) and Mayur Raj

College of Agricultural Information Technology,
Anand Agricultural University, Anand, Gujarat, India
abhichawda34@gmail.com, mraj@aau.in

Abstract. Role of Geographical Information is significant in agriculture and it can also play an important role in boosting agriculture based e-commerce. With majority of products available on internet, agricultural products still behind to be online. Lacking point behind this situation is the improper application of Geographical Information, as it would act as the backbone in this value chain. In spite of various steps are taken to acknowledge e-commerce related activities in the field of agriculture profit does not percolate to farmers. This can be accelerate using Geographical Information into the current stream of e-commerce. In majority of the scenario merchant act as intermediate which pilfers the seller's and buyer's share. This paper presents a merchant free platform Geographical Information Assisted e-Commerce (GIAE) which uses Geographical Information of the user and allow selling and purchasing based on rank of either party, so that both can gain maximum benefit.

Keywords: Geographical information · Agriculture · m-Commerce · e-Commerce · Agri-business

1 Introduction

India's e-commerce market was worth about $3.9 billion in 2009, it went up to $12.6 billion in 2013. In 2013, the e-retail segment was worth US$2.3 billion. About 70% of India's e-commerce market is travel related. According to Google India, there were 35 million online shoppers in India in 2014 Q1 and is expected to cross 100 million mark by end of year 2016.

This way e-commerce has the deepest roots in the today's dynamo. The effect of this dynamo has affected the agriculture sector as well. Today, India ranks second worldwide in farm output. Agriculture and allied sectors like forestry and fisheries accounted for 13.7% of the GDP (gross domestic product) in 2013 about 50% of the workforce (https://en.wikipedia.org/wiki/Agriculture_in_India#Infrastructure).

But somehow the major amount of agriculture product produce is being thrown as wastage. Reason beside this are various. One say it is lack of management, so other suggest it is improper functioning of the value chain. All these reasons results in low

S.C. Satapathy and A. Joshi (eds.), *Information and Communication Technology for Intelligent Systems (ICTIS 2017) - Volume 1*, Smart Innovation, Systems and Technologies 83, DOI 10.1007/978-3-319-63673-3_67

price of agricultural produce thus farmer forcefully sell them at low price or products are thrown away as wastage.

With the introduction of e-commerce and to boost this decay of agricultural product a new revolution of managing agricultural product online has taken a very stable shape across India as well as in whole world. But the stone in this fast moving path is the improper execution of selling and buying channel. Due to lack of proper Geographical locations of the seller and buyer various tries are being went resting.

Existing system of value chain follows the channel flow as: Seller to merchant and then merchant to buyer or vice versa (Fig. 1). Here the actual profit which was supposed to be made by either the seller or the buyer was actually gulped by the merchant. Thus the seller of agricultural commodity started taking less interest in this mode of marketing which can be seen by looking at failure of government project like Delhi Kissan Mandi etc. and various privately initiated projects like agro mall by WOTU.

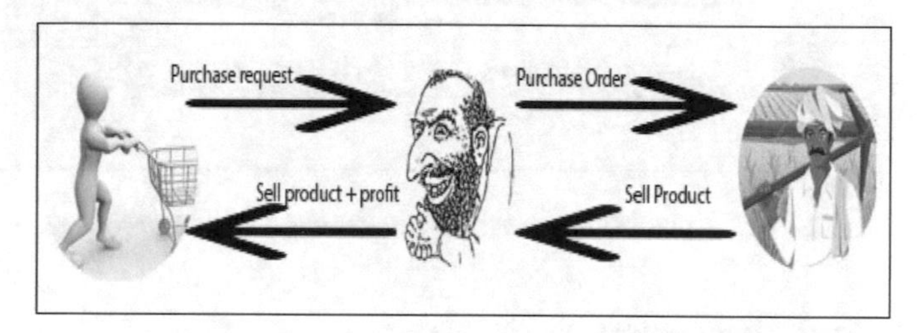

Fig. 1. Current process flow in selling or buying products

Research shows the need of a direct pathway between seller and the consumer is more fruitful. Deshpande and Biniwale (2006) have explained their views about regulated market and rural weekly market in their research paper entitled, “Restructuring Agriculture Marketing System: Karnataka Experience”. Agricultural marketing system of the primary level in Karnataka involves four board marketing channels i.e. (i) direct to consumer (ii) through private wholesalers and retailers (iii) through public agencies (regulated market) or co-operatives and (iv) through processors. Although the share for direct selling to consumer is lowest among the other which he thinks should be highest.

2 Research Objective

The main objective of this research was to create an online platform Geographical Information Assisted e-Commerce (GIAE) which using Geographical Information of the user allow selling and purchasing of agricultural products between them without any intermediate person.

3 Methodology

GIAE platform has been developed using MVC architecture for providing separation of concerns, easier integration with client side tool, search engine optimization and test driven development as a developer's point of view. From technical aspects, problem with the usage of any system is device on which it would be accessed. It can be smart phone or a personal computer. In simple terms the same website can be viewed in Desktop computer as well as in mobile with same ease that it help in providing responsiveness. Using this platform no intermediate person is required to cut the deal. Thus the channel flow in this system would be: Seller to buyer and vice versa (Fig. 2).

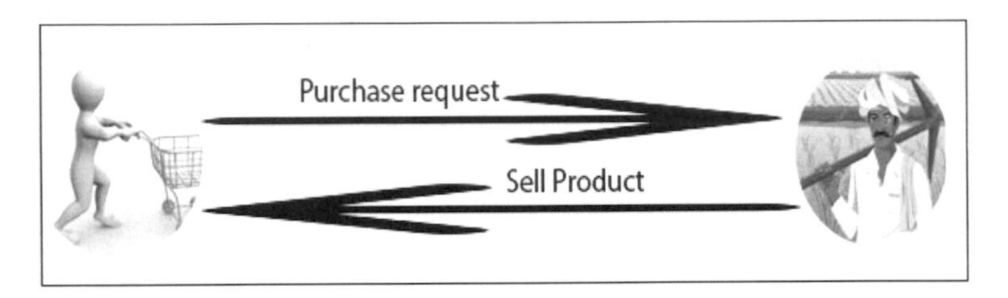

Fig. 2. GIAE platform process flow in selling or buying products

4 Results and Discussion

GIAE is a try to create a platform where a seller can directly sell his products to consumers while consumer can directly purchase from his interested farmer without any intermediate merchant. Customer can search for product of his interest from given various menus and options. Interesting feature which is the unique and most wanting is that search of products would be based on the Geographical Information possessed by the buyer and various sellers. In this way both the seller and the buyer would be happy using this platform. He can go for purchasing of the interested product then or he can save it in the cart for further future purchase. One more interesting feature added is the rating system. The customer can rate the farmer (seller) on the basis of past purchase experience. Also vice versa. When a customer finishes with the platform for purchasing, an email will be send with necessary purchase details like commodity with its requested amount and total price and simultaneously a message will be given to the seller of that respective commodity regarding this purchase request. (Customer can also note down address and phone number for further query and contact).

This system is divided in three main modules VIZ. Buying, Selling and Admin (Fig. 3).

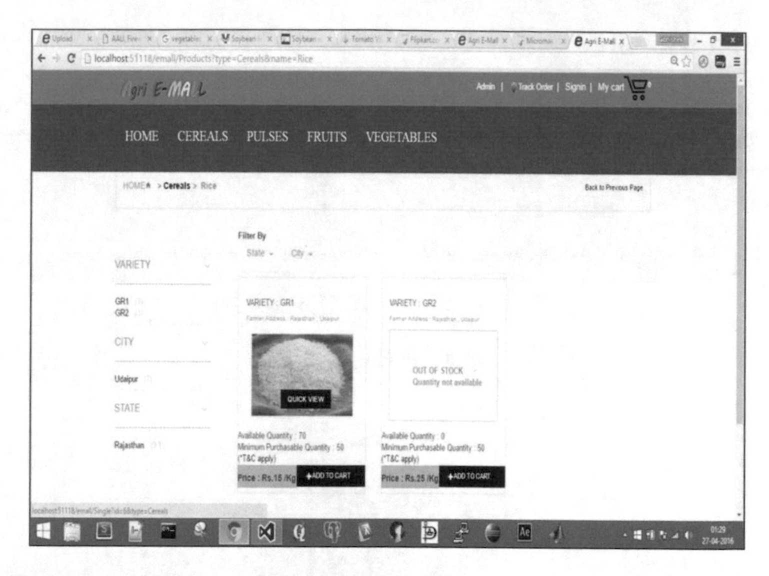

Fig. 3. Shows scenario when a particular category is selected by the buyer. Here filter are applied based on Geographical Information of user

4.1 Buying Module

- In this part of website customer can view various items provided by various farmer. He can view the price, variety and farmer name and address associate with a particular product.
- He can then put this product of his into cart for future purchase or can directly order products from the farmer of its interest by providing his address and quantity he required.
- A confirmation email and mobile message will be provided whenever an order is placed by the customer. Meanwhile a mobile message will be send to the appropriate farmer regarding the placed order. (An order will be placed only when customer is logged in while he can view products without logging in).

4.2 Selling Module

- In this part of website farmer can upload the information like category of product (cereals, pulses, fruits, vegetable etc.), name, its quantity and its image(s). This item thereon will be available for selling online.
- With this he can see his uploads and sales statistics like purchase request, successfully completed orders etc.
- For all his activities he will need to register first by providing his name, address, phone number and a password. He will then log in with phone number and his password.
- With this he can rate his customers also on the basis of his experience with that customer.

5 Admin Module

- Here in this part admin can see and analysis the ongoing activities on the websites like the total number of successful order done, the information regarding the uploaded information of the commodity by the seller.

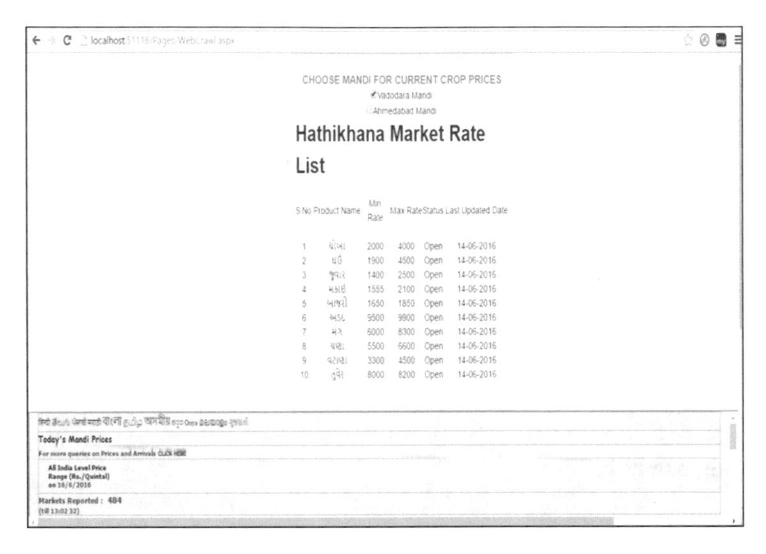

Fig. 4. Shows screen where online current market prices of various commodities are shown.

Table 1. Table showing a brief overview of situation existing in currently proposed or launched similar proposals/projects.

S. no.	GIAE	Other similar systems
1	No intermediate person(s)	Some have, some don't but without improper functioning
2	Searching on the basis of Geolocation	Common searching on the basis of searched keyword
3	Viewing of ongoing current market price possible and comparison of same with GIAE prices	Very few of such projects this functionality
4	Proper validation and authenticity of the user by admin side	Lack in proper validation and authenticity of the users. Various fake accounts with fake information is being created out
5	Mode of Payment is restricted to COD for the initial phase. Later on other mode of payment can be introduced	Variable mode of payment available
6	Delivery of consumable will be on mutual understanding of buyer and seller	Delivery part is taken by third party which again cut margin of either buyer or seller or both

- He can also add and delete the categories for the purchase dynamically from here. Also he can add or delete crops in this categories.

Besides this GIAE platform offers a very interesting feature of checking ongoing current market price of different commodity which will help a buyer to compare prices (Fig. 4 and Table 1).

6 Conclusion

Information technology has been creating an efficient and competitive platform for e-commerce especially using current GIS and GPS techniques. e-Commerce not only provides lower cost and easy to use alternative for businesses, but it also changes the scenario of product selling.

In this paper we have designed a web solution (platform) for online selling and buying of agriculture commodity providing a free place to sell product direct to the buyer without any intermediary based on their Geographical Information which could lead its success as compare with other same kind of platforms. Regular sell statics provided to seller and admin. After an order is being placed, email and mobile notifications are send to both seller and buyer. Freight service will not be provided by seller i.e. farmer, so buyer is responsible for making arrangement for goods pickup. In future freight service provider module can be incorporated. A new payment mode Cash on Goods pickup (COP) is used for this system. This payment mode will allow buyer to ensure quality & quantity of goods. At last it could be concluded that usage of Geographical Information with e-commerce and Web Technologies could bring the much awaited changes in the Agri-Market and Agri-Business.

References

1. Dagar, G.: Study of agriculture marketing. AIMA J. Manag. Res. **9**(2/4), 1–9 (2015)
2. Ehmke, C., Ernst, S., Hopkins, J., Luther, T.: The market for e-commerce services in agriculture, pp. 1–16 (2001)
3. Martinez, S.E.A.: Local food systems: concepts, impacts, and issues. ERR 97, U.S. Department of Agriculture, Economic, May 2010
4. Vadivela, A., Kiran, B.: Problems and prospects of agricultural marketing in India: an overview. Int. J. Agric. Food Sci. **3**(3), 108–118 (2013)
5. Abdullah, M., Hossain, M.R.: A new cooperative marketing strategy for agricultural products in Bangladesh. World Res. Bus. Rev. **3**, 130–144 (2013)
6. Wen, W.: A knowledge-based intelligent electronic commerce system for selling agricultural products. Comput. Electron. Agric. **57**(1), 33–46 (2007)

Performance Analysis of Video Watermarking in Transform Domain Using Differential Embedding

Rasika Rana(✉), Sharmelee Thangjam, and Sarvjit Singh

UIET, Panjab University, Chandigarh, India
rasika.rana@gmail.com, sharmeleeth@yahoo.com,
sarvjit100878@yahoo.co.in

Abstract. Digital Watermarking has been discussed as a potential solution against illegitimate use of videos over the globe. In this paper, blind hybrid video watermarking technique in transformation domain using differential embedding approach is analyzed. Discrete wavelet transforms and discrete cosine transform with differential embedding is used. The embedding and extraction of watermark is described and watermarks are examined in terms of invisibility and robustness respectively. The method performance in terms of peak signal to noise ratio and bit correction ratio is analyzed. The numerical results show the method has better performance when compared with DWT alone and DCT alone methods under noise and different video processing operations.

Keywords: Blind watermarking · Discrete wavelet transform · Discrete cosine transform · Differential embedding · Sub-sampling

1 Introduction

Effective growth in information technology has led to easy and fast distribution of digital data. But this progress has also raised the concern for content providers. Many are working to nullify the illegal video sharing and use. Digital video watermarking is believed to be a strong technique to give copyright and tamper protection to the owners. The unlicensed use and unlawful production of a patented video without the sanction of author is termed as video piracy. In video watermarking a watermark or a secret mark is embedded in the video. This watermark ensures the complete security of the video. The user can extract this embedded watermark and compare with the original watermark for security. To increase the capability of the watermarking technique a secret key is used [1]. This secret key decides the position inside the video where the watermark is embedded. The owner uses this key at the extraction time and thus makes the process unique.

When a video is being transferred over the communication channel or is processed for some other applications, its quality degrades. Noise and common processing operation like compression, rotation, filtering etc. alters the original watermark. This affects the embedded watermark. Hence, the watermark should be robust [1]. To hide the watermark in the video it must be invisible. Therefore a watermark has two

S.C. Satapathy and A. Joshi (eds.), *Information and Communication Technology for Intelligent Systems (ICTIS 2017) - Volume 1*, Smart Innovation, Systems and Technologies 83, DOI 10.1007/978-3-319-63673-3_68

important identities: robustness and invisibility. The redundancy present in the digital video frames makes way for the watermark embedding. These frames are made up of pixels with varying intensities. Some of these pixels contribute to the perceptual quality and other adds details to the image. If a watermark is added in perceptually significant coefficient, this result in better robustness as these coefficients do not change much after common video processing operations. But, the invisibility of the watermark is compromised. Thus, watermark's position in frames is crucial as these two properties conflict with each other [2].

Embedding of a watermark can be performed in spatial domain or in frequency domain. In spatial domain the pixel values of the frames of a video are directly get altered by the pixel values of secret image. Least Significant Bit (LSB) and Intermediate Significant Bit (ISB) are the examples of these methods. These techniques make invisible but less robust watermarks. In transform domain watermarking the watermark introduces changes in the frequency component of the pixels of the images [4]. DWT, DCT, DFT and FFT are the examples of such methods which provide better robustness than spatial domain. These techniques provide safe watermarks after different attacks. These transformations are used together by many authors to develop a better approach.

The method used in this paper has been used successfully in case of images [1] and hence been analyzed for video data in this paper. The method uses DCT of sub sampled vectors of LL-sub band coefficient obtained after taking DWT of each frame. Sub sampling applies different modifications to transform coefficients belonging to different sub images and thereby increases the space for watermarking [1]. Blind extraction of watermark sequence is carried out by taking the advantage of sub sampling. The method is compared against DWT alone and DCT alone method.

1.1 Related Work

A method using DWT and chaotic map is discussed by Somayyeh Mohammadi [2]. Logistic chaotic map is used to embed a logo image on video signal. The chaotic

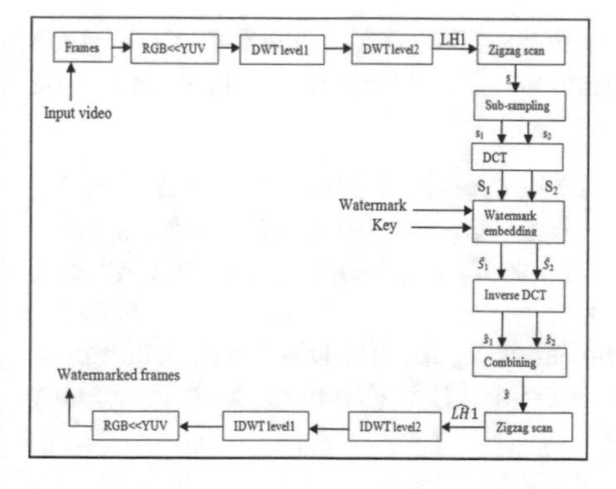

Fig. 1. Shows the embedding flowchart

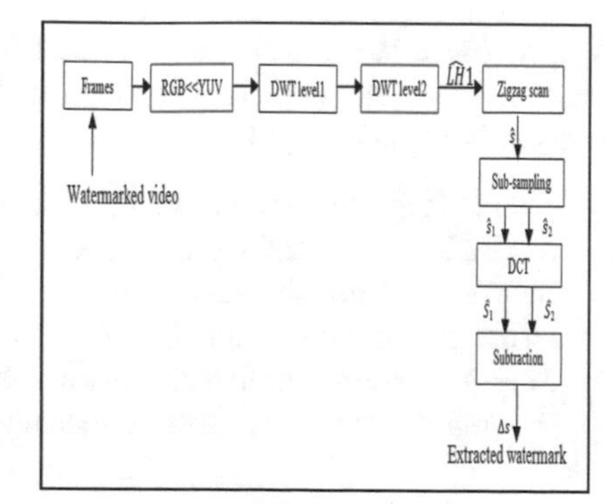

Fig. 2. Shows the extraction flowchart

parameters are used as keys. The watermark is embedded in I-frame as their existence in a video sequence is needed and they are least compressible frames. The results were verified by evaluating BER. A maximum of 50db PSNR is achieved. I frame is selected because their existence in a video is necessary and P & B are extremely compressed by motion compensation.

Md. Asikuzzaman, Jhangir and Mark R. [4] have given a blind and robust video watermarking scheme in DT CWT and SVD domain. The watermark is added in U channel of the frames of the video after applying 3-level CT DWT and the watermark is extracted from any level. Singular value decomposition is used to increase the stability. Video quality is maintained and robustness against downscaling is achieved. Pik Wah Chan and Michael R. Lyu [4] research is based on scene changing analysis and error correction code. In this watermarking scheme DWT is performed on video as well as on the secret image. The secret image is decomposed into different parts and thus each part is embedded in different scenes or frames. This means identical secret image is inserted in indistinguishable frames. This technique is robust against frame dropping. DCT-DWT-SVD algorithm is presented by C.D. Rawat and Sneha [5] and performance is evaluated by PSNR and correlation. The watermark is embedded in Y frame of 1-level DWT video. DCT is taken for LL band. Singular values of reference frame are modified in accordance to the singular values of watermark. Robustness achieved is good but the imperceptibility is poor.

2 Proposed Method

The watermarking process is done in two steps. First the watermark is embedded in the frames of the video without altering the visual properties. In second step the watermark is extracted without using the original video frames from the embedded video. The degree of closeness of the embedded and the extracted watermark decides the quality of technique used. Figures 1 and 2 shows the flowchart for the embedding and the extraction.

2.1 Process of Embedding

We choose a string of binary numbers as watermark W (i) to embed it in the video frames. The watermark is embedded in Y frames of YUV color matrix format. The process is explained as follows:

- Step 1: Represent the videos into frames. Let F_n denotes the frames of size M $\times$ M, n = 1, 2, 3 …n. and Convert every video frame from RGB to YUV color space.
- Step 2: Perform the steps from 4 to 14 for the Y_n i.e. brightness matrix of each frame of the video with size M $\times$ M $\times$ n.
- Step 3: Perform the first level DWT of the input Y_n matrix. DWT will decompose the matrix into LL (low frequency sub-band), LH (Vertical high frequency

sub-band), HL (Horizontal high frequency sub-band), and HH (High frequency sub-band):

$$[\mathrm{LL},\ \mathrm{LH},\ \mathrm{HL},\ \mathrm{HH}] = \mathrm{dwt}\ (\mathrm{Y_n}) \tag{1}$$

- Step 4: Perform the second level DWT of the LL sub band obtained from step 4. This will further decompose this sub band to LL1, LH1, HL1 and HH1 sub-bands, such that:

$$[\mathrm{LL1},\ \mathrm{LH1},\ \mathrm{HL1},\ \mathrm{HH}] = \mathrm{dwt}\ (\mathrm{LL}) \tag{2}$$

- Step 5: Convert LH21 to vector $s_n(m)$, m = 1, 2, 3, 4 …N, where N = M*M/16 after performing a zigzag scan on them.
- Step 6: The vector $s_n(m)$ is sub sampled in two correlated sub-vectors using the following criteria:

$$\mathrm{s_{n1}} = \mathrm{s_n}(2\mathrm{p}) \tag{3}$$

$$\mathrm{s_{n2}} = \mathrm{s_n}(2\mathrm{p}-1),\ \text{Where},\ \mathrm{p}\ = 1\ldots\mathrm{N}/2. \tag{4}$$

- Step 7: Apply DCT transformation on these to vectors, such that:

$$\mathrm{S_{n1}} = \ \mathrm{dct}\ (\mathrm{s_{n1}}) \tag{5}$$

$$\mathrm{S_{n2}} = \ \mathrm{dct}\ (\mathrm{s_{n2}}) \tag{6}$$

- Step 8: Add the watermark W (i) in the DCT transformed vectors using differential embedding technique. This watermark is a sequence of binary digits with length L, therefore i = 1, 2, 3…L. This will generate two new vectors with watermark hidden in them. $\propto$ Represents the strength of embedded watermark and is called as gain factor and i′ are the locations in the S_{n1} and S_{n2} vectors where watermark is inserted. These locations are purely random. So, $(\propto, i')$ is the unique key of the user. Here, the frames of the video are sub sampled into sub-frames. These sub frames are differently altered. Further, differential embedding is used in which the watermark is firstly added to a sub frame and subtracted from the other sub frame. This is used as both the frame vectors are highly correlated.

$$E_{n1}(i') = \frac{1}{2}[S_{n1}(i') + S_{n2}(i')] + \propto W(i) \tag{7}$$

$$E_{n2}(i') = \frac{1}{2}[S_{n1}(i') + S_{n2}(i')] - \propto W(i) \tag{8}$$

- Step 9: Perform the inverse DCT of E_{n1} and E_{n2} using these equations:

$$\hat{S}_{n1} = \mathrm{idct}(E_{n1}) \tag{9}$$

$$\hat{S}_{n2} = \mathrm{idct}(E_{n2}) \tag{10}$$

- Step 10: By performing the reverse operation in (3) and (4) combine the $\hat{S}_{n1}$ and $\hat{S}_{n2}$ vectors to get:

$$\hat{S}_{n1}(2p) = \hat{S}_{n1}(p) \tag{11}$$

$$\hat{S}_{n2}(2p - 1) = \hat{S}_{n2}(p), \text{ Where, } p = 1\ldots N/2. \tag{12}$$

- Step 11: Convert the modified vector $\hat{S}_n$ into modified matrix by using inverse of zigzag scan used in step 6. For frame regeneration take first level inverse DWT and then second level inverse DWT of this modified matrix. This will give the modified Y_n frames with watermark hidden inside them.
- Step 12: Convert YUV color space with modified brightness matrix to RGB color space for viewing purpose. Repeat this procedure for all frames.

2.2 Process of Extraction

The process of extraction includes the steps of embedding up to step 6. The two sub sampled vectors of embedded video Y_n frames will then be used for further extraction of watermark. The extraction depends on the fact that the two sub-sampled matrices i.e. E_{n1} and E_{n2} are highly correlated. So in extraction phase the subtraction of these two vectors will give the watermark inserted in embedding phase. Where i = 1, 2, 3…L. The watermark extracted here can then convert to binary by setting the value to one if the difference is greater than zero and to zero if difference is less than one.

$$\Delta e_n(i) = E_{n1}(i') - E_{n2}(i') = 2 \propto W(i) \tag{13}$$

2.3 DWT Alone and DCT Alone Methods

The flowchart of dwt alone is similar to the Fig. 1 except that in case of dwt alone there will be no dct block. Similarly DCT alone can also be used to embed watermark in video. The flowchart of DCT alone is same as of DWT alone except that in place of DWT, DCT is used. Analyzing DWT alone and DCT alone will show why we have combined both the transform.

3 Performance Analyses

3.1 Description of Parameters Evaluated

The effectiveness of the proposed technique is being evaluated and presented in this section. For this method the technique is applied on the sample video 'wildlife' in mp4 video format. The duration of this video is of seven seconds with 25 frames per second. Thus the video has 193 frames. The frame size is selected as 256 × 256 i.e. the frame height and width are equal. The watermark used is a binary string of 256 binary bits. To check the transparency of the watermark PSNR equation is:

$$MSE = \sum\nolimits_{i=1}^{M} \sum\nolimits_{j=1}^{N} |X(i, j) - X'(i, j)| \tag{14}$$

$$PSNR = 10 \log\left(255^2\right)/MSE \tag{15}$$

To calculate the similarity between the actual and the extracted watermark BCR i.e. Bit Correction Ratio is evaluated. PSNR gives the measure of imperceptibility and BCR gives the measure of robustness. If the watermarks are similar then the BCR will 100% [1]. This can be presented as:

$$BCR = \frac{1}{L} \sum\nolimits_{K=0}^{L-1} \overline{W(K) \oplus W'(\mathrm{K})} \times 100\% \tag{16}$$

3.2 Invisibility Test

As invisibility is judged by PSNR, so here PSNR values are being calculated. PSNR depends on the gain factor. The gain factor or the strength of watermark is of immense importance. It sets a tradeoff between invisibility and robustness. If alpha is high the similarity between the extracted and embedded watermark will be high but this will

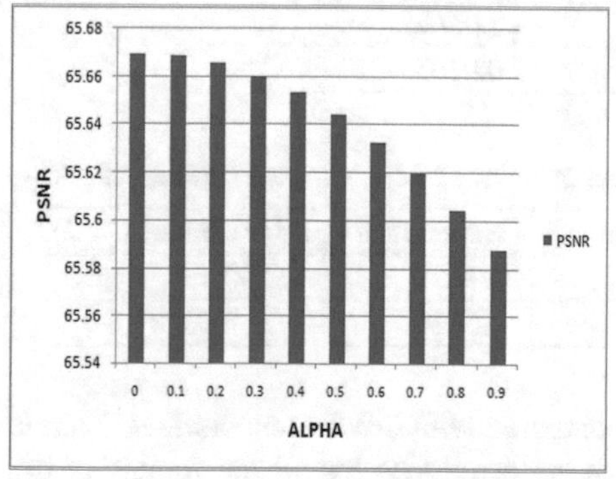

Fig. 3. PSNR values for different values of alpha.

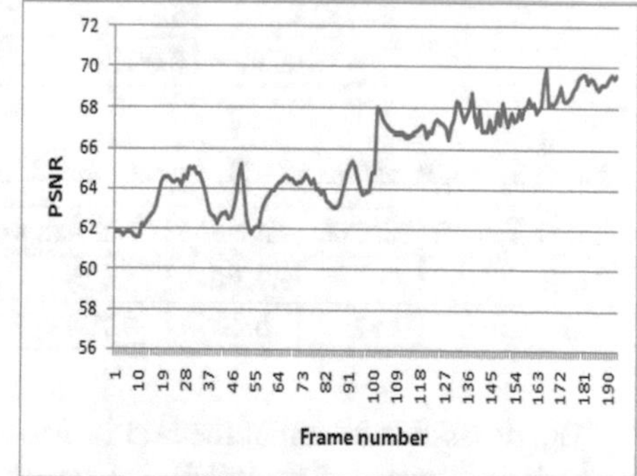

Fig. 4. PSNR values for all 193 frames of video.

give poor PSNR. A low alpha means high PSNR but low BCR. So, its value is precisely taken. Figure 3 shows the variation of PSNR with gain factor alpha. From this figure we can conclude that a good PSNR, alpha is between 0.2 and 0.3. For the rest of the evaluation the alpha is fixed at 0.2. Figure 4 shows the PSNR value for each frame. From this we can see that for all frames PSNR is above 60 db. These results are better than other techniques given in video watermarking techniques.

3.3 Robustness Test

Robustness means the ability of watermark to resist the geometrical attacks or the video processing attacks or the harm caused to the video by compression. Table 1 shows the numerical values of correlation and BCR when video is attacked by noise. The technique is checked against Gaussian noise, salt and pepper noise and speckle noise. The BCR values are close to 100%; this means the watermark is extracted successfully when attacked by noise and is robust against noise addition. Table 2 shows the values of BCR against geometrical attacks. The video is resized and rotated to some value to check the robustness of the method explained. The table verifies that BCR values are good and thus proves the robustness of method. Table 3 shows the BCR values of method under low pass filtering attacks. All the result shows that the BCR is around 70%. Hence, it is verified that the method has better robustness results for low pass filtering attacks.

Table 1. BCR values when video is attacked by noise

Noise	Gaussian var = 0.05	Gaussian var = 0.01	Salt pepper var = 0.05	Salt pepper var = 0.01	Speckle var = 0.05	Speckle var = 0.01
BCR	94.5313	97.2656	98.8281	99.2188	94.5313	99.6094
Correlation	0.9606	0.9783	0.9882	0.9921	0.9542	0.9941

Table 2. BCR values of differential embedding technique under geometrical attacks

Attacks	Resizing (two times)	Rotation (0.05°)
BCR	50	51.5625
Correlation	0.0062	0.0165

Table 3. BCR values of differential embedding technique under low pass filtering attacks

Filter(3 × 3)	Median	Average	Gaussian var = 1	Gaussian var = 1.5	Wiener (3 × 3)
BCR	75	69.1406	78.125	70.7031	93.75
Correlation	0.5254	0.4968	0.7045	0.5624	0.9318

The dwt-dct-differential method is also compared with DWT alone and DCT alone methods in Table 4. The PSNR values of these techniques for all the frames of the given video are shown and compared in Fig. 5. From the table and figure we can see that the above described technique has better results for the noise attacks and low pass

Table 4. Comparison of evaluated method with other techniques under noise

Noise	Gaussian var 0.05	Gaussian var 0.01	Salt pepper var 0.05	Salt pepper var0.01	Speckle var 0.05	Speckle var 0.01
DWT alone	92.9688	92.1875	92.1875	92.5781	89.4531	92.5781
DCT alone	92.5781	95.7031	91.0156	92.9688	91.4063	94.5313
Evaluated method	94.5313	97.2656	99.2188	98.8281	94.5313	99.6094

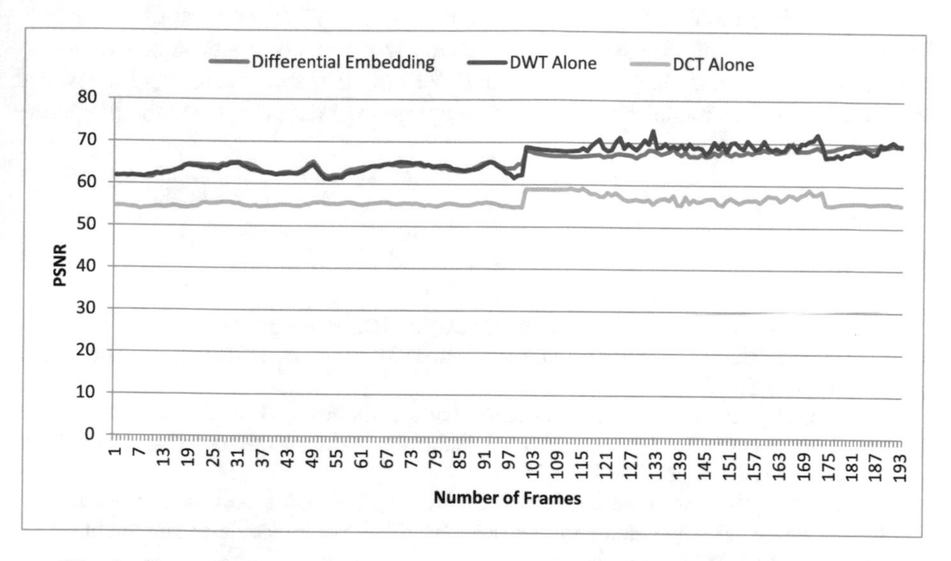

Fig. 5. Shows the comparison of evaluated technique with dwt alone and dct alone

Table 5. Comparison of evaluated method with other techniques under filtering

Filter (3 × 3)	Median	Average	Gaussian var = 1	Gaussian var 1.5	Wiener
DWT alone	53.9063	63.2813	72.2656	65.2344	91.0156
DCT alone	82.8125	80.4688	81.6406	80.0781	81.25
Evaluated method	75	69.1406	78.125	70.7031	93.75

filtering attacks. DCT only method shows good results when the frames are analyzed against filtering techniques but shows poor PSNR values. As invisibility and robustness are two important characteristics a watermark should possess, hence differential embedding with both DCT and DWT wins over DWT alone and DCT alone. The result of used method is 99% for salt and pepper noise and where as for dwt alone and dct alone it is only 92% and 93% respectively (Table 5).

4 Conclusion

In this paper the performance analysis of digital video watermarking based on transform domain and differential embedding is demonstrated. It is stated that a powerful technique must possess an invisible and robust watermark. A low gain factor, gives better invisibility but robustness becomes poor. It is observed that when alpha is fixed neither too low nor too high both invisibility and robustness could be achieved. The imperceptibility is judged in terms of PSNR and BCR gives the measure of robustness. The PSNR and BCR values are calculated and represented for noise attack, geometrical attacks and low pass filtering attacks. The PSNR values are above 65 db for all the frames which gives invisibility to the embedded watermark. It has been noticed that for the dwt-dct-differential method the BCR value is above 90% for most of the attacks. Thus a watermark can be successfully extracted when video is attacked. The performance of the method is also compared with DWT only method. The results suggest that the technique has potential for securing the copyright of the author and the illegal use of the videos.

References

1. Benoraira, A., Benmahammed, K., Boucenna, N.: Blind image watermarking technique based on differential embedding in DWT and DCT domains. EURASIP J. Adv. Signal Process. **1**, 1 (2015)
2. Mohammadi, S.: A novel video watermarking algorithm based on Chaotic maps in the transform domain. In: IEEE Artificial Intelligence and Signal Processing International Symposium (AISP), pp. 188–191 (2015)
3. Asikuzzaman, M., Alam, M.J., Pickering, M.R.: A blind and robust video watermarking scheme in the DT CWT and SVD Domain. In: IEEE Picture Coding Symposium (PCS), pp. 277–281 (2015)
4. Chan, P.W., Lyu, M.R., Chin, R.T.: A novel scheme for hybrid digital video watermarking; approach, evaluation, experiment. IEEE Trans. Circ. Syst. Video Technol. **15**(12), 1638–1649 (2005)
5. Rawat, C.D., Shivamkutty, S.M.: Digital watermarking of video using hybrid techniques. In: IEEE Advances in Communication and Computing Technologies International Conference (ICACACT), pp. 1–5 (2014)

Multi-agent Simulation Model for Sequence Generation for Specially Abled Learners

Jonita Roman[1(✉)], Devarshi R. Mehta[2], and Priti S. Sajja[3]

[1] Sardar Vallabhbhai Patel Institute of Technology, Vasad, Gujarat, India
jonita27@rediffmail.com
[2] GLC-ICT, Ahmedabad, Gujarat, India
Devarshi.mehta@glsuniversity.ac.in
[3] G.H. Patel P.G. Department of Computer Science and Technology, Anand, Gujarat, India
priti@pritisajja.info

Abstract. To help specially abled learners few Intelligent Tutoring Systems solutions exist. However, these tutoring systems lack in construction of individualized dynamic sequences and on the go learning content provision as per the learning ability of the learner. In order to overcome this limitation, there is a need of an intelligent system with a distributed approach that will handle this issue. This paper describes an architecture using Multi Agent Simulation model that where the agents interact with each other to cater to the individual needs of the specially abled learner. The paper suggests a simulation model where the agents such as Pedagogical agents, Intelligent Sequencing agent and Feedback with their detailed design.

Keywords: Multi-agent-based simulation · Intelligent Tutoring System · Specially abled learners · Pedagogy · Sequencing

1 Introduction

Agent based systems, being intelligent, dynamic, computational, having decision-making ability, autonomous, collaborative and many such attributes have gained a good place in various real time applications [1–3]. Various area of applications are Agriculture [2], Air Traffic Control [3], Biomedical Research [4], Crime Analysis [5], Ecology [6, 7], Economics [8], Energy Analysis [9–11], Infectious Diseases [12], Healthcare [13], Social Psychology [14] and many such. Intelligent tutoring systems are developed such that they adapt themselves to the learning needs of the learner [15]. Learning objects in digital form are already made available in the form of DLO (Digital Learning Objects) repository. Later on using these, a sequence is generated to cater to the learning needs of the learner. There are two main concerns when ITS is talked about: (i) Creation of DLOs and (ii) Sequencing of DLOs for learner needs [16–19]. ITS for specially abled learners also take these two parameter along with a few others such as the difficulty faced and the level of difficulty faced by the learner [20]. It is thus, an open area of research for designing an ITS such specially abled learners. Agent based simulations

S.C. Satapathy and A. Joshi (eds.), *Information and Communication Technology for Intelligent Systems (ICTIS 2017) - Volume 1*, Smart Innovation, Systems and Technologies 83, DOI 10.1007/978-3-319-63673-3_69

can come to aid for these learners [15, 21, 22]. In this research, we propose a simulation model, which, with the help of learning agents provide learning contents to the specially abled children.

2 Background Work

2.1 Need for Intelligent Tutoring System for Specially Abled Learners

Teaching learning process is highly affected by the use of various Learning Technologies in regular curriculum. This is also is equally true in the case of special needs learners, both with physical as well as mental difficulties [20]. Various assistive learning tools are devised and developed using several educational theories [19], in order to aid these learners and to cope up with their educational needs. Assistive technologies work for mainly two concerns: (i) Content of Learning as learning objects (ii) Sequencing of learning objects to get the best learning patterns. Learning objects creation is a challenging venture, especially for special needs learners, as the objects have to be created looking to the difficulties the learner faces [16–19]. Here we discuss about various methods applied for sequencing of learning objects. Competency based learning method was used by [23] for learning purposes. A dynamic learning path was formulated by defining relationship between competencies and meta-data by grouping the learners in various groups. The research is based on problem solving using Ant Colony optimization and Swarm Optimization techniques. In an ecological model proposed by [21], the concept of learning thorough peers was considered and depending on this, the learning objects were linked to the students. In continuation of the previous work, [21] proposed a learning model where the experiences with the learning objects of the students were maintained as different states in the form of initial and final states, after the experience of the learners was noted. A nature based, Ant Colony Optimization evolutionary technique was proposed by [24]. The level of the student was considered, and according to the levels noted, appropriate learning concepts relevant for the student were proposed. As per the change in student learning behavior, appropriately change in learning concepts would be provided to the students. An Ontology-based curriculum was developed [25] where a general sequencing knowledge base was created. The knowledge base had two basic components: (i) Ontology: used for representation of abstract views of content sequencing and course materials and (ii) Semantic rules were used for representation of relationship between individuals. Online learning intelligent tutoring system was developed by [15], which used natural language interface. It aimed to mimic the human knowledge for interaction. They named it Oscar CITS, that took input as natural language and detected the individual's learning style and accordingly adapted itself to create learning sequences.

2.2 Multi-agent Based Approach

Agent-based simulation (ABS), or agent-based modeling (ABM), is a computational and modeling framework that is used for simulating dynamic processes that involve autonomous agents. An agent is a generalized concept of people or group of people interacting with each other to represent certain processes. They are also involved in taking self-decisions and respond to order or change during the model simulation [26, 27]. Agent-based simulations gain a vast and broad applicability domain. Right from Agriculture [2], Air Traffic Control [3], Biomedical Research [4], Crime Analysis [5], Ecology [6, 7], Economics [8], Energy Analysis [9–11], Infectious Diseases [12], Healthcare [13], Social Psychology [14] and many such domains.

2.3 Objective and Need

Looking to the autonomous, dynamicity, collaborative nature, Agent based simulation model actually fits into the forming of learning model for specially abled learners through intelligent tutoring systems. These simulation models will have agents that will interact with the learners in a similar way a teacher or a mentor does and provides individualized learning patterns. Thus the main concern is to develop a model that uses human like modules that interact with the learners.

3 Process Model

3.1 General Process Model

A general agent based system comprises of various agents that interact with each other to achieve a target or a task. An interface agent generally interacts with various other agents and repositories. It is from this interaction, it obtains its dynamicity and autonomous behavior. The framework helps in sequence of interactions and simulations.
In the model proposed by this research, we have identified three agents:

i. Pedagogical Agent
ii. Sequencing Agent
iii. Feedback Agent

Each agent interacts with the others in order to meet the learning requirements of the specially abled learner (Fig. 1).

This multi-agent simulation model works in two phases.

The first phase: Phase I, takes the inputs related to the specially abled learner. Here the inputs are in the form of the learner information pertaining to the difficulty faced by the learner along with the level of difficulty faced. Apart from this the other personal information that lies in the repository also acts as an input. A replica of the pedagogical agent (from Phase II) is stored and is also considered for providing learning content. The inputs from the learners taken in the form of preferences, pedagogical structures help the sequencing agent in retrieving the apt content for being learnt.

The second phase: Phase II, maintains the repositories related to learning, practice and assessment related content. It also had frequently asked question module and popular content sequences.

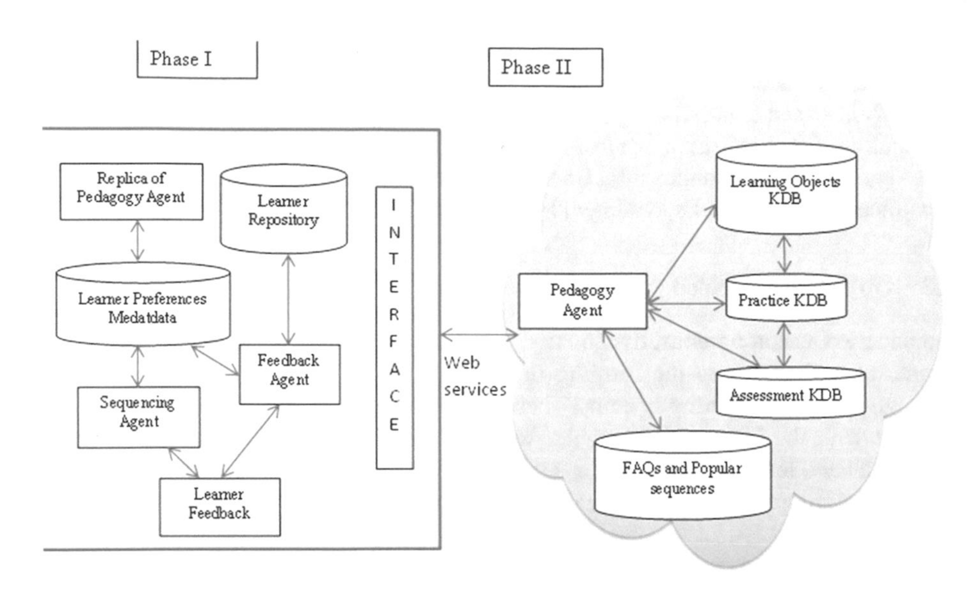

Fig. 1. Multi-agent architecture with interacting agents

Pedagogical Agent: Learning module is divided into three main parts: Learning, Practice and Assessment. Every learner goes through a cycle of Learn-Practice-Assess. Since every abled learner is associated with a different type of difficulty and level of difficulty, hence the number of iterations of this cycle varies from learner to learner. It should also be noted that some learners take the individual part for a number of times. As a sample, a learner may take four times the same learning content. This frequency of studying the same content repeatedly depends on the learner's preference and various other individual's personal parameters. The pedagogical agent is thus responsible to maintain this check of repetitive behavior of the learner. The pedagogical agent along with the sequencing agent, needs to see the learning pattern of the learner through the assessment results. Timely making the already learnt contents to be revised is also the responsibility of the pedagogical agent.

Sequencing Agent: On the basis of the preferences set and feedback obtained for the individual learner from the feedback and learning preferences and the content offered by the pedagogical agent, an individual sequence is maintained by the sequencing agent. This agent works in collaboration with pedagogical agent and the feedback agent. An important task performed by the sequencing agent is to maintain content sequences for each individual separately. Thus the sequencing agent is responsible for maintaining and provided individualized dynamic contents as per the learner needs.

Feedback Agent: The learning cycle (learn-practice-assess) taken by the learner for every learning content is unique. The feedback agent acts as a teacher or a mentor that provides feedback to the sequencing and pedagogy agents. Just a teacher, understands the need of the learner from the manual feedback of the learner, similarly this agent also takes the feedback as input and helps in providing appropriate learning sequences.

4 Conclusion

This paper proposes a multi-agent simulation model for providing intelligent tutoring systems for specially abled children. The model provides dynamic learning sequences on the basis of the interaction between the pedagogical, sequencing and feedback agents. The intelligent and autonomous behavior of the model is decided on the basis of the various parameters passed to the agents through the repositories present in the model. As a part of future enhancement, the simulation model can be compared with the existing standard models for quality assurance. As an application of this model, we can further enhance the learning repositories by adding content as agents that can develop learner specific learning objects.

References

1. Macal, C., North, M.: Introductory tutorial: agent-based modeling and simulation. Proceedings of the 2014 Winter Simulation Conference. IEEE Press (2014)
2. Leyk, S., Binder, C.R., Nuckols, J.R.: Spatial modeling of personalized exposure dynamics: the case of pesticide use in small-scale agricultural production landscapes of the developing world. Int. J. Health Geogr. **8**(1), 1 (2009)
3. Conway, S.R.: An Agent-Based Model for Analyzing Control Policies and the Dynamic Service-Time Performance of a Capacity-Constrained Air Traffic Management Facility. NASA Langley Research Center, Hampton (2006)
4. Folcik, V.A., An, G.C., Orosz, C.G.: The basic immune simulator: an agent-based model to study the interactions between innate and adaptive immunity. Theor. Biol. Med. Model. **4**(1), 1 (2007)
5. Malleson, N., Heppenstall, A., See, L.: Crime reduction through simulation: an agent-based model of burglary. Comput. Environ. Urban Syst. **34**(3), 236–250 (2010)
6. Grosman, P.D., et al.: Trade-off between road avoidance and attraction by roadside salt pools in moose: An agent-based model to assess measures for reducing moose-vehicle collisions. Ecol. Model. **222**(8), 1423–1435 (2011)
7. Mock, K.J., Testa, J.W.: An agent-based model of predator-prey relationships between transient killer whales and other marine mammals. University of Alaska Anchorage, Anchorage, AK, Technical report (2007)
8. Bookstaber, R.: Using agent-based models for analyzing threats to financial stability (2012)
9. Chen, J., Taylor, J.E., Wei, H.-H.: Modeling building occupant network energy consumption decision-making: the interplay between network structure and conservation. Energy Build. **47**, 515–524 (2012)
10. Jackson, J.: Improving energy efficiency and smart grid program analysis with agent-based end-use forecasting models. Energy Policy **38**(7), 3771–3780 (2010)
11. Wittmann, T.: Agent-Based Models of Energy Investment Decisions. Springer, Heidelberg (2008)
12. Parker, J., Epstein, J.M.: A distributed platform for global-scale agent-based models of disease transmission. ACM Trans. Model. Comput. Simul. **22**(1), 2 (2011)
13. Hunt, C.A., et al.: Agent-based modeling: a systematic assessment of use cases and requirements for enhancing pharmaceutical research and development productivity. Wiley Interdiscip.Rev. **5**(4), 461–480 (2013)

14. Smith, E.R., Conrey, F.R.: Agent-based modeling: a new approach for theory building in social psychology. Personal. Soc. Psychol. Rev. **11**(1), 87–104 (2007)
15. Latham, A., Keeley, C., McLean, D.: An adaptation algorithm for an intelligent natural language tutoring system. Comput. Educ. **71**, 97–110 (2014)
16. Chatterjee, P., Asoke, N.: Massive open online courses (MOOCs) in education—A case study in Indian context and vision to ubiquitous learning. MOOC, Innovation and Technology in Education (MITE), 2014 IEEE International Conference on. IEEE (2014)
17. Chakraborty, S., Devshri. R., Anupam, B.: Shikshak: an architecture for an intelligent tutoring system. In: Proceedings of 16th International Conference on Computers in Education, Taipei, Taiwan (2008)
18. Chakraborty, S., Roy, D., Basu, A.: Development of knowledge based intelligent tutoring system. Adv. Knowl. Based Syst. **1**, 74–100 (2010)
19. Sajadi, S.S., Khan, T.M.: An evaluation of constructivism for learners with ADHD: development of a constructivist pedagogy for special needs. In: European, Mediterranean & Middle Eastern Conference on Information Systems (2011)
20. Madaus, J.W., McKeown, K., Gelbar, N., Banerjee, M.: The online and blended learning experience: differences for students with and without learning disabilities and attention deficit/ hyperactivity disorder. Int. J. Res. Learn Disabil **1**(1), 21–36 (2012)
21. McCalla, G.I.: The search for adaptability, flexibility, and individualization: Approaches to curriculum in intelligent tutoring systems. Adaptive Learning Environments, pp. 91–121. Springer, Berlin (1992)
22. Champaign, J., Robin, C.: A model for content sequencing in intelligent tutoring systems based on the ecological approach and its validation through simulated students. In: FLAIRS Conference (2010)
23. Rodríguez-Fórtiz, M.J., et al.: Sc@ut: developing adapted communicators for special education. Proc. Soc. Behav. Sci. **1**(1), 1348–1352 (2009)
24. Sharma, R., Hema, B., Punam, B.: Adaptive content sequencing for e-learning courses using ant colony optimization. In: Proceedings of the International Conference on Soft Computing for Problem Solving (SocProS 2011), December 20–22, 2011. Springer, India (2012)
25. Chi, Yu-Liang: Ontology-based curriculum content sequencing system with semantic rules. Expert Syst. Appl. **36**(4), 7838–7847 (2009)
26. Gilbert, N., Troitzsch, K.: Simulation for the Social Scientist. McGraw-Hill, New York (2005)
27. Bonabeau, E.: Agent-based modeling: methods and techniques for simulating human systems. Proc. Natl. Acad. Sci. U.S.A. **99**(suppl 3), 7280–7287 (2002)

Analysis of Text Messages in Social Media to Investigate CyberPsycho Attack

Prashant Gupta(✉) and Manisha J. Nene

Department of Computer Science and Engineering,
Defence Institute of Advanced Technology (DIAT),
Pune, Maharashtra 411025, India
{prashant_mce15,mjnene}@diat.ac.in

Abstract. Social media proved to be a medium to document our daily lives. It is a virtual tool used by an individual or organizations for various communications and information exchanges. Data available on social media assists in solving crimes; like to identify people, locations, evidence gathering notifying public, community, soliciting crimes etc. It is desired to attain privacy & security on these platforms. Further, it is observed that the information exchange and forwards are persuasive messages. These messages are intended to attain economical, political, religious, regional gains. The study in this paper proposes a model to analysis persuasive messages on social media. The results of the analysis demonstrate that these messages are the potential contents which leads to cyberpsycho attacks. The work in this paper is a step towards a study of threat analysis which demonstrates the psychological health of society & cyberpsycho attacks.

Keywords: Cyberpsycho attack · Text analysis · Social media · Whatsapp messages · Persuasive messages

1 Introduction

Cyber space spreads its volume exponentially in terms of its users and user- generated contents. A variety of platforms on social networks are available in cyber space for its users, and provides ease of technology, connectivity and reachability with lesser amount of time and money. It is the place where top leader, celebrities, professionals, students, housewives, worker etc., can share his or her words without any physical appearance and hesitation. It is a place which is also used to spread facts and figures about any matter for making or breaking the trending news. Since contents are shared through connected and accepted users, one may believe the concerned matter.

Later on it is identified as *viral contents*. This imposes a challenge for analysis to the cyber experts; since they do not fall in the category of *privacy* or *security threat*.

P. Gupta and M.J. Nene are employees of Defence Institute of Psychological Research (DIPR), DRDO, MoD, Delhi and Defence Institute of Advanced Technology (DIAT), DRDO, MoD, Pune, respectively. Prashant Gupta is currently pursuing his post graduation in Computer Science & Engg from DIAT.

S.C. Satapathy and A. Joshi (eds.), *Information and Communication Technology for Intelligent Systems (ICTIS 2017) - Volume 1*, Smart Innovation, Systems and Technologies 83, DOI 10.1007/978-3-319-63673-3_70

However, there lies a sever impact of such messages which leads to an attack that influences the behavior patterns/change in the users who are the part of that social networks. The bigger impact of causes leads to *cyberpsycho attack;* the attacks use cyberspace to impact the thought process & psychology of the users of the social network.

In this attack, attacker make content more attractive and persuasive by using trending news, events with sentiments, and forward it on cyberspace and use social network of users to fulfill his/her motive or objective without any illegal access to person's device or information. To do this activity, an attacker need not include his/her identity but only need to apply emotions which will effect others to do same activity or perform as per the objective of the messages. Users who influenced from this *cyberpsycho attack* are motivated, influenced, as a result of this impact, users may spend their money, time and other cyber resources. The influence of this attack is so much that they take decisions on the basis of it without knowing the actuality. Some times contents are designed in a way that it looks to be authentic. The study in this paper describes the potential of messages with respect to sentiments and emotions that leads to *cyberpsycho attack*. Further, it sets motivation *to* researchers, analyst to acknowledge this domain by including technologies and psychological strategies together. The proposed work in this paper for the analysis of persuasive messages is a step towards an investigation of cyberpsycho attacks.

Outline of this paper: Sect. 2 describes persuasion with cyberpsycho effect and the methods which related to analysis of text and sentiments. Section 3 describes the motivation towards the proposed work. Our proposed work is explained along with data collection, platform and techniques used in the analysis of cyberpsycho attacks. Section 4 shows some result associated to analysis and describes observations and Sect. 5 summarizes and concludes the work.

2 Related Work

Strategies that are used to create funny short messages, taglines, interactive images, online campaigns, survey & polls, video, advertisements, etc., attract the targeted audience like customers, voters or people according to their objectives. Some of them are used to trigger cyber attacks. By making contents effective and impressive, they persuade audience to believe in it and act accordingly which leads to *cyberpsycho effect*. Persuasive computing technologies can influence people's attitudes and bring some constructive changes in many domains like marketing, health, safety, environment etc. [2]. The content-generator generates such persuasive contents which have some triggers to influence users which persuade them to share this content to others, recursively.

Researchers have proposed different methods to analyze the text messages of social media. In [3], the authors proposed *trendminer* an architecture for real time analysis of social media texts which is a open source framework to perform text analysis tasks on Social Media data. In [4], the authors proposed a framework for social media analysis in political context. In [5], the authors proposed a system that assigns scores indicating positive or negative opinion to each distinct entity in the text corpus in large-scale sentiment analysis for news and blogs. In [6], the authors proposed the basic design and architecture of text analysis for large scale news analysis. In [7], the authors proposes an automatic trend mining method based on knowledge integrating learning approach. In [8], the authors present a web service to analyze the contents shared by multiple online social network platforms.

The work done in [9–12] proposed methods, frameworks & approaches to analysis text messages, classification, detection of trending topics in social networks. In [13], the authors analyzed emotions and sociability of tweets based on demographics characteristics. In [14], the authors analyzed sentiments of trending topic in twitter to trace the precise date of an event by analyzing the number of Tweets and their sentiment scores. However, the analysis of cyberpsycho effects which leads to cyberpsycho attack remains and hence becomes the need of the present. To do this, the authors proposed an approach to analyze the tweets based on the trending topics and demonstrate the enabling phases of cyberpsycho attacks [15]. Further, the work in this paper is a step towards the study of threat, enabled using text messages, which predicts the psychological health of society & cyberpsycho attacks.

3 Proposed Work

This section describes the motivation for the proposed work, the proposed model, the parameters that leads to *cyberpsycho-effect* and the steps to analyze it.

3.1 Motivation

The persuasive strategies or psychology used in various domains like to increase productivity or improve services are acknowledged as quite impressive in changing people's behavior and attitude. The volume of messages flow throughout social-networks very quickly on the occurrence of an incident, the information is spread in no time. Everybody on social-networks are a kind of author, poet or publisher to spread the contents related to the occurrence of incident. This consumes cyber space, time and amount of data transactions. Hence, there is a need to study and analysis these messages and provide enough information about how human beings perceive and express information in the form of text to express their feelings and emotions, and how it will affect emotions, perception and decision taking ability of others.

3.2 Model & Parameters

The Fig. 1 represents the model to demonstrate interrelationships of the parameters of *cyberpsycho-effec*t in which persuasiveness made the contents on social media viral and were disseminated using different technology and social network services [1]. These social network services used persuasive technology to provide better services as well as to increase user involvement. Further it will drew some objective and laid psychologically impact on society.

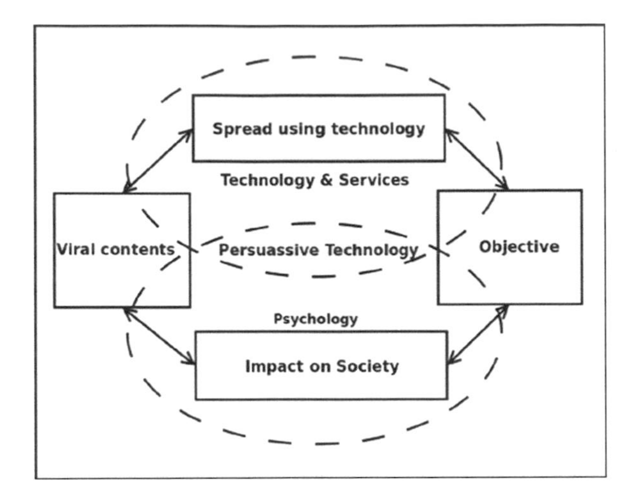

Fig. 1. Parameters of CyberPsycho effect.

The contents are decorated with some objectives that persuade human sentiments, that are forwarded through different social media platforms. These platforms are accessed using different computing devices. However, these forwarded messages with persuasive contents becomes viral contents may impact on one's perception or decisions and/or society by sentiments, emotions, or the facts that it includes.

3.3 Stages and Steps

There are two ways to analyze the cyberpsycho attacks. First, to analyze the content collected through a single machine or individual device of social network users and Second, to analyze the contents by collecting through servers of social network service providers. The work done in this paper, uses first way of analysis for contents collection.

To analyze the whatsapp messages as shown in Fig. 2, The steps start from the collection of data, here messages collected through *email chat history* feature of application. This will generate a text file which can be used as input for analyses framework for further processing. Here *Rstudio* used as content analysis framework in which r script is used to analyze the sentiment, emotions in the messages. Sample text messages used for analysis are regional or English language but excluding media like images etc. In content analysis framework, apart from the *R*, any other analysis framework can be used which analyze the text contents. The steps describe in this paper are general which can be applied in any social network platform as well as with any content analyses framework. The scope of the work in this paper restricted to analyze the emotions, sentiment of all forwarded messages with an objective to predict the occurrence cyberpsycho attack.

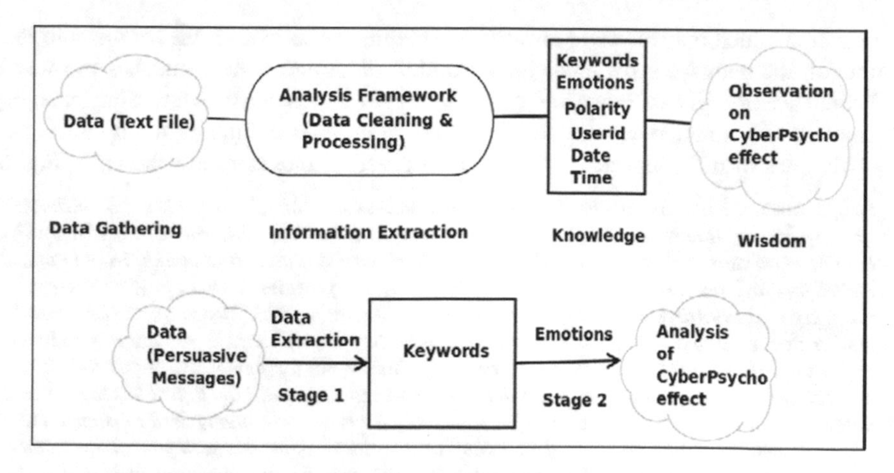

Fig. 2. Model for identification and analysis of CyberPsycho attacks.

4 Result and Analysis

Data sets used to analyze the cyberpsycho effect collected through author's WhatsApp account having all the messages including current trending news and events consist about the one year of databases. Data sets are in the form of texts messages, Phases involved during analysis process are as per phases in Fig. 2. analysis output which shown in Fig. 3 is generated using *RStudio*.

In content analysis framework, First, cleaning of input data is performed that involves removing punctuation, special symbols, white spaces etc., Second, by using *sentiment* package of *R* performs the emotions extraction from the collected text data and generate polarity graph similar to as shown in Figs. 3 and 4. In the sentiment and emotion extraction process the script matches the positive & negative words with pre-defined positive & negative matches as defined in words dictionary. Then it counts with respects to sentiments and make a list of relevant words that shows in Fig. 3. there are different kind of messages found which spreads positive and negative impact. Some of them are identified as neutral messages by the system but these also containing some

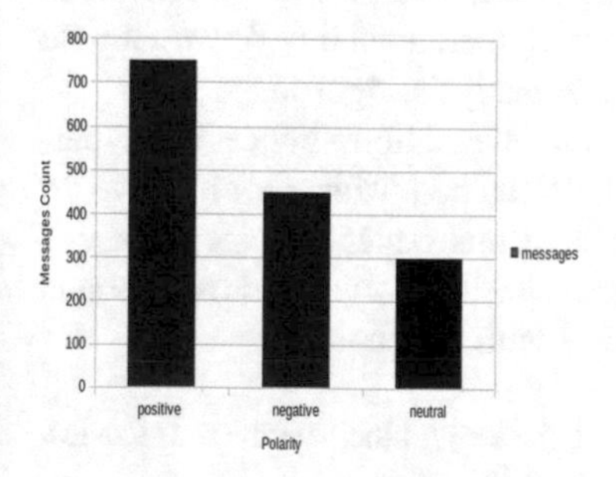

Fig. 3. Sentiments ANALYSIS of WhatsApp messages on polarity

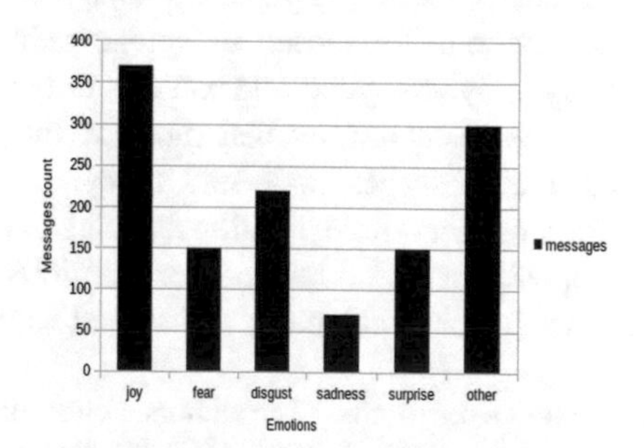

Fig. 4. Sentiment analysis of WhatsApp messages on emotions

information which may effective towards persuading perception. Same for the emotions extraction the package have predefined emotion dictionary which matches the words and count accordingly as shown in Fig. 4. The words which does not fall in any category of emotions are counted as other words & shown in *other* column in the figure.

Examples of most repeated words that observed in messages are shown in Fig. 5.

> ***Single word-*** *Share, Indian, hindu, modi, demonatisation, patanjli, pm, biwi, jio, history, groups, friends, family, pappu, forward, strike, surgical, market, new, namo, rss, isis, free.* ***Double word-*** *surgical strike, ache din, modi sarkar, new message, true indian, truly hindu, please forward, plz forward, create awareness, bad news, good news, don't buy, no reservations, try once.* ***Phrases & sentenses -*** *share if you like, share with friends, if you love your mother then read and share, share with your family, very urgent news, if you are true indian then share, if you are true hindu then share, important news for family, share this link fast, don't share if you are not human, don't delete need to share, forward to all, fwd this msg to as many as you can, fed as received, only for girls, only for boys, we actually need to change the system, issued by indian gov, issued by indian army, please share this with your friends and help the nation, share the truth, issued in public interest, send this message to all who matter to you, breaking news*

Fig. 5. Most repeatetive keywords in WhatsApp messages.

5 Conclusion

The study in this paper analyzes the text messages and demonstrates that these messages are the potential contents which leads to *cyberpsycho effect*.

The work in this paper explains the method to analysis the volume of sentiments in the contents forwarded or spread using popular social media mainly *WhatsAap* to analyze cyberpsycho attacks; emotions and sentiments analysis, rigorously.

The results show that most of the messages contain emotions hence these persuasive messages may lead to cyberpsycho attacks launched with an objective of attaining economical/political/social gains. The work in this paper is a step towards proposing a method to analyze emotions and sentiments that are forwarded using social networks or cyberspace; and intend to use technology with psychology.

Acknowledgements. The authors would like to thank the Director and the Members of Review Committee, DIPR, DRDO, MoD for the permission to publish this work.

References

1. Prashant, G., Nene, M.J.: CyberPsycho attacks: techniques, causes, effects and recommendations to end-users. Int. J. Comput. Appl. **156**, 11 (2016)
2. Cheng, R.: Persuasion strategies for computers as persuasive technologies. Department of Computer Science, University of Saskatchewan, Saskatoon (2003)
3. Preotiuc-Pietro, D., et al.: Trendminer: an architecture for real time analysis of social media text. In: Proceedings of the Workshop on Real-Time Analysis and Mining of Social Streams (2012)
4. Stieglitz, S., Dang-Xuan, L.: Social media and political communication: a social media analytics framework. Soc. Network Anal. Min **3**(4), 1277–1291 (2013)
5. Godbole, N., Srinivasaiah, M., Skiena, S.: Large-scale sentiment analysis for news and blogs. ICWSM **7**(21), 219–222 (2007)
6. Levon, L., Kechagias, D., Skiena, S.: Lydia: a system for large-scale news analysis. In: International Symposium on String Processing and Information Retrieval. Springer, Heidelberg (2005)
7. Olga, S.: Mining trends in texts on the web. In: Proceedings of the Doctoral Consortium of the 3rd Future Internet Symposium (2010)
8. Tiago, R., et al.: uTrack: track yourself! monitoring information on online social media. In: Proceedings of the 22nd International Conference on World Wide Web. ACM (2013)
9. Tim, A., et al.: Analysis and forecasting of trending topics in online media streams. In: Proceedings of the 21st ACM International Conference on Multimedia. ACM (2013)
10. Jey Han, L., Collier, N., Baldwin T.: On-line trend analysis with topic models:\# twitter trends detection topic model online In: COLING 2012 (2012)
11. Kathy, L., et al.: Twitter trending topic classification. In: 2011 IEEE 11th International Conference on Data Mining Workshops (ICDMW). IEEE (2011)
12. Raju, E., Sravanthi, K.: Analysis of social networks using the techniques of web mining. Int. J. **2**, 10 (2012)
13. Kristina, L., et al.: Emotions, Demographics and Sociability in Twitter (2016)
14. Dongjin, C., et al.: Tracing trending topics by analyzing the sentiment status of tweets. Comput. Sci. Info. Syst. **11**(1), 157–169 (2014)
15. Gupta, P., Nene, M. J.: An approach to analyze cyberpsycho attacks enabled using persuasive messages. In: 1st International Conference on Smart Computing & Informatics. Springer (in press, 2017)

K-Means Clustering with Neural Networks for ATM Cash Repository Prediction

Pankaj Kumar Jadwal[1(✉)], Sonal Jain[2], Umesh Gupta[3], and Prashant Khanna[4,5]

[1] JK Lakshmipat University, Jaipur, India
pankajjadwal@gmail.com
[2] Department of Computer Science Engineering, JK Lakshmipat University, Jaipur, India
sonaljain@jklu.edu.in
[3] Department of Mathematics, JK Lakshmipat University, Jaipur, India
umeshgupta@jklu.edu.in
[4] Wintec, Hamilton, New Zealand
perukhan@gmail.com
[5] Wintec, Hamilton, New Zealand

Abstract. Optimal forecasting of ATM cash repository in an optimal way is a complex task. This paper deals with cash demand forecasting of NN5 time series data using neural networks. NN5 reduced Dataset is a subsample of 11 time series of complete dataset of 111 daily time series drawn from homogeneous population of empirical cash demand time series. Main objective of this paper is to forecast cash demand forecasting of NN5 data with neural networks. Further, the same process is applied on clusters of ATMs. Discrete time wrapping is used as distance measure. Root mean square error has been calculated for such clustered group of ATMs and average is calculated. Root Mean Square error indicates applications of clustering before applying Neural Network increases precision in forecasting of ATM Cash Repository.

Keywords: Clustering · Forecasting · Neural networks-prediction

1 Introduction

Banking industry plays a major role in terms of an integral part of economy of any country. Today, Major Banks are competing with each other for getting large share of customer financial transactions. Banks want to have more and more customers, more and more customer transactions and retain them as long as possible in an optimal way. ATMs play a significant role in terms of cash flow between customers and banks. ATMs are the machines for customers to provide easy cash withdrawal in a hassle freeway. As the technology is getting better day by day, Number of customers in banks is also increasing in a rapid way so it is obvious that banking customers to ATMs ratio should be increased relatively. After installing and placing ATMs, cash demand forecasting is a crucial and important issue. Because if cash in an ATM is much higher than customer demand in that area, then this will keep the cash unused and if cash in

S.C. Satapathy and A. Joshi (eds.), *Information and Communication Technology for Intelligent Systems (ICTIS 2017) - Volume 1*, Smart Innovation, Systems and Technologies 83, DOI 10.1007/978-3-319-63673-3_71

that ATM is much lower, then it will lead towards customer dissatisfaction in terms of not having required cash in ATMs.

In this paper application of neural networks is discussed in concern with cash demand forecasting in ATMs. Forecasting of cash in ATMs should be more accurate for keeping both sides alive. Machine learning techniques especially neural networks can play an important role in terms of forecasting of cash in ATMs.

2 Review of Literature

Premchand Kumar et al. [3] presented two neural network models (daily model and weekly model) for cash predicting for a bank branch. Real data for three months regarding the cash withdrawal was collected from a Chandigarh (India) based branch of State Bank of India. The model is a three-layer feed-forward neural network and was trained using fast back propagation algorithm. Simutis et al. [8] has presented an optimized approach based on artificial neural network for cash demand forecasting in ATMs (automatic teller machines) in the network. Most authors proposed different ideas for minimizing the complexity of cash demand forecasting problem by performing either a data reduction of the input feature space before prediction [5] or by employing a local learning model [9]. For analysing NN5 data, Self-organizing fuzzy neural network approach [6], Multi-layer perceptron's [5], artificial neural network along with principle component analysis has been used. Combinations of Computational models and linear models has used for prediction of NN5 data.

Acuna along with authors [1] did a comparative study between NARX (non linear auto regressive models with exogenous variables) and NARMAX (Non linear auto regressive models with moving average) models developed with ANN and SVM for predicting cash demand in ATMs. Significant differences were not found between NARX and NARMAX for both ANN and SVM. Hence it was advised to choose simpler models, such as NARX and a user-friendly tool like ANN at least for this particular application. M. Ramakrishna Murthy et al. [4] proposed solutions for two major problems with k means clustering. First one is to choose k (number of clusters) and second one is the selection of initial centroids. R. Madhuri et al. [7] did cluster analysis on iris data set and cholesterol data set with incremental k means clustering and also applied modified k-Mode algorithm on Contact-Lens data set and post-operative data set and compared the results.

Venketesh Kamini et al. [10] have recommended the prediction of cash demand for groups of ATMs using clustering followed by neural network techniques to deal with problem of optimal availability of cash in ATMs. Authors applied Tailors Bhutina clustering algorithm on NN5 data then predicted ATM's cash demand using GRNN, MLFF, GMDH and WNN methods on different clusters. Venketesh Kamini et al. [11] used two types of approaches for cash prediction in ATMs. First approach was traditional and non-chaotic and second approach was non-traditional and chaotic. In the first method which was traditional and non-chaotic, missing values of the time series are filled and further de-seasonalization was done with a time lag of seven days as a part of pre-processing of NN5 data. Then following techniques Auto Regressive Integrated Moving Average (ARIMA), Wavelet Neural Network (WNN), Multi-Layer

Perceptron (MLP) and General Regression Neural Network (GRNN) were applied in the forecasting phase. In the second method which was non-traditional and chaotic, phase space was reconstructed using the chaotic parameters, embedding dimension and delay time using TISEAN tool. SMPAE was taken as evaluation parameter. GRNN yielded a SMAPE value of 14.71%.

For improving the predication accuracy of bank cash flow, a mixed model based on back propagation (BP) neural network and grey prediction method is proposed by Wang et al. [2] Experimental results proved that grey neural networks had the highest prediction precision and better generalization ability Table 1.

Table 1. Comparative analysis of related research work in terms of error and usage of clustering

Paper	Author	Type of neural network	Performance parameter	Clustering Used	Clustering algorithm
Cash Forecastingan application of Artificial neural Network in Finance	PremchandKumar & EktaWalia	a three-layer feed-forward neural network trained using back propagation algorithm	Error[4–5%)	No	
A flexible neural network for ATM cash demand forecasting	Simuitis et al.	a three layer feed-forward neural network was trained using Levenberg-Marquardt algorithm	Mean Square absolute Error [20–25%)	No	
Analysis of Time Series with Artificial Neural Networks	Raymundo A. &Gonealez-Grimaldo	General regression neural networks	Mean Square Error (30%)	No	
Cash demand forecasting for ATM using neural networks and support vector regression algorithms	Simuitis et al.	Three layer feed forward Artifical neural network	Mean absolute percentage errors (13–40%)	No	
Comparing NARX and NARMAX models using ANN and SVM for cash demand forecasting for ATM	Acuna et al.	NARX and NARMAX developed with SVM and NN	SMPAE (20%)	No	
Cash demand forecasting in	Venkatesh Kamini et al.		SMPAE (18.44%)	Yes	

(*continued*)

Table 1. (*continued*)

Paper	Author	Type of neural network	Performance parameter	Clustering Used	Clustering algorithm
ATMs by clustering and neural networks		MLFF, WNN, GRNN and GMDH			Tailor Bhutina Clustering algorithm
Cash demand forecasting in ATMs by clustering and neural networks	Venkatesh Kamini et al.	MLFF, WNN, GRNN and GMDH	SMPAE (23.16%)	No	
Chaotic Time Series Analysis with Neural Networks to forecast cash demand in ATMs	Venkatesh Kamini et al.	MLP, WNN, GRNN and ARIMA	SMPAE (14.71%)	No	
Time Series Prediction of Bank Cash Flow Based on grey neural network algorithm	Lie-sheng Wang, Chen-xu Ning, Wen-hua Cui	Back propagation neural network and grey model prediction method	Mean square error (6.92×10^{5}), Absolute Error [$5.23 \times 10^{5-}$], Relative Error [8.86%) and prediction accuracy (91.14%)	No	

3 Methodology and Dataset

In this paper, NN5 reduced dataset has been used which is publically available on UCI repository (http://www.neural-forecasting-Competition.com/NN5/datasets.html) for research perspective. It consists of cash withdrawal of 2 years at different ATMs at different locations in England. Figure 1 Represents flow diagram of proposed approach.

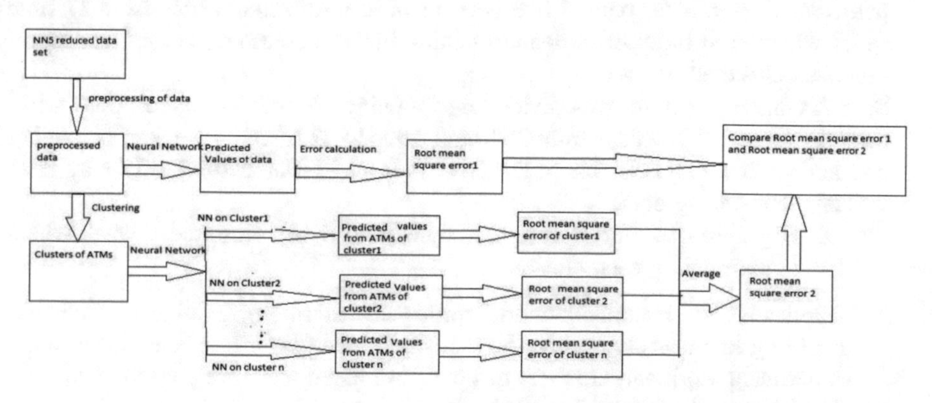

Fig. 1. Flow diagram of proposed approach

4 Algorithm Utilized: K Means Clustering with Neural Networks for ATM Cash Repository Prediction

Input: NN5 reduced dataset was taken as an input dataset.

Step 1: Pre-processing of dataset in terms of filling missing values by replacing that missing values by mean of that time series. Pre-process function was used in R for pre-processing of dataset.
Step 2: Dataset was normalized. Min-Max normalization was used for normalization of dataset. Min Max normalization was used for normalization of dataset. Table 2 shows normalized NN5 reduced dataset.

$$\text{Formula for Min} - \text{Max normalization} = (\text{value} - \text{Min})/\text{Max} - \text{Min}.$$

Step 3: Neural Network was applied on reduced, pre-processed and normalized data. In this paper, rprop + algorithm was used for training of network. Learning rate was assumed to be 0.5. One hidden layer had been taken in the experiment work and number of nodes in hidden layer was 8.
Step 4: Compute function was used to predict the values of testing data with help of the trained model. 10 fold cross validation was applied on testing data. Average of root mean square errors obtained from different ATMs was calculated and represented by RMSE1. Comparison of actual output and predicted output is represented in Fig. 1.
Step 5: Now instead of applying neural networks on individual time series, K Means clustering was applied on original dataset and numbers of clusters were assumed to be 3. Discrete time wrapping (DTW) was used as a distance measure between time series. DTW is a family of algorithms which compute the local stretch or compression to apply to the time axes of two time series in order to optimally map one (query) onto the other (reference). DTW produced the remaining cumulative distance between time series in NN5 dataset.
Step 6: Centroid obtained from each cluster, worked as a time series for that particular cluster. Thus dataset had been reduced into 3 time series from 11 time series where each time series denotes centroid of each cluster. Dataset was represented as clustered dataset.
Step 7: Clustered dataset was divided into training set and testing set. Cash withdrawal data of first 735 days from 28 March 1996 to 22 March was taken as training data and Cash withdrawal data of last 56 days from 23 March 1998 to 17 may 1998 was taken as testing data.
Step 8: Repetition of step 8 for each time series[i] of clustered dataset where $I = 1 - 3$ representing each cluster.

(a) Application of Neural Network. rprop+ algorithm for training of network. Learning rate was assumed to be 0.5. One hidden layer had been taken in the experiment work and numbers of nodes in hidden layer are chosen as 8.
(b) Prediction of Output of Test Records.

(c) Calculation of Root Mean Square Error for each clustered time series.

Step 9: RMSE2 = Average of errors obtained for all three time-series of clustered dataset

Step 10: Comparison of RMSE 1 obtained in step 4 and RMSE2 obtained in step 9.

5 Results and Discussion

Intermediate Values of the entire experiment has been showcased in Tables 2, 3, 4, Figs. 2 and 3. Final result is represented in terms of comparison of root mean square error between Neural Network with clustering and Neural Networks without clustering in Table 5. Prediction Error rate has been reduced when model is created using centroid of clusters.

Table 2. Normalized NN5 reduced dataset

Time series #1D	NN5-101	NN5-102	NN5-103	NN5-104	NN5-105	NN5-106	NN5-107	NN5-108	NN5-109	NN5-110	NN5-111
Starting day	77	77	77	77	77	77	77	77	77	77	77
Dataset	Reduced	Reduced	Reduced	Reduced	Reduced	Reduced	Reduced	Reduced	Reduced	Reduced	Reduced
18–03	0.52	0.19	0.16	0.19	0.22	0.14	0.17	0.17	0.12	0.21	0.13
19–03	0.55	0.23	0.15	0.09	0.26	0.10	0.21	0.22	0.18	0.19	0.16
20–03	0.61	0.30	0.22	0.33	0.31	0.30	0.22	0.24	0.30	0.22	0.26
21–03	0.65	0.39	0.34	0.35	0.52	0.44	0.36	0.44	0.45	0.31	0.36
22–03	0.12	0.49	0.25	0.44	0.51	0.45	0.46	0.34	0.39	0.43	0.29
23–03	0.41	0.11	0.17	0.10	0.34	0.14	0.13	0.16	0.20	0.26	0.07
24–03	0.12	0.25	0.28	0.18	0.30	0.26	0.22	0.20	0.15	0.30	0.13
25–03	0.14	0.20	0.11	0.14	0.20	0.24	0.20	0.19	0.18	0.23	0.14
26–03	0.21	0.20	0.20	0.18	0.26	0.28	0.21	0.20	0.19	0.22	0.18
27–03	0.42	0.32	0.16	0.33	0.32	0.03	0.30	0.32	0.28	0.25	0.32
28–03	0.57	0.52	0.39	0.49	0.63	0.80	0.45	0.49	0.49	0.38	0.49
29–03	0.35	0.60	0.46	0.49	0.79	0.58	0.48	0.46	0.44	0.44	0.38
30–03	0.18	0.11	0.25	0.14	0.46	0.13	0.15	0.17	0.17	0.33	0.14
31.–03	0.28	0.29	0.54	0.23	0.19	0.35	0.23	0.28	0.29	0.51	0.18
01–04	0.34	0.28	0.45	0.19	0.39	0.29	0.25	0.23	0.24	0.39	0.19
02–04	0.37	0.30	0.43	0.15	0.30	0.34	0.27	0.34	0.25	0.42	0.25
03–04	0.54	0.53	0.46	0.50	0.47	0.45	0.42	0.51	0.62	0.54	0.36
04–04	0.30	0.38	0.55	0.32	0.47	0.37	0.26	0.40	0.36	0.53	0.38
05–04	0.24	0.57	0.65	0.46	0.43	0.56	0.48	0.48	0.43	0.88	0.32
06–04	0.19	0.11	0.49	0.08	0.00	0.15	0.18	0.24	0.20	0.60	0.13

Table 3. RMSE for each time-series, (average RMSE1 = 8.50)

Time series	NN5101	NN5102	NN5103	NN5104	NN5105	NN5105	NN5107	NN5108	NN5109	NN5110	NN5111
Error (RMSE)	6.72	10.91	11.96	6.19	9.43	7.14	6.69	11.71	8.24	4.61	9.91

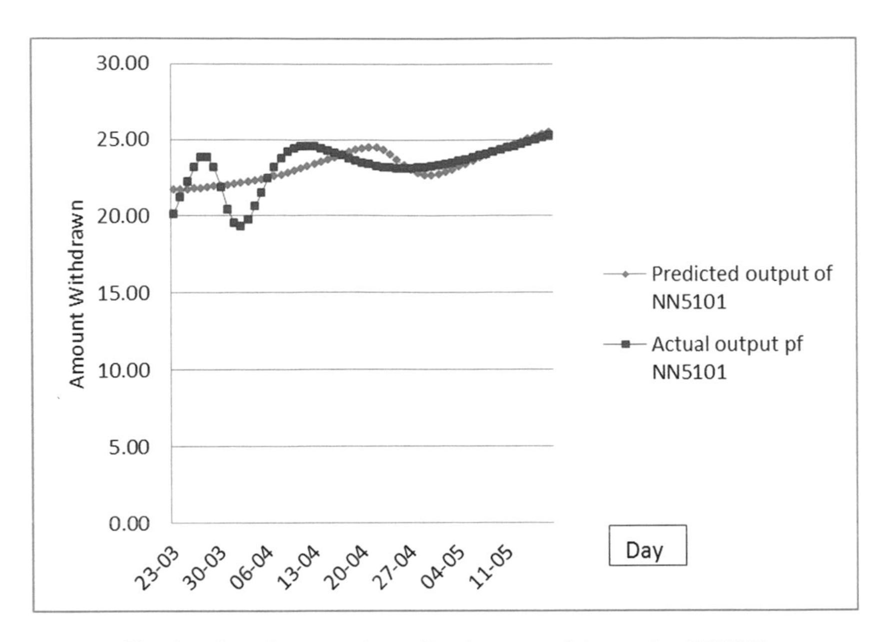

Fig. 2. Actual output v/s predicted output of time series NN5101

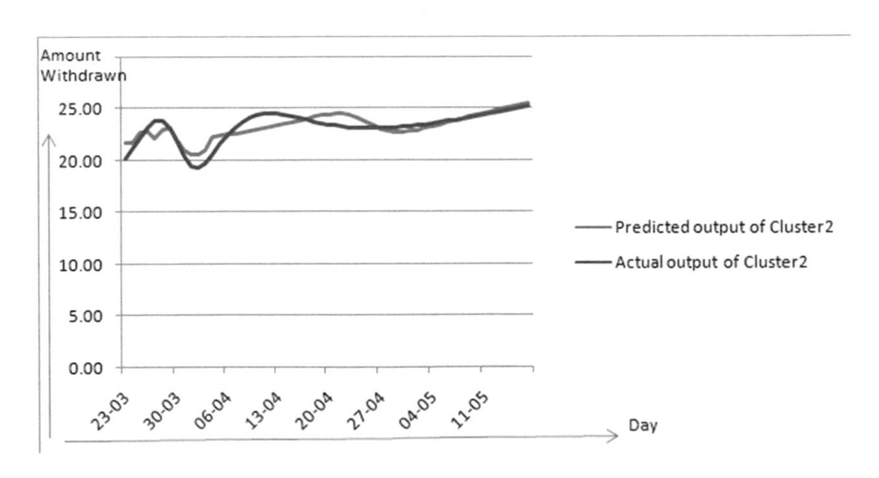

Fig. 3. Actual output v/s predicted output of Cluster2

Table 4. Root mean square error (RMSE) from different 3 clusters

Time series	Cluster 1	Cluster 2	Cluster 3
Error (RMSE)	7.91	4.61	7.21

Table 5. Comparison of Root mean square errors obtained from Neural Network with Clustering and Neural Network without clustering.

Error	Clustered time series	Non-clustered time series
RMSE	6.58	8.5

6 Conclusion and Future Work

Forecasting of ATM cash repository in an optimal way is a crucial task. After installing and placing ATMs, cash demand prediction is a crucial issue. Cash in the ATMs should be balanced in such a way that neither customer will be dissatisfied nor money will be blocked. Accurate forecasting is feasible using application of Neural Networks on the transaction data of ATMs. This paper showcases increase in accuracy by indicating reduction in error rate. Error rate was reduced when ATMs were clustered before application of Neural Networks. Training of machines is more accurate when data of ATMs with similar transactions were fed and leading to precise forecasting. Authors aim to apply various clustering techniques to further reduce error rate with Neural Networks as well use Support Vector Regression for Training of network as a part of future work.

References

1. Acuna, G., Ramirez, C., Curilem, M.: Comparing NARX and NARMAX models using ANN and SVM for cash demand forecasting for ATM. In: The 2012 International Joint Conference on Neural Networks (IJCNN), pp. 1–6. IEEE (2012)
2. Cai, W.Z., Wang, D.T., Wang, Y.S., Yang, Y., Gao, Z.L.: Study of short-term wind power forecasting based on adaptive grey prediction method. Appl. Mech. Mater. **734**, 697–700 (2015)
3. Kumar, P., Walia, E.: Cash forecasting: an application of artificial neural networks in finance. IJCSA **3**(1), 61–77 (2006)
4. Murty, M.R., Murthy, J.V.R., Prasad Reddy, P.V.G.D.: A new approach for finding appropriate number of clusters using SVD along with determining best initial centroids for K-means algorithm. Int. J. Enhanc. Res. Sci. Technol. Eng. **3**(3), 430–437 (2014)
5. Pasero, E., Raimondo, G., Ruffa, S.: MULP: a multi-layer perceptron application to long-term, out-of-sample time series prediction. Advances in Neural Networks, pp. 566–575. Springer, Berlin (2010)
6. Prasad, G., Leng, G., McGinnity, T.M., Coyle, D.: Online identification of self-organizing fuzzy neural networks for modelling time-varying complex systems. Evol. Intell. Syst: Methodol. Appl. **12**, 201 (2010)

7. Madhuri, R., Ramakrishna Murty, M., Murthy, J.V.R., Prasad Reddy, P.V.G.D.: Cluster analysis on different data sets using K-modes and K-prototype algorithms. In: Proceeding in AISC and Computing. Springer Series, vol. 249, pp. 137–144 (2014)
8. Simutis, R., Dilijonas, D., Bastina, L., Friman, J.: A flexible neural network for ATM cash demand forecasting. In: 6th WSEAS International Conference on Computational Intelligence, Man-Machine Systems and Cybernetics, pp. 162–167 (2007)
9. Teddy, S.D., Ng, S.K.: Forecasting ATM cash demands using a local learning model of cerebellar associative memory network. Int. J. Forecast. **27**(3), 760–776 (2011)
10. Venkatesh, K., Ravi, V., Prinzie, A., Van den Poel, D.: Cash demand forecasting in ATMs by clustering and neural networks. Eur. J. Oper. Res. **232**(2), 383–392 (2014)
11. Kamini, V., Ravi, V., Kumar, D.N.: Chaotic time series analysis with neural networks to forecast cash demand in ATMs. In: IEEE International Conference on Computational Intelligence and Computing Research (ICCIC), pp. 1–5 (2014)

Social Media User Ranking Based on Temporal Trust

Harish Kumar and Prabhjot Kaur(✉)

Computer Sceince and Engineering Branch,
U.I.E.T, Panjab University, Chandigarh, India
harishk@pu.ac.in, pavibti@gmail.com

Abstract. This paper proposes a methodology to compute trust and rank user on the basis of time, number and nature of interactions. Technique models trust based on two main factors: Engagement of user, temporal/behavioural factor based on difference between time of user reaction to that of any social media activity. It is evaluated using verified Facebook page of 'Panjab University' with 160,000+ users. Validation uses 985 active users. On analyzing the result it is found that the top scorers are from 'Public Relation Department' of university. Naïve Bayesian machine learning technique has classified the data more accurately as compared to SVM and Logistic regression. Accuracy of Naïve Bayesian is 77.54%, as the condition of independence of dataset is satisfied and degree of overlapping is null. This result states that proposed model provide efficient technique to rank users on social networks.

Keywords: Social networks · Social capital · Social trust · Social influence · Engagement trust · Behavioral trust · Propagational trust

1 Introduction

Use of social media is not confined to connecting people. Nowadays government bodies, organization, agencies etc are registered on social media to promote their services and encourage themselves to deal with dynamic demands of the user. Term Social networks was introduced in 1954 by J.A. Branes to represent social media as a graph, where nodes are entity, organization, agencies, communities or user of social media and edges represent relationship, interaction or behavior between two entities. Edges/relationship can be given weights by the user on basis of designed parameters [1]. These interactions in social network provides with qualitative and quantitative analysis of the user. Judgment of user being trustworthiness can be traced from these relationships. Interactions among entity are not random but measure to calculate *Trust*. Pattern in interactions among the entities in the social network corresponds to meaningful information which represents trust. Trust on social networking sites is highly influenced by social capital and influence. *Social capital* defines social ties, interactions, bonds and popularity with respect to entity in social environment [2]. It stands for social earning in social media. High the social demand of the entity higher is the social capital. It is the main measure in driving *social trust* [3]. *Social influence* describes the effect or influence of the social capital on the user. Influence can be positive or negative

S.C. Satapathy and A. Joshi (eds.), *Information and Communication Technology for Intelligent Systems (ICTIS 2017) - Volume 1*, Smart Innovation, Systems and Technologies 83, DOI 10.1007/978-3-319-63673-3_72

depending on social capital. Presented model in paper develops a formula based sentiment scoring of the interaction based on machine learning techniques (engagement trust), user behavior based on the user activities and participates and temporal factor of the interaction among the user of the page (behavioral trust).

Section 2.1 describes trust, its types and properties in social networking sites. Section 2.2 describes techniques for sentiment analysis that can be used in trust modeling of user. Section 3 describes the methodology to compute engagement trust with behavioral factor of the users. Section 4 measures result evaluation of obtained score of the user. Section 5 present conclusion and challenges in trust modeling in social media.

2 Trust and Sentiment Analysis in Social Network

2.1 Trust in Social Network

Trust in social networks is stated as a confidence measure that an entity will react in a particular manner despite of the prevailing opportunities in environment. Trust can be direct which is developed from personal behavior or it can or indirect which is incurred from second person's experience. Trust in social network develops on the basis of the three dimensions [1]: *Attitude*, reflects entities like, dislike, positive, negative view for certain objects or program. This attitude information can be derived from user's interaction in the social network. Second is *Behavior*, it implies the interaction pattern. The pattern of interaction among trusted entities will be frequent. Mean time in responses is negligible. The third is *experience*, which is can be considered an external factor in calculating trust in social networks.

Trust is present with different properties which are [1] *context specific*, *dynamic*, *propagative*, subjective, self enforcing and event sensitive. Trust is defined in different aspect related to several disciplines: *Calculative*, *Relational*, *Emotional*, *Institutional*, *Cognitive* and *Dispositional*. To measure trust, there are three trust evaluation models (i) G*raph based/network based model* where graphs are used for the representation of the social networks. (ii) *Interaction based model* where the message exchange plays an important role. Here reviews, feedbacks, promotion, comment, shares, likes, rating parameters are used. (iii) *Hybrid model* is the mixture of the above two models. Analyzing the property, aspects, measures and evaluation model for trust, trust scoring on networking sites is based on following characteristic properties of trust.

Engagement trust is defined as the involvement of the user with the entity (individual, organization, government body). Engagement of the user in the services represents its interest in the provider and it can be calculated through various parameters [4]. Both graph and interaction based model can be used to calculate engagement of the user. In calculating engagement trust interaction with semantic analysis is done. Interaction of the user on social networking sites are likes per post, comment done, clicks on post and information exchanged. These parameters are defined under the reach or audience engagement per post. For example Insights of facebook page provides complete engagement statistics based on each post in particular time period. It helps in defining engagement in social networking sites. For engagement main characteristic is opinion mining on the comment or information exchanges by the entity on social networking sites. Classification of the comments, feedbacks or message exchange in three classes: positive, negative or neutral done on the basis of various semantic analysis techniques using dictionary based or machine learning technique. Various researchers have given their perspective in improving efficiency of engagement trust through confidence measures, analyzing external factor, considering time factor etc. [5].

Popularity trust is stated purely on social ties and popularity. More the social capital high is popularity trust. Most efficient method to measure popularity is in-degree of number of friend requests, invitation send by the entity accepted, invitation came to the entity etc. [6]. How brand loyalty influence the user and thus affecting social capital of the entity increasing the profit of the organization [7].

Behavioral trust is completely based on interaction without semantic analysis of the context. It consider characteristic like how frequent, continuous, regular and number of interactions among the entities. Time plays important role in behavioral aspect of trust. There are two category of behavioral trust [8].

- *Conversational trust:* It is defines on three tuple sender, receiver and time. It works more efficiently on sorted list of time difference when the information is exchanged. Smoothing factor groups the conversation describing its pattern. Longer, equal and regular conversation implies more trust.
- *Propagational trust:* When entity A transmits information to B, and B to C in this case C has a propagational trust on A. Percentage of trust developed is defined on information propagated and manipulated A and C. This behavior in social networks corresponds to propagational trust. Table 1 lists properties of various trust model proposed in literature.

Table 1. Properties of existing trust models

Trust model	Properties
Strust (engagement, popularity) [4]	Dynamic, context specific, subjective
Behavioral based model [8]	Event sensitive
Trust computing model [9]	Propagative, composable
Behavioral Integration on web services [10]	Composable
Trust modeling (Epinion case study) [11]	Dynamic
Social trust [3]	Dynamic
Sunny [5]	Subjective

2.2 Semantic Analysis

Semantic focuses on bonding of the words, phrases, lines with each other. Semantic analysis states analyzing relationship in a productive manner. Semantic classification determines whether a document/sentence carry positive or negative opinion. It is basically done with the help of feedback, reviews, and ratings given by the user or through some criteria [12]. For example 'excellent work' signifies positive opinion, 'poor approach' negative opinion and 'what's up' neutral opinion of the user toward the service. The most important part of a speech is *verb* which gives a sentence power to express. This 'verb' is given importance in context mining. Sentiment analysis has two main techniques [13]:

(i) *Machine Learning Technique* predict label by first training the classifier on standardized dataset and then using for testing or predicting purpose. Techniques used in machine learning are: *Naïve Bayesian classification (*classifies on the basis of the probability of the word), *Support vector Machine classification* (classifies on the basis of co-variance matrix), *Logistic Regression Technique* (classifies text on the basis of the odds from the frequency table) and many others.
(ii) *Dictionary-based approaches* are based on lexicons and use a dictionary of words mapped to their semantic value. The lexicon of a language is its vocabulary. WordNet, SentiWorldNet etc represents an index of sentiment words, and it has the polarity information of the relevant word irrespective of whether it carries a positive sentiment or a negative one.

Measures in Sentiment Analysis: Data present on the social networking sites is raw and unstructured like gud → good etc. Main challenge in sentiment analysis of data in social networking sites are:

- pre-processing of the data before classification [14].
- To fetch structured data from unstructured data.
- Identifying relevant and irrelevant objects.
- Feature extraction of the data should be done in an efficient manner as same word can be expressed in different manner like congratulation/congrats/congo.
- Choosing a standardized dataset to classify opinion positive or negative based upon feature extracted from the dataset.

After following the above mentioned step the data is in structured form with all the meaningful content and ready for text mining with better and efficient results. Next step for semantic analysis is identifying the approach for text mining.

Bag of words means collection of the words present in the document with additional feature of frequency. But in text mining of social networking sites it consist of big corpus data from various trusted sites [13]. List of positive and negative words are created with the frequency which represent how positive/negative a word is trained from a corpus data and used in predicting new labels. Naïve Bayesian technique is used in classifying the data in required classes. Zero probability problem in bag of word is solved by Laplace smoothing that is frequency count 1 is added to all the words which are positive but not negative to calculate its negative score [12]. Some complexities with ambiguous terms like 'low price' and 'low quality' here low is positive and negative both. In these case after negation 'not' is attached like 'low not_quality' in bag of words to improve efficiency. N-Gram, extension of bag of words, where N is set of co-occurring words in a given window. This window is decided by 'N' [13].

POS (Part of speech) is a process of marking up a word in a text to a particular part of speech on the basis of context and definition that is relation within a sentence paragraph or phrase [12].

After examine trust and sentiment analysis models, time of creation of comment with opinion mining of comment is not practiced. Hence, quality and quantitative parameters both are analyzed in proposed methodology.

- Engagement trust: is calculated by number of likes and comment by the user (quantitative). Opinion mining of the comment in the proposed method uses sentiment analysis techniques (qualitative).
- Behavioral trust: is the time difference among responses of entities. This time difference with context analysis gives behavioral and engagement value of the user in accordance to a particular response (qualitative).

3 Method Proposed

In the proposed methodology trust is modeled on various actions of the users which account to their engagement on social networks. Temporal dataset is retrieved and formulated along with engagement metrics which include *posts, comments, likes* on the officially verified page of Panjab University, Chandigarh (India) along with the creation time within six month period that is from December, 2015 to May, 2016. Figure 1 represents work flow of technique modeled to simulate trust.

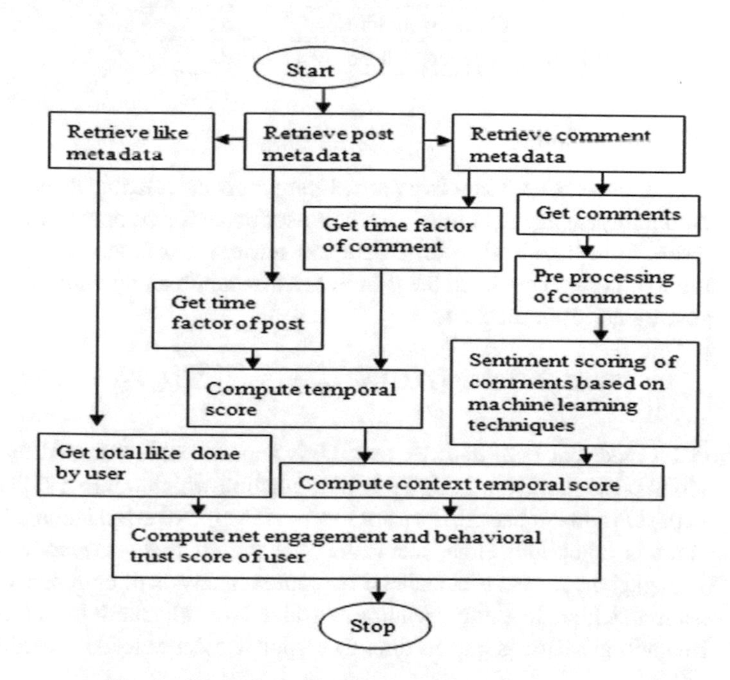

Fig. 1. Work flow of proposed trust modeling model

3.1 Database

Dataset had been fetched with the help of Facebook graph API tool and SDK library using asp.net from Facebook's Social Graph. Access token of page is used to access and retrieve data. Table 2 lays out the statistics of the dataset retrieved for trust simulation. *Like* by the user's depicts their willingness, interest, engagement and frequent analysis toward the new data/post posted on the page. Their steady interest in the events on the page increases their reach and trust in the organization. Thus frequency of *likes* is calculated through total like corresponding to build trust among the user and organization. In general, *comments* by the user are comparatively less than *likes* on a post. On the university page, if a post has 10,000 likes, the comment are about 40. This data state that comments are less preferred by the user but the users which enroll themselves in comment show their full involvement by providing feedbacks, reviews and interest. Thus comment corresponds to additional trust factor among the user toward the organization. Hence different scoring method is designed for *comment* in accordance to *likes.*

Table 2. Statistics of temporal dataset collected

Total no of posts	83
Total no of comments	1,605
Total no of likes	1,09,221
Total no of user who commented	985
Total no of user who likes	58,898
Maximum likes by an id/user	83
Maximum comments done by a user	39

3.2 Formula Formation

Trust modeling of users is split into two parts. First part is calculating *Total Like Count (TLC)* corresponding engagement trust based on user behavior computed from number of likes by users from ‘likedata’ table of database retrieved. It is quantitative measure corresponding to all users who liked the post in last six month along with their id. Total likes on a post by an id are fetched.

$$Total\ like\ count\ (TLC)\ of\ U_i = \sum_{j=1}^{n} L(U_i, P_j) \tag{1}$$

Equation 1 stated that *total likes* by user U_i is sum of value returned by function $L(U_i, P_j)$ on total posts retrieved. $L(U_i, P_j)$ is like function which return 1 if like is done by user U_i on post P_j and 0 if user U_i does not like post P_j. ‘n’ is the total number of posts.

Second part is compiling engagement and behavioral trust or *context temporal score (CTS).* Engagement trust is calculated by context analysis of comments based on naïve Bayesian machine learning techniques trained on standard set of sentiment sentences. Behavioral factor is gap in time to response to new feeds provided by the service provider.

$$CTS\ (U_i) = \sum_{k=1}^{n} \left(\sum_{j=1}^{J} T(U_i, C_j, P_k) * S(U_i, C_j) \right) \tag{2}$$

Equation 2 states function to calculate sentiment score $S(U_i, C_j)$, where function return the predicted value for comment based on the trained dataset by model used. S (U_i, C_j) return value from 0 to 1 based on the opinion analyzed by naïve Bayesian model. Temporal function $T(U_i, C_j, P_k)$, return the temporal score based on the

Table 3. Scoring of temporal factor (in hours/days)

Time difference (h)	Score	Time difference (days)	Score
0–0.02	15	0–0.015	10
0.02–0.5	13	0.015–0.523	7.5
0.5–1.5	11.5	0.523–1	5
1.5–5	9.5	1–5	3.5
5–15	8	5–10	2
15–30	6.5	10–20	1.5
30–60	5	>20	1
60–125	3.5	–	–
125–370	2	–	–
>370	1	–	–

frequency of the user to respond to the post on scale of 1–10 or 1–15 as stated in Table 3 based on the time domain that is in days or hours. Scaling is done in order to make clean classification based on behavioral factor. Temporal scoring of time gap is provided on analyzing the statistical measures mean, mode, median and standard deviation of data collected. Scaling factor is 1–15 in hours to fairly classify user's frequency and behavior with the post. Similarly in days, scaling is done between 1 and 10 to neatly classify users on their behavior in the domain looking on statistical measures. In $T(U_i, C_j, P_k)$, U_i is the user who wrote comment C_j on post P_k. From Eqs. 1 and 2, trust score of the user correspond to sum of *TLC and CTS.* Equation 3 represents proposed formula to simulate trust of user U_i.

$$\textit{Final score} = \sum\nolimits_{k=1}^{n} \left(\sum\nolimits_{j=1}^{J} T(U_i, C_j, P_k) * S(U_i, C_j) \right) + \sum\nolimits_{j=1}^{n} L(U_i, P_j) \quad (3)$$

4 Results

Different performance parameters are used to compute the efficiency of the model. Accuracy states total comment correctly classified in captured data. Precision defines the comments which are relevant to the class (positive, negative, neutral) and are retrieved. Recall is a measure of fraction of retrieved relevant document to the true positives. Tables 4 and 5 show the accuracy of the machine learning technique modeled on standardized dataset of sentiment sentences present in LightSide tool. Naïve Bayesian classifier with bag of words techniques in LightSide has better accuracy over SVM and Logistic Regression as the condition of data independence of dataset is satisfied and degree of overlapping of classes is small.

Tables 6 and 7 represents confusion matrix of the Naïve Bayesian model when labels are predicted on raw comments and pre-processed comments. Accuracy for positive classified sentences in Table 7 is 65.4%, negatively classified sentences 72% and accuracy of neutral sentences without pre-processing is 52.02%. After pre-processing of the data, Table 8 shows confusion matrix of comment classified by

Table 4. Performance measures of models trained

Models/performance measures		Precision	Recall	f-measure
NB	Positive	0.756	0.783	0.764
	Negative	0.79	0.766	0.777
SVM	Positive	0.724	0.735	0.724
	Negative	0.74	0.728	0.733
LR	Positive	0.752	0.762	0.754
	Negative	0.76	0.75	0.754

Table 5. Comparison of accuracy of machine learning techniques

Classification method	Naïve Bayesian	Logistic regression	SVM
Accuracy obtained	77.54	75.9	73.23

Table 6. Confusion matrix of text without preprocessing

Actual/predicted	Positive	Negative	Neutral
Positive	467	91	219
Negative	24	47	17
Neutral	217	314	206

Table 7. Confusion matrix of pre processed data

Actual/predicted	Positive	Negative	Neutral
Positive	467	55	255
Negative	17	50	11
Neutral	223	141	373

Naïve Bayesian corresponding to manually classification of comments. Accuracy obtained after pre-processing is 64.3% for positive, 85.8% for negative and 59.9% for neutral class. This implies that classification of sentences with respect to sentiment sentence has increased by 13.8% after pre-processing for negatively classified sentences and 7.9% for neutral class. This increase in efficiency is because the data is now in structured form.

Precision, Recall and F-score are the measure of sensitivity and specificity that are common in language technology research. In Table 8, Recall is 60.1% for the positive class which states the percentage of positive comments labeled in all the positive comments of dataset. In case of negative class it is 64.1% labeled as negative. Similarly, for neutral class it is 50.6%. Precision of positive class is 66%. This implies out of all the comments labeled as positive only 66% is validated as being positive. Similar case for negative out of all the comment labeled as negative only 81% is validated as being negative. Same goes for neutral class out of all the comment labeled neutral comment only 58.9% is validated as neutral. F-measure is average of both precision and recall.

Table 8. Performance measures of comment classes

Measure/classes	Positive	Negative	Neutral
Recall	0.601	0.641	0.506
Precision	0.66	0.819	0.583
F-score	0.623	0.719	0.590

Table 9. Comparison of the active user in six month

Ranked user in case of days	Same ranked user in case of hours	Difference in rank
A1	A1	0
A2	A2	0
A7	A10	3
A8	A7	1
A9	A9	0
A10	A8	2
A11	A25	14
A12	A14	2

F-score for positive class is 62.3%, negative 71.3% and neutral 59%. Accuracy obtained by the class is 64.3% for positive, 85.8% for negative, 59.9% for neutral.

Table 9 represent few instances all active users (985) scaling temporal factor on basis of days and hours. On analyzing the user it is found that top 5 ranker of the trust modeled score belong to the public relation department of the university. The users are made anonymous in order to retain their identity. Table 10 represents the standard deviation and mean difference among the rank differ of the user.

Table 10. Statistical measures in evaluating result

List/statistical measures	Mean	Standard deviation
Top 50	5.14	5.07
Top 100	11	14
All active users	117	114.5

Mean of the ranking difference of top 50 users is 5.14 implies that on an average the shifting of the rank user among the two lists is by 5 ranks. Standard deviation is 5.07 which states that the standard error among the ranking of list is by the rank of 5. Similarly in other two cases, standard deviation states standard error among the two lists computed. On analyzing sorted lists of 985 users maximum rank difference is 583 of single user whereas maximum user had difference in rank between 75–125. It implies that formula works for highly active user or debater and activist of the page.

5 Conclusion

Main contribution of paper is in measuring behavioral trust along with engagement using temporal factor. Main focus is to gather temporal database. It uses both qualitative and quantitative measures. Metrics used in formulating formula are actions for examples likes, comments, reactions or feedbacks. Text mining of comments is done along with behavioral factor focusing on interaction and dyadic relationship. Formula is developed using above mentioned metrics and evaluated using performance measures. Variation among top ten users is negligible. Maximum variation in user among top 50 is by five ranks, top 100 by 11 ranks and among all 985 users standard deviation is 114. Thus stating modeled technique is significant and efficient.

References

1. Sherchan, W., Nepal, S., Paris, C.: A survey of trust in social networks. J. ACM Comput. Surv. **45**(4), 47 (2013)
2. Claridge, T.: Social capital and natural resource management. http://www.socialcapitalresearch.com (2017). Accessed 30 Jan 2017
3. Caverlee J., Liu L., Webb S.: Socialtrust: tamper-resilient trust establishment in online communities. In: Proceedings of the 8th ACM/IEEE-CS Joint Conference on Digital Libraries, pp. 104–114 (2008)
4. Nepal S., Sherchan W., Paris C.: Strust: a trust model for social networks. In: Proceedings of the IEEE 10th International Conference on Trust, Security and Privacy in Computing and Communications, pp. 841–846 (2011)
5. Kuter U., Golbeck J.: Sunny: a new algorithm for trust inference in social networks using probabilistic confidence models. In: Proceedings of the International Conference of AAAI, vol. 7, pp. 1377–1382 (2007)
6. Ortega, F.J., Troyano, J.A., Cruz, F.L., Vallejo, C.G., EnríQuez, F.: Propagation of trust and distrust for the detection of trolls in a social network. J. Comput. Netw. **56**(12), 2884–2895 (2012)
7. Laroche, M., Habibi, M.R., Richard, M.O.: To be or not to be in social media: How brand loyalty is affected by social media? Int. J. Inf. Manag. **33**(1), 76–82 (2013)
8. Adali, S., Escriva, R., Goldberg, M.K., Hayvanovych, M., Magdon-Ismail, M., Szymanski, B.K., Wallace, W.A., Williams, G.: Measuring behavioral trust in social networks. In: Proceedings of the IEEE International Conference on Intelligence and Security Informatics (ISI), pp. 150–152 (2010)
9. Zuo Y., Hu W.C., O'Keefe, T.: Trust computing for social networking. In: Proceedings of the IEEE 6th International Conference on Information Technology, pp. 1534–1539 (2009)
10. Paradesi, S., Doshi, P., Swaika, S.: Integrating behavioral trust in web service compositions. In: ICWS, IEEE International Conference of Web Services, pp. 453–460 (2009)
11. Liu, H., Lim, E.P., Lauw, H.W., Le, M.T., Sun, A., Srivastava, J., Kim, Y.A.: Predicting trusts among users of online communities: an epinions case study. In: Proceedings of the 9th ACM Conference on Electronic commerce, pp. 310–319 (2008)
12. Ahkter, J.K., Soria, S.: Sentiment analysis: Facebook status messages. Stanford, CA (2010). Accessed 30 May 2016

13. Hamouda, S.B., Akaichi, J.: 'Social networks' text mining for sentiment classification: the case of Facebook' statuses updates in the 'Arabic Spring era'. Int. J. Appl. Innov. Eng. Manag. **2**(5), 470–478 (2013)
14. Hemalatha, I., Varma, G.S., Govardhan, A: Preprocessing the informal text for efficient sentiment analysis. Int. J. Emerg. Trends Technol. Comput. Sci. **1**(2), 58–61 (2012)

Computer Vision Based Real Time Lip Tracking for Person Authentication

Sumita Nainan(✉), Vaishali Kulkarni, and Aditya Srivastava

Department of Electronics and Telecommunication,
Mukesh Patel School of Technology Management and Engineering
(SVKM'S NMIMS), Mumbai, India
{sumita.nainan, vaishali.kulkarni}@nmims.edu,
adityasri612@gmail.com

Abstract. Automatic Person Recognition and Authentication faces a huge challenge especially in high risk environment. Multimodal Biometrics do address this issue up to a certain extent. Voice is the simplest and distinct single modality and lip movement is unique for each speaker. Dynamic movement of the lips if tracked online while a Speaker is speaking can authenticate a speaker in real time and the imposter can be identified thus making the biometric system robust and secure. This paper detects the lip movement of a Speaker, creates contours around the mouth region in real time and creates data base for a Speaker. Using the USB webcam, face detection was done from which the mouth region was segmented and contours were created around the lip area with the help of random points selected on both the upper and lower lips which were mathematically obtained. Based on these points of movement, a database was created for an individual. The time taken to create the speaker's lip contour database takes hardly a few seconds which is the advantage and requirement for Real time applications which we have achieved.

Keywords: Automatic Speaker Recognition (ASR) · Region of Interest (ROI) · Face detection · Lip movement tracking · Mathematical model

1 Introduction

Biometrics is a science of measuring and analysing biological data of human body by extracting a feature set from the acquired data and comparing this set against a template set in a data base.

With advances in the development of Hardware and Software in the field of Digital Signal Processing, Automatic Speaker Recognition System (ASR) over the years have eased the Identification/verification of a speaker in domains of voice dialling, banking, security systems employed for confidential information, remote access to information through Computers, data base accessing services, attendance systems and also for telephone shopping [1]. The challenge is in matching the performance of the human counterpart in terms of Accuracy and Speed.

With incorporation of state-of-the-art mobile phones as well as advanced digital imaging and sensing platforms where voice, GPS, temperature, acceleration, ECG/EEG

S.C. Satapathy and A. Joshi (eds.), *Information and Communication Technology for Intelligent Systems (ICTIS 2017) - Volume 1*, Smart Innovation, Systems and Technologies 83, DOI 10.1007/978-3-319-63673-3_73

sensors are available for sharing personal information, the requirement of the hour is to incorporate high privacy protection and security against identity theft or fraudulent use [2], hence selecting a unique biometric modality becomes important.

Conventionally physiological or behavioural characteristics chosen were voice, face, fingerprint, iris and signature. Unimodal Biometric Systems for Person Recognition however has its limitations, as the circumstances under which modalities are acquired affect the percentage accuracy of authentication. Noise, illumination, angle at which image has been acquired, age of the person, health conditions all add to increasing the error in person recognition, besides making the biometric system vulnerable to spoof and replay attacks. Multimodal Biometric system do address these problems to a large extent but the challenge is in how and which single modality can be selected to give a secure and robust Biometric system.

For effectively accurate system, liveness of Audio or Video information if made available can lead to real time Automatic Person recognition. Lip motion which is intrinsic to speech production can provide proof of liveness.

We have in our work used Lip movement detection for person Authentication.

The correlation between the acoustics of speech and the accompanying lip movement has been established in [3]. In case of visual person authentication systems, the articulatory movements of the lips, tongue as well as teeth provide a simple but dynamic proof of liveness and represents a speaking face [4]. Face detection and tracking is a problem frequently addressed in Computer vision as the applications involve human machine interaction, teleconferencing, person identification etc.

In this paper we are focusing on liveness detection of a speaker hence enabling Speaker recognition by first detecting the face. The mouth region has been considered as the region of interest ROI. Once the rectangle boundaries of the ROI were defined, contours were created around the LIP area automatically as the speaker speaks. Relative calculation of distance between the mouth's rectangle formed and random points on both upper and lower lips were calculated. We hence mathematically calculated the points of high and low movement for each frame thus establishing the dynamic lip movement. We use Matplotlib to draw patterns and saved the pattern in a database created for each person. Person Recognition and authentication can then be done using pattern recognition where new patterns created can be matched with the existing ones in the database.

The paper is structured as follows: work done by researchers have been discussed in Sect. 1. Section 2 highlights the methods used for face and lip detection. Work done has been detailed in Sect. 3 while results are discussed in Sect. 4 followed by conclusion and list of references.

1.1 Literature Review

In the field of human-machine interface, face recognition is becoming one of the pursued research spots. Lips are one of the main feature for face recognition [5]. Positioning and segmenting lips have been done using Matlab for feature extraction. YCbCr color space method has been used instead of traditional RGB color space algorithm which yielded higher accuracy with faster calculations. The Algorithm uses dynamic threshold to secondary position the lips, for localization of its shape for image segmentation and edge detection. Characteristics of the lips are then obtained. Uneven illumination although influences the computational complexity of lip feature recognition. Jun Shiraishi proposes [6] a novel optical flow-based lip reading method that includes head motion and employs facial features based ROIs instead of the rectangular ROI. With the surface movement of the skin, the Optical flow can be estimated. The method was tested on the utterance scenes in CUAVE database. A Lip Mouse which is a multimodal human computer interface [7] where lip movement, and gestures made with mouth can allow user to interact with the computer. Lip image segmentation and lip contour analysis was done after harr-like feature was used to detect the face. However in [8] the lip centroid was considered as the origin of the coordinates and five points fitted to the outer lip boundary made the method highly flexible and accurate. Researchers have also proposed lip contour tracking using level set evolution and re-initialization function once lip contours have been obtained [9]. Visual character recognition was performed using distance weighted k nearest neighbour algorithm. SVM has also been used for training and classification along with kalman filters to realize real time lip tracking and detection. Variance based Harr-Like feature and SVM prove much more efficient than using traditional Harr-Like features [10]. Comparative study to locate the mouth region using two color based techniques, i.e. hue and accumulated distance and chrominance technique proves the later on to be more efficient than the former while dual approach results in a much more robust system [11]. A hybrid approach to improve lip localization and tracking has been proposed in [12] where new color space transformation for enhancing lip segmentation is implemented. PCA is used to maximize the discrimination between lip and non lip color. Fusion of acoustic and visual modalities have warranted improvements in the robustness of person recognition biometric systems.

2 Face and Lip Detection

Lip movement of a person is a very significant visual component. Relevant information of the spoken words as well as the unique voice combined with it gives a very robust Multimodal Biometric system. For authenticating a person for real time high security applications lip movement detection and tracking resolves a lot of issues.

Limitations in determining the accuracy of recognition however is posed by the image capturing device, the illumination conditions and low discrimination in the color of the skin and lips. There are two basic methods of extracting features from the lips from the image sequences captured, 1. Image based approach and 2. Model based approach.

1. Image Based approach: this approach is based on PCA i.e. the Principal Component analysis method or the DCT method where the focus is on the intensity values of the pixels around the lips region [12]. These low dimensional features are usually used for Speech recognition used for Speech recognition. As other facial features from the jaw movement and the tongue are also extracted depending on the ROI established. It will be sensitive to scaling and illumination variations.

2. Model Based Approach: A lip model is created by the height and width parameters of the lips. The variations in this approach are the active contour model or the snake model, the active shape model and the deformable geometry model. This method reduces the processing time and provides better accuracy and is widely used for Lip tracking. Region Scalable fitting energy (RCEF) and Localized Active Contour model (LACM) are also other models [13] effectively used for Lip tracking.

2.1 Lip Tracking

Once face detection is done the task now is to detect the lip contours which when established from frame to frame, gives the liveness of the speech utterances. The mouth ROI once located, Lips can be tracked.

2.1.1 Snake Method

This is an active control model introduced by Kass et al. It is an energy minimizing technique where the user must first establish the initial snake points closer to the target. It then follows a line. The shape of the lips keep changing and hence the challenge here is to establish the initial contours. The snake algorithm that gives the energy [13] is as given in Eq. (1).

$$E = \int \left(\alpha(s)E_{cont} + \beta(s)E_{curve} + \gamma(s)E_{image}\right)ds \tag{1}$$

where α, β and Υ are the relative controls.

2.1.2 Region Scalable Fitting Energy Method

This method was introduced particularly for biomedical image segmentation but can be effectively used for lip movement tracking. The data fitting energy, drives the active contour towards object boundaries. The Eqs. (2) and (3) explain the lip tracking.

$$\frac{d\phi}{dt} = -\delta_\varepsilon(\phi)(\lambda_1 e_1 - \lambda_2 e_2) \tag{2}$$

The zero level which represents the lip contours decides that the contour belongs to a domain in this level set segmentation method.

$$e_i(x) = \int K_\sigma(y - x)|I(x) - f_i(x)|^2 dy, \quad i = 1, 2 \tag{3}$$

Variable K_σ = Gaussian kernel, $I(x)$ = image intensity

2.1.3 Localized Active Control Method (LACM)

Local image statistics are considered instead of global image statistics. Contours now can be used for segmenting objects which can have multiple features. The Eq. (4) is used for lip tracking [13].

$$\frac{d\phi}{dt}(x) = \delta\phi(x)\int_{\sigma} D(x,y)F(I(y), \phi(y)dy + K(\phi(x)) \quad (4)$$

where x, y are independent variables and represent a point in the domain Ω of the image. *I* is one image from the frame in 2D.

3 Implementation

In this paper we present a Mathematical approach to generation of a database by tracking lip movement for real time person recognition.

The flow diagram for the process is as shown in Fig. 1.

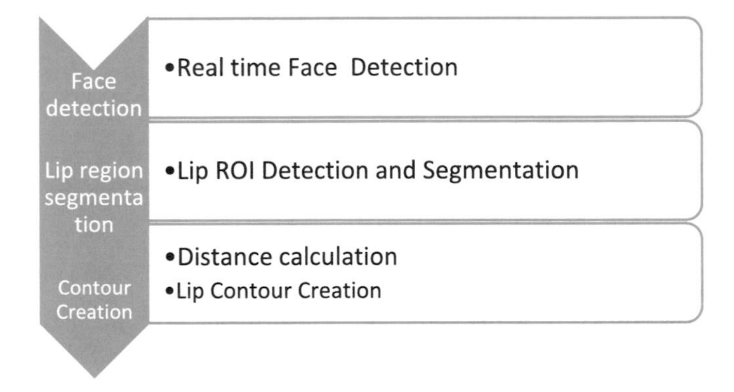

Fig. 1. Flowchart of the process

Step 1. Face Detection: Using VideoCapture option in OpenCv and using the harrcascade function for frontal face and mouth, face in real time was captured speaking a word and images were saved for an individual. Once face detection is complete, rectangle ROI for face area is created thus minimizing the size of the image shown in Fig. 2.

Step 2. Extraction of Mouth/Lips: This was the next step done to extract mouth from the face ROI. The rectangle ROI drawn around the face was divided into two equal halves horizontally and we used only the lower portion of the rectangle thus reducing the frame size and increasing the number of frames per second. Harr cascade mouth_mcs.xml file was used to detect the mouth and draw the rectangle around it.

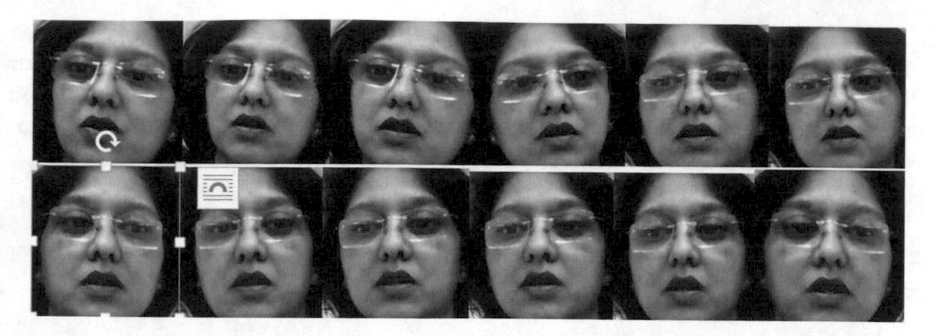

Fig. 2. Face detection and creation of ROI

Step 3. Contour Creation/Pattern Generation: The boundary once drawn was used as the upper limit and the lower limit of the mouth movement, which gave the flexibility for contour creation when a sentence/word was spoken. Canny edge detection was used to detect the edges. As the Speaker speaks the word 'hello' which was chosen for this experiment, for the lip movement accompanying that, we start drawing contour lines according to where the upper and lower lip have been moving thus generating a pattern, then this pattern is automatically extracted from the incoming image and saved to database of that specific person.

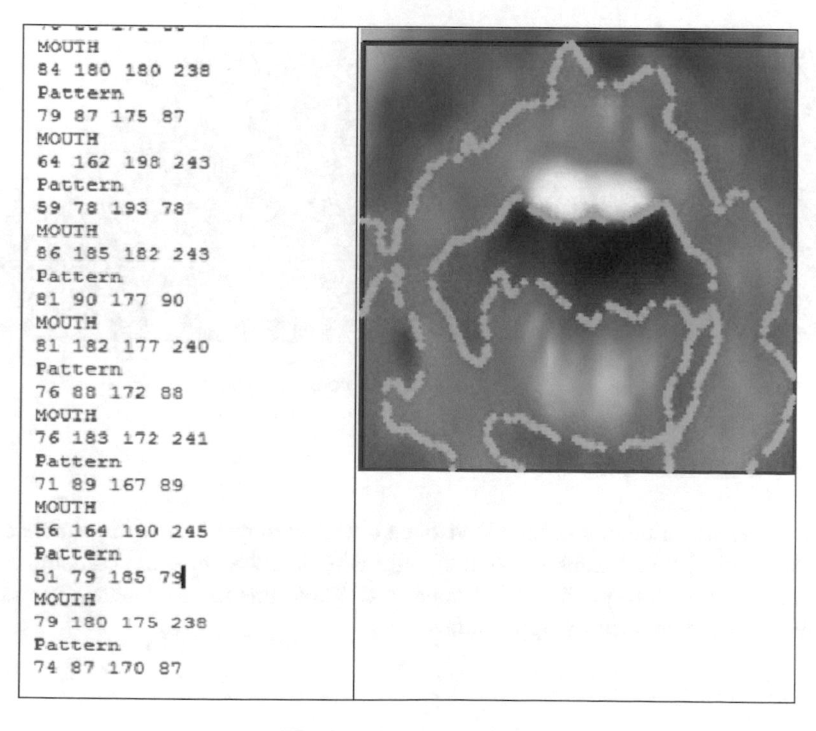

Fig. 3. Pattern generation

The Pattern is generated mathematically so that it does not depend on the location of mouth on the screen and remains same when the sentence is spoken again regardless the location of mouth. Each dot in the pattern is formed by taking relative distance from the starting of mouth both upper and sideways to the location of the lip at the particular word. The location can be easily calculated by simple distance formula using x, y, w, h as length, breadth, width and height of the location as in Fig. 3.

The contours predicted and drawn around the mouth using matplotlib in python as shown in Figs. 4 and 5. A database is created with the name and images of the lip for each individual. 20 to 25 frames per second were created and stored.

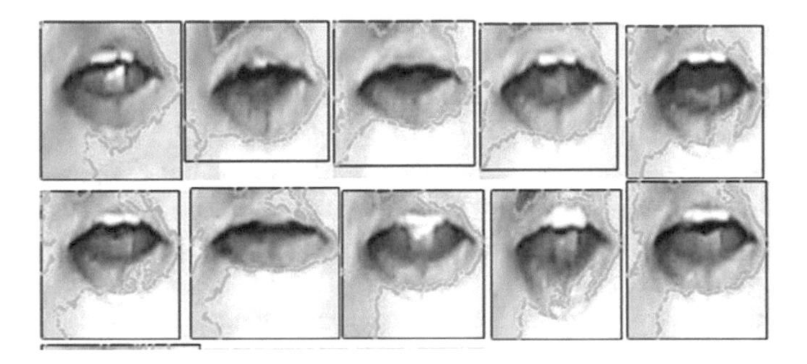

Fig. 4. Contour created around the lip movement

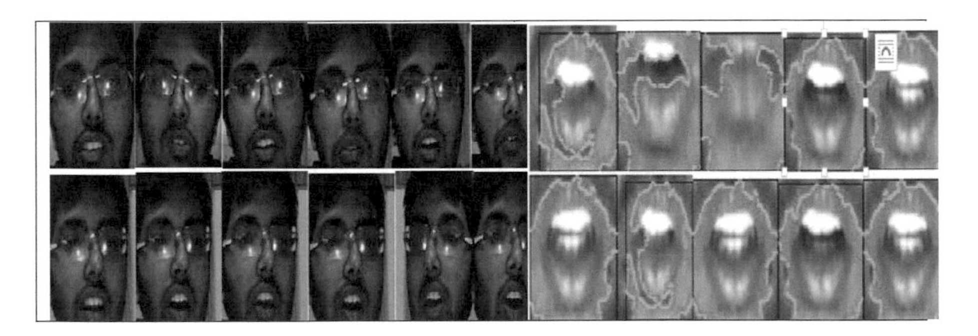

Fig. 5. Face detection and contour creation

4 Results and Discussion

Optimum results in minimum time have been obtained in real time using simple codes. As mathematical calculations are used to define the bounding box and contours around the lip movements are predicted it becomes a simple method to detect liveness of a speaker in any high security applications.

5 Conclusion

Voice as a single modality is widely used for Text Independent Speaker Recognition. As no single modality gives 100% accuracy, this work tries to simplify the process of acquiring images using a USB webcam and creating a database of the contours of the lip movement created. Applying a pattern matching algorithm we propose to take this work forward to achieve person recognition. The features acquired from the lips can also be fused with the features extracted from the voice of the same speaker and Multimodal biometric system can be implemented.

References

1. Nainan, S., Kulkarni, V.: A comparison of performance evaluation of ASR for noisy and enhanced signal using GMM. In: International Conference on Computing and Security Trends. IEEE, Pune (2016)
2. Wang, C., Xuebin, L.: A new spectral image assessment based on energy of structural distortion. In: IEEE International Conference on Image Analysis and Signal Processing (2009)
3. Matsui, T., Furui, S.: Concatenated phoneme models for text-variable speaker recognition. In: Proceedings International Conference on Acoustics, Speech and Signal Processing, ICSLP, pp. 391–394 (1993)
4. Chetty, G., Wagner, M.: Automated lip feature extraction for liveness verification in audio-video authentication. HCC laboratory University of Canberra, Australia (2004)
5. Shen, X.-G., Wu, W.: An algorithm of lips secondary positioning and feature extraction based on YCbCr color space. In: International Conference on advances in Mechanical Engineering and Industrial Informatics. AMEII (2015)
6. Shiraishi, J., Saitoh, T.: Optical flow based lip reading using non rectangular ROI and head motion reduction. IEEE (2015). 971-1-4799-6026-2
7. John Hubert, P., Sheeba, M.S.: Lip and head gesture recognition based PC interface using image processing. Biomed. J. Pharm. J. **8**(1), 77–82 (2015)
8. Gurumurthy, S., Tripathy, B.K.: Design and implementation of face recognition system in matlab using the features of lips. IJISA **4**(8), 30 (2012)
9. Mehrotra, H., Agrawal, G.: Automatic lip contour tracking and visual character recognition for computerized lip reading. Int. J. Electr. Comput. Energ. Electron. Commun. Eng. **3**(4), 62–71 (2009)
10. Wang, L. Wang, X.: Lip detection and tracking using variance based Harr-like features and Kalman filter. In: Fifth International Conference on Frontier of computer Science and Technology. IEEE (2010). 978-0-7695-4139-6/10
11. Craig, B., Harte, N.: Region of interest extraction using colour based methods on the CUAVE database ISSC, Dublin (2009)
12. Ooi, W.C., Jeon, C.: Effective lip localization and tracking and achieving multimodal speech recognition. In: International Conference on Multisensor Fusion and Integration for Intelligent Systems, Seoul Korea (2008)
13. Morade, S., Patnaik, S.: Automatic lip tracking and extraction of lip geometric features for lip reading. Int. J. Mach. Learn. Comput. **3**(2), 168 (2013)

Multi-focus Image Fusion Method Using 2D-Wavelet Analysis and PCA

Anil Singh[1], Vikrant Bhateja[1](✉), Ashutosh Singhal[1], and Suresh Chandra Satapathy[2]

[1] Department of Electronics and Communication Engineering, Shri Ramswaroop Memorial Group of Professional Colleges (SRMGPC), Lucknow 226028, Uttar Pradesh, India
lfsanilsinghgkp@gmail.com, bhateja.vikrant@gmail.com, ashutoshsinghal71@gmail.com

[2] Department of CSE, PVP Siddhartha Institute of Technology, Vijayawada, Andhra Pradesh, India
sureshsatapathy@gmail.com

Abstract. In digital cameras, optical lens suffers from a limited depth of focus; therefore, it is limited in providing sufficient information regarding the scene. Two images with different depth of focus can be used to produce an "all-in-focus" image in order to provide better description. The concept of merging two different multi-focus images having different focus content using the combination of Wavelet and Principal Component Analysis (PCA) is carried out in this paper. In the proposed multi-focus image fusion method, each of the registered source images are decomposed into approximation and detailed layers using Discrete Wavelet Transform (DWT). Later, average fusion rule is applied on approximation layer and PCA is applied on detail layer to preserve both spectral as well as spatial information. Image quality assessment of the fused image is made using Standard Deviation (SD), Fusion Factor (FF) and Entropy (E) which justifies effectiveness of proposed method.

Keywords: Multi-focus · Daubechies · Entropy · PCA

1 Introduction

Image fusion is a method of synthesizing different registered source images having complementary information. The aim is to produce a new image, which carries complementary as well as common features of individual images. Multi-focus image fusion can be applied for various applications involving medical as well as remote sensing images [1]. During image acquisition by digital camera, it is usually difficult to generate an image that contains significant depth of focus for particular objects. Because of this limitation, there is a need to form a composite image for relevant visual perception. When a scene captured by optical lens contains objects at different distances, the depth of focus is different on each objects, creating a set of pictures [2]. Thus, in the obtained image, objects near to lens are in "in-focus", but others which vary in distance from optical lens will be "out-focus" and thus blurred. Multi-focus image fusion aims at

S.C. Satapathy and A. Joshi (eds.), *Information and Communication Technology for Intelligent Systems (ICTIS 2017) - Volume 1*, Smart Innovation, Systems and Technologies 83, DOI 10.1007/978-3-319-63673-3_74

combining information of two or more complementary images (In-focus and Out-focus) of a scene and the obtained result has an "all-in-focus" image [3]. Multi-focus image fusion aims to minimize the insufficiency of details in in-focus or out-focus exposed region. Literature in the last two decades provides various studies discussing number of image multi-resolution decomposition methodologies such as Discrete Wavelet Transform (DWT) [4], Laplacian Pyramid (LP) [5, 6] and Complex Wavelet [7]. The various fusion mechanisms such as Principal Component Analysis (PCA) [8], pixel based image fusion techniques which includes maximum and minimum selection rule [9] and weighted averaging image fusion have been popularly used. Choi [10] worked on fast Intensity Hue Saturation (IHS) with a trade-off parameter in which standard deviation of fused image increases but the spectral resolution decreases. Liu [11] performed Brovey Transform (BT) for multi-focus fusion of images. Other transform based fusion approaches are based on Contourlet [12], Bandlet [13] and Curvelet [14]. Yang et al. [15] deployed wavelets for pixel level decomposition of source images along with average fusion rule. The algorithm used was simple; but the obtained results do not visually depicted the presence of object from corresponding registered source images. Godse et al. worked on wavelet fusion technique by using maximum intensity approach but the fused image suffered from contrast reduction and blurring [16]. Sadhasivam et al. [17] performed fusion using PCA by selecting maximum pixel intensity but the obtained fused image has low contrast and illumination. From the above discussion, it can be inferred that there is a need to remove the redundant information from the fused image so that quality of image is improved. Wavelet based approach remove the artifacts by preserving time and frequency information [15]. PCA based approach improves the redundant information and feature enhancement property in the fused image [4]. Therefore, it is assimilated that combination of wavelet and PCA can be better modelled to provide fusion of multi-focus images. In this paper, the decomposition of source images is carried out by using DWT and further average fusion rule is applied on approximate co-efficients and PCA is applied on detailed co-efficients. The proposed fusion results have been evaluated by using SD, FF and E as image quality metrics.

2 Proposed Multi-focus Fusion Method

The scope of the problem presented in this paper is to perform multi-focus image fusion by using combination of 2D-Wavelets and PCA. PCA is a method of extracting relevant information from non-linear data sets. It determines Eigen values and subsequently its Eigen vectors [18, 19]. By determining Eigen values, it projects optimum information from original image to Eigen values; which increases the covariance. It helps to suppress the redundant information and highlights the components having maximum information in the source image [20]. This technique is useful in dimensionality reduction, data representation, image compression etc.

2.1 2D-Wavelets

A wavelet can be coined as small wave that grows and decays in a very small time period. It represents the time–frequency transform of an image. It was introduced as a substitute for Short-time Fourier Transform (STFT) to improve resolution of an image in both time and frequency domain. The wavelet transform of an image is done by decomposing image into scaled (dilation) and shifted (translation) version of chosen 'mother' wavelet denoted by $\Psi p,s(x)$ [4]. The 'mother' wavelet is defined in the following Eq. 1. The term 'mother' implies that the function with different region of support is from one main function during transformation process.

$$\psi_{p,s}(t) = \frac{1}{\sqrt{p}}\psi\left(\frac{t-s}{p}\right) for\ (p, s \in R), p > 0 \tag{1}$$

$\psi(t)$ represents mother wavelet function, whereas p and s represents scale and translation parameters as given by Eq. 1.

$$p = p_o^j, s = lp_o^j s_o for\ (j, l \in Z) \tag{2}$$

Thus, the wavelet family can be represented by Eq. 3

$$\phi_{j,l}(t) = p_o^{-j/2}\phi(p_o^{-j}t - ls_0) for\ (j, l \in Z) \tag{3}$$

DWT applies two-channel filter bank on a source image to obtain different layers or sub-band co-efficents. The transform is non-redundant and invertible. It decomposes the source images into four layers namely HH, HL, LH, LL as shown in Fig. 1. The HH layer is approximate layer with maximum information content while other layer is called detail layer which contain information pertaining to edges and the details.

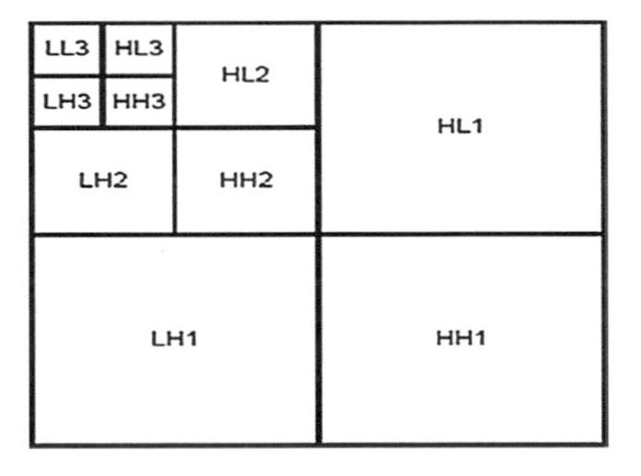

Fig. 1. Sub-band decomposition of source image using DWT [18].

2.2 Image Fusion Algorithm Based on PCA and Average Rule

The first pre-processing step involves conversion of source image from RGB scale to gray scale. After pre-processing decompose the in-focus and out-focus images (source images) using DWT. When DWT is applied on both the registered source images; a pair of detailed and approximate co-efficients are obtained. Average fusion technique is applied over the approximate co-efficients (in-focus and out-focus) whereas PCA is applied over the detailed co-efficients. Finally, the fusion of enhanced co-efficients

Table 1. Algorithm for proposed methodology using DWT, average and PCA fusion rule.

BEGIN

Step 1: Input source images.
Step 2: Decompose source images to approximate and detailed co-efficients by DWT.
Step 3: Apply Average fusion rule on approximate coefficients given by Eq. (4)

$$\xi(i,j) = \frac{\phi(i,j) + \psi(i,j)}{2} \quad (4)$$

Where: $\phi(i,j), \psi(i,j)$ are the approximate co-efficients of each of the source images and $\xi(i,j)$ is resulting mean co- efficient.
Step 4: Determine the column vector of Approximate and Detailed co-efficients
Step 5: Compute Covariance matrix for column vector obtained in Step 4.
Step 6: Determine Eigen vectors for Eigen values to Covariance matrix obtained in Step 5.
Step 7: Normalization of Column vector equivalent to larger Eigen value.
Step 8: Normalized Column vector and wavelet coefficient are multiplied.
Step 9: Reiterate the aforesaid process to all the coefficient of detailed co-efficients.
Step 10: Perform Inverse DWT of matrix obtained in Step 9.

END

takes place to combine the images for final reconstruction. The algorithm defining the proposed methodology is shown below in Table 1:

After the generation of final reconstructed image using the proposed fusion method; there are various performance parameters that are calculated: Standard Deviation (SD), Fusion Factor (FF) and Entropy (E) [4, 9]. Higher values of these metrics justifies the better worth of the proposed method.

3 Results and Discussions

Fusion of different multi-focus images is performed by decomposing it into approximation and detailed co-efficients by applying 2D-DWT and combining each co-efficients of corresponding source images by utilizing PCA and Average rule as fusion algorithms. The proposed fusion method is tested with 4 different sets of multi-focus images IMG-1(p1, q1, r1), similarly for IMG-2, IMG-3, IMG-4 to confirm the effectiveness. It can be inferred that in IMG-1 due to inward depth of focus of image (in-focus) the alphabets are clearly visible on the larger bottle but the smaller bottle and metal screw at the back is not visible as shown in Fig. 2(p1) whereas in Fig. 2(q1) due to outward depth of focus of image (out-focus) the smaller bottle and the metal screw

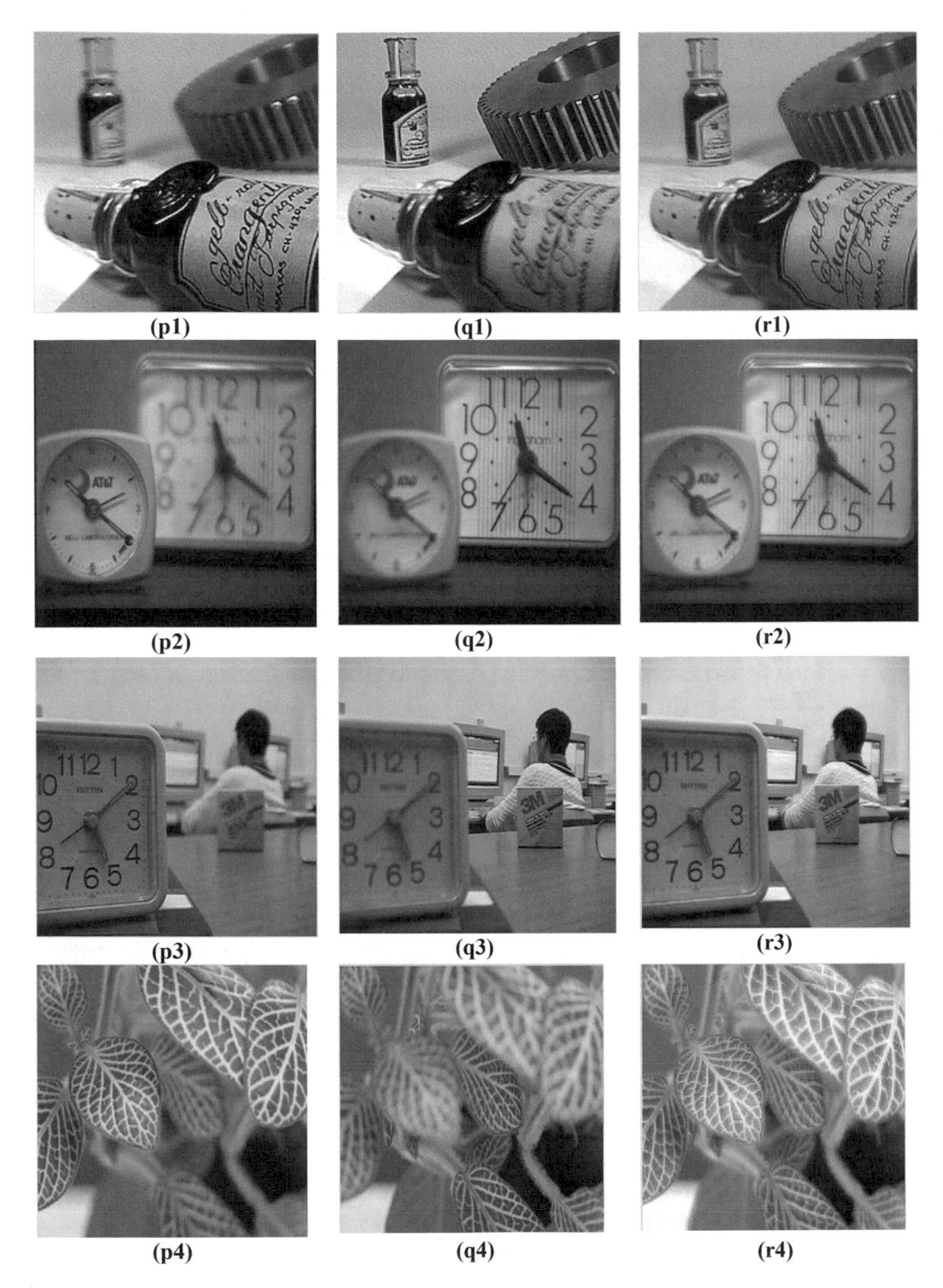

Fig. 2. An illustration of image fusion results for different sets of multi-focus images using proposed methodology: (p1)–(p4) input source images (in-focus), (q1)–(q4) input source images (out-focus), (r1)–(r4) output fused images.

are clearly visible but the alphabets on larger bottle are not clearly visible. Figure 2(r1) represents the fusion of in-focus and out-focus image from the proposed method which shows letters on larger bottle, smaller bottle and metal screw in fused image. High values of SD depict better quality of fused images in terms of contrast as shown in Table 2. This has been the methodology of utilizing the fused image to improve the spectral characteristics. Similarly, in IMG-2, Fig. 2(p2) small clock in the foreground clearly shows the letter and hands of clock whereas background clock is blurred while in Fig. 2(q2), the background clock can be clearly visualized while the foreground clock is not. The obtained fused images shows both the clocks which inferred the effectiveness of proposed fusion method. Figure 2 also depicts the other test input source images and the obtained fusion results.

The obtained results are evaluated for fusion quality using performance parameters such as E, FF, and SD. The overall impact is that the Entropy (E) of the fused image has been improved which shows an high amount of information content in the fused image. It is further verified through Fusion Factor (FF) shown in Table 2. Large values of E, FF and SD depicts better performance of proposed fusion approach for various source images. Hence, the proposed methodology is validated to show its worth in conserving the spectral as well as spatial features.

Table 2. Image quality computation using various fusion metrics.

Images	E	FF	SD
IMG-1	7.4680	5.5013	65.8902
IMG-2	7.3727	7.1426	49.3604
IMG-3	7.1971	7.4955	45.3861
IMG-4	7.4628	4.3866	41.4954

4 Conclusion

2D-wavelet based fusion method applicable to multi-focus images employing PCA as the subspace method is proposed in this paper. Due to time and frequency conservation property of wavelet both spatial and spectral features are extracted during fusion. PCA is a linear transformation of input image which basically results in orthogonal components. These components result in preservation of feature enhancement in the fused image. The same has been quantitatively validated using different fusion metrics pertaining to mutual information, structural and contrast content. Future work in this paper could be extended to usage of PCA in combination with other multi-scale technique Bandlet [13] and Contourlet [12, 21]. Therefore, multi-focus fusion method presented combines image with better description than individual image to produce an "all in focus" image.

References

1. Wang, Z., Ma, Y., Gu, J.: Multi-focus image fusion using PCNN. Pattern Recogn. **43**, 2003–2016 (2010)
2. Sinha, A., Bhateja, V., Sharma, A., Satapathy, S.C.: Bilateral filtering in wavelet domain for synthesis of flash and no-flash image pairs. In: International Conference on Information Systems Design and Intelligent Applications (INDIA-2015), Kalyani (W.B.), India, vol. 2, pp. 797–805, January 2015
3. Chai, Y., Li, H., Li, Z.: Multi-focus image fusion scheme using focused region detection and multi-resolution. Opt. Commun. **284**(19), 4376–4389 (2011)
4. Krishn, A., Bhateja, V., Himanshi, Sahu, A.: Medical image fusion using combination of PCA and wavelet analysis. In: International Conference on Advances in Computing, Communication and Informatics (ICACCI-2014), Gr. Noida (U.P.), India, pp. 986–991. IEEE, September 2014
5. Unser, M.: An improved least squares Laplacian pyramid for image compression. Signal Process. J. **27**(2), 187–203 (1992)
6. Sahu, A., Bhateja, V., Krishn, A., Himanshi: Medical image fusion with Laplacian pyramid. In: International Conference on Medical Imaging, m-Health & Emerging Communication Systems (MEDCom-2014), Gr. Noida (U.P.), pp. 448–453. IEEE, November 2014
7. Himanshi, Bhateja, V., Krishn, A., Sahu, A.: An improved medical image fusion approach using PCA and complex wavelets. In: International Conference on Medical Imaging, m-Health & Emerging Communication Systems (MEDCom-2014), Gr. Noida (U.P.), pp. 442–447. IEEE, November 2014)
8. He, C., Liu, Q., Li, H., Wang, H.: Multimodal medical image fusion based on IHS and PCA. In: Symposium on Security Detection and Information Processing, vol. 7, pp. 280–285. Elsevier (2010)
9. Himanshi, Bhateja, V., Krishn, A., Sahu, A.: Medical image fusion in Curvelet domain employing PCA and maximum selection rule. In: International Conference on Computers and Communication Technologies (IC3T-2015), Hyderabad, India, vol. 1, pp. 1–9. IEEE , July 2015
10. Choi, M.: A new intensity-hue-saturation fusion approach to image fusion with a trade-off parameter. IEEE Trans. Geosci. Remote Sens. **44**(6), 1672–1682 (2006)
11. Liu, J.G.: Smoothing filter-based intensity modulation: a spectral preserve image fusion technique for improving spatial details. Int. J. Remote Sens. **21**(18), 3461–3472 (2000)
12. Xiao-bo, Q., Wen, Y.J., Zhi, X.H., Zi-Qian, Z.: Image fusion algorithm based on spatial frequency-motivated pulse coupled neural networks in nonsubsampled contourlet transform domain. Acta Autom. Sinica **34**(12), 1508–1514 (2008)
13. Qu, X., Yan, J., Xie, G.: A novel image fusion algorithm based on bandelet transform. Chin. Opt. Lett. **5**(10), 569–572 (2007)
14. Li, S., Yang, B.: Multi-focus image fusion by combining Curvelet and wavelet transform. Pattern Recognit. Lett. J. **29**(9), 1295–1301 (2008)
15. Yang, B., Zing, Z.L., Zhao, H.: Review of pixel level image fusion. J. Shanghai Jiao Tong Univ. **15**, 6–12 (2010)
16. Godse, D.A., Bormane, D.S.: Wavelet based image fusion using pixel based maximum selection rule. Int. J. Eng. Sci. Technol. **3**(7), 5572–5578 (2011)
17. Sadhasivam, S.K., Keerthivasan, M.K., Muttan, S.: Implementation of max principle with PCA in image fusion for surveillance and navigation application. Electron. Lett. Comput. Vis. Image Anal. **10**(1), 1–10 (2011)

18. Moin, A., Bhateja, V., Srivastava, A.: Multispectral medical image fusion using PCA in wavelet domain, In: Proceedings of the (ACM-ICPS) Second International Conference on Information and Communication Technology for Competitive Strategies (ICTCS-2016), Udaipur, India, pp. 1–6, March 2016
19. Krishn, A., Bhateja, V., Himanshi, Sahu, A.: PCA based medical image fusion in ridgelet domain. In: International Conference on Frontiers in Intelligent Computing Theory and Applications (FICTA-2014), Bhubaneswar, India, vol. 328, pp. 475–482. Springer (2014)
20. Naidu, V.P.S., Raol, J.R.: Pixel-level image fusion using wavelets and principal component analysis. Def. Sci. J. **58**(3), 338–352 (2008)
21. Srivastava, A., Bhateja, V., Moin, A.: Combination of PCA and Contourlets for multispectral image fusion, In: International Conference on Data Engineering and Communication Technology (ICDECT-2016), Pune, India, vol. 2, pp. 1–8. Springer (2016)

Author Index

S.C. Satapathy and A. Joshi (eds.), *Information and Communication Technology for Intelligent Systems (ICTIS 2017) - Volume 1*, Smart Innovation, Systems and Technologies 83, DOI 10.1007/978-3-319-63673-3

Printed in the United States
By Bookmasters